SCHAUM'S OUTLINE OF
THEORY and
PROBLEMS of

D1257853

COMPLEX
VARIABLES

with an introduction to

CONFORMAL
MAPPING and
its applications

by
MURRAY R. SPIEGEL, Ph.D.
Professor of Mathematics
Rensselaer Polytechnic Institute

SCHAUM'S OUTLINE SERIES
McGRAW-HILL BOOK COMPANY
New York, St. Louis, San Francisco, Toronto, Sydney

ISBN 07-060230-1

20 21 22 23 24 25 26 27 28 29 30 SH SH 8 7 6

Preface

The theory of functions of a complex variable, also called for brevity complex variables or complex analysis, is one of the most beautiful as well as useful branches of mathematics. Although originating in an atmosphere of mystery, suspicion and distrust, as evidenced by the terms "imaginary" and "complex" present in the literature, it was finally placed on a sound foundation in the 19th century through the efforts of Cauchy, Riemann, Weierstrass, Gauss and other great mathematicians.

Today the subject is recognized as an essential part of the mathematical background of engineers, physicists, mathematicians and other scientists. From the theoretical viewpoint this is because many mathematical concepts become clarified and unified when examined in the light of complex variable theory. From the applied viewpoint the theory is of tremendous value in the solution of problems of heat flow, potential theory, fluid mechanics, electromagnetic theory, aerodynamics, elasticity and many other fields of science and engineering.

This book is designed for use as a supplement to all current standard texts or as a textbook for a formal course in complex variable theory and applications. It should also be of considerable value to those taking courses in mathematics, physics, aerodynamics, elasticity or any of the numerous other fields in which complex variable methods are employed.

Each chapter begins with a clear statement of pertinent definitions, principles and theorems together with illustrative and other descriptive material. This is followed by graded sets of solved and supplementary problems. The solved problems serve to illustrate and amplify the theory, bring into sharp focus those fine points without which the student continually feels himself on unsafe ground, and provide the repetition of basic principles so vital to effective learning. Numerous proofs of theorems and derivations of formulas are included among the solved problems. The large number of supplementary problems with answers serve as a complete review of the material in each chapter.

Topics covered include the algebra and geometry of complex numbers, complex differential and integral calculus, infinite series including Taylor and Laurent series, the theory of residues with applications to the evaluation of integrals and series, and conformal mapping with applications drawn from various fields. An added feature is the chapter on special topics which should prove useful as an introduction to some more advanced topics.

Considerably more material has been included here than can be covered in most first courses. This has been done to make the book more flexible, to provide a more useful book of reference and to stimulate further interest in the topics.

I wish to take this opportunity to thank the staff of the Schaum Publishing Company for their splendid cooperation.

M. R. SPIEGEL

Rensselaer Polytechnic Institute
July, 1964

CONTENTS

Page

Chapter *1*   COMPLEX NUMBERS   1
The real number system. Graphical representation of real numbers. The complex number system. Fundamental operations with complex numbers. Absolute value. Axiomatic foundations of the complex number system. Graphical representation of complex numbers. Polar form of complex numbers. De Moivre's theorem. Roots of complex numbers. Euler's formula. Polynomial equations. The nth roots of unity. Vector interpretation of complex numbers. Spherical representation of complex numbers. Stereographic projection. Dot and cross product. Complex conjugate coordinates. Point sets.

Chapter *2*   FUNCTIONS, LIMITS AND CONTINUITY   33
Variables and functions. Single and multiple-valued functions. Inverse functions. Transformations. Curvilinear coordinates. The elementary functions. Branch points and branch lines. Riemann surfaces. Limits. Theorems on limits. Infinity. Continuity. Continuity in a region. Theorems on continuity. Uniform continuity. Sequences. Limit of a sequence. Theorems on limits of sequences. Infinite series.

Chapter *3*   COMPLEX DIFFERENTIATION AND
THE CAUCHY-RIEMANN EQUATIONS   63
Derivatives. Analytic functions. Cauchy-Riemann equations. Harmonic functions. Geometric interpretation of the derivative. Differentials. Rules for differentiation. Derivatives of elementary functions. Higher order derivatives. L'Hospital's rule. Singular points. Orthogonal families. Curves. Applications to geometry and mechanics. Complex differential operators. Gradient, divergence, curl and Laplacian. Some identities involving gradient, divergence and curl.

Chapter *4*   COMPLEX INTEGRATION AND CAUCHY'S THEOREM   92
Complex line integrals. Real line integrals. Connection between real and complex line integrals. Properties of integrals. Change of variables. Simply and multiply-connected regions. Jordan curve theorem. Convention regarding traversal of a closed path. Green's theorem in the plane. Complex form of Green's theorem. Cauchy's theorem. The Cauchy-Goursat theorem. Morera's theorem. Indefinite integrals. Integrals of special functions. Some consequences of Cauchy's theorem.

Chapter *5*   CAUCHY'S INTEGRAL FORMULAS AND
RELATED THEOREMS   118
Cauchy's integral formulas. Some important theorems. Morera's theorem. Cauchy's inequality. Liouville's theorem. Fundamental theorem of algebra. Gauss' mean value theorem. Maximum modulus theorem. Minimum modulus theorem. The argument theorem. Rouché's theorem. Poisson's integral formulas for a circle. Poisson's integral formulas for a half plane.

CONTENTS

Chapter **6** **INFINITE SERIES. TAYLOR'S AND LAURENT SERIES**.... **139**

Sequences of functions. Series of functions. Absolute convergence. Uniform convergence of sequences and series. Power series. Some important theorems. General theorems. Theorems on absolute convergence. Special tests for convergence. Theorems on uniform convergence. Theorems on power series. Taylor's theorem. Some special series. Laurent's theorem. Classification of singularities. Entire functions. Meromorphic functions. Lagrange's expansion. Analytic continuation.

Chapter **7** **THE RESIDUE THEOREM.**
EVALUATION OF INTEGRALS AND SERIES **172**

Residues. Calculation of residues. The residue theorem. Evaluation of definite integrals. Special theorems used in evaluating integrals. The Cauchy principal value of integrals. Differentiation under the integral sign. Leibnitz's rule. Summation of series. Mittag-Leffler's expansion theorem. Some special expansions.

Chapter **8** **CONFORMAL MAPPING** **200**

Transformations or mappings. Jacobian of a transformation. Complex mapping functions. Conformal mapping. Riemann's mapping theorem. Fixed or invariant points of a transformation. Some general transformations. Translation. Rotation. Stretching. Inversion. Successive transformations. The linear transformation. The bilinear or fractional transformation. Mapping of a half plane on to a circle. The Schwarz-Christoffel transformation. Transformations of boundaries in parametric form. Some special mappings.

Chapter **9** **PHYSICAL APPLICATIONS OF CONFORMAL MAPPING**.... **232**

Boundary-value problems. Harmonic and conjugate functions. Dirichlet and Neumann problems. The Dirichlet problem for the unit circle. Poisson's formula. The Dirichlet problem for the half plane. Solutions to Dirichlet and Neumann problems by conformal mapping. Applications to fluid flow. Basic assumptions. The complex potential. Equipotential lines and streamlines. Sources and sinks. Some special flows. Flow around obstacles. Bernoulli's theorem. Theorems of Blasius. Applications to electrostatics. Coulomb's law. Electric field intensity. Electrostatic potential. Gauss' theorem. The complex electrostatic potential. Line charges. Conductors. Capacitance. Applications to heat flow. Heat flux. The complex temperature.

Chapter **10** **SPECIAL TOPICS** **265**

Analytic continuation. Schwarz's reflection principle. Infinite products. Absolute, conditional and uniform convergence of infinite products. Some important theorems on infinite products. Weierstrass' theorem for infinite products. Some special infinite products. The gamma function. Properties of the gamma function. The beta function. Differential equations. Solution of differential equations by contour integrals. Bessel functions. Legendre functions. The hypergeometric function. The zeta function. Asymptotic series. The method of steepest descents. Special asymptotic expansions. Elliptic functions.

INDEX **307**

Chapter 1

Complex Numbers

THE REAL NUMBER SYSTEM

The number system as we know it today is a result of gradual development as indicated in the following list.

1. **Natural numbers** $1, 2, 3, 4, \ldots$, also called *positive integers*, were first used in counting. The symbols varied with the times, e.g. the Romans used I, II, III, IV, $\ldots$. If a and b are natural numbers, the *sum* $a + b$ and *product* $a \cdot b$, $(a)(b)$ or ab are also natural numbers. For this reason the set of natural numbers is said to be *closed* under the operations of *addition* and *multiplication* or to satisfy the *closure property* with respect to these operations.

2. **Negative integers and zero,** denoted by $-1, -2, -3, \ldots$ and 0 respectively, arose to permit solutions of equations such as $x + b = a$ where a and b are any natural numbers. This leads to the operation of *subtraction*, or *inverse of addition*, and we write $x = a - b$.

   The set of positive and negative integers and zero is called the set of *integers* and is closed under the operations of addition, multiplication and subtraction.

3. **Rational numbers** or *fractions* such as $\frac{3}{4}, -\frac{8}{3}, \ldots$ arose to permit solutions of equations such as $bx = a$ for all integers a and b where $b \neq 0$. This leads to the operation of *division* or *inverse of multiplication*, and we write $x = a/b$ or $a \div b$ [called the *quotient* of a and b] where a is the *numerator* and b is the *denominator*.

   The set of integers is a part or *subset* of the rational numbers, since integers correspond to rational numbers a/b where $b = 1$.

   The set of rational numbers is closed under the operations of addition, subtraction, multiplication and division, so long as division by zero is excluded.

4. **Irrational numbers** such as $\sqrt{2} = 1.41423\cdots$ and $\pi = 3.14159\cdots$ are numbers which are not rational, i.e. cannot be expressed as a/b where a and b are integers and $b \neq 0$.

The set of rational and irrational numbers is called the set of *real* numbers. It is assumed that the student is already familiar with the various operations on real numbers.

GRAPHICAL REPRESENTATION OF REAL NUMBERS

Real numbers can be represented by points on a line called the *real axis*, as indicated in Fig. 1-1. The point corresponding to zero is called the *origin*.

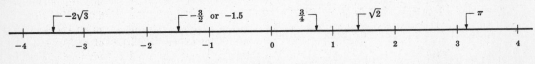

Fig. 1-1

Conversely, to each point on the line there is one and only one real number. If a point A corresponding to a real number a lies to the right of a point B corresponding to a real number b, we say that a is *greater than* b or b is *less than* a and write respectively $a > b$ or $b < a$.

The set of all values of x such that $a < x < b$ is called an *open interval* on the real axis while $a \leqq x \leqq b$, which also includes the endpoints a and b, is called a *closed interval*. The symbol x, which can stand for any of a set of real numbers, is called a *real variable*.

The *absolute value* of a real number a, denoted by $|a|$, is equal to a if $a > 0$, to $-a$ if $a < 0$ and to 0 if $a = 0$. The distance between two points a and b on the real axis is $|a - b|$.

THE COMPLEX NUMBER SYSTEM

There is no real number x which satisfies the polynomial equation $x^2 + 1 = 0$. To permit solutions of this and similar equations, the set of *complex numbers* is introduced.

We can consider a *complex number* as having the form $a + bi$ where a and b are real numbers and i, which is called the *imaginary unit*, has the property that $i^2 = -1$. If $z = a + bi$, then a is called the *real part* of z and b is called the *imaginary part* of z and are denoted by Re $\{z\}$ and Im $\{z\}$ respectively. The symbol z, which can stand for any of a set of complex numbers, is called a *complex variable*.

Two complex numbers $a + bi$ and $c + di$ are *equal* if and only if $a = c$ and $b = d$. We can consider real numbers as a subset of the set of complex numbers with $b = 0$. Thus the complex numbers $0 + 0i$ and $-3 + 0i$ represent the real numbers 0 and -3 respectively. If $a = 0$, the complex number $0 + bi$ or bi is called a *pure imaginary number*.

The *complex conjugate*, or briefly *conjugate*, of a complex number $a + bi$ is $a - bi$. The complex conjugate of a complex number z is often indicated by $\bar{z}$ or z^*.

FUNDAMENTAL OPERATIONS WITH COMPLEX NUMBERS

In performing operations with complex numbers we can proceed as in the algebra of real numbers, replacing i^2 by -1 when it occurs.

1. *Addition*
$$(a + bi) + (c + di) = a + bi + c + di = (a + c) + (b + d)i$$

2. *Subtraction*
$$(a + bi) - (c + di) = a + bi - c - di = (a - c) + (b - d)i$$

3. *Multiplication*
$$(a + bi)(c + di) = ac + adi + bci + bdi^2 = (ac - bd) + (ad + bc)i$$

4. *Division*
$$\frac{a + bi}{c + di} = \frac{a + bi}{c + di} \cdot \frac{c - di}{c - di} = \frac{ac - adi + bci - bdi^2}{c^2 - d^2 i^2}$$
$$= \frac{ac + bd + (bc - ad)i}{c^2 + d^2} = \frac{ac + bd}{c^2 + d^2} + \frac{bc - ad}{c^2 + d^2} i$$

ABSOLUTE VALUE

The *absolute value* or *modulus* of a complex number $a + bi$ is defined as $|a + bi| = \sqrt{a^2 + b^2}$.

Example: $|-4 + 2i| = \sqrt{(-4)^2 + (2)^2} = \sqrt{20} = 2\sqrt{5}$

If $z_1, z_2, z_3, \ldots, z_m$ are complex numbers, the following properties hold.

1. $|z_1 z_2| = |z_1||z_2|$ or $|z_1 z_2 \cdots z_m| = |z_1||z_2| \cdots |z_m|$

2. $\left|\dfrac{z_1}{z_2}\right| = \dfrac{|z_1|}{|z_2|}$ if $z_2 \neq 0$

3. $|z_1 + z_2| \leqq |z_1| + |z_2|$ or $|z_1 + z_2 + \cdots + z_m| \leqq |z_1| + |z_2| + \cdots + |z_m|$

4. $|z_1 + z_2| \geqq |z_1| - |z_2|$ or $|z_1 - z_2| \geqq |z_1| - |z_2|$

AXIOMATIC FOUNDATIONS OF THE COMPLEX NUMBER SYSTEM

From a strictly logical point of view it is desirable to define a complex number as an ordered pair (a, b) of real numbers a and b subject to certain operational definitions which turn out to be equivalent to those above. These definitions are as follows, where all letters represent real numbers.

A. Equality $(a, b) = (c, d)$ if and only if $a = c$, $b = d$

B. Sum $(a, b) + (c, d) = (a + c,\ b + d)$

C. Product $(a, b) \cdot (c, d) = (ac - bd,\ ad + bc)$

$$m(a, b) = (ma, mb)$$

From these we can show [Problem 14] that $(a, b) = a(1, 0) + b(0, 1)$ and we associate this with $a + bi$ where i is the symbol for $(0, 1)$ and has the property that $i^2 = (0, 1)(0, 1) = (-1, 0)$ [which can be considered equivalent to the real number -1] and $(1, 0)$ can be considered equivalent to the real number 1. The ordered pair $(0, 0)$ corresponds to the real number 0.

From the above we can prove that if z_1, z_2, z_3 belong to the set S of complex numbers, then

1. $z_1 + z_2$ and $z_1 z_2$ belong to S Closure law
2. $z_1 + z_2 = z_2 + z_1$ Commutative law of addition
3. $z_1 + (z_2 + z_3) = (z_1 + z_2) + z_3$ Associative law of addition
4. $z_1 z_2 = z_2 z_1$ Commutative law of multiplication
5. $z_1(z_2 z_3) = (z_1 z_2)z_3$ Associative law of multiplication
6. $z_1(z_2 + z_3) = z_1 z_2 + z_1 z_3$ Distributive law
7. $z_1 + 0 = 0 + z_1 = z_1$, $1 \cdot z_1 = z_1 \cdot 1 = z_1$, 0 is called the *identity with respect to addition*, 1 is called the *identity with respect to multiplication*.
8. For any complex number z_1 there is a unique number z in S such that $z + z_1 = 0$; z is called the *inverse of z_1 with respect to addition* and is denoted by $-z_1$.
9. For any $z_1 \ne 0$ there is a unique number z in S such that $z_1 z = z z_1 = 1$; z is called the *inverse of z_1 with respect to multiplication* and is denoted by z_1^{-1} or $1/z_1$.

In general any set, such as S, whose members satisfy the above is called a *field*.

GRAPHICAL REPRESENTATION OF COMPLEX NUMBERS

If real scales are chosen on two mutually perpendicular axes $X'OX$ and $Y'OY$ [called the *x and y axes* respectively] as in Fig. 1-2, we can locate any point, in the plane determined by these lines, by the ordered pair of real numbers (x, y) called *rectangular coordinates* of the point. Examples of the location of such points are indicated by P, Q, R, S and T in Fig. 1-2.

Since a complex number $x + iy$ can be considered as an ordered pair of real numbers, we can represent such numbers by points in an xy plane called the *complex plane* or *Argand diagram*. The complex number represented by P, for example, could then be read as either $(3, 4)$ or $3 + 4i$. To each complex number there corresponds one and only one point in the plane, and conversely to each point in the plane there corresponds one and only one complex number. Because of this we often refer to the complex number z as the *point z*.

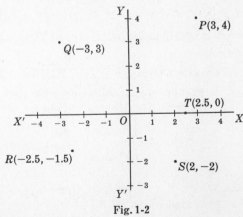

Fig. 1-2

Sometimes we refer to the x and y axes as the *real* and *imaginary* axes respectively and to the complex plane as the z *plane*. The distance between two points $z_1 = x_1 + iy_1$ and $z_2 = x_2 + iy_2$ in the complex plane is given by $|z_1 - z_2| = \sqrt{(x_1 - x_2)^2 + (y_1 - y_2)^2}$.

POLAR FORM OF COMPLEX NUMBERS

If P is a point in the complex plane corresponding to the complex number (x, y) or $x + iy$, then we see from Fig. 1-3 that

$$x = r \cos \theta, \quad y = r \sin \theta$$

where $r = \sqrt{x^2 + y^2} = |x + iy|$ is called the *modulus* or *absolute value* of $z = x + iy$ [denoted by *mod z* or $|z|$]; and θ, called the *amplitude* or *argument* of $z = x + iy$ [denoted by *arg z*], is the angle which line OP makes with the positive x axis.

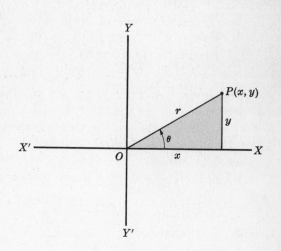

Fig. 1-3

It follows that

$$z = x + iy = r(\cos \theta + i \sin \theta) \qquad (1)$$

which is called the *polar form* of the complex number, and r and θ are called *polar coordinates*. It is sometimes convenient to write the abbreviation cis θ for $\cos \theta + i \sin \theta$.

For any complex number $z \neq 0$ there corresponds only one value of θ in $0 \leqq \theta < 2\pi$. However, any other interval of length 2π, for example $-\pi < \theta \leqq \pi$, can be used. Any particular choice, decided upon in advance, is called the *principal range*, and the value of θ is called its *principal value*.

DE MOIVRE'S THEOREM

If $z_1 = x_1 + iy_1 = r_1 (\cos \theta_1 + i \sin \theta_1)$ and $z_2 = x_2 + iy_2 = r_2 (\cos \theta_2 + i \sin \theta_2)$, we can show that [see Problem 19]

$$z_1 z_2 = r_1 r_2 \{\cos (\theta_1 + \theta_2) + i \sin (\theta_1 + \theta_2)\} \qquad (2)$$

$$\frac{z_1}{z_2} = \frac{r_1}{r_2} \{\cos (\theta_1 - \theta_2) + i \sin (\theta_1 - \theta_2)\} \qquad (3)$$

A generalization of (2) leads to

$$z_1 z_2 \cdots z_n = r_1 r_2 \cdots r_n \{\cos (\theta_1 + \theta_2 + \cdots + \theta_n) + i \sin (\theta_1 + \theta_2 + \cdots + \theta_n)\} \qquad (4)$$

and if $z_1 = z_2 = \cdots = z_n = z$ this becomes

$$z^n = \{r(\cos \theta + i \sin \theta)\}^n = r^n (\cos n\theta + i \sin n\theta) \qquad (5)$$

which is often called *De Moivre's theorem*.

ROOTS OF COMPLEX NUMBERS

A number w is called an nth *root* of a complex number z if $w^n = z$, and we write $w = z^{1/n}$. From De Moivre's theorem we can show that if n is a positive integer,

$$z^{1/n} = \{r(\cos \theta + i \sin \theta)\}^{1/n}$$

$$= r^{1/n} \left\{ \cos \left(\frac{\theta + 2k\pi}{n} \right) + i \sin \left(\frac{\theta + 2k\pi}{n} \right) \right\} \qquad k = 0, 1, 2, \ldots, n-1 \qquad (6)$$

from which it follows that there are n different values for $z^{1/n}$, i.e. n different nth roots of z, provided $z \neq 0$.

EULER'S FORMULA

By assuming that the infinite series expansion $e^x = 1 + x + x^2/2! + x^3/3! + \cdots$ of elementary calculus holds when $x = i\theta$, we can arrive at the result

$$e^{i\theta} = \cos\theta + i\sin\theta \qquad e = 2.71828\ldots \qquad (7)$$

which is called *Euler's formula*. It is more convenient, however, simply to take (7) as a definition of $e^{i\theta}$. In general, we define

$$e^z = e^{x+iy} = e^x e^{iy} = e^x(\cos y + i\sin y) \qquad (8)$$

In the special case where $y = 0$ this reduces to e^x.

Note that in terms of (7) De Moivre's theorem essentially reduces to $(e^{i\theta})^n = e^{in\theta}$.

POLYNOMIAL EQUATIONS

Often in practice we require solutions of polynomial equations having the form

$$a_0 z^n + a_1 z^{n-1} + a_2 z^{n-2} + \cdots + a_{n-1}z + a_n = 0 \qquad (9)$$

where $a_0 \neq 0, a_1, \ldots, a_n$ are given complex numbers and n is a positive integer called the *degree* of the equation. Such solutions are also called *zeros* of the polynomial on the left of (9) or *roots of the equation*.

A very important theorem called the *fundamental theorem of algebra* [to be proved in Chapter 5] states that every polynomial equation of the form (9) has at least one root which is complex. From this we can show that it has in fact n complex roots, some or all of which may be identical.

If $z_1, z_2, \ldots, z_n$ are the n roots, (9) can be written

$$a_0(z - z_1)(z - z_2) \cdots (z - z_n) = 0 \qquad (10)$$

which is called the *factored form* of the polynomial equation. Conversely if we **can write** (9) in the form (10), we can easily determine the roots.

THE nth ROOTS OF UNITY

The solutions of the equation $z^n = 1$ where n is a positive integer are called the nth *roots of unity* and are given by

$$z = \cos 2k\pi/n + i\sin 2k\pi/n = e^{2k\pi i/n} \qquad k = 0, 1, 2, \ldots, n-1 \qquad (11)$$

If we let $\omega = \cos 2\pi/n + i\sin 2\pi/n = e^{2\pi i/n}$, the n roots are $1, \omega, \omega^2, \ldots, \omega^{n-1}$. Geometrically they represent the n vertices of a regular polygon of n sides inscribed in a circle of radius one with center at the origin. This circle has the equation $|z| = 1$ and is often called the *unit circle*.

VECTOR INTERPRETATION OF COMPLEX NUMBERS

A complex number $z = x + iy$ can be considered as a vector OP whose *initial point* is the origin O and whose *terminal point* P is the point (x, y) as in Fig. 1-4. We sometimes call $OP = x + iy$ the *position vector* of P. Two vectors having the same *length* or *magnitude* and *direction* but different initial points, such as OP and AB in Fig. 1-4, are considered equal. Hence we write $OP = AB = x + iy$.

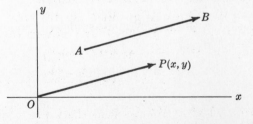

Fig. 1-4

Addition of complex numbers corresponds to the *parallelogram law* for addition of vectors [see Fig. 1-5]. Thus to add the complex numbers z_1 and z_2, we complete the parallelogram $OABC$ whose sides OA and OC correspond to z_1 and z_2. The diagonal OB of this parallelogram corresponds to $z_1 + z_2$. See Problem 5.

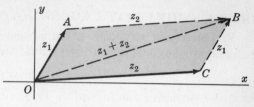

Fig. 1-5

SPHERICAL REPRESENTATION OF COMPLEX NUMBERS. STEREOGRAPHIC PROJECTION

Let $\mathcal{P}$ [Fig. 1-6] be the complex plane and consider a unit sphere $\mathcal{S}$ [radius one] tangent to $\mathcal{P}$ at $z = 0$. The diameter NS is perpendicular to $\mathcal{P}$ and we call points N and S the *north* and *south poles* of $\mathcal{S}$. Corresponding to any point A on $\mathcal{P}$ we can construct line NA intersecting $\mathcal{S}$ at point A'. Thus to each point of the complex plane $\mathcal{P}$ there corresponds one and only one point of the sphere $\mathcal{S}$, and we can represent any complex number by a point on the sphere. For completeness we say that the point N itself corresponds to the "point at infinity" of the plane. The set of all points of the complex plane including the point at infinity is called the *entire complex plane*, the *entire z plane*, or the *extended complex plane*.

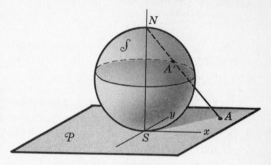

Fig. 1-6

The above method for mapping the plane on to the sphere is called *stereographic projection*. The sphere is sometimes called the *Riemann sphere*.

DOT AND CROSS PRODUCT

Let $z_1 = x_1 + iy_1$ and $z_2 = x_2 + iy_2$ be two complex numbers [vectors]. The *dot product* [also called the *scalar product*] of z_1 and z_2 is defined by

$$z_1 \circ z_2 = |z_1||z_2| \cos\theta = x_1 x_2 + y_1 y_2 = \text{Re}\{\bar{z}_1 z_2\} = \tfrac{1}{2}\{\bar{z}_1 z_2 + z_1 \bar{z}_2\} \tag{12}$$

where θ is the angle between z_1 and z_2 which lies between 0 and π.

The *cross product* of z_1 and z_2 is defined by

$$z_1 \times z_2 = |z_1||z_2| \sin\theta = x_1 y_2 - y_1 x_2 = \text{Im}\{\bar{z}_1 z_2\} = \frac{1}{2i}\{\bar{z}_1 z_2 - z_1 \bar{z}_2\} \tag{13}$$

Clearly,

$$\bar{z}_1 z_2 = (z_1 \circ z_2) + i(z_1 \times z_2) = |z_1||z_2| e^{i\theta} \tag{14}$$

If z_1 and z_2 are non-zero, then

1. A necessary and sufficient condition that z_1 and z_2 be perpendicular is that $z_1 \circ z_2 = 0$.

2. A necessary and sufficient condition that z_1 and z_2 be parallel is that $z_1 \times z_2 = 0$.

3. The magnitude of the projection of z_1 on z_2 is $|z_1 \circ z_2|/|z_2|$.

4. The area of a parallelogram having sides z_1 and z_2 is $|z_1 \times z_2|$.

COMPLEX CONJUGATE COORDINATES

A point in the complex plane can be located by rectangular coordinates (x, y) or polar coordinates (r, θ). Many other possibilities exist. One such possibility uses the fact that $x = \frac{1}{2}(z + \bar{z})$, $y = \frac{1}{2i}(z - \bar{z})$ where $z = x + iy$. The coordinates $(z, \bar{z})$ which locate a point are called *complex conjugate coordinates* or briefly *conjugate coordinates* of the point [see Problems 43 and 44].

POINT SETS

Any collection of points in the complex plane is called a (*two-dimensional*) point set, and each point is called a *member* or *element* of the set. The following fundamental definitions are given here for reference.

1. **Neighborhoods.** A *delta, or δ, neighborhood* of a point z_0 is the set of all points z such that $|z - z_0| < \delta$ where δ is any given positive number. A *deleted δ neighborhood* of z_0 is a neighborhood of z_0 in which the point z_0 is omitted, i.e. $0 < |z - z_0| < \delta$.

2. **Limit Points.** A point z_0 is called a *limit point, cluster point,* or *point of accumulation* of a point set S if every deleted δ neighborhood of z_0 contains points of S.

    Since δ can be any positive number, it follows that S must have infinitely many points. Note that z_0 may or may not belong to the set S.

3. **Closed Sets.** A set S is said to be *closed* if every limit point of S belongs to S, i.e. if S contains all its limit points. For example, the set of all points z such that $|z| \leqq 1$ is a closed set.

4. **Bounded Sets.** A set S is called *bounded* if we can find a constant M such that $|z| < M$ for every point z in S. An *unbounded set* is one which is not bounded. A set which is both bounded and closed is sometimes called *compact*.

5. **Interior, Exterior and Boundary Points.** A point z_0 is called an *interior point* of a set S if we can find a δ neighborhood of z_0 all of whose points belong to S. If every δ neighborhood of z_0 contains points belonging to S and also points not belonging to S, then z_0 is called a *boundary point*. If a point is not an interior or boundary point of a set S, it is an *exterior point* of S.

6. **Open Sets.** An *open set* is a set which consists only of interior points. For example, the set of points z such that $|z| < 1$ is an open set.

7. **Connected Sets.** An open set S is said to be *connected* if any two points of the set can be joined by a path consisting of straight line segments (i.e. a *polygonal path*) all points of which are in S.

8. **Open Regions or Domains.** An open connected set is called an *open region* or *domain*.

9. **Closure of a Set.** If to a set S we add all the limit points of S, the new set is called the *closure* of S and is a closed set.

10. **Closed Regions.** The closure of an open region or domain is called a *closed region*.

11. **Regions.** If to an open region or domain we add some, all or none of its limit points, we obtain a set called a *region*. If all the limit points are added, the region is *closed*; if none are added, the region is *open*. In this book whenever we use the word *region* without qualifying it, we shall mean *open region* or *domain*.

12. **Union and Intersection of Sets.** A set consisting of all points belonging to set S_1 or set S_2 or to both sets S_1 and S_2 is called the *union* of S_1 and S_2 and is denoted by $S_1 + S_2$ or $S_1 \cup S_2$.

   A set consisting of all points belonging to both sets S_1 and S_2 is called the *intersection* of S_1 and S_2 and is denoted by $S_1 S_2$ or $S_1 \cap S_2$.

13. **Complement of a Set.** A set consisting of all points which do not belong to S is called the *complement* of S and is denoted by $\widetilde{S}$.

14. **Null Sets and Subsets.** It is convenient to consider a set consisting of no points at all. This set is called the *null set* and is denoted by $\emptyset$. If two sets S_1 and S_2 have no points in common (in which case they are called *disjoint* or *mutually exclusive sets*), we can indicate this by writing $S_1 \cap S_2 = \emptyset$.

   Any set formed by choosing some, all or none of the points of a set S is called a *subset* of S. If we exclude the case where all points of S are chosen, the set is called a *proper subset* of S.

15. **Countability of a Set.** If the members or elements of a set can be placed into a one to one correspondence with the natural numbers $1, 2, 3, \ldots$, the set is called *countable* or *denumerable*; otherwise it is *non-countable* or *non-denumerable*.

The following are two important theorems on point sets.

1. **Weierstrass-Bolzano Theorem.** Every bounded infinite set has at least one limit point.

2. **Heine-Borel Theorem.** Let S be a compact set each point of which is contained in one or more of the open sets $A_1, A_2, \ldots$ [which are then said to *cover* S]. Then there exists a finite number of the sets $A_1, A_2, \ldots$ which will cover S.

Solved Problems

FUNDAMENTAL OPERATIONS WITH COMPLEX NUMBERS

1. Perform each of the indicated operations.

   (a) $(3 + 2i) + (-7 - i) = 3 - 7 + 2i - i = -4 + i$

   (b) $(-7 - i) + (3 + 2i) = -7 + 3 - i + 2i = -4 + i$

   The results (a) and (b) illustrate the *commutative law of addition*.

   (c) $(8 - 6i) - (2i - 7) = 8 - 6i - 2i + 7 = 15 - 8i$

   (d) $(5 + 3i) + \{(-1 + 2i) + (7 - 5i)\} = (5 + 3i) + \{-1 + 2i + 7 - 5i\} = (5 + 3i) + (6 - 3i) = 11$

   (e) $\{(5 + 3i) + (-1 + 2i)\} + (7 - 5i) = \{5 + 3i - 1 + 2i\} + (7 - 5i) = (4 + 5i) + (7 - 5i) = 11$

   The results (d) and (e) illustrate the *associative law of addition*.

   (f) $(2 - 3i)(4 + 2i) = 2(4 + 2i) - 3i(4 + 2i) = 8 + 4i - 12i - 6i^2 = 8 + 4i - 12i + 6 = 14 - 8i$

   (g) $(4 + 2i)(2 - 3i) = 4(2 - 3i) + 2i(2 - 3i) = 8 - 12i + 4i - 6i^2 = 8 - 12i + 4i + 6 = 14 - 8i$

   The results (f) and (g) illustrate the *commutative law of multiplication*.

   (h) $(2 - i)\{(-3 + 2i)(5 - 4i)\} = (2 - i)\{-15 + 12i + 10i - 8i^2\}$
   $$= (2 - i)(-7 + 22i) = -14 + 44i + 7i - 22i^2 = 8 + 51i$$

   (i) $\{(2 - i)(-3 + 2i)\}(5 - 4i) = \{-6 + 4i + 3i - 2i^2\}(5 - 4i)$
   $$= (-4 + 7i)(5 - 4i) = -20 + 16i + 35i - 28i^2 = 8 + 51i$$

   The results (h) and (i) illustrate the *associative law of multiplication*.

(j) $(-1+2i)\{(7-5i)+(-3+4i)\} = (-1+2i)(4-i) = -4+i+8i-2i^2 = -2+9i$

     Another method. $(-1+2i)\{(7-5i)+(-3+4i)\} = (-1+2i)(7-5i)+(-1+2i)(-3+4i)$

$$= \{-7+5i+14i-10i^2\}+\{3-4i-6i+8i^2\}$$

$$= (3+19i)+(-5-10i) = -2+9i$$

     This illustrates the *distributive law*.

(k) $\dfrac{3-2i}{-1+i} = \dfrac{3-2i}{-1+i}\cdot\dfrac{-1-i}{-1-i} = \dfrac{-3-3i+2i+2i^2}{1-i^2} = \dfrac{-5-i}{2} = -\dfrac{5}{2}-\dfrac{1}{2}i$

     Another method. By definition, $(3-2i)/(-1+i)$ is that number $a+bi$, where a and b are real, such that $(-1+i)(a+bi) = -a-b+(a-b)i = 3-2i$. Then $-a-b = 3$, $a-b = -2$ and solving simultaneously, $a = -5/2$, $b = -1/2$ or $a+bi = -5/2-i/2$.

(l) $\dfrac{5+5i}{3-4i}+\dfrac{20}{4+3i} = \dfrac{5+5i}{3-4i}\cdot\dfrac{3+4i}{3+4i}+\dfrac{20}{4+3i}\cdot\dfrac{4-3i}{4-3i}$

$$= \dfrac{15+20i+15i+20i^2}{9-16i^2}+\dfrac{80-60i}{16-9i^2} = \dfrac{-5+35i}{25}+\dfrac{80-60i}{25} = 3-i$$

(m) $\dfrac{3i^{30}-i^{19}}{2i-1} = \dfrac{3(i^2)^{15}-(i^2)^9 i}{2i-1} = \dfrac{3(-1)^{15}-(-1)^9 i}{-1+2i}$

$$= \dfrac{-3+i}{-1+2i}\cdot\dfrac{-1-2i}{-1-2i} = \dfrac{3+6i-i-2i^2}{1-4i^2} = \dfrac{5+5i}{5} = 1+i$$

2.   If $z_1 = 2+i$, $z_2 = 3-2i$ and $z_3 = -\dfrac{1}{2}+\dfrac{\sqrt{3}}{2}i$, evaluate each of the following.

(a) $|3z_1-4z_2| = |3(2+i)-4(3-2i)| = |6+3i-12+8i|$

$$= |-6+11i| = \sqrt{(-6)^2+(11)^2} = \sqrt{157}$$

(b) $z_1^3 - 3z_1^2 + 4z_1 - 8 = (2+i)^3 - 3(2+i)^2 + 4(2+i) - 8$

$$= \{(2)^3 + 3(2)^2(i) + 3(2)(i)^2 + i^3\} - 3(4+4i+i^2) + 8 + 4i - 8$$

$$= 8 + 12i - 6 - i - 12 - 12i + 3 + 8 + 4i - 8 = -7 + 3i$$

(c) $(\bar{z}_3)^4 = \left(\overline{-\dfrac{1}{2}+\dfrac{\sqrt{3}}{2}i}\right)^4 = \left(-\dfrac{1}{2}-\dfrac{\sqrt{3}}{2}i\right)^4 = \left[\left(-\dfrac{1}{2}-\dfrac{\sqrt{3}}{2}i\right)^2\right]^2$

$$= \left[\dfrac{1}{4}+\dfrac{\sqrt{3}}{2}i+\dfrac{3}{4}i^2\right]^2 = \left(-\dfrac{1}{2}+\dfrac{\sqrt{3}}{2}i\right)^2 = \dfrac{1}{4}-\dfrac{\sqrt{3}}{2}i+\dfrac{3}{4}i^2 = -\dfrac{1}{2}-\dfrac{\sqrt{3}}{2}i$$

(d) $\left|\dfrac{2z_2+z_1-5-i}{2z_1-z_2+3-i}\right|^2 = \left|\dfrac{2(3-2i)+(2+i)-5-i}{2(2+i)-(3-2i)+3-i}\right|^2$

$$= \left|\dfrac{3-4i}{4+3i}\right|^2 = \dfrac{|3-4i|^2}{|4+3i|^2} = \dfrac{(\sqrt{(3)^2+(-4)^2})^2}{(\sqrt{(4)^2+(3)^2})^2} = 1$$

3.   Find real numbers x and y such that $3x+2iy-ix+5y = 7+5i$.

     The given equation can be written as $3x+5y+i(2y-x) = 7+5i$. Then equating real and imaginary parts, $3x+5y = 7$, $2y-x = 5$. Solving simultaneously, $x = -1$, $y = 2$.

4.   Prove: (a) $\overline{z_1+z_2} = \bar{z}_1+\bar{z}_2$,   (b) $|z_1z_2| = |z_1|\,|z_2|$.

     Let $z_1 = x_1+iy_1$, $z_2 = x_2+iy_2$.   Then

(a) $\overline{z_1+z_2} = \overline{x_1+iy_1+x_2+iy_2} = \overline{x_1+x_2+i(y_1+y_2)}$

$$= x_1+x_2-i(y_1+y_2) = x_1-iy_1+x_2-iy_2 = \overline{x_1+iy_1}+\overline{x_2+iy_2} = \bar{z}_1+\bar{z}_2$$

(b) $|z_1z_2| = |(x_1+iy_1)(x_2+iy_2)| = |x_1x_2-y_1y_2+i(x_1y_2+y_1x_2)|$

$$= \sqrt{(x_1x_2-y_1y_2)^2+(x_1y_2+y_1x_2)^2} = \sqrt{(x_1^2+y_1^2)(x_2^2+y_2^2)} = \sqrt{x_1^2+y_1^2}\sqrt{x_2^2+y_2^2} = |z_1|\,|z_2|$$

     Another method.

$$|z_1z_2|^2 = (z_1z_2)(\overline{z_1z_2}) = z_1z_2\bar{z}_1\bar{z}_2 = (z_1\bar{z}_1)(z_2\bar{z}_2) = |z_1|^2|z_2|^2 \quad \text{or} \quad |z_1z_2| = |z_1|\,|z_2|$$

     where we have used the fact that the conjugate of a product of two complex numbers is equal to the product of their conjugates (see Problem 55).

GRAPHICAL REPRESENTATION OF COMPLEX NUMBERS. VECTORS

5. Perform the indicated operations both analytically and graphically:

(a) $(3 + 4i) + (5 + 2i)$, (b) $(6 - 2i) - (2 - 5i)$, (c) $(-3 + 5i) + (4 + 2i) + (5 - 3i) + (-4 - 6i)$.

(a) *Analytically.* $(3 + 4i) + (5 + 2i) = 3 + 5 + 4i + 2i = 8 + 6i$

Graphically. Represent the two complex numbers by points P_1 and P_2 respectively as in Fig. 1-7 below. Complete the parallelogram with OP_1 and OP_2 as adjacent sides. Point P represents the sum, $8 + 6i$, of the two given complex numbers. Note the similarity with the parallelogram law for addition of vectors OP_1 and OP_2 to obtain vector OP. For this reason it is often convenient to consider a complex number $a + bi$ as a vector having *components* a and b in the directions of the positive x and y axes respectively.

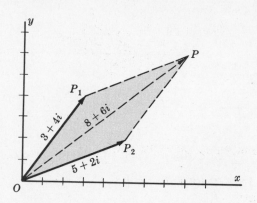

Fig. 1-7

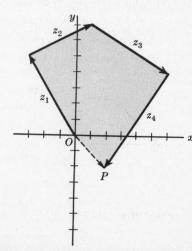

Fig. 1-8

(b) *Analytically.* $(6 - 2i) - (2 - 5i) = 6 - 2 - 2i + 5i = 4 + 3i$

Graphically. $(6 - 2i) - (2 - 5i) = 6 - 2i + (-2 + 5i)$. We now add $6 - 2i$ and $(-2 + 5i)$ as in part (a). The result is indicated by OP in Fig. 1-8 above.

(c) *Analytically.*

$(-3 + 5i) + (4 + 2i) + (5 - 3i) + (-4 - 6i) = (-3 + 4 + 5 - 4) + (5i + 2i - 3i - 6i) = 2 - 2i$

Graphically. Represent the numbers to be added by z_1, z_2, z_3, z_4 respectively. These are shown graphically in Fig. 1-9. To find the required sum proceed as shown in Fig. 1-10. At the terminal point of vector z_1 construct vector z_2. At the terminal point of z_2 construct vector z_3, and at the terminal point of z_3 construct vector z_4. The required sum, sometimes called the *resultant*, is obtained by constructing the vector OP from the initial point of z_1 to the terminal point of z_4, i.e. $OP = z_1 + z_2 + z_3 + z_4 = 2 - 2i$.

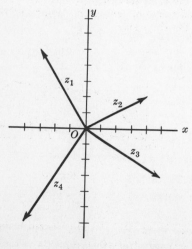

Fig. 1-9

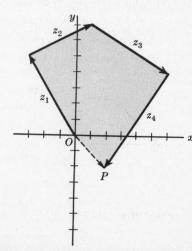

Fig. 1-10

6. If z_1 and z_2 are two given complex numbers (vectors) as in Fig. 1-11, construct graphically

(a) $3z_1 - 2z_2$ (b) $\frac{1}{2}z_2 + \frac{5}{3}z_1$

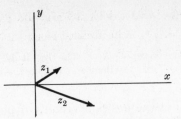

Fig. 1-11

(a) In Fig. 1-12 below, $OA = 3z_1$ is a vector having length 3 times vector z_1 and the same direction.

$OB = -2z_2$ is a vector having length 2 times vector z_2 and the opposite direction.

Then vector $OC = OA + OB = 3z_1 - 2z_2$.

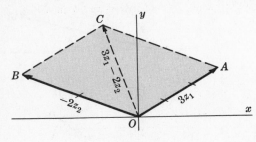

Fig. 1-12

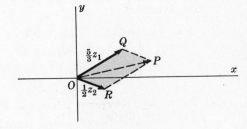

Fig. 1-13

(b) The required vector (complex number) is represented by OP in Fig. 1-13 above.

7. Prove   (a) $|z_1 + z_2| \leqq |z_1| + |z_2|$,   (b) $|z_1 + z_2 + z_3| \leqq |z_1| + |z_2| + |z_3|$,   (c) $|z_1 - z_2| \geqq |z_1| - |z_2|$ and give a graphical interpretation.

(a) *Analytically.*   Let $z_1 = x_1 + iy_1$, $z_2 = x_2 + iy_2$. Then we must show that

$$\sqrt{(x_1 + x_2)^2 + (y_1 + y_2)^2} \leqq \sqrt{x_1^2 + y_1^2} + \sqrt{x_2^2 + y_2^2}$$

Squaring both sides, this will be true if

$$(x_1 + x_2)^2 + (y_1 + y_2)^2 \leqq x_1^2 + y_1^2 + 2\sqrt{(x_1^2 + y_1^2)(x_2^2 + y_2^2)} + x_2^2 + y_2^2$$

i.e. if $$x_1 x_2 + y_1 y_2 \leqq \sqrt{(x_1^2 + y_1^2)(x_2^2 + y_2^2)}$$

or if (squaring both sides again)

$$x_1^2 x_2^2 + 2x_1 x_2 y_1 y_2 + y_1^2 y_2^2 \leqq x_1^2 x_2^2 + x_1^2 y_2^2 + y_1^2 x_2^2 + y_1^2 y_2^2$$

or $$2x_1 x_2 y_1 y_2 \leqq x_1^2 y_2^2 + y_1^2 x_2^2$$

But this is equivalent to $(x_1 y_2 - x_2 y_1)^2 \geqq 0$ which is true. Reversing the steps, which are reversible, proves the result.

Graphically.   The result follows graphically from the fact that $|z_1|$, $|z_2|$, $|z_1 + z_2|$ represent the lengths of the sides of a triangle (see Fig. 1-14) and that the sum of the lengths of two sides of a triangle is greater than or equal to the length of the third side.

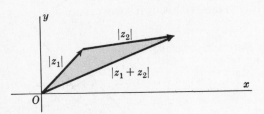

Fig. 1-14

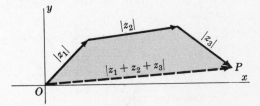

Fig. 1-15

(b) *Analytically.*   By part (a),

$$|z_1 + z_2 + z_3| = |z_1 + (z_2 + z_3)| \leqq |z_1| + |z_2 + z_3| \leqq |z_1| + |z_2| + |z_3|$$

Graphically.   The result is a consequence of the geometric fact that in a plane a straight line is the shortest distance between two points O and P (see Fig. 1-15).

(c) *Analytically.* By part (a), $|z_1| = |z_1 - z_2 + z_2| \leqq |z_1 - z_2| + |z_2|$. Then $|z_1 - z_2| \geqq |z_1| - |z_2|$. An equivalent result obtained on replacing z_2 by $-z_2$ is $|z_1 + z_2| \geqq |z_1| - |z_2|$.

Graphically. The result is equivalent to the statement that a side of a triangle has length greater than or equal to the difference in lengths of the other two sides.

8. Let the position vectors of points $A(x_1, y_1)$ and $B(x_2, y_2)$ be represented by z_1 and z_2 respectively. (a) Represent the vector AB as a complex number. (b) Find the distance between points A and B.

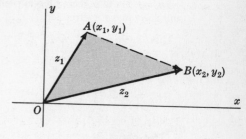

(a) From Fig. 1-16, $OA + AB = OB$ or
$$AB = OB - OA = z_2 - z_1$$
$$= (x_2 + iy_2) - (x_1 + iy_1)$$
$$= (x_2 - x_1) + i(y_2 - y_1)$$

Fig. 1-16

(b) The distance between points A and B is given by
$$|AB| = |(x_2 - x_1) + i(y_2 - y_1)| = \sqrt{(x_2 - x_1)^2 + (y_2 - y_1)^2}$$

9. Let $z_1 = x_1 + iy_1$ and $z_2 = x_2 + iy_2$ represent two non-collinear or non-parallel vectors. If a and b are real numbers (scalars) such that $az_1 + bz_2 = 0$, prove that $a = 0$ and $b = 0$.

The given condition $az_1 + bz_2 = 0$ is equivalent to $a(x_1 + iy_1) + b(x_2 + iy_2) = 0$ or $ax_1 + bx_2 + i(ay_1 + by_2) = 0$. Then $ax_1 + bx_2 = 0$ and $ay_1 + by_2 = 0$. These equations have the simultaneous solution $a = 0$, $b = 0$ if $y_1/x_1 \neq y_2/x_2$, i.e. if the vectors are non-collinear or non-parallel vectors.

10. Prove that the diagonals of a parallelogram bisect each other.

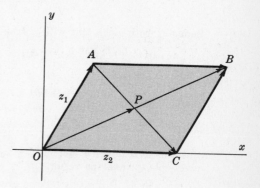

Let $OABC$ [Fig. 1-17] be the given parallelogram with diagonals intersecting at P.

Since $z_1 + AC = z_2$, $AC = z_2 - z_1$. Then $AP = m(z_2 - z_1)$ where $0 \leqq m \leqq 1$.

Since $OB = z_1 + z_2$, $OP = n(z_1 + z_2)$ where $0 \leqq n \leqq 1$.

But $OA + AP = OP$, i.e. $z_1 + m(z_2 - z_1) = n(z_1 + z_2)$ or $(1 - m - n)z_1 + (m - n)z_2 = 0$. Hence by Problem 9, $1 - m - n = 0$, $m - n = 0$ or $m = \frac{1}{2}$, $n = \frac{1}{2}$ and so P is the midpoint of both diagonals.

Fig. 1-17

11. Find an equation for the straight line which passes through two given points $A(x_1, y_1)$ and $B(x_2, y_2)$.

Let $z_1 = x_1 + iy_1$ and $z_2 = x_2 + iy_2$ be the position vectors of A and B respectively. Let $z = x + iy$ be the position vector of any point P on the line joining A and B.

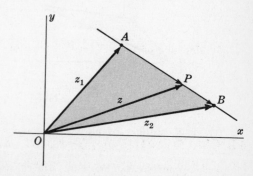

From Fig. 1-18,
$OA + AP = OP$ or $z_1 + AP = z$, i.e. $AP = z - z_1$
$OA + AB = OB$ or $z_1 + AB = z_2$, i.e. $AB = z_2 - z_1$

Since AP and AB are collinear, $AP = t AB$ or $z - z_1 = t(z_2 - z_1)$ where t is real, and the required equation is

$$z = z_1 + t(z_2 - z_1) \quad \text{or} \quad z = (1 - t)z_1 + tz_2$$

Fig. 1-18

Using $z_1 = x_1 + iy_1$, $z_2 = x_2 + iy_2$ and $z = x + iy$, this can be written

$$x - x_1 = t(x_2 - x_1), \quad y - y_1 = t(y_2 - y_1) \qquad \text{or} \qquad \frac{x - x_1}{x_2 - x_1} = \frac{y - y_1}{y_2 - y_1}$$

The first two are called *parametric equations* of the line and t is the parameter; the second is called the equation of the line in *standard form*.

Another method. Since AP and PB are collinear, we have for real numbers m and n:

$$mAP = nPB \qquad \text{or} \qquad m(z - z_1) = n(z_2 - z)$$

Solving, $\qquad z = \dfrac{mz_1 + nz_2}{m + n} \qquad$ or $\qquad x = \dfrac{mx_1 + nx_2}{m + n}, \quad y = \dfrac{my_1 + ny_2}{m + n}$

which is called the *symmetric form*.

12. Let $A(1, -2)$, $B(-3, 4)$, $C(2, 2)$ be the three vertices of triangle ABC. Find the length of the median from C to the side AB.

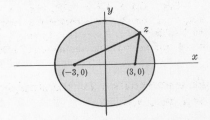

> The position vectors of A, B and C are given by $z_1 = 1 - 2i$, $z_2 = -3 + 4i$ and $z_3 = 2 + 2i$ respectively. Then from Fig. 1-19,
>
> $AC = z_3 - z_1 = 2 + 2i - (1 - 2i) = 1 + 4i$
> $BC = z_3 - z_2 = 2 + 2i - (-3 + 4i) = 5 - 2i$
> $AB = z_2 - z_1 = -3 + 4i - (1 - 2i) = -4 + 6i$
> $AD = \frac{1}{2}AB = \frac{1}{2}(-4 + 6i) = -2 + 3i \quad$ since D is the midpoint of AB.
>
> $AC + CD = AD \qquad$ or $\qquad CD = AD - AC = -2 + 3i - (1 + 4i) = -3 - i$.
>
> Then the length of median CD is $\quad |CD| = |-3 - i| = \sqrt{10}$.

Fig. 1-19

13. Find an equation for (a) a circle of radius 4 with center at $(-2, 1)$, (b) an ellipse with major axis of length 10 and foci at $(-3, 0)$ and $(3, 0)$.

> (a) The center can be represented by the complex number $-2 + i$. If z is any point on the circle [Fig. 1-20], the distance from z to $-2 + i$ is
>
> $$|z - (-2 + i)| = 4$$
>
> Then $|z + 2 - i| = 4$ is the required equation. In rectangular form this is given by
>
> $$|(x + 2) + i(y - 1)| = 4, \quad \text{i.e.} \quad (x + 2)^2 + (y - 1)^2 = 16$$

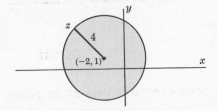

Fig. 1-20

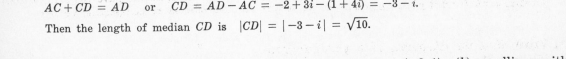

Fig. 1-21

> (b) The sum of the distances from any point z on the ellipse [Fig. 1-21] to the foci must equal 10. Hence the required equation is
>
> $$|z + 3| + |z - 3| = 10$$
>
> In rectangular form this reduces to $\quad x^2/25 + y^2/16 = 1 \quad$ (see Problem 74).

AXIOMATIC FOUNDATIONS OF COMPLEX NUMBERS

14. Use the definition of a complex number as an ordered pair of real numbers and the definitions on Page 3 to prove that $(a, b) = a(1, 0) + b(0, 1)$ where $(0, 1)(0, 1) = (-1, 0)$.

From the definitions of sum and product on Page 3, we have

$$(a, b) \;=\; (a, 0) + (0, b) \;=\; a(1, 0) + b(0, 1)$$

where

$$(0, 1)(0, 1) \;=\; (0 \cdot 0 - 1 \cdot 1, \; 0 \cdot 1 + 1 \cdot 0) \;=\; (-1, 0)$$

By identifying $(1, 0)$ with 1 and $(0, 1)$ with i, we see that $(a, b) = a + bi$.

15. If $z_1 = (a_1, b_1)$, $z_2 = (a_2, b_2)$ and $z_3 = (a_3, b_3)$, prove the distributive law: $z_1(z_2 + z_3) = z_1 z_2 + z_1 z_3$.

We have
$$
\begin{aligned}
z_1(z_2 + z_3) \;&=\; (a_1, b_1)\{(a_2, b_2) + (a_3, b_3)\} \;=\; (a_1, b_1)(a_2 + a_3, \; b_2 + b_3) \\
&=\; \{a_1(a_2 + a_3) - b_1(b_2 + b_3), \; a_1(b_2 + b_3) + b_1(a_2 + a_3)\} \\
&=\; (a_1 a_2 - b_1 b_2 + a_1 a_3 - b_1 b_3, \; a_1 b_2 + b_1 a_2 + a_1 b_3 + b_1 a_3) \\
&=\; (a_1 a_2 - b_1 b_2, \; a_1 b_2 + b_1 a_2) + (a_1 a_3 - b_1 b_3, \; a_1 b_3 + b_1 a_3) \\
&=\; (a_1, b_1)(a_2, b_2) + (a_1, b_1)(a_3, b_3) \;=\; z_1 z_2 + z_1 z_3
\end{aligned}
$$

POLAR FORM OF COMPLEX NUMBERS

16. Express each of the following complex numbers in polar form.

(a) $2 + 2\sqrt{3}\,i$

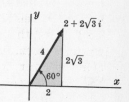

Modulus or absolute value, $r = |2 + 2\sqrt{3}\,i| = \sqrt{4 + 12} = 4$.

Amplitude or argument, $\theta = \sin^{-1} 2\sqrt{3}/4 = \sin^{-1} \sqrt{3}/2 = 60° = \pi/3$ (radians).

Then
$$
\begin{aligned}
2 + 2\sqrt{3}\,i \;&=\; r(\cos\theta + i\sin\theta) \;=\; 4(\cos 60° + i\sin 60°) \\
&=\; 4(\cos\pi/3 + i\sin\pi/3)
\end{aligned}
$$

Fig. 1-22

The result can also be written as 4 cis $\pi/3$ or, using Euler's formula, as $4e^{\pi i/3}$.

(b) $-5 + 5i$

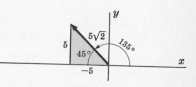

$$r = |-5 + 5i| = \sqrt{25 + 25} = 5\sqrt{2}$$
$$\theta = 180° - 45° = 135° = 3\pi/4 \text{ (radians)}$$

Then
$$
\begin{aligned}
-5 + 5i \;&=\; 5\sqrt{2}\,(\cos 135° + i\sin 135°) \\
&=\; 5\sqrt{2} \text{ cis } 3\pi/4 \;=\; 5\sqrt{2}\,e^{3\pi i/4}
\end{aligned}
$$

Fig. 1-23

(c) $-\sqrt{6} - \sqrt{2}\,i$

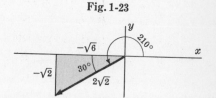

$$r = |-\sqrt{6} - \sqrt{2}\,i| = \sqrt{6 + 2} = 2\sqrt{2}$$
$$\theta = 180° + 30° = 210° = 7\pi/6 \text{ (radians)}$$

Then
$$
\begin{aligned}
-\sqrt{6} - \sqrt{2}\,i \;&=\; 2\sqrt{2}\,(\cos 210° + i\sin 210°) \\
&=\; 2\sqrt{2} \text{ cis } 7\pi/6 \;=\; 2\sqrt{2}\,e^{7\pi i/6}
\end{aligned}
$$

Fig. 1-24

(d) $-3i$

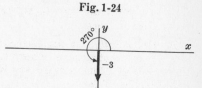

$$r = |-3i| = |0 - 3i| = \sqrt{0 + 9} = 3$$
$$\theta = 270° = 3\pi/2 \text{ (radians)}$$

Then
$$
\begin{aligned}
-3i \;&=\; 3(\cos 3\pi/2 + i\sin 3\pi/2) \\
&=\; 3 \text{ cis } 3\pi/2 \;=\; 3e^{3\pi i/2}
\end{aligned}
$$

Fig. 1-25

17. Graph each of the following: (a) $6(\cos 240° + i\sin 240°)$,   (b) $4e^{3\pi i/5}$,   (c) $2e^{-\pi i/4}$.

(a)
$$6(\cos 240° + i\sin 240°) \;=\; 6 \text{ cis } 240° \;=\; 6 \text{ cis } 4\pi/3 \;=\; 6\,e^{4\pi i/3}$$
can be represented graphically by OP in Fig. 1-26 below.

If we start with vector OA, whose magnitude is 6 and whose direction is that of the positive x axis, we can obtain OP by rotating OA counterclockwise through an angle of 240°. In general, $re^{i\theta}$ is equivalent to a vector obtained by rotating a vector of magnitude r and direction that of the positive x axis, counterclockwise through an angle θ.

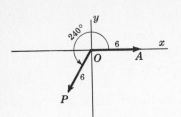

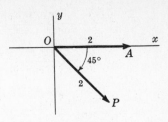

Fig. 1-26   Fig. 1-27   Fig. 1-28

(b) $4\,e^{3\pi i/5} = 4(\cos 3\pi/5 + i \sin 3\pi/5) = 4(\cos 108° + i \sin 108°)$

is represented by OP in Fig. 1-27 above.

(c) $2\,e^{-\pi i/4} = 2\{\cos (-\pi/4) + i \sin (-\pi/4)\} = 2\{\cos (-45°) + i \sin (-45°)\}$

   This complex number can be represented by vector OP in Fig. 1-28 above. This vector can be obtained by starting with vector OA, whose magnitude is 2 and whose direction is that of the positive x axis, and rotating it counterclockwise through an angle of $-45°$ (which is the same as rotating it *clockwise* through an angle of $45°$).

18. A man travels 12 miles northeast, 20 miles 30° west of north, and then 18 miles 60° south of west. Determine (a) analytically and (b) graphically how far and in what direction he is from his starting point.

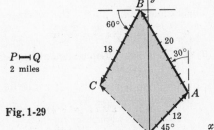

Fig. 1-29

(a) *Analytically.* Let O be the starting point (see Fig. 1-29). Then the successive displacements are represented by vectors OA, AB and BC. The result of all three displacements is represented by the vector

$$OC = OA + AB + BC$$

Now
$$OA = 12(\cos 45° + i \sin 45°) = 12\,e^{\pi i/4}$$
$$AB = 20\{\cos (90° + 30°) + i \sin (90° + 30°)\} = 20\,e^{2\pi i/3}$$
$$BC = 18\{\cos (180° + 60°) + i \sin (180° + 60°)\} = 18\,e^{4\pi i/3}$$

Then
$$OC = 12\,e^{\pi i/4} + 20\,e^{2\pi i/3} + 18\,e^{4\pi i/3}$$
$$= \{12 \cos 45° + 20 \cos 120° + 18 \cos 240°\} + i\{12 \sin 45° + 20 \sin 120° + 18 \sin 240°\}$$
$$= \{(12)(\sqrt{2}/2) + (20)(-1/2) + (18)(-1/2)\} + i\{(12)(\sqrt{2}/2) + (20)(\sqrt{3}/2) + (18)(-\sqrt{3}/2)\}$$
$$= (6\sqrt{2} - 19) + (6\sqrt{2} + \sqrt{3})i$$

If $r(\cos \theta + i \sin \theta) = 6\sqrt{2} - 19 + (6\sqrt{2} + \sqrt{3})i$, then $r = \sqrt{(6\sqrt{2} - 19)^2 + (6\sqrt{2} + \sqrt{3})^2} =$ 14.7 approximately, and $\theta = \cos^{-1}(6\sqrt{2} - 19)/r = \cos^{-1}(-.717) = 135°49'$ approximately.

   Thus the man is 14.7 miles from his starting point in a direction $135°49' - 90° = 45°49'$ west of north.

(b) *Graphically.* Using a convenient unit of length such as PQ in Fig. 1-29 which represents 2 miles, and a protractor to measure angles, construct vectors OA, AB and BC. Then by determining the number of units in OC and the angle which OC makes with the y axis, we obtain the approximate results of (a).

DE MOIVRE'S THEOREM

19. If $z_1 = r_1(\cos \theta_1 + i \sin \theta_1)$ and $z_2 = r_2(\cos \theta_2 + i \sin \theta_2)$,   prove:

(a) $z_1 z_2 = r_1 r_2 \{\cos (\theta_1 + \theta_2) + i \sin (\theta_1 + \theta_2)\}$

(b) $\dfrac{z_1}{z_2} = \dfrac{r_1}{r_2} \{\cos (\theta_1 - \theta_2) + i \sin (\theta_1 - \theta_2)\}$.

(a) $z_1 z_2 = \{r_1(\cos \theta_1 + i \sin \theta_1)\}\{r_2(\cos \theta_2 + i \sin \theta_2)\}$
$$= r_1 r_2\{(\cos \theta_1 \cos \theta_2 - \sin \theta_1 \sin \theta_2) + i(\sin \theta_1 \cos \theta_2 + \cos \theta_1 \sin \theta_2)\}$$
$$= r_1 r_2\{\cos (\theta_1 + \theta_2) + i \sin (\theta_1 + \theta_2)\}$$

$(b)\ \dfrac{z_1}{z_2}\ =\ \dfrac{r_1(\cos\theta_1 + i\sin\theta_1)}{r_2(\cos\theta_2 + i\sin\theta_2)}\cdot\dfrac{(\cos\theta_2 - i\sin\theta_2)}{(\cos\theta_2 - i\sin\theta_2)}$

$\qquad\qquad =\ \dfrac{r_1}{r_2}\left\{\dfrac{(\cos\theta_1\cos\theta_2 + \sin\theta_1\sin\theta_2)\ +\ i(\sin\theta_1\cos\theta_2 - \cos\theta_1\sin\theta_2)}{\cos^2\theta_2 + \sin^2\theta_2}\right\}$

$\qquad\qquad =\ \dfrac{r_1}{r_2}\{\cos(\theta_1 - \theta_2) + i\sin(\theta_1 - \theta_2)\}$

In terms of Euler's formula $e^{i\theta} = \cos\theta + i\sin\theta$, the results state that if $z_1 = r_1 e^{i\theta_1}$ and $z_2 = r_2 e^{i\theta_2}$, then $z_1 z_2 = r_1 r_2\, e^{i(\theta_1 + \theta_2)}$ and $\dfrac{z_1}{z_2} = \dfrac{r_1 e^{i\theta_1}}{r_2 e^{i\theta_2}} = \dfrac{r_1}{r_2}\, e^{i(\theta_1 - \theta_2)}$.

20. Prove De Moivre's theorem: $(\cos\theta + i\sin\theta)^n = \cos n\theta + i\sin n\theta$ where n is any positive integer.

We use the *principle of mathematical induction*. Assume that the result is true for the particular positive integer k, i.e. assume $(\cos\theta + i\sin\theta)^k = \cos k\theta + i\sin k\theta$. Then multiplying both sides by $\cos\theta + i\sin\theta$, we find

$(\cos\theta + i\sin\theta)^{k+1} = (\cos k\theta + i\sin k\theta)(\cos\theta + i\sin\theta) = \cos(k+1)\theta + i\sin(k+1)\theta$

by Problem 19. Thus *if* the result is true for $n = k$, then it is also true for $n = k+1$. But since the result is clearly true for $n = 1$, it must also be true for $n = 1+1 = 2$ and $n = 2+1 = 3$, etc., and so must be true for all positive integers.

The result is equivalent to the statement $(e^{i\theta})^n = e^{ni\theta}$.

21. Prove the identities: $(a)\ \cos 5\theta = 16\cos^5\theta - 20\cos^3\theta + 5\cos\theta$; $(b)\ (\sin 5\theta)/(\sin\theta) = 16\cos^4\theta - 12\cos^2\theta + 1$, if $\theta \neq 0, \pm\pi, \pm 2\pi, \ldots$.

We use the *binomial formula*

$$(a+b)^n = a^n + \tbinom{n}{1}a^{n-1}b + \tbinom{n}{2}a^{n-2}b^2 + \cdots + \tbinom{n}{r}a^{n-r}b^r + \cdots + b^n$$

where the coefficients $\tbinom{n}{r} = \dfrac{n!}{r!\,(n-r)!}$, also denoted by $_nC_r$, are called the *binomial coefficients*. The number $n!$ or *factorial* n, is defined as the product $1\cdot 2\cdot 3\cdots n$ and we define $0! = 1$.

From Problem 20, with $n = 5$, and the binomial formula,

$\cos 5\theta + i\sin 5\theta\ =\ (\cos\theta + i\sin\theta)^5$

$\qquad =\ \cos^5\theta\ +\ \tbinom{5}{1}(\cos^4\theta)(i\sin\theta)\ +\ \tbinom{5}{2}(\cos^3\theta)(i\sin\theta)^2$
$\qquad\qquad +\ \tbinom{5}{3}(\cos^2\theta)(i\sin\theta)^3\ +\ \tbinom{5}{4}(\cos\theta)(i\sin\theta)^4\ +\ (i\sin\theta)^5$

$\qquad =\ \cos^5\theta\ +\ 5i\cos^4\theta\sin\theta\ -\ 10\cos^3\theta\sin^2\theta$
$\qquad\qquad -\ 10i\cos^2\theta\sin^3\theta\ +\ 5\cos\theta\sin^4\theta\ +\ i\sin^5\theta$

$\qquad =\ \cos^5\theta\ -\ 10\cos^3\theta\sin^2\theta\ +\ 5\cos\theta\sin^4\theta$
$\qquad\qquad +\ i(5\cos^4\theta\sin\theta - 10\cos^2\theta\sin^3\theta + \sin^5\theta)$

Hence

$(a)\qquad \cos 5\theta\ =\ \cos^5\theta\ -\ 10\cos^3\theta\sin^2\theta\ +\ 5\cos\theta\sin^4\theta$

$\qquad\qquad\quad =\ \cos^5\theta\ -\ 10\cos^3\theta\,(1 - \cos^2\theta)\ +\ 5\cos\theta\,(1 - \cos^2\theta)^2$

$\qquad\qquad\quad =\ 16\cos^5\theta\ -\ 20\cos^3\theta\ +\ 5\cos\theta$

and

$(b)\qquad \sin 5\theta\ =\ 5\cos^4\theta\sin\theta\ -\ 10\cos^2\theta\sin^3\theta\ +\ \sin^5\theta$

or

$\qquad\qquad \dfrac{\sin 5\theta}{\sin\theta}\ =\ 5\cos^4\theta\ -\ 10\cos^2\theta\sin^2\theta\ +\ \sin^4\theta$

$\qquad\qquad\qquad =\ 5\cos^4\theta\ -\ 10\cos^2\theta\,(1 - \cos^2\theta)\ +\ (1 - \cos^2\theta)^2$

$\qquad\qquad\qquad =\ 16\cos^4\theta\ -\ 12\cos^2\theta\ +\ 1$

provided $\sin\theta \neq 0$, i.e. $\theta \neq 0, \pm\pi, \pm 2\pi, \ldots$.

22. Show that $(a)\ \cos\theta = \dfrac{e^{i\theta} + e^{-i\theta}}{2}$, $(b)\ \sin\theta = \dfrac{e^{i\theta} - e^{-i\theta}}{2i}$.

We have $\qquad (1)\ e^{i\theta} = \cos\theta + i\sin\theta$, $\qquad (2)\ e^{-i\theta} = \cos\theta - i\sin\theta$

(a) Adding (1) and (2), $e^{i\theta} + e^{-i\theta} = 2\cos\theta$ or $\cos\theta = \dfrac{e^{i\theta} + e^{-i\theta}}{2}$

(b) Subtracting (2) from (1), $e^{i\theta} - e^{-i\theta} = 2i\sin\theta$ or $\sin\theta = \dfrac{e^{i\theta} - e^{-i\theta}}{2i}$

23. Prove the identities (a) $\sin^3\theta = \frac{3}{4}\sin\theta - \frac{1}{4}\sin 3\theta$, (b) $\cos^4\theta = \frac{1}{8}\cos 4\theta + \frac{1}{2}\cos 2\theta + \frac{3}{8}$.

(a) $\sin^3\theta = \left(\dfrac{e^{i\theta} - e^{-i\theta}}{2i}\right)^3 = \dfrac{(e^{i\theta} - e^{-i\theta})^3}{8i^3} = -\dfrac{1}{8i}\{(e^{i\theta})^3 - 3(e^{i\theta})^2(e^{-i\theta}) + 3(e^{i\theta})(e^{-i\theta})^2 - (e^{-i\theta})^3\}$

$= -\dfrac{1}{8i}(e^{3i\theta} - 3e^{i\theta} + 3e^{-i\theta} - e^{-3i\theta}) = \dfrac{3}{4}\left(\dfrac{e^{i\theta} - e^{-i\theta}}{2i}\right) - \dfrac{1}{4}\left(\dfrac{e^{3i\theta} - e^{-3i\theta}}{2i}\right)$

$= \dfrac{3}{4}\sin\theta - \dfrac{1}{4}\sin 3\theta$

(b) $\cos^4\theta = \left(\dfrac{e^{i\theta} + e^{-i\theta}}{2}\right)^4 = \dfrac{(e^{i\theta} + e^{-i\theta})^4}{16}$

$= \dfrac{1}{16}\{(e^{i\theta})^4 + 4(e^{i\theta})^3(e^{-i\theta}) + 6(e^{i\theta})^2(e^{-i\theta})^2 + 4(e^{i\theta})(e^{-i\theta})^3 + (e^{-i\theta})^4\}$

$= \dfrac{1}{16}(e^{4i\theta} + 4e^{2i\theta} + 6 + 4e^{-2i\theta} + e^{-4i\theta}) = \dfrac{1}{8}\left(\dfrac{e^{4i\theta} + e^{-4i\theta}}{2}\right) + \dfrac{1}{2}\left(\dfrac{e^{2i\theta} + e^{-2i\theta}}{2}\right) + \dfrac{3}{8}$

$= \dfrac{1}{8}\cos 4\theta + \dfrac{1}{2}\cos 2\theta + \dfrac{3}{8}$

24. Given a complex number (vector) z, interpret geometrically $ze^{i\alpha}$ where α is real.

Let $z = re^{i\theta}$ be represented graphically by vector OA in Fig. 1-30. Then

$$ze^{i\alpha} = re^{i\theta} \cdot e^{i\alpha} = re^{i(\theta + \alpha)}$$

is the vector represented by OB.

Hence multiplication of a vector z by $e^{i\alpha}$ amounts to rotating z counterclockwise through angle α. We can consider $e^{i\alpha}$ as an *operator* which acts on z to produce this rotation.

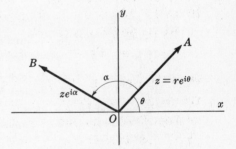

Fig. 1-30

25. Prove: $e^{i\theta} = e^{i(\theta + 2k\pi)}$, $k = 0, \pm 1, \pm 2, \ldots$.

$e^{i(\theta + 2k\pi)} = \cos(\theta + 2k\pi) + i\sin(\theta + 2k\pi) = \cos\theta + i\sin\theta = e^{i\theta}$

26. Evaluate each of the following.

(a) $[3(\cos 40° + i\sin 40°)][4(\cos 80° + i\sin 80°)] = 3 \cdot 4[\cos(40° + 80°) + i\sin(40° + 80°)]$

$= 12(\cos 120° + i\sin 120°)$

$= 12\left(-\dfrac{1}{2} + \dfrac{\sqrt{3}}{2}i\right) = -6 + 6\sqrt{3}\,i$

(b) $\dfrac{(2\text{ cis }15°)^7}{(4\text{ cis }45°)^3} = \dfrac{128\text{ cis }105°}{64\text{ cis }135°} = 2\text{ cis }(105° - 135°)$

$= 2[\cos(-30°) + i\sin(-30°)] = 2[\cos 30° - i\sin 30°] = \sqrt{3} - i$

(c) $\left(\dfrac{1 + \sqrt{3}\,i}{1 - \sqrt{3}\,i}\right)^{10} = \left\{\dfrac{2\text{ cis }(60°)}{2\text{ cis }(-60°)}\right\}^{10} = (\text{cis }120°)^{10} = \text{cis }1200° = \text{cis }120° = -\dfrac{1}{2} + \dfrac{\sqrt{3}}{2}i$

Another method.

$\left(\dfrac{1 + \sqrt{3}\,i}{1 - \sqrt{3}\,i}\right)^{10} = \left(\dfrac{2e^{\pi i/3}}{2e^{-\pi i/3}}\right)^{10} = (e^{2\pi i/3})^{10} = e^{20\pi i/3}$

$= e^{6\pi i}e^{2\pi i/3} = (1)[\cos(2\pi/3) + i\sin(2\pi/3)] = -\dfrac{1}{2} + \dfrac{\sqrt{3}}{2}i$

27. Prove that   (*a*) $\arg(z_1 z_2) = \arg z_1 + \arg z_2$,   (*b*) $\arg(z_1/z_2) = \arg z_1 - \arg z_2$, stating appropriate conditions of validity.

Let $z_1 = r_1(\cos\theta_1 + i\sin\theta_1)$, $z_2 = r_2(\cos\theta_2 + i\sin\theta_2)$.   Then $\arg z_1 = \theta_1$, $\arg z_2 = \theta_2$.

(*a*) Since $z_1 z_2 = r_1 r_2\{\cos(\theta_1 + \theta_2) + i\sin(\theta_1 + \theta_2)\}$, $\arg(z_1 z_2) = \theta_1 + \theta_2 = \arg z_1 + \arg z_2$.

(*b*) Since $\dfrac{z_1}{z_2} = \dfrac{r_1}{r_2}\{\cos(\theta_1 - \theta_2) + i\sin(\theta_1 - \theta_2)\}$, $\arg\left(\dfrac{z_1}{z_2}\right) = \theta_1 - \theta_2 = \arg z_1 - \arg z_2$.

Since there are many possible values for $\theta_1 = \arg z_1$ and $\theta_2 = \arg z_2$, we can only say that the two sides in the above equalities are equal for *some* values of $\arg z_1$ and $\arg z_2$. They may not hold even if principal values are used.

ROOTS OF COMPLEX NUMBERS

28. (*a*) Find all values of z for which $z^5 = -32$,   and   (*b*) locate these values in the complex plane.

(*a*) In polar form, $-32 = 32\{\cos(\pi + 2k\pi) + i\sin(\pi + 2k\pi)\}$, $k = 0, \pm1, \pm2, \cdots$.

Let $z = r(\cos\theta + i\sin\theta)$.   Then by De Moivre's theorem,

$$z^5 = r^5(\cos 5\theta + i\sin 5\theta) = 32\{\cos(\pi + 2k\pi) + i\sin(\pi + 2k\pi)\}$$

and so $r^5 = 32$, $5\theta = \pi + 2k\pi$,   from which $r = 2$, $\theta = (\pi + 2k\pi)/5$.   Hence

$$z = 2\left\{\cos\left(\frac{\pi + 2k\pi}{5}\right) + i\sin\left(\frac{\pi + 2k\pi}{5}\right)\right\}$$

If $k = 0$, $z = z_1 = 2(\cos\pi/5 + i\sin\pi/5)$.

If $k = 1$, $z = z_2 = 2(\cos 3\pi/5 + i\sin 3\pi/5)$.

If $k = 2$, $z = z_3 = 2(\cos 5\pi/5 + i\sin 5\pi/5) = -2$.

If $k = 3$, $z = z_4 = 2(\cos 7\pi/5 + i\sin 7\pi/5)$.

If $k = 4$, $z = z_5 = 2(\cos 9\pi/5 + i\sin 9\pi/5)$.

By considering $k = 5, 6, \ldots$ as well as negative values, $-1, -2, \ldots$, repetitions of the above five values of z are obtained. Hence these are the only *solutions* or *roots* of the given equation. These five roots are called the *fifth roots of -32* and are collectively denoted by $(-32)^{1/5}$. In general, $a^{1/n}$ represents the nth roots of a and there are n such roots.

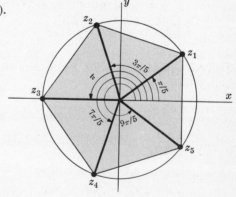

(*b*) The values of z are indicated in Fig. 1-31. Note that they are equally spaced along the circumference of a circle with center at the origin and radius 2. Another way of saying this is that the roots are represented by the vertices of a regular polygon.

Fig. 1-31

29. Find each of the indicated roots and locate them graphically.

(*a*) $(-1 + i)^{1/3}$

$$-1 + i = \sqrt{2}\{\cos(3\pi/4 + 2k\pi) + i\sin(3\pi/4 + 2k\pi)\}$$

$$(-1 + i)^{1/3} = 2^{1/6}\left\{\cos\left(\frac{3\pi/4 + 2k\pi}{3}\right) + i\sin\left(\frac{3\pi/4 + 2k\pi}{3}\right)\right\}$$

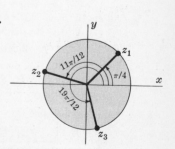

If $k = 0$, $z_1 = 2^{1/6}(\cos\pi/4 + i\sin\pi/4)$.

If $k = 1$, $z_2 = 2^{1/6}(\cos 11\pi/12 + i\sin 11\pi/12)$.

If $k = 2$, $z_3 = 2^{1/6}(\cos 19\pi/12 + i\sin 19\pi/12)$.

These are represented graphically in Fig. 1-32.

Fig. 1-32

(b) $(-2\sqrt{3} - 2i)^{1/4}$

$$-2\sqrt{3} - 2i = 4\{\cos(7\pi/6 + 2k\pi) + i\sin(7\pi/6 + 2k\pi)\}$$

$$(-2\sqrt{3} - 2i)^{1/4} = 4^{1/4}\left\{\cos\left(\frac{7\pi/6 + 2k\pi}{4}\right) + i\sin\left(\frac{7\pi/6 + 2k\pi}{4}\right)\right\}$$

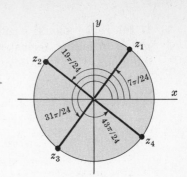

If $k = 0$, $z_1 = \sqrt{2}(\cos 7\pi/24 + i\sin 7\pi/24)$.

If $k = 1$, $z_2 = \sqrt{2}(\cos 19\pi/24 + i\sin 19\pi/24)$.

If $k = 2$, $z_3 = \sqrt{2}(\cos 31\pi/24 + i\sin 31\pi/24)$.

If $k = 3$, $z_4 = \sqrt{2}(\cos 43\pi/24 + i\sin 43\pi/24)$.

These are represented graphically in Fig. 1-33.

Fig. 1-33

30. Find the square roots of $-15 - 8i$.

Method 1.

$-15 - 8i = 17\{\cos(\theta + 2k\pi) + i\sin(\theta + 2k\pi)\}$ where $\cos\theta = -15/17$, $\sin\theta = -8/17$.

Then the square roots of $-15 - 8i$ are

$$\sqrt{17}(\cos\theta/2 + i\sin\theta/2) \tag{1}$$

and

$$\sqrt{17}\{\cos(\theta/2 + \pi) + i\sin(\theta/2 + \pi)\} = -\sqrt{17}(\cos\theta/2 + i\sin\theta/2) \tag{2}$$

Now

$$\cos\theta/2 = \pm\sqrt{(1 + \cos\theta)/2} = \pm\sqrt{(1 - 15/17)/2} = \pm 1/\sqrt{17}$$

$$\sin\theta/2 = \pm\sqrt{(1 - \cos\theta)/2} = \pm\sqrt{(1 + 15/17)/2} = \pm 4/\sqrt{17}$$

Since θ is an angle in the third quadrant, $\theta/2$ is an angle in the second quadrant. Hence $\cos\theta/2 = -1/\sqrt{17}$, $\sin\theta/2 = 4/\sqrt{17}$ and so from (1) and (2) the required square roots are $-1 + 4i$ and $1 - 4i$. As a check note that $(-1 + 4i)^2 = (1 - 4i)^2 = -15 - 8i$.

Method 2.

Let $p + iq$, where p and q are real, represent the required square roots. Then

$$(p + iq)^2 = p^2 - q^2 + 2pqi = -15 - 8i \qquad \text{or} \qquad (3)\ p^2 - q^2 = -15, \quad (4)\ pq = -4$$

Substituting $q = -4/p$ from (4) into (3), it becomes $p^2 - 16/p^2 = -15$ or $p^4 + 15p^2 - 16 = 0$, i.e. $(p^2 + 16)(p^2 - 1) = 0$ or $p^2 = -16$, $p^2 = 1$. Since p is real, $p = \pm 1$. From (4) if $p = 1$, $q = -4$; if $p = -1$, $q = 4$. Thus the roots are $-1 + 4i$ and $1 - 4i$.

POLYNOMIAL EQUATIONS

31. Solve the *quadratic equation* $az^2 + bz + c = 0$, $a \neq 0$.

Transposing c and dividing by $a \neq 0$, $\qquad z^2 + \dfrac{b}{a}z = -\dfrac{c}{a}$

Adding $\left(\dfrac{b}{2a}\right)^2$ [completing the square], $\qquad z^2 + \dfrac{b}{a}z + \left(\dfrac{b}{2a}\right)^2 = -\dfrac{c}{a} + \left(\dfrac{b}{2a}\right)^2$

Then $\qquad \left(z + \dfrac{b}{2a}\right)^2 = \dfrac{b^2 - 4ac}{4a^2}$

Taking square roots, $\qquad z + \dfrac{b}{2a} = \dfrac{\pm\sqrt{b^2 - 4ac}}{2a}$

Hence $\qquad z = \dfrac{-b \pm \sqrt{b^2 - 4ac}}{2a}$

32. Solve the equation $z^2 + (2i - 3)z + 5 - i = 0$.

From Problem 31, $a = 1$, $b = 2i - 3$, $c = 5 - i$ and so the solutions are

$$z = \frac{-b \pm \sqrt{b^2 - 4ac}}{2a} = \frac{-(2i - 3) \pm \sqrt{(2i - 3)^2 - 4(1)(5 - i)}}{2(1)} = \frac{3 - 2i \pm \sqrt{-15 - 8i}}{2}$$

$$= \frac{3 - 2i \pm (1 - 4i)}{2} = 2 - 3i \text{ or } 1 + i$$

using the fact that the square roots of $-15 - 8i$ are $\pm(1 - 4i)$ [see Problem 30]. These are found to satisfy the given equation.

33. If the real rational number p/q (where p and q have no common factor except ± 1, i.e. p/q is in lowest terms) satisfies the polynomial equation $a_0 z^n + a_1 z^{n-1} + \cdots + a_n = 0$ where $a_0, a_1, \ldots, a_n$ are integers, show that p and q must be factors of a_n and a_0 respectively.

Substituting $z = p/q$ in the given equation and multiplying by q^n yields

$$a_0 p^n + a_1 p^{n-1} q + \cdots + a_{n-1} p q^{n-1} + a_n q^n = 0 \tag{1}$$

Dividing by p and transposing the last term,

$$a_0 p^{n-1} + a_1 p^{n-2} q + \cdots + a_{n-1} q^{n-1} = -\frac{a_n q^n}{p} \tag{2}$$

Since the left side of (2) is an integer, so also is the right side. But since p has no factor in common with q, it cannot divide q^n and so must divide a_n.

Similarly on dividing (1) by q and transposing the first term, we find that q must divide a_0.

34. Solve $6z^4 - 25z^3 + 32z^2 + 3z - 10 = 0$.

The integer factors of 6 and -10 are respectively $\pm 1, \pm 2, \pm 3, \pm 6$ and $\pm 1, \pm 2, \pm 5, \pm 10$. Hence by Prob. 33 the possible rational solutions are $\pm 1, \pm 1/2, \pm 1/3, \pm 1/6, \pm 2, \pm 2/3, \pm 5, \pm 5/2, \pm 5/3, \pm 5/6, \pm 10, \pm 10/3$.

By trial we find that $z = -1/2$ and $z = 2/3$ are solutions, and so the polynomial $(2z+1)(3z-2) = 6z^2 - z - 2$ is a factor of $6z^4 - 25z^3 + 32z^2 + 3z - 10$, the other factor being $z^2 - 4z + 5$ as found by long division. Hence

$$6z^4 - 25z^3 + 32z^2 + 3z - 10 = (6z^2 - z - 2)(z^2 - 4z + 5) = 0$$

The solutions of $z^2 - 4z + 5 = 0$ are [see Problem 31]

$$z = \frac{4 \pm \sqrt{16 - 20}}{2} = \frac{4 \pm \sqrt{-4}}{2} = \frac{4 \pm 2i}{2} = 2 \pm i$$

Then the solutions are $-1/2,\ 2/3,\ 2+i,\ 2-i$.

35. Prove that the sum and product of all the roots of $a_0 z^n + a_1 z^{n-1} + \cdots + a_n = 0$ where $a_0 \neq 0$, are $-a_1/a_0$ and $(-1)^n a_n/a_0$ respectively.

If $z_1, z_2, \ldots, z_n$ are the n roots, the equation can be written in factored form as

$$a_0(z - z_1)(z - z_2) \cdots (z - z_n) = 0$$

Direct multiplication shows that

$$a_0 \{ z^n - (z_1 + z_2 + \cdots + z_n) z^{n-1} + \cdots + (-1)^n z_1 z_2 \cdots z_n \} = 0$$

It follows that $-a_0(z_1 + z_2 + \cdots + z_n) = a_1$ and $a_0 (-1)^n z_1 z_2 \cdots z_n = a_n$, from which

$$z_1 + z_2 + \cdots + z_n = -a_1/a_0, \qquad z_1 z_2 \cdots z_n = (-1)^n a_n/a_0$$

as required.

36. If $p + qi$ is a root of $a_0 z^n + a_1 z^{n-1} + \cdots + a_n = 0$ where $a_0 \neq 0$, $a_1, \ldots, a_n$, p and q are real, prove that $p - qi$ is also a root.

Let $p + qi = re^{i\theta}$ in polar form. Since this satisfies the equation,

$$a_0 r^n e^{in\theta} + a_1 r^{n-1} e^{i(n-1)\theta} + \cdots + a_{n-1} re^{i\theta} + a_n = 0$$

Taking the conjugate of both sides

$$a_0 r^n e^{-in\theta} + a_1 r^{n-1} e^{-i(n-1)\theta} + \cdots + a_{n-1} re^{-i\theta} + a_n = 0$$

we see that $re^{-i\theta} = p - qi$ is also a root. The result does not hold if $a_0, \ldots, a_n$ are not all real (see Problem 32).

The theorem is often expressed in the statement: The zeros of a polynomial with real coefficients occur in conjugate pairs.

THE nth ROOTS OF UNITY

37. Find all the 5th roots of unity.

$$z^5 = 1 = \cos 2k\pi + i \sin 2k\pi = e^{2k\pi i} \quad \text{where} \quad k = 0, \pm1, \pm2, \ldots$$

Then
$$z = \cos\frac{2k\pi}{5} + i \sin\frac{2k\pi}{5} = e^{2k\pi i/5}$$

where it is sufficient to use $k = 0, 1, 2, 3, 4$ since all other values of k lead to repetition.

Thus the roots are $1, e^{2\pi i/5}, e^{4\pi i/5}, e^{6\pi i/5}, e^{8\pi i/5}$. If we call $e^{2\pi i/5} = \omega$, these can be denoted by $1, \omega, \omega^2, \omega^3, \omega^4$.

38. If $n = 2, 3, 4, \ldots$, prove that

(a)
$$\cos\frac{2\pi}{n} + \cos\frac{4\pi}{n} + \cos\frac{6\pi}{n} + \cdots + \cos\frac{2(n-1)\pi}{n} = -1$$

(b)
$$\sin\frac{2\pi}{n} + \sin\frac{4\pi}{n} + \sin\frac{6\pi}{n} + \cdots + \sin\frac{2(n-1)\pi}{n} = 0$$

Consider the equation $z^n - 1 = 0$ whose solutions are the nth roots of unity,
$$1, \quad e^{2\pi i/n}, \quad e^{4\pi i/n}, \quad e^{6\pi i/n}, \quad \ldots, \quad e^{2(n-1)\pi i/n}$$

By Problem 35 the sum of these roots is zero. Then
$$1 + e^{2\pi i/n} + e^{4\pi i/n} + e^{6\pi i/n} + \cdots + e^{2(n-1)\pi i/n} = 0$$
i.e.,
$$\left\{ 1 + \cos\frac{2\pi}{n} + \cos\frac{4\pi}{n} + \cdots + \cos\frac{2(n-1)\pi}{n} \right\} + i\left\{ \sin\frac{2\pi}{n} + \sin\frac{4\pi}{n} + \cdots + \sin\frac{2(n-1)\pi}{n} \right\} = 0$$
from which the required results follow.

DOT AND CROSS PRODUCT

39. If $z_1 = 3 - 4i$ and $z_2 = -4 + 3i$, find (a) $z_1 \circ z_2$, (b) $z_1 \times z_2$.

(a) $z_1 \circ z_2 = \text{Re}\{\bar{z}_1 z_2\} = \text{Re}\{(3 + 4i)(-4 + 3i)\} = \text{Re}\{-24 - 7i\} = -24$

Another method. $z_1 \circ z_2 = (3)(-4) + (-4)(3) = -24$

(b) $z_1 \times z_2 = \text{Im}\{\bar{z}_1 z_2\} = \text{Im}\{(3 + 4i)(-4 + 3i)\} = \text{Im}\{-24 - 7i\} = -7$

Another method. $z_1 \times z_2 = (3)(3) - (-4)(-4) = -7$

40. Find the acute angle between the vectors in Problem 39.

From Problem 39(a), we have $\cos\theta = \dfrac{z_1 \circ z_2}{|z_1|\,|z_2|} = \dfrac{-24}{|3 - 4i|\,|-4 + 3i|} = \dfrac{-24}{25} = -.96$.

Then the acute angle is $\cos^{-1} .96 = 16°16'$ approximately.

41. Prove that the area of a parallelogram having sides z_1 and z_2 is $|z_1 \times z_2|$.

Area of parallelogram [Fig. 1-34]

$= \text{(base)(height)}$

$= (|z_2|)(|z_1| \sin\theta)$

$= |z_1|\,|z_2| \sin\theta = |z_1 \times z_2|$

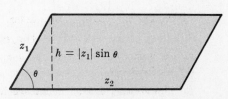

Fig. 1-34

42. Find the area of a triangle with vertices at $A(x_1, y_1)$, $B(x_2, y_2)$ and $C(x_3, y_3)$.

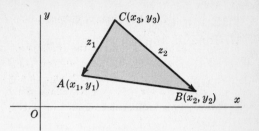

Fig. 1-35

The vectors from C to A and B [Fig. 1-35] are respectively given by

$$z_1 = (x_1 - x_3) + i(y_1 - y_3),$$
$$z_2 = (x_2 - x_3) + i(y_2 - y_3)$$

Since the area of a triangle with sides z_1 and z_2 is half the area of the corresponding parallelogram, we have by Problem 41:

$$
\begin{aligned}
\text{Area of triangle} &= \tfrac{1}{2}\,|\,z_1 \times z_2\,| = \tfrac{1}{2}\,|\,\text{Im}\,\{[(x_1 - x_3) - i(y_1 - y_3)][(x_2 - x_3) + i(y_2 - y_3)]\}\,| \\
&= \tfrac{1}{2}\,|\,(x_1 - x_3)(y_2 - y_3) - (y_1 - y_3)(x_2 - x_3)\,| \\
&= \tfrac{1}{2}\,|\,x_1 y_2 - y_1 x_2 + x_2 y_3 - y_2 x_3 + x_3 y_1 - y_3 x_1\,| \\
&= \tfrac{1}{2}\,\left|\,\begin{vmatrix} x_1 & y_1 & 1 \\ x_2 & y_2 & 1 \\ x_3 & y_3 & 1 \end{vmatrix}\,\right|
\end{aligned}
$$

in determinant form.

COMPLEX CONJUGATE COORDINATES

43. Express each equation in terms of conjugate coordinates: (a) $2x + y = 5$, (b) $x^2 + y^2 = 36$.

(a) Since $z = x + iy$, $\bar{z} = x - iy$, $x = \dfrac{z + \bar{z}}{2}$, $y = \dfrac{z - \bar{z}}{2i}$. Then $2x + y = 5$ becomes

$$2\left(\frac{z + \bar{z}}{2}\right) + \left(\frac{z - \bar{z}}{2i}\right) = 5 \qquad \text{or} \qquad (2i + 1)z + (2i - 1)\bar{z} = 10i$$

The equation represents a straight line in the z plane.

(b) **Method 1.** The equation is $(x + iy)(x - iy) = 36$ or $z\bar{z} = 36$.

Method 2. Substitute $x = \dfrac{z + \bar{z}}{2}$, $y = \dfrac{z - \bar{z}}{2i}$ in $x^2 + y^2 = 36$ to obtain $z\bar{z} = 36$.

The equation represents a circle in the z plane of radius 6 with center at the origin.

44. Prove that the equation of any circle or line in the z plane can be written as $\alpha z\bar{z} + \beta z + \bar{\beta}\bar{z} + \gamma = 0$ where α and γ are real constants while β may be a complex constant.

The general equation of a circle in the xy plane can be written

$$A(x^2 + y^2) + Bx + Cy + D = 0$$

which in conjugate coordinates becomes

$$Az\bar{z} + B\left(\frac{z + \bar{z}}{2}\right) + C\left(\frac{z - \bar{z}}{2i}\right) + D = 0 \qquad \text{or} \qquad Az\bar{z} + \left(\frac{B}{2} + \frac{C}{2i}\right)z + \left(\frac{B}{2} - \frac{C}{2i}\right)\bar{z} + D = 0$$

Calling $A = \alpha$, $\dfrac{B}{2} + \dfrac{C}{2i} = \beta$ and $D = \gamma$, the required result follows.

In the special case $A = \alpha = 0$, the circle degenerates into a line.

POINT SETS

45. Given the point set $S: \{i, \tfrac{1}{2}i, \tfrac{1}{3}i, \tfrac{1}{4}i, \ldots\}$ or briefly $\{i/n\}$. (a) Is S bounded? (b) What are its limit points, if any? (c) Is S closed? (d) What are its interior and boundary points? (e) Is S open? (f) Is S connected? (g) Is S an open region or domain? (h) What is the closure of S? (i) What is the complement of S? (j) Is S countable? (k) Is S compact? (l) Is the closure of S compact?

(a) S is bounded since for every point z in S, $|z| < 2$ [for example], i.e. all points of S lie inside a circle of radius 2 with center at the origin.

(b) Since every deleted neighborhood of $z = 0$ contains points of S, a limit point is $z = 0$. It is the only limit point.

Note that since S is bounded and infinite the Weierstrass-Bolzano theorem predicts *at least one* limit point.

(c) S is not closed since the limit point $z = 0$ does not belong to S.

(d) Every δ neighborhood of any point i/n [i.e. every circle of radius δ with center at i/n] contains points which belong to S and points which do not belong to S. Thus every point of S, as well as the point $z = 0$, is a boundary point. S has *no* interior points.

(e) S does not consist of any interior points. Hence it cannot be open. Thus S is neither open nor closed.

(f) If we join any two points of S by a polygonal path, there are points on this path which do not belong to S. Thus S is not connected.

(g) Since S is not an open connected set, it is not an open region or domain.

(h) The closure of S consists of the set S together with the limit point zero, i.e. $\{0, i, \frac{1}{2}i, \frac{1}{3}i, \ldots\}$.

(i) The complement of S is the set of all points not belonging to S, i.e. all points $z \neq i, i/2, i/3, \ldots$.

(j) There is a one to one correspondence between the elements of S and the natural numbers $1, 2, 3, \ldots$ as indicated below.

$$i \qquad \tfrac{1}{2}i \qquad \tfrac{1}{3}i \qquad \tfrac{1}{4}i \qquad \cdots$$
$$\updownarrow \qquad \updownarrow \qquad \updownarrow \qquad \updownarrow$$
$$1 \qquad 2 \qquad 3 \qquad 4 \qquad \cdots$$

Hence S is countable.

(k) S is bounded but not closed. Hence it is not compact.

(l) The closure of S is bounded and closed and so is compact.

46. Given the point sets $A = \{3, -i, 4, 2+i, 5\}$, $B = \{-i, 0, -1, 2+i\}$, $C = \{-\sqrt{2}\,i, \frac{1}{2}, 3\}$. Find (a) $A + B$ or $A \cup B$, (b) AB or $A \cap B$, (c) AC or $A \cap C$, (d) $A(B+C)$ or $A \cap (B \cup C)$, (e) $AB + AC$ or $(A \cap B) \cup (A \cap C)$, (f) $A(BC)$ or $A \cap (B \cap C)$.

(a) $A + B = A \cup B$ consists of points belonging either to A or B or both and is given by $\{3, -i, 4, 2+i, 5, 0, -1\}$.

(b) AB or $A \cap B$ consists of points belonging to both A and B and is given by $\{-i, 2+i\}$.

(c) AC or $A \cap C = \{3\}$, consisting of only the member 3.

(d) $B + C$ or $B \cup C = \{-i, 0, -1, 2+i, -\sqrt{2}\,i, \frac{1}{2}, 3\}$.

Hence $A(B+C)$ or $A \cap (B \cup C) = \{3, -i, 2+i\}$, consisting of points belonging to both A and $B + C$.

(e) $AB = \{-i, 2+i\}$, $AC = \{3\}$ from parts (b) and (c). Hence $AB + AC = \{-i, 2+i, 3\}$.

From this and the result of (d) we see that $A(B+C) = AB + AC$ or $A \cap (B \cup C) = (A \cap B) \cup (A \cap C)$, which illustrates the fact that A, B, C satisfy the *distributive law*. We can show that sets exhibit many of the properties valid in the algebra of numbers. This is of great importance in theory and application.

(f) $BC = B \cap C = \emptyset$, the *null set*, since there are no points common to both B and C. Hence $A(BC) = \emptyset$ also.

MISCELLANEOUS PROBLEMS

47. A number is called an *algebraic number* if it is a solution of a polynomial equation $a_0 z^n + a_1 z^{n-1} + \cdots + a_{n-1} z + a_n = 0$ where $a_0, a_1, \ldots, a_n$ are integers. Prove that (a) $\sqrt{3} + \sqrt{2}$ and (b) $\sqrt[3]{4} - 2i$ are algebraic numbers.

(a) Let $z = \sqrt{3} + \sqrt{2}$ or $z - \sqrt{2} = \sqrt{3}$. Squaring, $z^2 - 2\sqrt{2}\,z + 2 = 3$ or $z^2 - 1 = 2\sqrt{2}\,z$. Squaring again, $z^4 - 2z^2 + 1 = 8z^2$ or $z^4 - 10z^2 + 1 = 0$, a polynomial equation with integer coefficients having $\sqrt{3} + \sqrt{2}$ as a root. Hence $\sqrt{3} + \sqrt{2}$ is an algebraic number.

(b) Let $z = \sqrt[3]{4} - 2i$ or $z + 2i = \sqrt[3]{4}$. Cubing, $z^3 + 3z^2(2i) + 3z(2i)^2 + (2i)^3 = 4$ or $z^3 - 12z - 4 = i(8 - 6z^2)$. Squaring, $z^6 + 12z^4 - 8z^3 + 48z^2 + 96z + 80 = 0$, a polynomial equation with integer coefficients having $\sqrt[3]{4} - 2i$ as a root. Hence $\sqrt[3]{4} - 2i$ is an algebraic number.

Numbers which are not algebraic, i.e. do not satisfy any polynomial equation with integer coefficients, are called *transcendental numbers*. It has been proved that the numbers $\pi = 3.14159\ldots$ and $e = 2.71828\ldots$ are transcendental. However, it is still not yet known whether numbers such as $e\pi$ or $e + \pi$, for example, are transcendental or not.

48. Represent graphically the set of values of z for which   (a) $\left|\dfrac{z-3}{z+3}\right| = 2$,   (b) $\left|\dfrac{z-3}{z+3}\right| < 2$.

(a) The given equation is equivalent to $|z-3| = 2|z+3|$ or, if $z = x+iy$, $|x+iy-3| = 2|x+iy+3|$, i.e.,

$$\sqrt{(x-3)^2 + y^2} = 2\sqrt{(x+3)^2 + y^2}$$

Squaring and simplifying, this becomes

$x^2 + y^2 + 10x + 9 = 0$ or $(x+5)^2 + y^2 = 16$

i.e. $|z+5| = 4$, a circle of radius 4 with center at $(-5, 0)$ as shown in Fig. 1-36.

Geometrically, any point P on this circle is such that the distance from P to point $B(3,0)$ is twice the distance from P to point $A(-3,0)$.

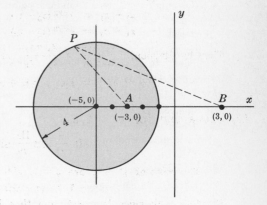

Fig. 1-36

Another method.

$\left|\dfrac{z-3}{z+3}\right| = 2$ is equivalent to

$\left(\dfrac{z-3}{z+3}\right)\left(\dfrac{\bar{z}-3}{\bar{z}+3}\right) = 4$ or $z\bar{z} + 5\bar{z} + 5z + 9 = 0$

i.e. $(z+5)(\bar{z}+5) = 16$ or $|z+5| = 4$.

(b) The given inequality is equivalent to $|z-3| < 2|z+3|$ or $\sqrt{(x-3)^2 + y^2} < 2\sqrt{(x+3)^2 + y^2}$. Squaring and simplifying, this becomes $x^2 + y^2 + 10x + 9 > 0$ or $(x+5)^2 + y^2 > 16$, i.e. $|z+5| > 4$.

The required set thus consists of all points external to the circle of Fig. 1-36.

49. Given the sets A and B represented by $|z-1| < 2$ and $|z-2i| < 1.5$ respectively. Represent geometrically   (a) $A \cap B$ or AB,   (b) $A \cup B$ or $A+B$.

The required sets of points are shown shaded in Figures 1-37 and 1-38 respectively.

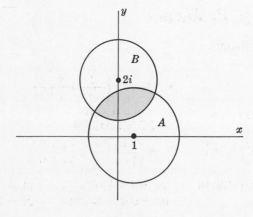

Fig. 1-37

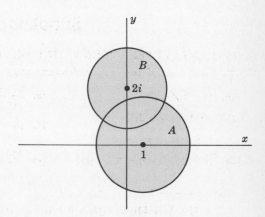

Fig. 1-38

50. Solve $z^2(1 - z^2) = 16$.

Method 1. The equation can be written $z^4 - z^2 + 16 = 0$, i.e. $z^4 + 8z^2 + 16 - 9z^2 = 0$, $(z^2 + 4)^2 - 9z^2 = 0$ or $(z^2 + 4 + 3z)(z^2 + 4 - 3z) = 0$. Then the required solutions are the solutions of $z^2 + 3z + 4 = 0$ and $z^2 - 3z + 4 = 0$, or $-\dfrac{3}{2} \pm \dfrac{\sqrt{7}}{2}i$ and $\dfrac{3}{2} \pm \dfrac{\sqrt{7}}{2}i$.

Method 2. Letting $w = z^2$, the equation can be written $w^2 - w + 16 = 0$ and $w = \dfrac{1}{2} \pm \dfrac{3}{2}\sqrt{7}\,i$. To obtain solutions of $z^2 = \dfrac{1}{2} \pm \dfrac{3}{2}\sqrt{7}\,i$, the methods of Problem 30 can be used.

51. If z_1, z_2, z_3 represent vertices of an equilateral triangle, prove that

$$z_1^2 + z_2^2 + z_3^2 = z_1 z_2 + z_2 z_3 + z_3 z_1$$

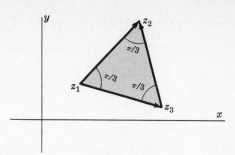

Fig. 1-39

From Fig. 1-39 we see that

$$z_2 - z_1 = e^{\pi i/3}(z_3 - z_1)$$
$$z_1 - z_3 = e^{\pi i/3}(z_2 - z_3)$$

Then by division, $\dfrac{z_2 - z_1}{z_1 - z_3} = \dfrac{z_3 - z_1}{z_2 - z_3}$ or

$$z_1^2 + z_2^2 + z_3^2 = z_1 z_2 + z_2 z_3 + z_3 z_1$$

52. Prove that for $m = 2, 3, \ldots$

$$\sin\frac{\pi}{m}\sin\frac{2\pi}{m}\sin\frac{3\pi}{m}\cdots\sin\frac{(m-1)\pi}{m} = \frac{m}{2^{m-1}}$$

The roots of $z^m = 1$ are $z = 1,\ e^{2\pi i/m},\ e^{4\pi i/m},\ \ldots,\ e^{2(m-1)\pi i/m}$. Then we can write

$$z^m - 1 = (z-1)(z - e^{2\pi i/m})(z - e^{4\pi i/m})\cdots(z - e^{2(m-1)\pi i/m})$$

Dividing both sides by $z - 1$ and then letting $z = 1$ [realizing that $(z^m - 1)/(z - 1) = 1 + z + z^2 + \cdots + z^{m-1}$] we find

$$m = (1 - e^{2\pi i/m})(1 - e^{4\pi i/m})\cdots(1 - e^{2(m-1)\pi i/m}) \tag{1}$$

Taking the complex conjugate of both sides of (1) yields

$$m = (1 - e^{-2\pi i/m})(1 - e^{-4\pi i/m})\cdots(1 - e^{-2(m-1)\pi i/m}) \tag{2}$$

Multiplying (1) by (2) using $(1 - e^{2k\pi i/m})(1 - e^{-2k\pi i/m}) = 2 - 2\cos(2k\pi/m)$, we have

$$m^2 = 2^{m-1}\left(1 - \cos\frac{2\pi}{m}\right)\left(1 - \cos\frac{4\pi}{m}\right)\cdots\left(1 - \cos\frac{2(m-1)\pi}{m}\right) \tag{3}$$

Since $1 - \cos(2k\pi/m) = 2\sin^2(k\pi/m)$, (3) becomes

$$m^2 = 2^{2m-2}\sin^2\frac{\pi}{m}\sin^2\frac{2\pi}{m}\cdots\sin^2\frac{(m-1)\pi}{m} \tag{4}$$

Then taking the positive square root of both sides yields the required result.

Supplementary Problems

FUNDAMENTAL OPERATIONS WITH COMPLEX NUMBERS

53. Perform each of the indicated operations:

(a) $(4 - 3i) + (2i - 8)$

(b) $3(-1 + 4i) - 2(7 - i)$

(c) $(3 + 2i)(2 - i)$

(d) $(i - 2)\{2(1 + i) - 3(i - 1)\}$

(e) $\dfrac{2 - 3i}{4 - i}$

(f) $(4 + i)(3 + 2i)(1 - i)$

(g) $\dfrac{(2 + i)(3 - 2i)(1 + 2i)}{(1 - i)^2}$

(h) $(2i - 1)^2\left\{\dfrac{4}{1 - i} + \dfrac{2 - i}{1 + i}\right\}$

(i) $\dfrac{i^4 + i^9 + i^{16}}{2 - i^5 + i^{10} - i^{15}}$

(j) $3\left(\dfrac{1 + i}{1 - i}\right)^2 - 2\left(\dfrac{1 - i}{1 + i}\right)^3$

Ans. (a) $-4 - i$ (c) $8 + i$ (e) $11/17 - (10/17)i$ (g) $-15/2 + 5i$ (i) $2 + i$
  (b) $-17 + 14i$ (d) $-9 + 7i$ (f) $21 + i$ (h) $-11/2 - (23/2)i$ (j) $-3 - 2i$

54. If $z_1 = 1 - i$, $z_2 = -2 + 4i$, $z_3 = \sqrt{3} - 2i$, evaluate each of the following:

(a) $z_1^2 + 2z_1 - 3$

(b) $|2z_2 - 3z_1|^2$

(c) $(z_3 - \bar{z}_3)^5$

(d) $|z_1\bar{z}_2 + z_2\bar{z}_1|$

(e) $\left|\dfrac{z_1 + z_2 + 1}{z_1 - z_2 + i}\right|$

(f) $\dfrac{1}{2}\left(\dfrac{z_3}{\bar{z}_3} + \dfrac{\bar{z}_3}{z_3}\right)$

(g) $\overline{(z_2 + z_3)(z_1 - z_3)}$

(h) $|z_1^2 + \bar{z}_2^2|^2 + |\bar{z}_3^2 - z_2^2|^2$

(i) $\operatorname{Re}\{2z_1^3 + 3z_2^2 - 5z_3^2\}$

(j) $\operatorname{Im}\{z_1 z_2/z_3\}$

Ans. (a) $-1 - 4i$ (c) $1024i$ (e) $3/5$ (g) $-7 + 3\sqrt{3} + \sqrt{3}\,i$ (i) -35
  (b) 170 (d) 12 (f) $-1/7$ (h) $765 + 128\sqrt{3}$ (j) $(6\sqrt{3} + 4)/7$

55. Prove that   (a) $(\overline{z_1 z_2}) = \bar{z}_1 \bar{z}_2$,   (b) $(\overline{z_1 z_2 z_3}) = \bar{z}_1 \bar{z}_2 \bar{z}_3$.   Generalize these results.

56. Prove that   (a) $(\overline{z_1/z_2}) = \bar{z}_1/\bar{z}_2$,   (b) $|z_1/z_2| = |z_1|/|z_2|$ if $z_2 \neq 0$.

57. Find real numbers x and y such that $2x - 3iy + 4ix - 2y - 5 - 10i = (x + y + 2) - (y - x + 3)i$.
    $Ans.$ $x = 1$, $y = -2$

58. Prove that   (a) $\text{Re}\{z\} = (z + \bar{z})/2$,   (b) $\text{Im}\{z\} = (z - \bar{z})/2i$.

59. Prove that if the product of two complex numbers is zero then at least one of the numbers must be zero.

60. If $w = 3iz - z^2$ and $z = x + iy$, find $|w|^2$ in terms of x and y.
    $Ans.$ $x^4 + y^4 + 2x^2y^2 - 6x^2y - 6y^3 + 9x^2 + 9y^2$

GRAPHICAL REPRESENTATION OF COMPLEX NUMBERS. VECTORS.

61. Perform the indicated operations both analytically and graphically.

    (a) $(2 + 3i) + (4 - 5i)$ (c) $3(1 + 2i) - 2(2 - 3i)$ (e) $\frac{1}{2}(4 - 3i) + \frac{3}{2}(5 + 2i)$
    (b) $(7 + i) - (4 - 2i)$ (d) $3(1 + i) + 2(4 - 3i) - (2 + 5i)$

    $Ans.$ (a) $6 - 2i$, (b) $3 + 3i$, (c) $-1 + 12i$, (d) $9 - 8i$, (e) $19/2 + (3/2)i$

62. If z_1, z_2 and z_3 are the vectors indicated in Fig. 1-40, construct graphically:

    (a) $2z_1 + z_3$ (c) $z_1 + (z_2 + z_3)$ (e) $\frac{1}{3}z_2 - \frac{3}{4}z_1 + \frac{2}{3}z_3$
    (b) $(z_1 + z_2) + z_3$ (d) $3z_1 - 2z_2 + 5z_3$

63. If $z_1 = 4 - 3i$ and $z_2 = -1 + 2i$, obtain graphically and analytically   (a) $|z_1 + z_2|$, (b) $|z_1 - z_2|$, (c) $\bar{z}_1 - \bar{z}_2$, (d) $|2\bar{z}_1 - 3\bar{z}_2 - 2|$.

    $Ans.$ (a) $\sqrt{10}$, (b) $5\sqrt{2}$, (c) $5 + 5i$, (d) 15

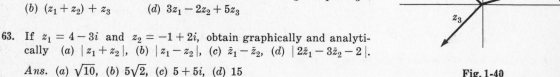

Fig. 1-40

64. The position vectors of points A, B and C of triangle ABC are given by $z_1 = 1 + 2i$, $z_2 = 4 - 2i$ and $z_3 = 1 - 6i$ respectively. Prove that ABC is an isosceles triangle and find the lengths of the sides.
    $Ans.$ $5, 5, 8$

65. Let z_1, z_2, z_3, z_4 be the position vectors of the vertices for quadrilateral $ABCD$. Prove that $ABCD$ is a parallelogram if and only if $z_1 - z_2 - z_3 + z_4 = 0$.

66. If the diagonals of a quadrilateral bisect each other, prove that the quadrilateral is a parallelogram.

67. Prove that the medians of a triangle meet in a point.

68. Let $ABCD$ be a quadrilateral and E, F, G, H the midpoints of the sides. Prove that $EFGH$ is a parallelogram.

69. In parallelogram $ABCD$, point E bisects side AD. Prove that the point where BE meets AC trisects AC.

70. The position vectors of points A and B are $2 + i$ and $3 - 2i$ respectively. (a) Find an equation for line AB. (b) Find an equation for the line perpendicular to AB at its midpoint.

    $Ans.$ (a) $z - (2 + i) = t(1 - 3i)$ or $x = 2 + t$, $y = 1 - 3t$ or $3x + y = 7$
           (b) $z - (5/2 - i/2) = t(3 + i)$ or $x = 3t + 5/2$, $y = t - 1/2$ or $x - 3y = 4$

71. Describe and graph the locus represented by each of the following: (a) $|z - i| = 2$,   (b) $|z + 2i| + |z - 2i| = 6$,   (c) $|z - 3| - |z + 3| = 4$,   (d) $z(\bar{z} + 2) = 3$,   (e) $\text{Im}\{z^2\} = 4$.

    $Ans.$ (a) circle, (b) ellipse, (c) hyperbola, (d) $z = 1$ and $z = -3$, (e) hyperbola

72. Find an equation for   (a) a circle of radius 2 with center at $(-3, 4)$,   (b) an ellipse with foci at $(0, 2)$ and $(0, -2)$ whose major axis has length 10.

    $Ans.$ (a) $|z + 3 - 4i| = 2$ or $(x + 3)^2 + (y - 4)^2 = 4$,   (b) $|z + 2i| + |z - 2i| = 10$

73. Describe graphically the region represented by each of the following:

   (a) $1 < |z + i| \leq 2$, (b) Re $\{z^2\} > 1$, (c) $|z + 3i| > 4$, (d) $|z + 2 - 3i| + |z - 2 + 3i| < 10$.

74. Show that the ellipse $|z + 3| + |z - 3| = 10$ can be expressed in rectangular form as $x^2/25 + y^2/16 = 1$ [see Problem 13(b)].

AXIOMATIC FOUNDATIONS OF COMPLEX NUMBERS

75. Use the definition of a complex number as an ordered pair of real numbers to prove that if the product of two complex numbers is zero then at least one of the numbers must be zero.

76. Prove the commutative laws with respect to (a) addition, (b) multiplication.

77. Prove the associative laws with respect to (a) addition, (b) multiplication.

78. (a) Find real numbers x and y such that $(c, d) \cdot (x, y) = (a, b)$ where $(c, d) \neq (0, 0)$.
   (b) How is (x, y) related to the result for division of complex numbers given on Page 2?

79. Prove that
$$(\cos \theta_1, \sin \theta_1)(\cos \theta_2, \sin \theta_2) \cdots (\cos \theta_n, \sin \theta_n) = (\cos [\theta_1 + \theta_2 + \cdots + \theta_n], \ \sin [\theta_1 + \theta_2 + \cdots + \theta_n])$$

80. (a) How would you define $(a, b)^{1/n}$ where n is a positive integer?
   (b) Determine $(a, b)^{1/2}$ in terms of a and b.

POLAR FORM OF COMPLEX NUMBERS

81. Express each of the following complex numbers in polar form.
   (a) $2 - 2i$, (b) $-1 + \sqrt{3}\,i$, (c) $2\sqrt{2} + 2\sqrt{2}\,i$, (d) $-i$, (e) -4, (f) $-2\sqrt{3} - 2i$, (g) $\sqrt{2}\,i$, (h) $\sqrt{3}/2 - 3i/2$.
   Ans. (a) $2\sqrt{2}$ cis $315°$ or $2\sqrt{2}\,e^{7\pi i/4}$, (b) 2 cis $120°$ or $2e^{2\pi i/3}$, (c) 4 cis $45°$ or $4e^{\pi i/4}$, (d) cis $270°$ or $e^{3\pi i/2}$,
   (e) 4 cis $180°$ or $4e^{\pi i}$, (f) 4 cis $210°$ or $4e^{7\pi i/6}$, (g) $\sqrt{2}$ cis $90°$ or $\sqrt{2}\,e^{\pi i/2}$, (h) $\sqrt{3}$ cis $300°$ or $\sqrt{3}\,e^{5\pi i/3}$.

82. Show that $2 + i = \sqrt{5}\,e^{i \tan^{-1}(1/2)}$.

83. Express in polar form: (a) $-3 - 4i$, (b) $1 - 2i$.
   Ans. (a) $5\,e^{i(\pi + \tan^{-1} 4/3)}$, (b) $\sqrt{5}\,e^{-i \tan^{-1} 2}$

84. Graph each of the following and express in rectangular form.
   (a) $6(\cos 135° + i \sin 135°)$, (b) 12 cis $90°$, (c) 4 cis $315°$, (d) $2e^{5\pi i/4}$, (e) $5e^{7\pi i/6}$, (f) $3e^{-2\pi i/3}$.
   Ans. (a) $-3\sqrt{2} + 3\sqrt{2}\,i$, (b) $12i$, (c) $2\sqrt{2} - 2\sqrt{2}\,i$, (d) $-\sqrt{2} - \sqrt{2}\,i$, (e) $-5\sqrt{3}/2 - (5/2)i$, (f) $-3\sqrt{3}/2 - (3/2)i$

85. An airplane travels 150 miles southeast, 100 miles due west, 225 miles $30°$ north of east, and then 200 miles northeast. Determine (a) analytically and (b) graphically how far and in what direction it is from its starting point.   *Ans.* 375 miles, $23°$ north of east (approx.)

86. Three forces as shown in Fig. 1-41 act in a plane on an object placed at O. Determine (a) graphically and (b) analytically what force is needed to prevent the object from moving. [This force is sometimes called the *equilibrant*.]

87. Prove that on the circle $z = Re^{i\theta}$, $|e^{iz}| = e^{-R \sin \theta}$.

88. (a) Prove that $r_1 e^{i\theta_1} + r_2 e^{i\theta_2} = r_3 e^{i\theta_3}$ where
$$r_3 = \sqrt{r_1^2 + r_2^2 + 2r_1 r_2 \cos(\theta_1 - \theta_2)}$$
$$\theta_3 = \tan^{-1}\left(\frac{r_1 \sin \theta_1 + r_2 \sin \theta_2}{r_1 \cos \theta_1 + r_2 \cos \theta_2}\right)$$

   (b) Generalize the result in (a).

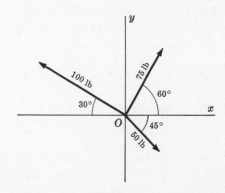

Fig. 1-41

DE MOIVRE'S THEOREM

89. Evaluate each of the following:

(a) $(5 \text{ cis } 20°)(3 \text{ cis } 40°)$

(b) $(2 \text{ cis } 50°)^6$

(c) $\dfrac{(8 \text{ cis } 40°)^3}{(2 \text{ cis } 60°)^4}$

(d) $\dfrac{(3e^{\pi i/6})(2e^{-5\pi i/4})(6e^{5\pi i/3})}{(4e^{2\pi i/3})^2}$

(e) $\left(\dfrac{\sqrt{3}-i}{\sqrt{3}+i}\right)^4 \left(\dfrac{1+i}{1-i}\right)^5$

Ans. (a) $15/2 + (15\sqrt{3}/2)i$, (b) $32 - 32\sqrt{3}\,i$, (c) $-16 - 16\sqrt{3}\,i$, (d) $3\sqrt{3}/2 - (3\sqrt{3}/2)i$, (e) $-\sqrt{3}/2 - (1/2)i$

90. Prove that (a) $\sin 3\theta = 3\sin\theta - 4\sin^3\theta$, (b) $\cos 3\theta = 4\cos^3\theta - 3\cos\theta$.

91. Prove that the solutions of $z^4 - 3z^2 + 1 = 0$ are given by $z = 2\cos 36°, 2\cos 72°, 2\cos 216°, 2\cos 252°$.

92. Show that (a) $\cos 36° = (\sqrt{5}+1)/4$, (b) $\cos 72° = (\sqrt{5}-1)/4$. [*Hint:* Use Problem 91.]

93. Prove that (a) $\dfrac{\sin 4\theta}{\sin\theta} = 8\cos^3\theta - 4\cos\theta = 2\cos 3\theta + 2\cos\theta$

(b) $\cos 4\theta = 8\sin^4\theta - 8\sin^2\theta + 1$

94. Prove De Moivre's theorem for (a) negative integers, (b) rational numbers.

ROOTS OF COMPLEX NUMBERS

95. Find each of the indicated roots and locate them graphically.

(a) $(2\sqrt{3} - 2i)^{1/2}$, (b) $(-4 + 4i)^{1/5}$, (c) $(2 + 2\sqrt{3}\,i)^{1/3}$, (d) $(-16i)^{1/4}$, (e) $(64)^{1/6}$, (f) $(i)^{2/3}$.

Ans. (a) $2 \text{ cis } 165°$, $2 \text{ cis } 345°$. (b) $\sqrt[5]{2} \text{ cis } 27°$, $\sqrt[5]{2} \text{ cis } 99°$, $\sqrt[5]{2} \text{ cis } 171°$, $\sqrt[5]{2} \text{ cis } 243°$, $\sqrt[5]{2} \text{ cis } 315°$.
(c) $\sqrt[3]{4} \text{ cis } 20°$, $\sqrt[3]{4} \text{ cis } 140°$, $\sqrt[3]{4} \text{ cis } 260°$. (d) $2 \text{ cis } 67.5°$, $2 \text{ cis } 157.5°$, $2 \text{ cis } 247.5°$, $2 \text{ cis } 337.5°$.
(e) $2 \text{ cis } 0°$, $2 \text{ cis } 60°$, $2 \text{ cis } 120°$, $2 \text{ cis } 180°$, $2 \text{ cis } 240°$, $2 \text{ cis } 300°$. (f) $\text{cis } 60°$, $\text{cis } 180°$, $\text{cis } 300°$.

96. Find all the indicated roots and locate them in the complex plane.

(a) cube roots of 8, (b) square roots of $4\sqrt{2} + 4\sqrt{2}\,i$, (c) fifth roots of $-16 + 16\sqrt{3}\,i$, (d) sixth roots of $-27i$.

Ans. (a) $2 \text{ cis } 0°$, $2 \text{ cis } 120°$, $2 \text{ cis } 240°$. (b) $\sqrt{8} \text{ cis } 22.5°$, $\sqrt{8} \text{ cis } 202.5°$. (c) $2 \text{ cis } 48°$, $2 \text{ cis } 120°$, $2 \text{ cis } 192°$, $2 \text{ cis } 264°$, $2 \text{ cis } 336°$. (d) $\sqrt{3} \text{ cis } 45°$, $\sqrt{3} \text{ cis } 105°$, $\sqrt{3} \text{ cis } 165°$, $\sqrt{3} \text{ cis } 225°$, $\sqrt{3} \text{ cis } 285°$, $\sqrt{3} \text{ cis } 345°$.

97. Solve the equations (a) $z^4 + 81 = 0$, (b) $z^6 + 1 = \sqrt{3}\,i$.

Ans. (a) $3 \text{ cis } 45°$, $3 \text{ cis } 135°$, $3 \text{ cis } 225°$, $3 \text{ cis } 315°$

(b) $\sqrt[6]{2} \text{ cis } 40°$, $\sqrt[6]{2} \text{ cis } 100°$, $\sqrt[6]{2} \text{ cis } 160°$, $\sqrt[6]{2} \text{ cis } 220°$, $\sqrt[6]{2} \text{ cis } 280°$, $\sqrt[6]{2} \text{ cis } 340°$

98. Find the square roots of (a) $5 - 12i$, (b) $8 + 4\sqrt{5}\,i$.

Ans. (a) $3 - 2i$, $-3 + 2i$. (b) $\sqrt{10} + \sqrt{2}\,i$, $-\sqrt{10} - \sqrt{2}\,i$

99. Find the cube roots of $-11 - 2i$. *Ans.* $1 + 2i$, $\frac{1}{2} - \sqrt{3} + (1 + \frac{1}{2}\sqrt{3})i$, $-\frac{1}{2} - \sqrt{3} + (\frac{1}{2}\sqrt{3} - 1)i$

POLYNOMIAL EQUATIONS

100. Solve the following equations, obtaining all roots: (a) $5z^2 + 2z + 10 = 0$, (b) $z^2 + (i-2)z + (3-i) = 0$.

Ans. (a) $(-1 \pm 7i)/5$, (b) $1 + i$, $1 - 2i$

101. Solve $z^5 - 2z^4 - z^3 + 6z - 4 = 0$. *Ans.* $1, 1, 2, -1 \pm i$

102. (a) Find all the roots of $z^4 + z^2 + 1 = 0$ and (b) locate them in the complex plane.

Ans. $\frac{1}{2}(1 \pm i\sqrt{3})$, $\frac{1}{2}(-1 \pm i\sqrt{3})$

103. Prove that the sum of the roots of $a_0 z^n + a_1 z^{n-1} + a_2 z^{n-2} + \cdots + a_n = 0$ where $a_0 \neq 0$ taken r at a time is $(-1)^r a_r/a_0$ where $0 < r < n$.

104. Find two numbers whose sum is 4 and whose product is 8. *Ans.* $2 + 2i$, $2 - 2i$

THE nth ROOTS OF UNITY

105. Find all the (a) fourth roots, (b) seventh roots of unity and exhibit them graphically.

Ans. (a) $e^{2\pi ik/4} = e^{\pi ik/2},\ k = 0,1,2,3$ (b) $e^{2\pi ik/7},\ k = 0,1,\ldots,6$

106. (a) Prove that $1 + \cos 72° + \cos 144° + \cos 216° + \cos 288° = 0$.
   (b) Give a graphical interpretation of the result in (a).

107. Prove that $\cos 36° + \cos 72° + \cos 108° + \cos 144° = 0$ and interpret graphically.

108. Prove that the sum of the products of all the nth roots of unity taken $2,3,4,\ldots,(n-1)$ at a time is zero.

109. Find all roots of $(1+z)^5 = (1-z)^5$.

Ans. $0,\ (\omega-1)/(\omega+1),\ (\omega^2-1)/(\omega^2+1),\ (\omega^3-1)/(\omega^3+1),\ (\omega^4-1)/(\omega^4+1),$ where $\omega = e^{2\pi i/5}$

THE DOT AND CROSS PRODUCT

110. If $z_1 = 2+5i$ and $z_2 = 3-i$, find (a) $z_1 \circ z_2$, (b) $z_1 \times z_2$, (c) $z_2 \circ z_1$, (d) $z_2 \times z_1$, (e) $|z_1 \circ z_2|$, (f) $|z_2 \circ z_1|$, (g) $|z_1 \times z_2|$, (h) $|z_2 \times z_1|$.

Ans. (a) 1, (b) −17, (c) 1, (d) 17, (e) 1, (f) 1, (g) 17, (h) 17

111. Prove that (a) $z_1 \circ z_2 = z_2 \circ z_1$, (b) $z_1 \times z_2 = -z_2 \times z_1$.

112. If $z_1 = r_1 e^{i\theta_1}$ and $z_2 = r_2 e^{i\theta_2}$, prove that (a) $z_1 \circ z_2 = r_1 r_2 \cos(\theta_2 - \theta_1)$, (b) $z_1 \times z_2 = r_1 r_2 \sin(\theta_2 - \theta_1)$.

113. Prove that (a) $z_1 \circ (z_2 + z_3) = z_1 \circ z_2 + z_1 \circ z_3$, (b) $z_1 \times (z_2 + z_3) = z_1 \times z_2 + z_1 \times z_3$.

114. Find the area of a triangle having vertices at $-4-i,\ 1+2i,\ 4-3i$.   Ans. 17

115. Find the area of a quadrilateral having vertices at $(2,-1),\ (4,3),\ (-1,2)$ and $(-3,-2)$.   Ans. 18

CONJUGATE COORDINATES

116. Describe each of the following loci expressed in terms of conjugate coordinates $z, \bar{z}$.
   (a) $z\bar{z} = 16$, (b) $z\bar{z} - 2z - 2\bar{z} + 8 = 0$, (c) $z + \bar{z} = 4$, (d) $\bar{z} = z + 6i$.

Ans. (a) $x^2 + y^2 = 16$, (b) $x^2 + y^2 - 4x + 8 = 0$, (c) $x = 2$, (d) $y = -3$

117. Write each of the following equations in terms of conjugate coordinates.
   (a) $(x-3)^2 + y^2 = 9$, (b) $2x - 3y = 5$, (c) $4x^2 + 16y^2 = 25$.

Ans. (a) $(z-3)(\bar{z}-3) = 9$, (b) $(2i-3)z + (2i+3)\bar{z} = 10i$, (c) $3(z^2 + \bar{z}^2) - 10z\bar{z} + 25 = 0$

POINT SETS

118. Let S be the set of all points $a + bi$, where a and b are rational numbers, which lie inside the square shown shaded in Fig. 1-42. (a) Is S bounded? (b) What are the limit points of S, if any? (c) Is S closed? (d) What are its interior and boundary points? (e) Is S open? (f) Is S connected? (g) Is S an open region or domain? (h) What is the closure of S? (i) What is the complement of S? (j) Is S countable? (k) Is S compact? (l) Is the closure of S compact?

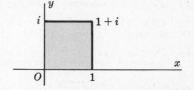

Fig. 1-42

Ans. (a) Yes. (b) Every point inside or on the boundary of the square is a limit point. (c) No. (d) All points of the square are boundary points; there are no interior points. (e) No. (f) No. (g) No. (h) The closure of S is the set of all points inside and on the boundary of the square. (i) The complement of S is the set of all points which are not equal to $a + bi$ when a and b [where $0 < a < 1,\ 0 < b < 1$] are rational. (j) Yes. (k) No. (l) Yes.

119. Answer Problem 118 if S is the set of all points inside the square.

Ans. (a) Yes. (b) Every point inside or on the square is a limit point. (c) No. (d) Every point inside is an interior point, while every point on the boundary is a boundary point. (e) Yes. (f) Yes. (g) Yes. (h) The closure of S is the set of all points inside and on the boundary of the square. (i) The complement of S is the set of all points exterior to the square or on its boundary. (j) No. (k) No. (l) Yes.

120. Answer Problem 118 if S is the set of all points inside or on the square.

  Ans. (a) Yes. (b) Every point of S is a limit point. (c) Yes. (d) Every point inside the square is an interior point, while every point on the boundary is a boundary point. (e) No. (f) Yes. (g) No. (h) S itself. (i) All points exterior to the square. (j) No. (k) Yes. (l) Yes.

121. Given the point sets $A = \{1, i, -i\}$, $B = \{2, 1, -i\}$, $C = \{i, -i, 1+i\}$, $D = \{0, -i, 1\}$. Find:
  (a) $A + (B + C)$ or $A \cup (B \cup C)$, (b) $AC + BD$ or $(A \cap C) \cup (B \cap D)$, (c) $(A + C)(B + D)$ or $(A \cup C) \cap (B \cup D)$.
  Ans. (a) $\{2, 1, -i, i, 1+i\}$, (b) $\{1, i, -i\}$, (c) $\{1, -i\}$

122. If A, B, C and D are any point sets, prove that (a) $A + B = B + A$, (b) $AB = BA$, (c) $A + (B + C) = (A + B) + C$, (d) $A(BC) = (AB)C$, (e) $A(B + C) = AB + AC$. Give equivalent results using the notations $\cap$ and $\cup$. Discuss how these can be used to define an algebra of sets.

123. If A, B and C are the point sets defined by $|z + i| < 3$, $|z| < 5$, $|z + 1| < 4$, represent graphically each of the following:
  (a) $A \cap B \cap C$, (b) $A \cup B \cup C$, (c) $A \cap B \cup C$, (d) $C(A + B)$, (d) $(A \cup B) \cap (B \cup C)$, (e) $AB + BC + CA$, (f) $A\widetilde{B} + B\widetilde{C} + C\widetilde{A}$.

124. Prove that the complement of a set S is open or closed according as S is closed or open.

125. If $S_1, S_2, \ldots, S_n$ are open sets, prove that $S_1 + S_2 + \cdots + S_n$ is open.

126. If a limit point of a set does not belong to the set, prove that it must be a boundary point of the set.

MISCELLANEOUS PROBLEMS

127. Let $ABCD$ be a parallelogram. Prove that $(AC)^2 + (BD)^2 = (AB)^2 + (BC)^2 + (CD)^2 + (DA)^2$.

128. Explain the fallacy: $-1 = \sqrt{-1}\sqrt{-1} = \sqrt{(-1)(-1)} = \sqrt{1} = 1$. Hence $1 = -1$.

129. (a) Show that the equation $z^4 + a_1 z^3 + a_2 z^2 + a_3 z + a_4 = 0$ where a_1, a_2, a_3, a_4 are real constants different from zero, has a pure imaginary root if $a_3^2 + a_1^2 a_4 = a_1 a_2 a_3$.

  (b) Is the converse of (a) true?

130. (a) Prove that $\cos^n \phi = \dfrac{1}{2^{n-1}} \left\{ \cos n\phi + n \cos(n-2)\phi + \dfrac{n(n-1)}{2!} \cos(n-4)\phi + \cdots + R_n \right\}$

  where $R_n = \begin{cases} \dfrac{n!}{[(n-1)/2]!\,[(n+1)/2]!} \cos \phi & \text{if } n \text{ is odd} \\[2ex] \dfrac{n!}{2\,[(n/2)!]^2} & \text{if } n \text{ is even} \end{cases}$

  (b) Derive a similar result for $\sin^n \phi$.

131. If $z = 6e^{\pi i/3}$, evaluate $|e^{iz}|$.   *Ans.* $e^{-3\sqrt{3}}$

132. Show that for any real numbers p and m, $e^{2mi \cot^{-1} p} \left\{ \dfrac{pi + 1}{pi - 1} \right\}^m = 1$.

133. If $P(z)$ is any polynomial in z with real coefficients, prove that $\overline{P(z)} = P(\bar{z})$.

134. If z_1, z_2 and z_3 are collinear, prove that there exist real constants α, β, γ, not all zero, such that $\alpha z_1 + \beta z_2 + \gamma z_3 = 0$ where $\alpha + \beta + \gamma = 0$.

135. Given the complex number z, represent geometrically (a) $\bar{z}$, (b) $-z$, (c) $1/z$, (d) z^2.

136. Given any two complex numbers z_1 and z_2 not equal to zero, show how to represent graphically using only ruler and compass (a) $z_1 z_2$, (b) z_1/z_2, (c) $z_1^2 + z_2^2$, (d) $z_1^{1/2}$, (e) $z_2^{3/4}$.

137. Prove that an equation for a line passing through the points z_1 and z_2 is given by
$$\arg \{(z - z_1)/(z_2 - z_1)\} = 0$$

138. If $z = x + iy$, prove that $|x| + |y| \leq \sqrt{2}\,|x + iy|$.

139. Is the converse to Problem 51 true? Justify your answer.

140. Find an equation for the circle passing through the points $1-i,\ 2i,\ 1+i$.
    Ans. $|z+1| = \sqrt{5}$ or $(x+1)^2 + y^2 = 5$

141. Show that the locus of z such that $|z-a|\,|z+a| = a^2$, $a > 0$ is a *lemniscate* as shown in Fig. 1-43.

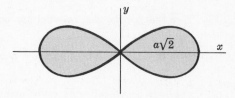

142. Let $p_n = a_n^2 + b_n^2$, $n = 1, 2, 3, \ldots$ where a_n and b_n are positive integers. Prove that for every positive integer M we can always find positive integers A and B such that $p_1 p_2 \cdots p_M = A^2 + B^2$. [Example: If $5 = 2^2 + 1^2$ and $25 = 3^2 + 4^2$, then $5 \cdot 25 = 2^2 + 11^2$.]

Fig. 1-43

143. Prove that

(a) $\cos\theta + \cos(\theta + \alpha) + \cdots + \cos(\theta + n\alpha) = \dfrac{\sin\frac{1}{2}(n+1)\alpha}{\sin\frac{1}{2}\alpha} \cos(\theta + \tfrac{1}{2}n\alpha)$

(b) $\sin\theta + \sin(\theta + \alpha) + \cdots + \sin(\theta + n\alpha) = \dfrac{\sin\frac{1}{2}(n+1)\alpha}{\sin\frac{1}{2}\alpha} \sin(\theta + \tfrac{1}{2}n\alpha)$

144. Prove that *(a)* $\mathrm{Re}\,\{z\} > 0$ and *(b)* $|z-1| < |z+1|$ are equivalent statements.

145. A wheel of radius 4 feet [Fig. 1-44] is rotating counterclockwise about an axis through its center at 30 revolutions per minute. (a) Show that the position and velocity of any point P on the wheel are given respectively by $4e^{i\pi t}$ and $4\pi i e^{i\pi t}$, where t is the time in seconds measured from the instant when P was on the positive x axis. (b) Find the position and velocity when $t = 2/3$ and $t = 15/4$.

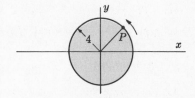

Fig. 1-44

146. Prove that for any integer $m > 1$,
$$(z+a)^{2m} - (z-a)^{2m} = 4maz \prod_{k=1}^{m-1} \{z^2 + a^2 \cot^2(k\pi/2m)\}$$
where $\displaystyle\prod_{k=1}^{m-1}$ denotes the product of all the factors indicated from $k=1$ to $m-1$.

147. If points P_1 and P_2, represented by z_1 and z_2 respectively, are such that $|z_1 + z_2| = |z_1 - z_2|$, prove that (a) z_1/z_2 is a pure imaginary number, (b) $\angle P_1 O P_2 = 90°$.

148. Prove that for any integer $m > 1$,
$$\cot\frac{\pi}{2m}\,\cot\frac{2\pi}{2m}\,\cot\frac{3\pi}{2m}\cdots\cot\frac{(m-1)\pi}{2m} = 1$$

149. Prove and generalize: (a) $\csc^2(\pi/7) + \csc^2(2\pi/7) + \csc^2(4\pi/7) = 2$
                                    (b) $\tan^2(\pi/16) + \tan^2(3\pi/16) + \tan^2(5\pi/16) + \tan^2(7\pi/16) = 28$

150. If masses m_1, m_2, m_3 are located at points z_1, z_2, z_3 respectively, prove that the center of mass is given by
$$\hat{z} = \frac{m_1 z_1 + m_2 z_2 + m_3 z_3}{m_1 + m_2 + m_3}$$
Generalize to n masses.

151. Find that point on the line joining points z_1 and z_2 which divides it in the ratio $p:q$.
    Ans. $(qz_1 + pz_2)/(q + p)$

152. Show that an equation for a circle passing through 3 points z_1, z_2, z_3 is given by
$$\left(\frac{z - z_1}{z - z_2}\right)\Big/\left(\frac{z_3 - z_1}{z_3 - z_2}\right) = \left(\frac{\bar{z} - \bar{z}_1}{\bar{z} - \bar{z}_2}\right)\Big/\left(\frac{\bar{z}_3 - \bar{z}_1}{\bar{z}_3 - \bar{z}_2}\right)$$

153. Prove that the medians of a triangle with vertices at z_1, z_2, z_3 intersect in the point $\frac{1}{3}(z_1 + z_2 + z_3)$.

154. Prove that the rational numbers between 0 and 1 are countable.

[*Hint.* Arrange the numbers as $0, \frac{1}{2}, \frac{1}{3}, \frac{2}{3}, \frac{1}{4}, \frac{3}{4}, \frac{1}{5}, \frac{2}{5}, \frac{3}{5}, \ldots$.]

155. Prove that all the real rational numbers are countable.

156. Prove that the irrational numbers between 0 and 1 are not countable.

157. Represent graphically the set of values of z for which (a) $|z| > |z-1|$, (b) $|z+2| > 1 + |z-2|$.

158. Show that (a) $\sqrt[3]{2} + \sqrt{3}$ and (b) $2 - \sqrt{2}\,i$ are algebraic numbers.

159. Prove that $\sqrt{2} + \sqrt{3}$ is an irrational number.

160. Let $ABCD\cdots PQ$ represent a regular polygon of n sides inscribed in a circle of unit radius. Prove that the product of the lengths of the diagonals $AC, AD, \ldots, AP$ is $\frac{1}{4}n \csc^2(\pi/n)$.

161. Prove that if $\sin\theta \neq 0$,

(a) $\dfrac{\sin n\theta}{\sin\theta} = 2^{n-1} \prod\limits_{k=1}^{n-1} \{\cos\theta - \cos(k\pi/n)\}$

(b) $\dfrac{\sin(2n+1)\theta}{\sin\theta} = (2n+1) \prod\limits_{k=1}^{n} \left\{1 - \dfrac{\sin^2\theta}{\sin^2 k\pi/(2n+1)}\right\}.$

162. Prove $\cos 2n\theta = (-1)^n \prod\limits_{k=1}^{n} \left\{1 - \dfrac{\cos^2\theta}{\cos^2(2k-1)\pi/4n}\right\}.$

163. If the product of two complex numbers z_1 and z_2 is real and different from zero, prove that there exists a real number p such that $z_1 = p\bar{z}_2$.

164. If z is any point on the circle $|z-1| = 1$, prove that $\arg(z-1) = 2\arg z = \frac{2}{3}\arg(z^2 - z)$ and give a geometrical interpretation.

165. Prove that under suitable restrictions (a) $z^m z^n = z^{m+n}$, (b) $(z^m)^n = z^{mn}$.

166. Prove (a) $\operatorname{Re}\{z_1 z_2\} = \operatorname{Re}\{z_1\}\operatorname{Re}\{z_2\} - \operatorname{Im}\{z_1\}\operatorname{Im}\{z_2\}$

(b) $\operatorname{Im}\{z_1 z_2\} = \operatorname{Re}\{z_1\}\operatorname{Im}\{z_2\} + \operatorname{Im}\{z_1\}\operatorname{Re}\{z_2\}.$

167. Find the area of the polygon with vertices at $2+3i, 3+i, -2-4i, -4-i, -1+2i$. *Ans.* 47/2

168. Let $a_1, a_2, \ldots, a_n$ and $b_1, b_2, \ldots, b_n$ be any complex numbers. Prove *Schwarz's inequality*,

$$\left| \sum_{k=1}^{n} a_k b_k \right|^2 \;\leq\; \left(\sum_{k=1}^{n} |a_k|^2 \right)\left(\sum_{k=1}^{n} |b_k|^2 \right)$$

Chapter 2

Functions, Limits and Continuity

VARIABLES AND FUNCTIONS

A symbol, such as z, which can stand for any one of a set of complex numbers is called a *complex variable*.

If to each value which a complex variable z can assume there corresponds one or more values of a complex variable w, we say that w is a *function* of z and write $w = f(z)$ or $w = G(z)$, etc. The variable z is sometimes called an *independent variable*, while w is called a *dependent variable*. The *value of a function* at $z = a$ is often written $f(a)$. Thus if $f(z) = z^2$, then $f(2i) = (2i)^2 = -4$.

SINGLE AND MULTIPLE-VALUED FUNCTIONS

If only one value of w corresponds to each value of z, we say that w is a *single-valued function* of z or that $f(z)$ is single-valued. If more than one value of w corresponds to each value of z, we say that w is a *multiple-valued* or *many-valued* function of z.

A multiple-valued function can be considered as a collection of single-valued functions, each member of which is called a *branch* of the function. It is customary to consider one particular member as a *principal branch* of the multiple-valued function and the value of the function corresponding to this branch as the *principal value*.

Example 1: If $w = z^2$, then to each value of z there is only one value of w. Hence $w = f(z) = z^2$ is a single-valued function of z.

Example 2: If $w = z^{1/2}$, then to each value of z there are two values of w. Hence $w = f(z) = z^{1/2}$ is a multiple-valued (in this case two-valued) function of z.

Whenever we speak of *function* we shall, unless otherwise stated, assume *single-valued function*.

INVERSE FUNCTIONS

If $w = f(z)$, then we can also consider z as a function of w, written $z = g(w) = f^{-1}(w)$. The function f^{-1} is often called the *inverse* function corresponding to f. Thus $w = f(z)$ and $w = f^{-1}(z)$ are *inverse functions* of each other.

TRANSFORMATIONS

If $w = u + iv$ (where u and v are real) is a single-valued function of $z = x + iy$ (where x and y are real), we can write $u + iv = f(x + iy)$. By equating real and imaginary parts this is seen to be equivalent to

$$u = u(x, y), \qquad v = v(x, y) \tag{1}$$

Thus given a point (x, y) in the z plane, such as P in Fig. 2-1 below, there corresponds a point (u, v) in the w plane, say P' in Fig. 2-2 below. The set of equations (1) [or the equivalent, $w = f(z)$] is called a *transformation*. We say that point P is *mapped* or *transformed* into point P' by means of the transformation and call P' the *image* of P.

33

Example: If $w = z^2$, then $u + iv = (x + iy)^2 = x^2 - y^2 + 2ixy$ and the transformation is $u = x^2 - y^2$, $v = 2xy$.   The image of a point $(1, 2)$ in the z plane is the point $(-3, 4)$ in the w plane.

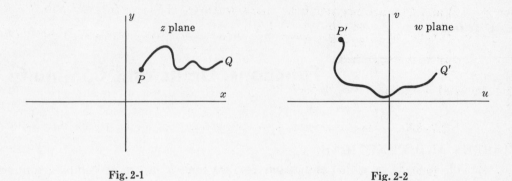

Fig. 2-1   Fig. 2-2

In general, under a transformation, a set of points such as those on curve PQ of Fig. 2-1 is mapped into a corresponding set of points, called the *image*, such as those on curve $P'Q'$ in Fig. 2-2.   The particular characteristics of the image depend of course on the type of function $f(z)$, which is sometimes called a *mapping function*.   If $f(z)$ is multiple-valued, a point (or curve) in the z plane is mapped in general into more than one point (or curve) in the w plane.

CURVILINEAR COORDINATES

Given the transformation $w = f(z)$ or, equivalently, $u = u(x, y)$, $v = v(x, y)$,   we call (x, y) the rectangular coordinates corresponding to a point P in the z plane   and (u, v) the *curvilinear coordinates* of P.

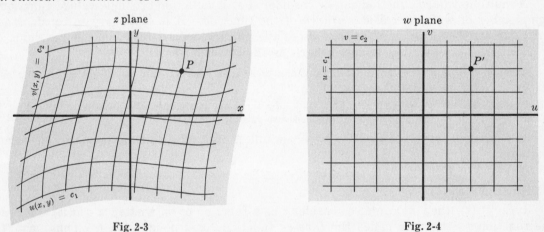

Fig. 2-3   Fig. 2-4

The curves $u(x, y) = c_1$, $v(x, y) = c_2$, where c_1 and c_2 are constants, are called *coordinate curves* [see Fig. 2-3] and each pair of these curves intersects in a point.   These curves map into mutually orthogonal lines in the w plane [see Fig. 2-4].

THE ELEMENTARY FUNCTIONS

1. Polynomial Functions are defined by

$$w = a_0 z^n + a_1 z^{n-1} + \cdots + a_{n-1} z + a_n = P(z) \tag{2}$$

where $a_0 \neq 0, a_1, \ldots, a_n$ are complex constants and n is a positive integer called the *degree* of the polynomial $P(z)$.

The transformation $w = az + b$ is called a *linear transformation*.

2. Rational Algebraic Functions are defined by

$$w = \frac{P(z)}{Q(z)} \tag{3}$$

where $P(z)$ and $Q(z)$ are polynomials. We sometimes call (3) a *rational transformation*. The special case $w = \frac{az+b}{cz+d}$ where $ad - bc \neq 0$ is often called a *bilinear* or *fractional linear transformation*.

3. Exponential Functions are defined by

$$w = e^z = e^{x+iy} = e^x(\cos y + i \sin y) \tag{4}$$

where $e = 2.71828\ldots$ is the *natural base of logarithms*. If a is real and positive, we define

$$a^z = e^{z \ln a} \tag{5}$$

where $\ln a$ is the *natural logarithm of* a. This reduces to (4) if $a = e$.

Complex exponential functions have properties similar to those of real exponential functions. For example, $e^{z_1} \cdot e^{z_2} = e^{z_1 + z_2}$, $e^{z_1}/e^{z_2} = e^{z_1 - z_2}$.

4. Trigonometric Functions. We define the trigonometric or circular functions $\sin z$, $\cos z$, etc., in terms of exponential functions as follows.

$$\sin z = \frac{e^{iz} - e^{-iz}}{2i} \qquad\qquad \cos z = \frac{e^{iz} + e^{-iz}}{2}$$

$$\sec z = \frac{1}{\cos z} = \frac{2}{e^{iz} + e^{-iz}} \qquad\qquad \csc z = \frac{1}{\sin z} = \frac{2i}{e^{iz} - e^{-iz}}$$

$$\tan z = \frac{\sin z}{\cos z} = \frac{e^{iz} - e^{-iz}}{i(e^{iz} + e^{-iz})} \qquad\qquad \cot z = \frac{\cos z}{\sin z} = \frac{i(e^{iz} + e^{-iz})}{e^{iz} - e^{-iz}}$$

Many of the properties familiar in the case of real trigonometric functions also hold for the complex trigonometric functions. For example, we have

$$\sin^2 z + \cos^2 z = 1 \qquad 1 + \tan^2 z = \sec^2 z \qquad 1 + \cot^2 z = \csc^2 z$$

$$\sin(-z) = -\sin z \qquad \cos(-z) = \cos z \qquad \tan(-z) = -\tan z$$

$$\sin(z_1 \pm z_2) = \sin z_1 \cos z_2 \pm \cos z_1 \sin z_2$$

$$\cos(z_1 \pm z_2) = \cos z_1 \cos z_2 \mp \sin z_1 \sin z_2$$

$$\tan(z_1 \pm z_2) = \frac{\tan z_1 \pm \tan z_2}{1 \mp \tan z_1 \tan z_2}$$

5. Hyperbolic Functions are defined as follows:

$$\sinh z = \frac{e^z - e^{-z}}{2} \qquad\qquad \cosh z = \frac{e^z + e^{-z}}{2}$$

$$\operatorname{sech} z = \frac{1}{\cosh z} = \frac{2}{e^z + e^{-z}} \qquad\qquad \operatorname{csch} z = \frac{1}{\sinh z} = \frac{2}{e^z - e^{-z}}$$

$$\tanh z = \frac{\sinh z}{\cosh z} = \frac{e^z - e^{-z}}{e^z + e^{-z}} \qquad\qquad \coth z = \frac{\cosh z}{\sinh z} = \frac{e^z + e^{-z}}{e^z - e^{-z}}$$

The following properties hold:

$$\cosh^2 z - \sinh^2 z = 1 \qquad 1 - \tanh^2 z = \operatorname{sech}^2 z \qquad \coth^2 z - 1 = \operatorname{csch}^2 z$$

$$\sinh(-z) = -\sinh z \qquad \cosh(-z) = \cosh z \qquad \tanh(-z) = -\tanh z$$

$$\sinh(z_1 \pm z_2) = \sinh z_1 \cosh z_2 \pm \cosh z_1 \sinh z_2$$

$$\cosh(z_1 \pm z_2) = \cosh z_1 \cosh z_2 \pm \sinh z_1 \sinh z_2$$

$$\tanh(z_1 \pm z_2) = \frac{\tanh z_1 \pm \tanh z_2}{1 \pm \tanh z_1 \tanh z_2}$$

The following relations exist between the trigonometric or circular functions and the hyperbolic functions:

$$\sin iz = i\sinh z \qquad \cos iz = \cosh z \qquad \tan iz = i\tanh z$$
$$\sinh iz = i\sin z \qquad \cosh iz = \cos z \qquad \tanh iz = i\tan z$$

6. **Logarithmic Functions.** If $z = e^w$, then we write $w = \ln z$, called the *natural logarithm* of z. Thus the natural logarithmic function is the inverse of the exponential function and can be defined by

$$w = \ln z = \ln r + i(\theta + 2k\pi) \qquad k = 0, \pm 1, \pm 2, \ldots$$

where $z = re^{i\theta} = re^{i(\theta + 2k\pi)}$. Note that $\ln z$ is a multiple-valued (in this case infinitely-many-valued) function. The *principal-value* or *principal branch* of $\ln z$ is sometimes defined as $\ln r + i\theta$ where $0 \le \theta < 2\pi$. However, any other interval of length 2π can be used, e.g. $-\pi < \theta \le \pi$, etc.

The logarithmic function can be defined for real bases other than e. Thus if $z = a^w$, then $w = \log_a z$ where $a > 0$ and $a \ne 0, 1$. In this case $z = e^{w \ln a}$ and so $w = (\ln z)/(\ln a)$.

7. **Inverse Trigonometric Functions.** If $z = \sin w$, then $w = \sin^{-1} z$ is called the *inverse sine of z* or *arc sine of z*. Similarly we define other inverse trigonometric or circular functions $\cos^{-1} z$, $\tan^{-1} z$, etc. These functions, which are multiple-valued, can be expressed in terms of natural logarithms as follows. In all cases we omit an additive constant $2k\pi i$, $k = 0, \pm 1, \pm 2, \ldots$, in the logarithm.

$$\sin^{-1} z = \frac{1}{i} \ln\left(iz + \sqrt{1 - z^2}\right) \qquad \csc^{-1} z = \frac{1}{i} \ln\left(\frac{i + \sqrt{z^2 - 1}}{z}\right)$$

$$\cos^{-1} z = \frac{1}{i} \ln\left(z + \sqrt{z^2 - 1}\right) \qquad \sec^{-1} z = \frac{1}{i} \ln\left(\frac{1 + \sqrt{1 - z^2}}{z}\right)$$

$$\tan^{-1} z = \frac{1}{2i} \ln\left(\frac{1 + iz}{1 - iz}\right) \qquad \cot^{-1} z = \frac{1}{2i} \ln\left(\frac{z + i}{z - i}\right)$$

8. **Inverse Hyperbolic Functions.** If $z = \sinh w$ then $w = \sinh^{-1} z$ is called the *inverse hyperbolic sine of z*. Similarly we define other inverse hyperbolic functions $\cosh^{-1} z$, $\tanh^{-1} z$, etc. These functions, which are multiple-valued, can be expressed in terms of natural logarithms as follows. In all cases we omit an additive constant $2k\pi i$, $k = 0, \pm 1, \pm 2, \ldots$, in the logarithm.

$$\sinh^{-1} z = \ln\left(z + \sqrt{z^2 + 1}\right) \qquad \operatorname{csch}^{-1} z = \ln\left(\frac{1 + \sqrt{z^2 + 1}}{z}\right)$$

$$\cosh^{-1} z = \ln\left(z + \sqrt{z^2 - 1}\right) \qquad \operatorname{sech}^{-1} z = \ln\left(\frac{1 + \sqrt{1 - z^2}}{z}\right)$$

$$\tanh^{-1} z = \frac{1}{2} \ln\left(\frac{1 + z}{1 - z}\right) \qquad \coth^{-1} z = \frac{1}{2} \ln\left(\frac{z + 1}{z - 1}\right)$$

9. **The Function z^α,** where α may be complex, is defined as $e^{\alpha \ln z}$. Similarly if $f(z)$ and $g(z)$ are two given functions of z, we can define $f(z)^{g(z)} = e^{g(z) \ln f(z)}$. In general such functions are multiple-valued.

10. **Algebraic and Transcendental Functions.** If w is a solution of the polynomial equation

$$P_0(z)w^n + P_1(z)w^{n-1} + \cdots + P_{n-1}(z)w + P_n(z) = 0 \tag{6}$$

where $P_0 \ne 0, P_1(z), \ldots, P_n(z)$ are polynomials in z and n is a positive integer, then $w = f(z)$ is called an *algebraic function* of z.

Example: $w = z^{1/2}$ is a solution of the equation $w^2 - z = 0$ and so is an algebraic function of z.

Any function which cannot be expressed as a solution of (*6*) is called a *transcendental function*. The logarithmic, trigonometric and hyperbolic functions and their corresponding inverses are examples of transcendental functions.

The functions considered in 1-9 above, together with functions derived from them by a finite number of operations involving addition, subtraction, multiplication, division and roots are called *elementary functions*.

BRANCH POINTS AND BRANCH LINES

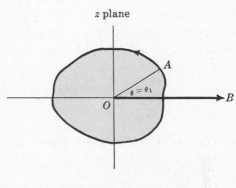

z plane

Fig. 2-5

Suppose that we are given the function $w = z^{1/2}$. Suppose further that we allow z to make a complete circuit (counterclockwise) around the origin starting from point A [Fig. 2-5]. We have $z = re^{i\theta}$, $w = \sqrt{r}\, e^{i\theta/2}$ so that at A, $\theta = \theta_1$ and $w = \sqrt{r}\, e^{i\theta_1/2}$. After a complete circuit back to A, $\theta = \theta_1 + 2\pi$ and $w = \sqrt{r}\, e^{i(\theta_1 + 2\pi)/2} = -\sqrt{r}\, e^{i\theta_1/2}$. Thus we have not achieved the same value of w with which we started. However, by making a second complete circuit back to A, i.e. $\theta = \theta_1 + 4\pi$, $w = \sqrt{r}\, e^{i(\theta_1 + 4\pi)/2} = \sqrt{r}\, e^{i\theta_1/2}$ and we then do obtain the same value of w with which we started.

We can describe the above by stating that if $0 \leq \theta < 2\pi$ we are on one branch of the multiple-valued function $z^{1/2}$, while if $2\pi \leq \theta < 4\pi$ we are on the other branch of the function.

It is clear that each branch of the function is single-valued. In order to keep the function single-valued, we set up an artificial barrier such as OB where B is at infinity [although any other line from O can be used] which we agree not to cross. This barrier [drawn heavy in the figure] is called a *branch line* or *branch cut*, and point O is called a *branch point*. It should be noted that a circuit around any point other than $z = 0$ does not lead to different values; thus $z = 0$ is the only finite branch point.

RIEMANN SURFACES

There is another way to achieve the purpose of the branch line described above. To see this we imagine that the z plane consists of two sheets superimposed on each other. We now cut the sheets along OB and imagine that the lower edge of the bottom sheet is joined to the upper edge of the top sheet. Then starting in the bottom sheet and making one complete circuit about O we arrive in the top sheet. We must now imagine the other cut edges joined together so that by continuing the circuit we go from the top sheet back to the bottom sheet.

The collection of two sheets is called a *Riemann surface* corresponding to the function $z^{1/2}$. Each sheet corresponds to a branch of the function and on each sheet the function is single-valued.

The concept of Riemann surfaces has the advantage in that the various values of multiple-valued functions are obtained in a continuous fashion.

The ideas are easily extended. For example, for the function $z^{1/3}$ the Riemann surface has 3 sheets; for $\ln z$ the Riemann surface has infinitely many sheets.

LIMITS

Let $f(z)$ be defined and single-valued in a neighborhood of $z = z_0$ with the possible exception of $z = z_0$ itself (i.e. in a deleted δ neighborhood of z_0). We say that the number l

is the *limit* of $f(z)$ as z approaches z_0 and write $\lim_{z \to z_0} f(z) = l$ if for any positive number ϵ (however small) we can find some positive number δ (usually depending on ϵ) such that $|f(z) - l| < \epsilon$ whenever $0 < |z - z_0| < \delta$.

In such case we also say that $f(z)$ approaches l as z approaches z_0 and write $f(z) \to l$ as $z \to z_0$. The limit must be independent of the manner in which z approaches z_0.

Geometrically, if z_0 is a point in the complex plane, then $\lim_{z \to z_0} f(z) = l$ if the difference in absolute value between $f(z)$ and l can be made as small as we wish by choosing points z sufficiently close to z_0 (excluding $z = z_0$ itself).

Example: Let $f(z) = \begin{cases} z^2 & z \neq i \\ 0 & z = i \end{cases}$. Then as z gets closer to i (i.e. z approaches i), $f(z)$ gets closer to $i^2 = -1$. We thus *suspect* that $\lim_{z \to i} f(z) = -1$. To *prove* this we must see whether the above definition of limit is satisfied. For this proof see Problem 23.

Note that $\lim_{z \to i} f(z) \neq f(i)$, i.e. the limit of $f(z)$ as $z \to i$ is not the same as the value of $f(z)$ at $z = i$, since $f(i) = 0$ by definition. The limit would in fact be -1 even if $f(z)$ were not defined at $z = i$.

When the limit of a function exists it is unique, i.e. it is the only one (see Problem 26). If $f(z)$ is multiple-valued, the limit as $z \to z_0$ may depend on the particular branch.

THEOREMS ON LIMITS

If $\lim_{z \to z_0} f(z) = A$ and $\lim_{z \to z_0} g(z) = B$, then

1. $\lim_{z \to z_0} \{f(z) + g(z)\} = \lim_{z \to z_0} f(z) + \lim_{z \to z_0} g(z) = A + B$

2. $\lim_{z \to z_0} \{f(z) - g(z)\} = \lim_{z \to z_0} f(z) - \lim_{z \to z_0} g(z) = A - B$

3. $\lim_{z \to z_0} \{f(z)\, g(z)\} = \left\{\lim_{z \to z_0} f(z)\right\}\left\{\lim_{z \to z_0} g(z)\right\} = AB$

4. $\lim_{z \to z_0} \dfrac{f(z)}{g(z)} = \dfrac{\lim_{z \to z_0} f(z)}{\lim_{z \to z_0} g(z)} = \dfrac{A}{B}$ if $B \neq 0$

INFINITY

By means of the transformation $w = 1/z$ the point $z = 0$ (i.e. the origin) is mapped into $w = \infty$, called the *point at infinity in the w plane*. Similarly we denote by $z = \infty$ *the point at infinity in the z plane*. To consider the behavior of $f(z)$ at $z = \infty$, it suffices to let $z = 1/w$ and examine the behavior of $f(1/w)$ at $w = 0$.

We say that $\lim_{z \to \infty} f(z) = l$ or $f(z)$ approaches l as z approaches infinity, if for any $\epsilon > 0$ we can find $M > 0$ such that $|f(z) - l| < \epsilon$ whenever $|z| > M$.

We say that $\lim_{z \to z_0} f(z) = \infty$ or $f(z)$ approaches infinity as z approaches z_0, if for any $N > 0$ we can find $\delta > 0$ such that $|f(z)| > N$ whenever $0 < |z - z_0| < \delta$.

CONTINUITY

Let $f(z)$ be defined and single-valued in a neighborhood of $z = z_0$ as well as at $z = z_0$ (i.e. in a δ neighborhood of z_0). The function $f(z)$ is said to be *continuous* at $z = z_0$ if $\lim_{z \to z_0} f(z) = f(z_0)$. Note that this implies three conditions which must be met in order that $f(z)$ be continuous at $z = z_0$:

1. $\lim\limits_{z \to z_0} f(z) = l$ must exist

2. $f(z_0)$ must exist, i.e. $f(z)$ is defined at z_0

3. $l = f(z_0)$

Equivalently, if $f(z)$ is continuous at z_0 we can write this in the suggestive form $\lim\limits_{z \to z_0} f(z) = f\left(\lim\limits_{z \to z_0} z\right)$.

Example 1:   If $f(z) = \begin{cases} z^2 & z \neq i \\ 0 & z = i \end{cases}$, then from the Example on Page 38, $\lim\limits_{z \to i} f(z) = -1$. But $f(i) = 0$.

Hence $\lim\limits_{z \to i} f(z) \neq f(i)$ and the function is not continuous at $z = i$.

Example 2:   If $f(z) = z^2$ for all z, then $\lim\limits_{z \to i} f(z) = f(i) = -1$ and $f(z)$ is continuous at $z = i$.

Points in the z plane where $f(z)$ fails to be continuous are called *discontinuities* of $f(z)$, and $f(z)$ is said to be *discontinuous* at these points. If $\lim\limits_{z \to z_0} f(z)$ exists but is not equal to $f(z_0)$, we call z_0 a *removable discontinuity* since by redefining $f(z_0)$ to be the same as $\lim\limits_{z \to z_0} f(z)$ the function becomes continuous.

Alternative to the above definition of continuity, we can define $f(z)$ as continuous at $z = z_0$ if for any $\epsilon > 0$ we can find $\delta > 0$ such that $|f(z) - f(z_0)| < \epsilon$ whenever $|z - z_0| < \delta$. Note that this is simply the definition of limit with $l = f(z_0)$ and removal of the restriction that $z \neq z_0$.

To examine the continuity of $f(z)$ at $z = \infty$, we place $z = 1/w$ and examine the continuity of $f(1/w)$ at $w = 0$.

CONTINUITY IN A REGION

A function $f(z)$ is said to be *continuous in a region* if it is continuous at all points of the region.

THEOREMS ON CONTINUITY

Theorem 1. If $f(z)$ and $g(z)$ are continuous at $z = z_0$, so also are the functions $f(z) + g(z)$, $f(z) - g(z)$, $f(z)\,g(z)$ and $\dfrac{f(z)}{g(z)}$, the last only if $g(z_0) \neq 0$. Similar results hold for continuity in a region.

Theorem 2. Among the functions continuous in every finite region are (a) all polynomials, (b) e^z, (c) $\sin z$ and $\cos z$.

Theorem 3. If $w = f(z)$ is continuous at $z = z_0$ and $z = g(\zeta)$ is continuous at $\zeta = \zeta_0$ and if $z_0 = g(\zeta_0)$, then the function $w = f[g(\zeta)]$, called a *function of a function* or *composite function*, is continuous at $\zeta = \zeta_0$. This is simetimes briefly stated as: A continuous function of a continuous function is continuous.

Theorem 4. If $f(z)$ is continuous in a closed region, it is bounded in the region; i.e. there exists a constant M such that $|f(z)| < M$ for all points z of the region.

Theorem 5. If $f(z)$ is continuous in a region, then the real and imaginary parts of $f(z)$ are also continuous in the region.

UNIFORM CONTINUITY

Let $f(z)$ be continuous in a region. Then by definition at each point z_0 of the region and for any $\epsilon > 0$, we can find $\delta > 0$ (which will in general depend on both ϵ and the particular point z_0) such that $|f(z) - f(z_0)| < \epsilon$ whenever $|z - z_0| < \delta$. If we can find δ depending on ϵ but not on the particular point z_0, we say that $f(z)$ is *uniformly continuous* in the region.

Alternatively, $f(z)$ is uniformly continuous in a region if for any $\epsilon > 0$ we can find $\delta > 0$ such that $|f(z_1) - f(z_2)| < \epsilon$ whenever $|z_1 - z_2| < \delta$ where z_1 and z_2 are any two points of the region.

Theorem. If $f(z)$ is continuous in a *closed* region, it is uniformly continuous there.

SEQUENCES

A function of a positive integral variable, designated by $f(n)$ or u_n, where $n = 1, 2, 3, \ldots$, is called a *sequence*. Thus a sequence is a set of numbers $u_1, u_2, u_3, \ldots$ in a definite order of arrangement and formed according to a definite rule. Each number in the sequence is called a *term* and u_n is called the nth *term*. The sequence $u_1, u_2, u_3, \ldots$ is also designated briefly by $\{u_n\}$. The sequence is called *finite* or *infinite* according as there are a finite number of terms or not. Unless otherwise specified, we shall consider infinite sequences only.

Example 1:   The set of numbers $i, i^2, i^3, \ldots, i^{100}$ is a finite sequence; the nth term is given by $u_n = i^n$, $n = 1, 2, \ldots, 100$.

Example 2:   The set of numbers $1 + i, \dfrac{(1+i)^2}{2!}, \dfrac{(1+i)^3}{3!}, \ldots$ is an infinite sequence; the nth term is given by $u_n = (1+i)^n/n!$, $n = 1, 2, 3, \ldots$.

LIMIT OF A SEQUENCE

A number l is called the *limit* of an infinite sequence $u_1, u_2, u_3, \ldots$ if for any positive number ϵ we can find a positive number N depending on ϵ such that $|u_n - l| < \epsilon$ for all $n > N$. In such case we write $\lim\limits_{n \to \infty} u_n = l$. If the limit of a sequence exists, the sequence is called *convergent*; otherwise it is called *divergent*. A sequence can converge to only one limit, i.e. if a limit exists it is unique.

A more intuitive but unrigorous way of expressing this concept of limit is to say that a sequence $u_1, u_2, u_3, \ldots$ has a limit l if the successive terms get "closer and closer" to l. This is often used to provide a "guess" as to the value of the limit, after which the definition is applied to see if the guess is really correct.

THEOREMS ON LIMITS OF SEQUENCES

If $\lim\limits_{n \to \infty} a_n = A$ and $\lim\limits_{n \to \infty} b_n = B$, then

1. $\lim\limits_{n \to \infty} (a_n + b_n) = \lim\limits_{n \to \infty} a_n + \lim\limits_{n \to \infty} b_n = A + B$

2. $\lim\limits_{n \to \infty} (a_n - b_n) = \lim\limits_{n \to \infty} a_n - \lim\limits_{n \to \infty} b_n = A - B$

3. $\lim\limits_{n \to \infty} (a_n b_n) = \left(\lim\limits_{n \to \infty} a_n\right)\left(\lim\limits_{n \to \infty} b_n\right) = AB$

4. $\lim\limits_{n \to \infty} \dfrac{a_n}{b_n} = \dfrac{\lim\limits_{n \to \infty} a_n}{\lim\limits_{n \to \infty} b_n} = \dfrac{A}{B}$ if $B \neq 0$

Further discussion of sequences is given in Chapter 6.

INFINITE SERIES

Let $u_1, u_2, u_3, \ldots$ be a given sequence.

Form a new sequence $S_1, S_2, S_3, \ldots$ defined by

$$S_1 = u_1, \quad S_2 = u_1 + u_2, \quad S_3 = u_1 + u_2 + u_3, \quad \cdots, \quad S_n = u_1 + u_2 + \cdots + u_n$$

where S_n, called the nth *partial sum*, is the sum of the first n terms of the sequence $\{u_n\}$

The sequence $S_1, S_2, S_3, \ldots$ is symbolized by

$$u_1 + u_2 + u_3 + \cdots = \sum_{n=1}^{\infty} u_n$$

which is called an *infinite series*. If $\lim\limits_{n \to \infty} S_n = S$ exists, the series is called *convergent* and S is its *sum*; otherwise the series is called *divergent*. A necessary condition that a series converge is $\lim\limits_{n \to \infty} u_n = 0$; however, this is not sufficient (see Problems 40 and 150).

Further discussion of infinite series is given in Chapter 6.

Solved Problems

FUNCTIONS AND TRANSFORMATIONS

1. Let $w = f(z) = z^2$. Find the values of w which correspond to (a) $z = -2 + i$ and (b) $z = 1 - 3i$, and show how the correspondence can be represented graphically.

(a) $w = f(-2 + i) = (-2 + i)^2 = 4 - 4i + i^2 = 3 - 4i$

(b) $w = f(1 - 3i) = (1 - 3i)^2 = 1 - 6i + 9i^2 = -8 - 6i$

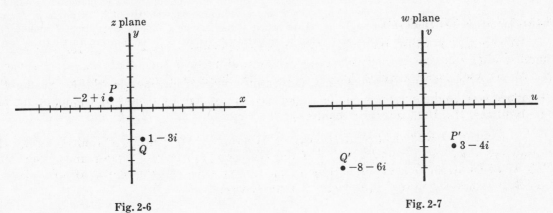

Fig. 2-6   Fig. 2-7

The point $z = -2 + i$, represented by point P in the z plane of Fig. 2-6, has the *image point* $w = 3 - 4i$ represented by P' in the w plane of Fig. 2-7. We say that P is *mapped* into P' by means of the *mapping function* or *transformation* $w = z^2$. Similarly, $z = 1 - 3i$ [point Q of Fig. 2-6] is mapped into $w = -8 - 6i$ [point Q' of Fig. 2-7]. To each point in the z plane there corresponds one and only one point (image) in the w plane, so that w is a single-valued function of z.

2. Show that the line joining the points P and Q in the z plane of Problem 1 [Fig. 2-6] is mapped by $w = z^2$ into a curve joining points $P'Q'$ [Fig. 2-7] and determine the equation of this curve.

Points P and Q have coordinates $(-2, 1)$ and $(1, -3)$. Then the parametric equations of the line joining these points are given by

$$\frac{x - (-2)}{1 - (-2)} = \frac{y - 1}{-3 - 1} = t \qquad \text{or} \qquad x = 3t - 2, \ y = 1 - 4t$$

The equation of the line PQ can be represented by $z = 3t - 2 + i(1 - 4t)$. The curve in the w plane into which this line is mapped has the equation

$$w = z^2 = \{3t - 2 + i(1 - 4t)\}^2 = (3t - 2)^2 - (1 - 4t)^2 + 2(3t - 2)(1 - 4t)i$$
$$= 3 - 4t - 7t^2 + (-4 + 22t - 24t^2)i$$

Then since $w = u + iv$, the parametric equations of the image curve are given by

$$u = 3 - 4t - 7t^2, \quad v = -4 + 22t - 24t^2$$

By assigning various values to the parameter t, this curve may be graphed.

3. A point P moves in a counterclockwise direction around a circle in the z plane having center at the origin and radius 1. If the mapping function is $w = z^3$, show that when P makes one complete revolution the image P' of P in the w plane makes three complete revolutions in a counterclockwise direction on a circle having center at the origin and radius 1.

Let $z = re^{i\theta}$. Then on the circle $|z| = 1$ [Fig. 2-8], $r = 1$ and $z = e^{i\theta}$. Hence $w = z^3 = (e^{i\theta})^3 = e^{3i\theta}$. Letting (ρ, ϕ) denote polar coordinates in the w plane, we have $w = \rho e^{i\phi} = e^{3i\theta}$ so that $\rho = 1$, $\phi = 3\theta$.

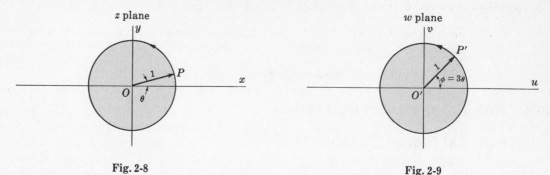

Fig. 2-8   Fig. 2-9

Since $\rho = 1$, it follows that the image point P' moves on a circle in the w plane of radius 1 and center at the origin [Fig. 2-9]. Also, when P moves counterclockwise through an angle θ, P' moves counterclockwise through an angle 3θ. Thus when P makes one complete revolution, P' makes three complete revolutions. In terms of vectors it means that vector $O'P'$ is rotating three times as fast as vector OP.

4. If c_1 and c_2 are any real constants, determine the set of all points in the z plane which map into the lines (a) $u = c_1$, (b) $v = c_2$ in the w plane by means of the mapping function $w = z^2$. Illustrate by considering the cases $c_1 = 2, 4, -2, -4$ and $c_2 = 2, 4, -2, -4$.

We have $w = u + iv = z^2 = (x + iy)^2 = x^2 - y^2 + 2ixy$ so that $u = x^2 - y^2$, $v = 2xy$. Then lines $u = c_1$ and $v = c_2$ in the w plane correspond respectively to hyperbolas $x^2 - y^2 = c_1$ and $2xy = c_2$ in the z plane as indicated in Figures 2-10 and 2-11.

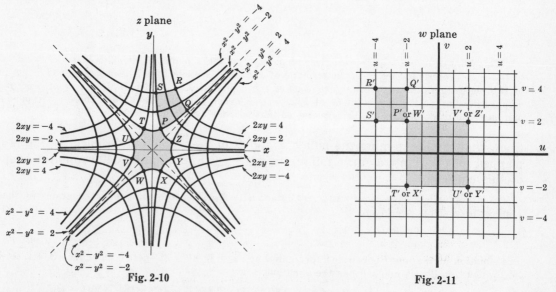

Fig. 2-10   Fig. 2-11

5. Referring to Problem 4, determine: (a) the image of the region in the first quadrant bounded by $x^2 - y^2 = -2$, $xy = 1$, $x^2 - y^2 = -4$ and $xy = 2$; (b) the image of the region in the z plane bounded by all the branches of $x^2 - y^2 = 2$, $xy = 1$, $x^2 - y^2 = -2$ and $xy = -1$; (c) the curvilinear coordinates of that point in the xy plane whose rectangular coordinates are $(2, -1)$.

(a) The region in the z plane is indicated by the shaded portion $PQRS$ of Fig. 2-10. This region maps into the required image region $P'Q'R'S'$ shown shaded in Fig. 2-11. It should be noted that curve $PQRSP$ is traversed in a counterclockwise direction and the image curve $P'Q'R'S'P'$ is also traversed in a counterclockwise direction.

(b) The region in the z plane is indicated by the shaded portion $PTUVWXYZ$ of Fig. 2-10. This region maps into the required image region $P'T'U'V'$ shown shaded in Fig. 2-11.

It is of interest to note that when the boundary of the region $PTUVWXYZ$ is traversed only once, the boundary of the image region $P'T'U'V'$ is traversed twice. This is due to the fact that the eight points P and W, T and X, U and Y, V and Z of the z plane map into the four points P' or W', T' or X', U' or Y', V' or Z' respectively.

However, when the boundary of region $PQRS$ is traversed only once, the boundary of the image region is also traversed only once. The difference is due to the fact that in traversing the curve $PTUVWXYZP$ we are encircling the origin $z = 0$, whereas when we are traversing the curve $PQRSP$ we are not encircling the origin.

(c) $u = x^2 - y^2 = (2)^2 - (-1)^2 = 3$, $v = 2xy = 2(2)(-1) = -4$. Then the curvilinear coordinates are $u = 3$, $v = -4$.

MULTIPLE-VALUED FUNCTIONS

6. Let $w^5 = z$ and suppose that corresponding to the particular value $z = z_1$ we have $w = w_1$. (a) If we start at the point z_1 in the z plane [see Fig. 2-12] and make one complete circuit counterclockwise around the origin, show that the value of w on returning to z_1 is $w_1 e^{2\pi i/5}$. (b) What are the values of w on returning to z_1, after $2, 3, \ldots$ complete circuits around the origin? (c) Discuss parts (a) and (b) if the paths do not enclose the origin.

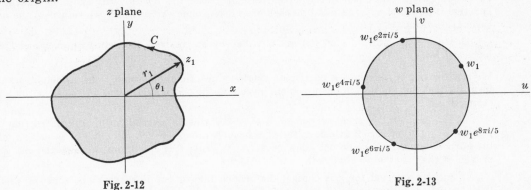

Fig. 2-12   Fig. 2-13

(a) We have $z = re^{i\theta}$, so that $w = z^{1/5} = r^{1/5} e^{i\theta/5}$. If $r = r_1$ and $\theta = \theta_1$, then $w_1 = r_1^{1/5} e^{i\theta_1/5}$.

As θ increases from θ_1 to $\theta_1 + 2\pi$, which is what happens when one complete circuit counterclockwise around the origin is made, we find

$$w = r_1^{1/5} e^{i(\theta_1 + 2\pi)/5} = r_1^{1/5} e^{i\theta_1/5} e^{2\pi i/5} = w_1 e^{2\pi i/5}$$

(b) After 2 complete circuits around the origin, we find

$$w = r_1^{1/5} e^{i(\theta_1 + 4\pi)/5} = r_1^{1/5} e^{i\theta_1/5} e^{4\pi i/5} = w_1 e^{4\pi i/5}$$

Similarly after 3 and 4 complete circuits around the origin, we find

$$w = w_1 e^{6\pi i/5} \qquad \text{and} \qquad w = w_1 e^{8\pi i/5}$$

After 5 complete circuits the value of w is $w_1 e^{10\pi i/5} = w_1$, so that the original value of w is obtained after 5 revolutions about the origin. Thereafter the cycle is repeated [see Fig. 2-13].

Another method. Since $w^5 = z$, we have $\arg z = 5 \arg w$ from which

$$\text{Change in } \arg w = \tfrac{1}{5}(\text{Change in } \arg z)$$

Then if $\arg z$ increases by $2\pi, 4\pi, 6\pi, 8\pi, 10\pi, \ldots$, $\arg w$ increases by $2\pi/5, 4\pi/5, 6\pi/5, 8\pi/5, 2\pi, \ldots$ leading to the same results obtained in (a) and (b).

(c) If the path does not enclose the origin then the increase in $\arg z$ is zero and so the increase in $\arg w$ is also zero. In this case the value of w is w_1, regardless of the number of circuits made.

7.   (a) In the preceding problem explain why we can consider w as a collection of five single-valued functions of z.

    (b) Explain geometrically the relationship between these single-valued functions.

    (c) Show geometrically how we can restrict ourselves to a particular single-valued function.

(a) Since $w^5 = z = re^{i\theta} = re^{i(\theta + 2k\pi)}$ where k is an integer, we have

$$w = r^{1/5} e^{i(\theta + 2k\pi)/5} = r^{1/5} \{\cos (\theta + 2k\pi)/5 + i \sin (\theta + 2k\pi)/5\}$$

and so w is a five-valued function of z, the five values being given by $k = 0, 1, 2, 3, 4$.

Equivalently, we can consider w as a collection of five single-valued functions, called *branches* of the multiple-valued function, by properly restricting θ. Thus, for example, we can write

$$w = r^{1/5} (\cos \theta/5 + i \sin \theta/5)$$

where we take the five possible intervals for θ given by $0 \leq \theta < 2\pi$, $2\pi \leq \theta < 4\pi$, ..., $8\pi \leq \theta < 10\pi$, all other such intervals producing repetitions of these.

The first interval, $0 \leq \theta < 2\pi$, is sometimes called the *principal range* of θ and corresponds to the *principal branch* of the multiple-valued function.

Other intervals for θ of length 2π can also be taken; for example, $-\pi \leq \theta < \pi$, $\pi \leq \theta < 3\pi$, etc., the first of these being taken as the principal range.

(b) We start with the (principal) branch

$$w = r^{1/5} (\cos \theta/5 + i \sin \theta/5) \qquad \text{where } 0 \leq \theta < 2\pi$$

After one complete circuit about the origin in the z plane, θ increases by 2π to give another branch of the function. After another complete circuit about the origin, still another branch of the function is obtained until all five branches have been found, after which we return to the original (principal) branch.

Because different values of $f(z)$ are obtained by successively encircling $z = 0$, we call $z = 0$ a *branch point*.

(c) We can restrict ourselves to a particular single-valued function, usually the principal branch, by insuring that not more than one complete circuit about the branch point is made, i.e. by suitably restricting θ.

In the case of the principal range $0 \leq \theta < 2\pi$, this is accomplished by constructing a cut, indicated by OA in Fig. 2-14 below, called a *branch cut* or *branch line*, on the positive real axis, the purpose being that we do not allow ourselves to cross this cut (if we do cross the cut, another branch of the function is obtained).

If another interval for θ is chosen, the branch line or cut is taken to be some other line in the z plane emanating from the branch point.

For some purposes, as we shall see later, it is useful to consider the curve of Fig. 2-15 of which Fig. 2-14 is a limiting case.

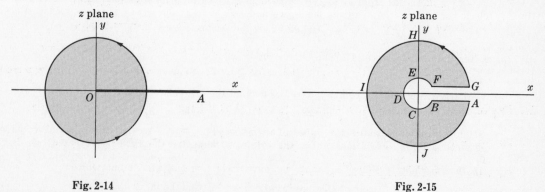

Fig. 2-14   Fig. 2-15

THE ELEMENTARY FUNCTIONS

8.   Prove that   (a) $e^{z_1} \cdot e^{z_2} = e^{z_1 + z_2}$,   (b) $|e^z| = e^x$,   (c) $e^{z + 2k\pi i} = e^z$, $k = 0, \pm 1, \pm 2, \ldots$.

(a) By definition $e^z = e^x (\cos y + i \sin y)$ where $z = x + iy$. Then if $z_1 = x_1 + iy_1$ and $z_2 = x_2 + iy_2$,

$$e^{z_1} \cdot e^{z_2} = e^{x_1}(\cos y_1 + i \sin y_1) \cdot e^{x_2}(\cos y_2 + i \sin y_2)$$
$$= e^{x_1} \cdot e^{x_2}(\cos y_1 + i \sin y_1)(\cos y_2 + i \sin y_2)$$
$$= e^{x_1+x_2}\{\cos(y_1 + y_2) + i \sin(y_1 + y_2)\} = e^{z_1+z_2}$$

(b) $$|e^z| = |e^x(\cos y + i \sin y)| = |e^x||\cos y + i \sin y| = e^x \cdot 1 = e^x$$

(c) By part (a), $$e^{z+2k\pi i} = e^z e^{2k\pi i} = e^z(\cos 2k\pi + i \sin 2k\pi) = e^z$$

This shows that the function e^z has *period* $2k\pi i$. In particular, it has period $2\pi i$.

9. Prove:

(a) $\sin^2 z + \cos^2 z = 1$ 　　　　　　　　(c) $\sin(z_1 + z_2) = \sin z_1 \cos z_2 + \cos z_1 \sin z_2$

(b) $e^{iz} = \cos z + i \sin z$, $e^{-iz} = \cos z - i \sin z$ 　(d) $\cos(z_1 + z_2) = \cos z_1 \cos z_2 - \sin z_1 \sin z_2$

By definition, $\sin z = \dfrac{e^{iz} - e^{-iz}}{2i}$, $\cos z = \dfrac{e^{iz} + e^{-iz}}{2}$.   Then

(a) $\sin^2 z + \cos^2 z = \left(\dfrac{e^{iz} - e^{-iz}}{2i}\right)^2 + \left(\dfrac{e^{iz} + e^{-iz}}{2}\right)^2$

$$= -\left(\frac{e^{2iz} - 2 + e^{-2iz}}{4}\right) + \left(\frac{e^{2iz} + 2 + e^{-2iz}}{4}\right) = 1$$

(b) 　　　　(1) $e^{iz} - e^{-iz} = 2i \sin z$, 　　(2) $e^{iz} + e^{-iz} = 2 \cos z$

Adding (1) and (2): $2e^{iz} = 2 \cos z + 2i \sin z$ and $e^{iz} = \cos z + i \sin z$

Subtracting (1) from (2): $2e^{-iz} = 2 \cos z - 2i \sin z$ and $e^{-iz} = \cos z - i \sin z$

(c) $\sin(z_1 + z_2) = \dfrac{e^{i(z_1+z_2)} - e^{-i(z_1+z_2)}}{2i} = \dfrac{e^{iz_1} \cdot e^{iz_2} - e^{-iz_1} \cdot e^{-iz_2}}{2i}$

$$= \frac{(\cos z_1 + i \sin z_1)(\cos z_2 + i \sin z_2) - (\cos z_1 - i \sin z_1)(\cos z_2 - i \sin z_2)}{2i}$$

$$= \sin z_1 \cos z_2 + \cos z_1 \sin z_2$$

(d) $\cos(z_1 + z_2) = \dfrac{e^{i(z_1+z_2)} + e^{-i(z_1+z_2)}}{2} = \dfrac{e^{iz_1} \cdot e^{iz_2} + e^{-iz_1} \cdot e^{-iz_2}}{2}$

$$= \frac{(\cos z_1 + i \sin z_1)(\cos z_2 + i \sin z_2) + (\cos z_1 - i \sin z_1)(\cos z_2 - i \sin z_2)}{2}$$

$$= \cos z_1 \cos z_2 - \sin z_1 \sin z_2$$

10. Prove that the zeros of (a) $\sin z$ and (b) $\cos z$ are all real and find them.

(a) If $\sin z = \dfrac{e^{iz} - e^{-iz}}{2i} = 0$, then $e^{iz} = e^{-iz}$ or $e^{2iz} = 1 = e^{2k\pi i}$, $k = 0, \pm 1, \pm 2, \ldots$.

Hence $2iz = 2k\pi i$ and $z = k\pi$, i.e. $z = 0, \pm\pi, \pm 2\pi, \pm 3\pi, \ldots$ are the zeros.

(b) If $\cos z = \dfrac{e^{iz} + e^{-iz}}{2} = 0$, then $e^{iz} = -e^{-iz}$ or $e^{2iz} = -1 = e^{(2k+1)\pi i}$, $k = 0, \pm 1, \pm 2, \ldots$.

Hence $2iz = (2k+1)\pi i$ and $z = (k + \frac{1}{2})\pi$, i.e. $z = \pm\pi/2, \pm 3\pi/2, \pm 5\pi/2, \ldots$ are the zeros.

11. Prove that (a) $\sin(-z) = -\sin z$, (b) $\cos(-z) = \cos z$, (c) $\tan(-z) = -\tan z$.

(a) $\sin(-z) = \dfrac{e^{i(-z)} - e^{-i(-z)}}{2i} = \dfrac{e^{-iz} - e^{iz}}{2i} = -\left(\dfrac{e^{iz} - e^{-iz}}{2i}\right) = -\sin z$

(b) $\cos(-z) = \dfrac{e^{i(-z)} + e^{-i(-z)}}{2} = \dfrac{e^{-iz} + e^{iz}}{2} = \dfrac{e^{iz} + e^{-iz}}{2} = \cos z$

(c) $\tan(-z) = \dfrac{\sin(-z)}{\cos(-z)} = \dfrac{-\sin z}{\cos z} = -\tan z$,   using (a) and (b).

Functions of z having the property that $f(-z) = -f(z)$ are called *odd functions*, while those for which $f(-z) = f(z)$ are called *even functions*. Thus $\sin z$ and $\tan z$ are odd functions, while $\cos z$ is an even function.

12. Prove: (a) $1 - \tanh^2 z = \text{sech}^2 z$ (c) $\cos iz = \cosh z$

               (b) $\sin iz = i \sinh z$ (d) $\sin (x + iy) = \sin x \cosh y + i \cos x \sinh y$

   (a) By definition, $\cosh z = \dfrac{e^z + e^{-z}}{2}$, $\sinh z = \dfrac{e^z - e^{-z}}{2}$.   Then

$$\cosh^2 z - \sinh^2 z = \left(\frac{e^z + e^{-z}}{2}\right)^2 - \left(\frac{e^z - e^{-z}}{2}\right)^2 = \frac{e^{2z} + 2 + e^{-2z}}{4} - \frac{e^{2z} - 2 + e^{-2z}}{4} = 1$$

     Dividing by $\cosh^2 z$, $\dfrac{\cosh^2 z - \sinh^2 z}{\cosh^2 z} = \dfrac{1}{\cosh^2 z}$ or $1 - \tanh^2 z = \text{sech}^2 z$.

   (b) $\sin iz = \dfrac{e^{i(iz)} - e^{-i(iz)}}{2i} = \dfrac{e^{-z} - e^z}{2i} = i\left(\dfrac{e^z - e^{-z}}{2}\right) = i \sinh z$

   (c) $\cos iz = \dfrac{e^{i(iz)} + e^{-i(iz)}}{2} = \dfrac{e^{-z} + e^z}{2} = \dfrac{e^z + e^{-z}}{2} = \cosh z$

   (d)   From Problem 9(c) and parts (b) and (c), we have

$$\sin (x + iy) = \sin x \cos iy + \cos x \sin iy = \sin x \cosh y + i \cos x \sinh y$$

13. (a) If $z = e^w$ where $z = r(\cos \theta + i \sin \theta)$ and $w = u + iv$, show that $u = \ln r$ and $v = \theta + 2k\pi$, $k = 0, \pm1, \pm2, \ldots$ so that $w = \ln z = \ln r + i(\theta + 2k\pi)$.   (b) Determine the values of $\ln (1 - i)$.   What is the principal value?

   (a) Since $z = r(\cos \theta + i \sin \theta) = e^w = e^{u+iv} = e^u(\cos v + i \sin v)$, we have on equating real and imaginary parts,

          (1) $e^u \cos v = r \cos \theta$ (2) $e^u \sin v = r \sin \theta$

     Squaring (1) and (2) and adding, we find $e^{2u} = r^2$ or $e^u = r$ and $u = \ln r$. Then from (1) and (2), $r \cos v = r \cos \theta, r \sin v = r \sin \theta$ from which $v = \theta + 2k\pi$. Hence $w = u + iv = \ln r + i(\theta + 2k\pi)$.

     If $z = e^w$, we say that $w = \ln z$. We thus see that $\ln z = \ln r + i(\theta + 2k\pi)$. An equivalent way of saying the same thing is to write $\ln z = \ln r + i\theta$ where θ can assume infinitely many values which differ by 2π.

     Note that *formally* $\ln z = \ln (re^{i\theta}) = \ln r + i\theta$ using laws of real logarithms familiar from elementary mathematics.

   (b) Since $1 - i = \sqrt{2}\, e^{7\pi i/4 + 2k\pi i}$, we have $\ln (1 - i) = \ln \sqrt{2} + \left(\dfrac{7\pi i}{4} + 2k\pi i\right) = \dfrac{1}{2} \ln 2 + \dfrac{7\pi i}{4} + 2k\pi i$.

     The principal value is $\dfrac{1}{2} \ln 2 + \dfrac{7\pi i}{4}$ obtained by letting $k = 0$.

14. Prove that $f(z) = \ln z$ has a branch point at $z = 0$.

     We have $\ln z = \ln r + i\theta$. Suppose that we start at some point $z_1 \neq 0$ in the complex plane for which $r = r_1, \theta = \theta_1$ so that $\ln z_1 = \ln r_1 + i\theta_1$ [see Fig. 2-16]. Then after making one complete circuit about the origin in the positive or counterclockwise direction, we find on returning to z_1 that $r = r_1$, $\theta = \theta_1 + 2\pi$ so that $\ln z_1 = \ln r_1 + i(\theta_1 + 2\pi)$. Thus we are on another branch of the function, and so $z = 0$ is a branch point.

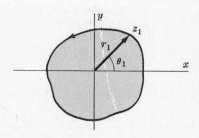

     Further complete circuits about the origin lead to other branches and (unlike the case of functions such as $z^{1/2}$ or $z^{1/5}$) we *never* return to the same branch.

Fig. 2-16

     It follows that $\ln z$ is an infinitely many-valued function of z with infinitely many branches. That particular branch of $\ln z$ which is real when z is real and positive is called the *principal branch*. To obtain this branch we require that $\theta = 0$ when $z > 0$. To accomplish this we can take $\ln z = \ln r + i\theta$ where θ is chosen so that $0 \leqq \theta < 2\pi$ or $-\pi \leqq \theta < \pi$, etc.

     As a generalization we note that $\ln (z - a)$ has a branch point at $z = a$.

15. Consider the transformation $w = \ln z$. Show that (a) circles with center at the origin in the z plane are mapped into lines parallel to the v axis in the w plane, (b) lines or *rays* emanating from the origin in the z plane are mapped into lines parallel to the u axis in the w plane, (c) the z plane is mapped into a strip of width 2π in the w plane. Illustrate the results graphically.

We have $w = u + iv = \ln z = \ln r + i\theta$ so that $u = \ln r, \ v = \theta$.

Choose the principal branch as $w = \ln r + i\theta$ where $0 \le \theta < 2\pi$.

(a) Circles with center at the origin and radius a have the equation $|z| = r = a$. These are mapped into lines in the w plane whose equations are $u = \ln a$. In Figures 2-17 and 2-18 the circles and lines corresponding to $a = 1/2, 1, 3/2, 2$ are indicated.

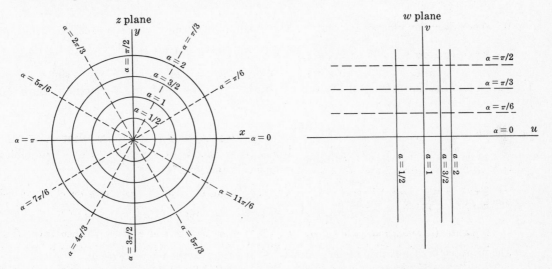

Fig. 2-17   Fig. 2-18

(b) Lines or rays emanating from the origin in the z plane (dashed in Fig. 2-17) have the equation $\theta = \alpha$. These are mapped into lines in the w plane (dashed in Fig. 2-18) whose equations are $v = \alpha$. We have shown the corresponding lines for $\alpha = 0, \pi/6, \pi/3$ and $\pi/2$.

(c) Corresponding to any given point P in the z plane defined by $z \ne 0$ and having polar coordinates (r, θ) where $0 \le \theta < 2\pi$, $r > 0$, there is a point P' in the strip of width 2π shown shaded in Fig. 2-20. Thus the z plane is mapped into this strip. The point $z = 0$ is mapped into a point of this strip sometimes called the *point at infinity*.

If θ is such that $2\pi \le \theta < 4\pi$, the z plane is mapped into the strip $2\pi \le v < 4\pi$ of Fig. 2-20. Similarly, we obtain the other strips shown in Fig. 2-20.

It follows that given any point $z \ne 0$ in the z plane, there are infinitely many image points in the w plane corresponding to it.

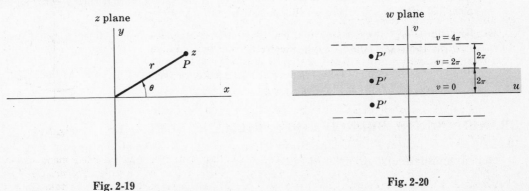

Fig. 2-19   Fig. 2-20

It should be noted that if we had taken θ such that $-\pi \le \theta < \pi$, $\pi \le \theta < 3\pi$, etc., the strips of Fig. 2-20 would be shifted vertically a distance π.

16. If we choose the principal branch of $\sin^{-1} z$ to be that one for which $\sin^{-1} 0 = 0$, prove that

$$\sin^{-1} z = \frac{1}{i} \ln\left(iz + \sqrt{1 - z^2}\right)$$

If $w = \sin^{-1} z$, then $z = \sin w = \dfrac{e^{iw} - e^{-iw}}{2i}$ from which

$$e^{iw} - 2iz - e^{-iw} = 0 \qquad \text{or} \qquad e^{2iw} - 2ize^{iw} - 1 = 0$$

Solving, $\qquad e^{iw} = \dfrac{2iz \pm \sqrt{4 - 4z^2}}{2} = iz \pm \sqrt{1 - z^2} = iz + \sqrt{1 - z^2}$

since $\pm\sqrt{1 - z^2}$ is implied by $\sqrt{1 - z^2}$. Now $e^{iw} = e^{i(w - 2k\pi)}$, $k = 0, \pm 1, \pm 2, \ldots$ so that

$$e^{i(w - 2k\pi)} = iz + \sqrt{1 - z^2} \qquad \text{or} \qquad w = 2k\pi + \frac{1}{i} \ln\left(iz + \sqrt{1 - z^2}\right)$$

The branch for which $w = 0$ when $z = 0$ is obtained by taking $k = 0$ from which we find, as required,

$$w = \sin^{-1} z = \frac{1}{i} \ln\left(iz + \sqrt{1 - z^2}\right)$$

17. If we choose the principal branch of $\tanh^{-1} z$ to be that one for which $\tanh^{-1} 0 = 0$, prove that

$$\tanh^{-1} z = \frac{1}{2} \ln\left(\frac{1 + z}{1 - z}\right)$$

If $w = \tanh^{-1} z$, then $z = \tanh w = \dfrac{\sinh w}{\cosh w} = \dfrac{e^w - e^{-w}}{e^w + e^{-w}}$ from which

$$(1 - z)e^w = (1 + z)e^{-w} \qquad \text{or} \qquad e^{2w} = (1 + z)/(1 - z)$$

Since $e^{2w} = e^{2(w - k\pi i)}$, we have

$$e^{2(w - k\pi i)} = \frac{1 + z}{1 - z} \qquad \text{or} \qquad w = k\pi i + \frac{1}{2} \ln\left(\frac{1 + z}{1 - z}\right)$$

The principal branch is the one for which $k = 0$ and leads to the required result.

18. (a) If $z = re^{i\theta}$, prove that $z^i = e^{-(\theta + 2k\pi)}\{\cos(\ln r) + i \sin(\ln r)\}$ where $k = 0, \pm 1, \pm 2, \ldots$.

(b) If z is a point on the unit circle with center at the origin, prove that z^i represents infinitely many real numbers and determine the principal value.

(c) Find the principal value of i^i.

(a) By definition, $\quad z^i = e^{i \ln z} = e^{i\{\ln r + i(\theta + 2k\pi)\}}$

$$= e^{i \ln r - (\theta + 2k\pi)} = e^{-(\theta + 2k\pi)}\{\cos(\ln r) + i \sin(\ln r)\}$$

The principal branch of the many-valued function $f(z) = z^i$ is obtained by taking $k = 0$ and is given by $e^{-\theta}\{\cos(\ln r) + i \sin(\ln r)\}$ where we can choose θ such that $0 \leqq \theta < 2\pi$.

(b) If z is any point on the unit circle with center at the origin, then $|z| = r = 1$. Hence by part (a), since $\ln r = 0$, we have $z^i = e^{-(\theta + 2k\pi)}$ which represents infinitely many real numbers. The principal value is $e^{-\theta}$ where we choose θ such that $0 \leqq \theta < 2\pi$.

(c) By definition, $\quad i^i = e^{i \ln i} = e^{i\{i(\pi/2 + 2k\pi)\}} = e^{-(\pi/2 + 2k\pi)}$ since $i = e^{i(\pi/2 + 2k\pi)}$ and $\ln i = i(\pi/2 + 2k\pi)$. The principal value is given by $e^{-\pi/2}$.

Another method. By part (b), since $z = i$ lies on the unit circle with center at the origin and since $\theta = \pi/2$, the principal value is $e^{-\pi/2}$.

BRANCH POINTS, BRANCH LINES, RIEMANN SURFACES

19. Let $w = f(z) = (z^2 + 1)^{1/2}$. (a) Show that $z = \pm i$ are branch points of $f(z)$. (b) Show that a complete circuit around both branch points produces no change in the branches of $f(z)$.

(a) We have $w = (z^2 + 1)^{1/2} = \{(z - i)(z + i)\}^{1/2}$. Then $\arg w = \frac{1}{2} \arg(z - i) + \frac{1}{2} \arg(z + i)$ so that

$$\text{Change in } \arg w = \tfrac{1}{2}\{\text{Change in } \arg(z - i)\} + \tfrac{1}{2}\{\text{Change in } \arg(z + i)\}$$

Let C [Fig. 2-21] be a closed curve enclosing the point i but not the point $-i$. Then as point z goes once counterclockwise around C,

$$\text{Change in } \arg(z-i) \; = \; 2\pi, \qquad \text{Change in } \arg(z+i) \; = \; 0$$

so that
$$\text{Change in } \arg w \; = \; \pi$$

Hence w does not return to its original value, i.e. a change in branches has occurred. Since a complete circuit about $z=i$ alters the branches of the function, $z=i$ is a branch point. Similarly if C is a closed curve enclosing the point $-i$ but not i, we can show that $z=-i$ is a branch point.

Another method.

Let $z-i = r_1 e^{i\theta_1}, \; z+i = r_2 e^{i\theta_2}$. Then

$$w \; = \; \{r_1 r_2 e^{i(\theta_1+\theta_2)}\}^{1/2} \; = \; \sqrt{r_1 r_2}\, e^{i\theta_1/2}\, e^{i\theta_2/2}$$

Suppose we start with a particular value of z corresponding to $\theta_1 = \alpha_1$ and $\theta_2 = \alpha_2$. Then $w = \sqrt{r_1 r_2}\, e^{i\alpha_1/2}\, e^{i\alpha_2/2}$. As z goes once counterclockwise around i, θ_1 increases to $\alpha_1 + 2\pi$ while θ_2 remains the same, i.e. $\theta_2 = \alpha_2$. Hence

$$w \; = \; \sqrt{r_1 r_2}\, e^{i(\alpha_1+2\pi)/2}\, e^{i\alpha_2/2}$$
$$= \; -\sqrt{r_1 r_2}\, e^{i\alpha_1/2}\, e^{i\alpha_2/2}$$

showing that we do not obtain the original value of w, i.e. a change of branches has occurred, showing that $z=i$ is a branch point.

(b) If C encloses both branch points $z = \pm i$ as in Fig. 2-22, then as point z goes counterclockwise around C,

$$\text{Change in } \arg(z-i) \; = \; 2\pi$$
$$\text{Change in } \arg(z+i) \; = \; 2\pi$$

so that
$$\text{Change in } \arg w \; = \; 2\pi$$

Hence a complete circuit around both branch points produces no change in the branches.

Another method.

In this case, referring to the second method of part (a), θ_1 increases from α_1 to $\alpha_1 + 2\pi$ while θ_2 increases from α_2 to $\alpha_2 + 2\pi$. Thus

$$w \; = \; \sqrt{r_1 r_2}\, e^{i(\alpha_1+2\pi)/2}\, e^{i(\alpha_2+2\pi)/2} \; = \; \sqrt{r_1 r_2}\, e^{i\alpha_1/2}\, e^{i\alpha_2/2}$$

and no change in branch is observed.

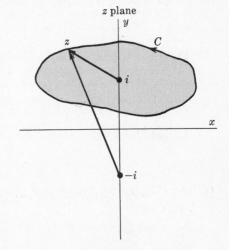

Fig. 2-21

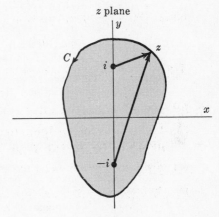

Fig. 2-22

20. Determine branch lines for the function of Problem 19.

The branch lines can be taken as those indicated heavy in either of Figures 2-23, 2-24. In both cases, by not crossing these heavy lines we insure the single-valuedness of the function.

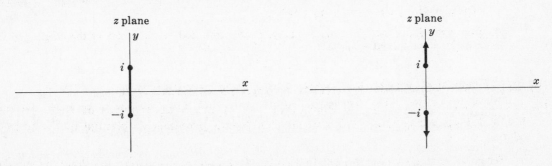

Fig. 2-23

Fig. 2-24

21. Discuss the Riemann surface for the function of Problem 19.

We can have different Riemann surfaces corresponding to Fig. 2-23 or 2-24 of Problem 20. Referring to Fig. 2-23, for example, we imagine that the z plane consists of two sheets superimposed on each other and cut along the branch line. Opposite edges of the cut are then joined, forming the Riemann surface. On making one complete circuit around $z = i$, we start on one branch and wind up on the other. However, if we make one circuit about both $z = i$ and $z = -i$, we do not change branches at all. This agrees with the results of Problem 19.

22. Discuss the Riemann surface for the function $f(z) = \ln z$ [see Problem 14].

In this case we imagine the z plane to consist of infinitely many sheets superimposed on each other and cut along a branch line emanating from the origin $z = 0$. We then connect each cut edge to the opposite cut edge of an adjacent sheet. Then every time we make a circuit about $z = 0$ we are on another sheet corresponding to a different branch of the function. The collection of sheets is the Riemann surface. In this case, unlike Problems 6 and 7, successive circuits never bring us back to the original branch.

LIMITS

23. (a) If $f(z) = z^2$, prove that $\lim\limits_{z \to z_0} f(z) = z_0^2$.

(b) Find $\lim\limits_{z \to z_0} f(z)$ if $f(z) = \begin{cases} z^2 & z \neq z_0 \\ 0 & z = z_0 \end{cases}$.

(a) We must show that given any $\epsilon > 0$ we can find δ (depending in general on ϵ) such that $|z^2 - z_0^2| < \epsilon$ whenever $0 < |z - z_0| < \delta$.

If $\delta \leq 1$, then $0 < |z - z_0| < \delta$ implies that

$$|z^2 - z_0^2| = |z - z_0||z + z_0| < \delta|z - z_0 + 2z_0| < \delta\{|z - z_0| + |2z_0|\} < \delta(1 + 2|z_0|)$$

Take δ as 1 or $\epsilon/(1 + 2|z_0|)$, whichever is smaller. Then we have $|z^2 - z_0^2| < \epsilon$ whenever $|z - z_0| < \delta$, and the required result is proved.

(b) There is no difference between this problem and that in part (a), since in both cases we exclude $z = z_0$ from consideration. Hence $\lim\limits_{z \to z_0} f(z) = z_0^2$. Note that the limit of $f(z)$ as $z \to z_0$ has nothing whatsoever to do with the value of $f(z)$ at z_0.

24. Interpret Problem 23 geometrically.

(a) The equation $w = f(z) = z^2$ defines a transformation or mapping of points of the z plane into points of the w plane. In particular let us suppose that point z_0 is mapped into $w_0 = z_0^2$.

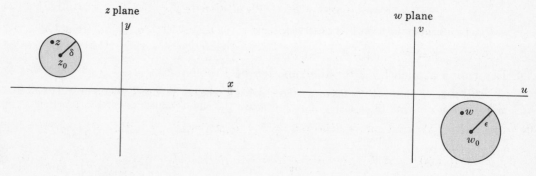

Fig. 2-25	Fig. 2-26

In Problem 23(a) we prove that given any $\epsilon > 0$ we can find $\delta > 0$ such that $|w - w_0| < \epsilon$ whenever $|z - z_0| < \delta$. Geometrically this means that if we wish w to be inside a circle of radius ϵ [see Fig. 2-26] we must choose δ (depending on ϵ) so that z lies inside a circle of radius δ. According to Problem 23(a) this is certainly accomplished if δ is the smaller of 1 and $\epsilon/(1 + 2|z_0|)$.

(b) In Problem 23(a), $w = w_0 = z_0^2$ is the image of $z = z_0$. However, in Problem 23(b), $w = 0$ is the image of $z = z_0$. Except for this, the geometric interpretation is identical with that given in part (a).

25. Prove that $\lim\limits_{z \to i} \dfrac{3z^4 - 2z^3 + 8z^2 - 2z + 5}{z - i} = 4 + 4i$.

We must show that for any $\epsilon > 0$ we can find $\delta > 0$ such that

$$\left| \frac{3z^4 - 2z^3 + 8z^2 - 2z + 5}{z - i} - (4 + 4i) \right| < \epsilon \qquad \text{when} \qquad 0 < |z - i| < \delta$$

Since $z \neq i$, we can write

$$\frac{3z^4 - 2z^3 + 8z^2 - 2z + 5}{z - i} = \frac{[3z^3 - (2 - 3i)z^2 + (5 - 2i)z + 5i][z - i]}{z - i}$$

$$= 3z^3 - (2 - 3i)z^2 + (5 - 2i)z + 5i$$

on cancelling the common factor $z - i \neq 0$.

Then we must show that for any $\epsilon > 0$, we can find $\delta > 0$ such that

$$|3z^3 - (2 - 3i)z^2 + (5 - 2i)z - 4 + i| < \epsilon \qquad \text{when} \qquad 0 < |z - i| < \delta$$

If $\delta \leqq 1$, then $0 < |z - i| < \delta$ implies

$$
\begin{aligned}
|3z^3 - (2 - 3i)z^2 + (5 - 2i)z - 4 + i| &= |z - i| \, |3z^2 + (6i - 2)z - 1 - 4i| \\
&= |z - i| \, |3(z - i + i)^2 + (6i - 2)(z - i + i) - 1 - 4i| \\
&= |z - i| \, |3(z - i)^2 + (12i - 2)(z - i) - 10 - 6i| \\
&< \delta \{ 3 |z - i|^2 + |12i - 2| \, |z - i| + |-10 - 6i| \} \\
&< \delta (3 + 13 + 12) = 28\delta
\end{aligned}
$$

Taking δ as the smaller of 1 and $\epsilon/28$, the required result follows.

THEOREMS ON LIMITS

26. If $\lim\limits_{z \to z_0} f(z)$ exists, prove that it must be unique.

We must show that if $\lim\limits_{z \to z_0} f(z) = l_1$ and $\lim\limits_{z \to z_0} f(z) = l_2$, then $l_1 = l_2$.

By hypothesis, given any $\epsilon > 0$, we can find $\delta > 0$ such that

$$|f(z) - l_1| < \epsilon/2 \qquad \text{when} \qquad 0 < |z - z_0| < \delta$$
$$|f(z) - l_2| < \epsilon/2 \qquad \text{when} \qquad 0 < |z - z_0| < \delta$$

Then

$$|l_1 - l_2| = |l_1 - f(z) + f(z) - l_2| \leqq |l_1 - f(z)| + |f(z) - l_2| < \epsilon/2 + \epsilon/2 = \epsilon$$

i.e. $|l_1 - l_2|$ is less than any positive number ϵ (however small) and so must be zero. Thus $l_1 = l_2$.

27. If $\lim\limits_{z \to z_0} g(z) = B \neq 0$, prove that there exists $\delta > 0$ such that

$$|g(z)| > \tfrac{1}{2}|B| \qquad \text{for} \qquad 0 < |z - z_0| < \delta$$

Since $\lim\limits_{z \to z_0} g(z) = B$, we can find δ such that $|g(z) - B| < \tfrac{1}{2}|B|$ for $0 < |z - z_0| < \delta$.

Writing $B = B - g(z) + g(z)$, we have

$$|B| \leqq |B - g(z)| + |g(z)| < \tfrac{1}{2}|B| + |g(z)|$$

i.e. $|B| < \tfrac{1}{2}|B| + |g(z)|$ from which $|g(z)| > \tfrac{1}{2}|B|$

28. Given $\lim\limits_{z \to z_0} f(z) = A$ and $\lim\limits_{z \to z_0} g(z) = B$, prove that (a) $\lim\limits_{z \to z_0} [f(z) + g(z)] = A + B$,

(b) $\lim\limits_{z \to z_0} f(z)\,g(z) = AB$, (c) $\lim\limits_{z \to z_0} \dfrac{1}{g(z)} = \dfrac{1}{B}$ if $B \neq 0$, (d) $\lim\limits_{z \to z_0} \dfrac{f(z)}{g(z)} = \dfrac{A}{B}$ if $B \neq 0$.

(a) We must show that for any $\epsilon > 0$ we can find $\delta > 0$ such that

$$|[f(z) + g(z)] - (A + B)| < \epsilon \qquad \text{when} \qquad 0 < |z - z_0| < \delta$$

We have

$$|[f(z) + g(z)] - (A + B)| = |[f(z) - A] + [g(z) - B]| \leqq |f(z) - A| + |g(z) - B| \qquad (1)$$

By hypothesis, given $\epsilon > 0$ we can find $\delta_1 > 0$ and $\delta_2 > 0$ such that

$$|f(z) - A| < \epsilon/2 \qquad \text{when} \qquad 0 < |z - z_0| < \delta_1 \qquad (2)$$
$$|g(z) - B| < \epsilon/2 \qquad \text{when} \qquad 0 < |z - z_0| < \delta_2 \qquad (3)$$

Then from (1), (2) and (3),

$$| [f(z) + g(z)] - (A + B) | \; < \; \epsilon/2 + \epsilon/2 \; = \; \epsilon \qquad \text{when} \qquad 0 < | z - z_0 | < \delta$$

where δ is chosen as the smaller of δ_1 and δ_2.

(b) We have
$$\begin{aligned}
| f(z) \, g(z) - AB | \; &= \; | f(z)\{g(z) - B\} + B\{f(z) - A\} | \\
&\leqq \; |f(z)| \, | g(z) - B | + |B| \, | f(z) - A | \\
&\leqq \; |f(z)| \, | g(z) - B | + (|B| + 1) \, | f(z) - A |
\end{aligned}$$ (4)

Since $\lim\limits_{z \to z_0} f(z) = A$, we can find δ_1 such that $| f(z) - A | < 1$ for $0 < | z - z_0 | < \delta_1$.
Hence by inequalities 4, Page 2,

$$| f(z) - A | \; \geqq \; |f(z)| - |A|, \quad \text{i.e.} \quad 1 \geqq |f(z)| - |A| \quad \text{or} \quad |f(z)| \leqq |A| + 1$$

i.e. $|f(z)| < P$ where P is a positive constant.

Since $\lim\limits_{z \to z_0} g(z) = B$, given $\epsilon > 0$ we can find $\delta_2 > 0$ such that $| g(z) - B | \; < \; \epsilon/2P$ for
$0 < | z - z_0 | < \delta_2$.

Since $\lim\limits_{z \to z_0} f(z) = A$, given $\epsilon > 0$ we can find $\delta_3 > 0$ such that $| f(z) - A | \; < \; \dfrac{\epsilon}{2(|B| + 1)}$
for $0 < | z - z_0 | < \delta_3$.

Using these in (4), we have

$$| f(z) \, g(z) - AB | \; < \; P \cdot \frac{\epsilon}{2P} + (|B| + 1) \cdot \frac{\epsilon}{2(|B| + 1)} \; = \; \epsilon$$

for $0 < | z - z_0 | < \delta$ where δ is the smaller of $\delta_1, \delta_2, \delta_3$, and the proof is complete.

(c) We must show that for any $\epsilon > 0$ we can find $\delta > 0$ such that

$$\left| \frac{1}{g(z)} - \frac{1}{B} \right| \; = \; \frac{| g(z) - B |}{|B| \, |g(z)|} \; < \; \epsilon \qquad \text{when} \qquad 0 < | z - z_0 | < \delta$$ (5)

By hypothesis, given any $\epsilon > 0$ we can find $\delta_1 > 0$ such that

$$| g(z) - B | \; < \; \tfrac{1}{2} |B|^2 \epsilon \qquad \text{when} \qquad 0 < | z - z_0 | < \delta_1$$

By Problem 27, since $\lim\limits_{z \to z_0} g(z) = B \neq 0$, we can find $\delta_2 > 0$ such that

$$|g(z)| > \tfrac{1}{2} |B| \qquad \text{when} \qquad 0 < | z - z_0 | < \delta_2$$

Then if δ is the smaller of δ_1 and δ_2, we can write

$$\left| \frac{1}{g(z)} - \frac{1}{B} \right| \; = \; \frac{| g(z) - B |}{|B| \, |g(z)|} \; < \; \frac{\tfrac{1}{2} |B|^2 \epsilon}{|B| \cdot \tfrac{1}{2} |B|} \; = \; \epsilon \qquad \text{whenever} \qquad 0 < | z - z_0 | < \delta$$

and the required result is proved.

(d) From parts (b) and (c),

$$\lim_{z \to z_0} \frac{f(z)}{g(z)} \; = \; \lim_{z \to z_0} \left\{ f(z) \cdot \frac{1}{g(z)} \right\} \; = \; \lim_{z \to z_0} f(z) \cdot \lim_{z \to z_0} \frac{1}{g(z)} \; = \; A \cdot \frac{1}{B} \; = \; \frac{A}{B}$$

This can also be proved directly [see Problem 145].

Note. In the proof of (a) we have used the results $| f(z) - A | < \epsilon/2$ and $| g(z) - B | < \epsilon/2$, so that the final result would come out to be $| f(z) + g(z) - (A + B) | < \epsilon$. Of course the proof would be *just as valid* if we had used 2ϵ [or any other positive multiple of ϵ] in place of ϵ. Similar remarks hold for the proofs of (b), (c) and (d).

29. Evaluate each of the following using theorems on limits.

(a)
$$\begin{aligned}
\lim_{z \to 1+i} (z^2 - 5z + 10) \; &= \; \lim_{z \to 1+i} z^2 + \lim_{z \to 1+i} (-5z) + \lim_{z \to 1+i} 10 \\
&= \; \left(\lim_{z \to 1+i} z \right)\left(\lim_{z \to 1+i} z \right) + \left(\lim_{z \to 1+i} -5 \right)\left(\lim_{z \to 1+i} z \right) + \lim_{z \to 1+i} 10 \\
&= \; (1 + i)(1 + i) - 5(1 + i) + 10 \; = \; 5 - 3i
\end{aligned}$$

In practice the intermediate steps are omitted.

(b)
$$\lim_{z \to -2i} \frac{(2z + 3)(z - 1)}{z^2 - 2z + 4} \; = \; \frac{\lim\limits_{z \to -2i} (2z + 3) \; \lim\limits_{z \to -2i} (z - 1)}{\lim\limits_{z \to -2i} (z^2 - 2z + 4)} \; = \; \frac{(3 - 4i)(-2i - 1)}{4i} \; = \; -\frac{1}{2} + \frac{11}{4} \, i$$

(c) $\displaystyle\lim_{z \to 2e^{\pi i/3}} \frac{z^3 + 8}{z^4 + 4z^2 + 16}$

In this case the limits of the numerator and denominator are each zero and the theorems on limits fail to apply.

However, by obtaining the factors of the polynomials, we see that

$$\lim_{z \to 2e^{\pi i/3}} \frac{z^3 + 8}{z^4 + 4z^2 + 16} = \lim_{z \to 2e^{\pi i/3}} \frac{(z+2)(z - 2e^{\pi i/3})(z - 2e^{5\pi i/3})}{(z - 2e^{\pi i/3})(z - 2e^{2\pi i/3})(z - 2e^{4\pi i/3})(z - 2e^{5\pi i/3})}$$

$$= \lim_{z \to 2e^{\pi i/3}} \frac{(z+2)}{(z - 2e^{2\pi i/3})(z - 2e^{4\pi i/3})} = \frac{e^{\pi i/3} + 1}{2(e^{\pi i/3} - e^{2\pi i/3})(e^{\pi i/3} - e^{4\pi i/3})}$$

$$= \frac{3}{8} - \frac{\sqrt{3}}{8} i$$

Another method. Since $z^6 - 64 = (z^2 - 4)(z^4 + 4z^2 + 16)$, the problem is equivalent to finding

$$\lim_{z \to 2e^{\pi i/3}} \frac{(z^2 - 4)(z^3 + 8)}{z^6 - 64} = \lim_{z \to 2e^{\pi i/3}} \frac{z^2 - 4}{z^3 - 8} = \frac{e^{2\pi i/3} - 1}{2(e^{\pi i} - 1)} = \frac{3}{8} - \frac{\sqrt{3}}{8} i$$

30. Prove that $\displaystyle\lim_{z \to 0} \frac{\bar{z}}{z}$ does not exist.

If the limit is to exist it must be independent of the manner in which z approaches the point 0.

Let $z \to 0$ along the x axis. Then $y = 0$, and $z = x + iy = x$ and $\bar{z} = x - iy = x$, so that the required limit is

$$\lim_{x \to 0} \frac{x}{x} = 1$$

Let $z \to 0$ along the y axis. Then $x = 0$, and $z = x + iy = iy$ and $\bar{z} = x - iy = -iy$, so that the required limit is

$$\lim_{y \to 0} \frac{-iy}{iy} = -1$$

Since the two approaches do not give the same answer, the limit does not exist.

CONTINUITY

31. (a) Prove that $f(z) = z^2$ is continuous at $z = z_0$.

(b) Prove that $f(z) = \begin{cases} z^2 & z \neq z_0 \\ 0 & z = z_0 \end{cases}$, where $z_0 \neq 0$, is discontinuous at $z = z_0$.

(a) By Problem 23(a), $\displaystyle\lim_{z \to z_0} f(z) = f(z_0) = z_0^2$ and so $f(z)$ is continuous at $z = z_0$.

Another method. We must show that given any $\epsilon > 0$, we can find $\delta > 0$ (depending on ϵ) such that $|f(z) - f(z_0)| = |z^2 - z_0^2| < \epsilon$ when $|z - z_0| < \delta$. The proof patterns that given in Problem 23(a).

(b) By Problem 23(b), $\displaystyle\lim_{z \to z_0} f(z) = z_0^2$, but $f(z_0) = 0$. Hence $\displaystyle\lim_{z \to z_0} f(z) \neq f(z_0)$ and so $f(z)$ is discontinuous at $z = z_0$ if $z_0 \neq 0$.

If $z_0 = 0$, then $f(z) = 0$; and since $\displaystyle\lim_{z \to z_0} f(z) = 0 = f(0)$, we see that the function is continuous.

32. Is the function $f(z) = \dfrac{3z^4 - 2z^3 + 8z^2 - 2z + 5}{z - i}$ continuous at $z = i$?

$f(i)$ does not exist, i.e. $f(z)$ is not defined at $z = i$. Thus $f(z)$ is not continuous at $z = i$.

By redefining $f(z)$ so that $f(i) = \displaystyle\lim_{z \to i} f(z) = 4 + 4i$ (see Problem 25), it becomes continuous at $z = i$. In such case we call $z = i$ a *removable discontinuity*.

33. Prove that if $f(z)$ and $g(z)$ are continuous at $z = z_0$, so also are

$$\text{(a)} \ \ f(z) + g(z), \qquad \text{(b)} \ \ f(z) \, g(z), \qquad \text{(c)} \ \ \frac{f(z)}{g(z)} \ \text{if} \ g(z_0) \neq 0$$

These results follow at once from Problem 28 by taking $A = f(z_0)$, $B = g(z_0)$ and rewriting $0 < |z - z_0| < \delta$ as $|z - z_0| < \delta$, i.e. *including* $z = z_0$.

34. Prove that $f(z) = z^2$ is continuous in the region $|z| \leqq 1$.

Let z_0 be any point in the region $|z| \leqq 1$. By Problem 23(a), $f(z)$ is continuous at z_0. Thus $f(z)$ is continuous in the region since it is continuous at any point of the region.

35. For what values of z are each of the following functions continuous?

(a) $f(z) = \dfrac{z}{z^2+1} = \dfrac{z}{(z-i)(z+i)}$. Since the denominator is zero when $z = \pm i$, the function is continuous everywhere except $z = \pm i$.

(b) $f(z) = \csc z = \dfrac{1}{\sin z}$. By Problem 10(a), $\sin z = 0$ for $z = 0, \pm\pi, \pm 2\pi, \ldots$. Hence $f(z)$ is continuous everywhere except at these points.

UNIFORM CONTINUITY

36. Prove that $f(z) = z^2$ is uniformly continuous in the region $|z| < 1$.

We must show that given any $\epsilon > 0$, we can find $\delta > 0$ such that $|z^2 - z_0^2| < \epsilon$ when $|z - z_0| < \delta$, where δ depends *only* on ϵ and not on the particular point z_0 of the region.

If z and z_0 are any points in $|z| < 1$, then

$$|z^2 - z_0^2| \; = \; |z + z_0| \, |z - z_0| \; \leqq \; \{|z| + |z_0|\} \, |z - z_0| \; < \; 2\,|z - z_0|$$

Thus if $|z - z_0| < \delta$, it follows that $|z^2 - z_0^2| < 2\delta$. Choosing $\delta = \epsilon/2$, we see that $|z^2 - z_0^2| < \epsilon$ when $|z - z_0| < \delta$, where δ depends only on ϵ and not on z_0. Hence $f(z) = z^2$ is uniformly continuous in the region.

37. Prove that $f(z) = 1/z$ is not uniformly continuous in the region $|z| < 1$.

Method 1.

Suppose that $f(z)$ is uniformly continuous in the region. Then for any $\epsilon > 0$ we should be able to find δ, say between 0 and 1, such that $|f(z) - f(z_0)| < \epsilon$ when $|z - z_0| < \delta$ for all z and z_0 in the region.

Let $z = \delta$ and $z_0 = \dfrac{\delta}{1+\epsilon}$. Then $|z - z_0| = \left| \delta - \dfrac{\delta}{1+\epsilon} \right| = \dfrac{\epsilon}{1+\epsilon}\delta < \delta$.

However, $\left| \dfrac{1}{z} - \dfrac{1}{z_0} \right| = \left| \dfrac{1}{\delta} - \dfrac{1+\epsilon}{\delta} \right| = \dfrac{\epsilon}{\delta} > \epsilon$ (since $0 < \delta < 1$).

Thus we have a contradiction, and it follows that $f(z) = 1/z$ cannot be uniformly continuous in the region.

Method 2.

Let z_0 and $z_0 + \zeta$ be any two points of the region such that $|z_0 + \zeta - z_0| = |\zeta| = \delta$. Then

$$|f(z_0) - f(z_0 + \zeta)| \; = \; \left| \dfrac{1}{z_0} - \dfrac{1}{z_0 + \zeta} \right| \; = \; \dfrac{|\zeta|}{|z_0|\,|z_0 + \zeta|} \; = \; \dfrac{\delta}{|z_0|\,|z_0 + \zeta|}$$

can be made larger than any positive number by choosing z_0 sufficiently close to 0. Hence the function cannot be uniformly continuous in the region.

SEQUENCES AND SERIES

38. Investigate the convergence of the sequences

(a) $u_n = \dfrac{i^n}{n}$, $n = 1, 2, 3, \ldots$,   (b) $u_n = \dfrac{(1+i)^n}{n}$.

(a) The first few terms of the sequence are $i, \dfrac{i^2}{2}, \dfrac{i^3}{3}, \dfrac{i^4}{4}, \dfrac{i^5}{5}$, etc., or $i, -\dfrac{1}{2}, \dfrac{-i}{3}, \dfrac{1}{4}, \dfrac{i}{5}, \ldots$. On plotting the corresponding points in the z plane, we suspect that the limit is zero. To prove this we must show that

$$|u_n - l| \; = \; |i^n/n - 0| \; < \; \epsilon \qquad \text{when} \qquad n > N \tag{1}$$

Now $|i^n/n - 0| = |i^n/n| = |i|^n/n = 1/n < \epsilon$ when $n > 1/\epsilon$

Let us choose $N = 1/\epsilon$. Then we see that (1) is true, and so the sequence converges to zero.

(b) Consider $\left|\dfrac{u_{n+1}}{u_n}\right| = \left|\dfrac{(1+i)^{n+1}/(n+1)}{(1+i)^n/n}\right| = \dfrac{n}{n+1}|1+i| = \dfrac{n\sqrt{2}}{n+1}$.

For all $n \geqq 10$ (for example), we have $\dfrac{n\sqrt{2}}{n+1} > \dfrac{6}{5} = 1.2$. Thus $|u_{n+1}| > 1.2|u_n|$ for $n > 10$, i.e. $|u_{11}| > 1.2\,|u_{10}|,\ |u_{12}| > 1.2\,|u_{11}| > (1.2)^2\,|u_{10}|$, and in general $|u_n| > (1.2)^{n-10}\,|u_{10}|$. It follows that $|u_n|$ can be made larger than any preassigned positive number (no matter how large) and thus the limit of $|u_n|$ cannot exist, and consequently the limit of u_n cannot exist. Thus the sequence diverges.

39. If $\lim\limits_{n\to\infty} a_n = A$ and $\lim\limits_{n\to\infty} b_n = B$, prove that $\lim\limits_{n\to\infty}(a_n + b_n) = A + B$.

By definition, given ϵ we can find N such that
$$|a_n - A| < \epsilon/2,\ |b_n - B| < \epsilon/2 \quad \text{for } n > N$$
Then for $n > N$,
$$|(a_n + b_n) - (A+B)| = |(a_n - A) + (b_n - B)| \leqq |a_n - A| + |b_n - B| < \epsilon$$
which proves the result.

It is seen that this parallels the proof for limits of functions [Problem 28].

40. Prove that if a series $u_1 + u_2 + u_3 + \cdots$ is to converge, we must have $\lim\limits_{n\to\infty} u_n = 0$.

If S_n is the sum of the first n terms of the series, then $S_{n+1} = S_n + u_n$. Hence if $\lim\limits_{n\to\infty} S_n$ exists and equals S, we have $\lim\limits_{n\to\infty} S_{n+1} = \lim\limits_{n\to\infty} S_n + \lim\limits_{n\to\infty} u_n$ or $S = S + \lim\limits_{n\to\infty} u_n$, i.e. $\lim\limits_{n\to\infty} u_n = 0$.

Conversely, however, if $\lim\limits_{n\to\infty} u_n = 0$ the series may or may not converge. See Problem 150.

41. Prove that $1 + z + z^2 + z^3 + \cdots = \dfrac{1}{1-z}$ if $|z| < 1$.

Let $\qquad\qquad S_n = 1 + z + z^2 + \cdots + z^{n-1}$

Then $\qquad\quad zS_n = \quad\ \ z + z^2 + \cdots + z^{n-1} + z^n$

Subtracting, $\quad (1-z)S_n = 1 - z^n \qquad\qquad \text{or} \qquad S_n = \dfrac{1-z^n}{1-z}$

If $|z| < 1$, then we suspect that $\lim\limits_{n\to\infty} z^n = 0$. To prove this we must show that given any $\epsilon > 0$ we can find N such that $|z^n - 0| < \epsilon$ for all $n > N$. The result is certainly true if $z = 0$; hence we can consider $z \neq 0$.

Now $|z^n| = |z|^n < \epsilon$ when $n \ln|z| < \ln \epsilon$ or $n > (\ln \epsilon)/(\ln |z|) = N$ [since if $|z| < 1$, $\ln |z|$ is negative]. We have therefore found the required N, and $\lim\limits_{n\to\infty} z^n = 0$.

Thus $\quad 1 + z + z^2 + \cdots = \lim\limits_{n\to\infty} S_n = \lim\limits_{n\to\infty}\dfrac{1-z^n}{1-z} = \dfrac{1-0}{1-z} = \dfrac{1}{1-z}$.

The series
$$a + az + az^2 + \cdots = \dfrac{a}{1-z}$$
is called a *geometric series* with first term equal to a and ratio z, and its sum is $a/(1-z)$ provided $|z| < 1$.

MISCELLANEOUS PROBLEMS

42. Let $w = (z^2 + 1)^{1/2}$. (a) If $w = 1$ when $z = 0$, and z describes the curve C_1 shown in Fig. 2-27 below, find the value of w when $z = 1$. (b) If z describes the curve C_2 shown in Fig. 2-28 below, is the value of w when $z = 1$ the same as that obtained in (a)?

(a) The branch points of $w = f(z) = (z^2 + 1)^{1/2} = \{(z-i)(z+i)\}^{1/2}$ are at $z = \pm i$ by Problem 19.

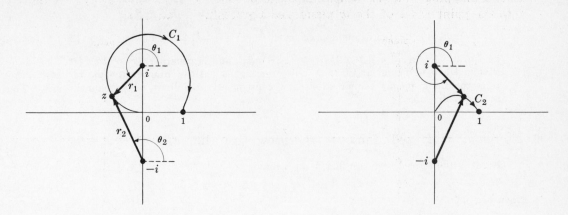

Fig. 2-27   Fig. 2-28

Let   (1) $z - i = r_1 e^{i\theta_1}$,   (2) $z + i = r_2 e^{i\theta_2}$.   Then since θ_1 and θ_2 are determined only within integer multiples of $2\pi i$, we can write

$$w = \sqrt{r_1 r_2}\, e^{i(\theta_1 + \theta_2)/2}\, e^{2k\pi i/2} = \sqrt{r_1 r_2}\, e^{i(\theta_1 + \theta_2)/2}\, e^{k\pi i} \tag{3}$$

Referring to Fig. 2-27 [or by using the equations (1) and (2)] we see that when z is at 0, $r_1 = 1$, $\theta_1 = 3\pi/2$ and $r_2 = 1$, $\theta_2 = \pi/2$. Since $w = 1$ at $z = 0$, we have from (3), $1 = e^{(k+1)\pi i}$ and we choose $k = -1$ [or $1, -3, \ldots$]. Then

$$w = -\sqrt{r_1 r_2}\, e^{i(\theta_1 + \theta_2)/2}$$

As z traverses C_1 from 0 to 1, r_1 changes from 1 to $\sqrt{2}$, θ_1 changes from $3\pi/2$ to $-\pi/4$, r_2 changes from 1 to $\sqrt{2}$, θ_2 changes from $\pi/2$ to $\pi/4$. Then

$$w = -\sqrt{(\sqrt{2})(\sqrt{2})}\; e^{i(-\pi/4 + \pi/4)/2} = -\sqrt{2}$$

(b) As in part (a), $w = -\sqrt{r_1 r_2}\, e^{i(\theta_1 + \theta_2)/2}$.   Referring to Fig. 2-28 we see that as z traverses C_2, r_1 changes from 1 to $\sqrt{2}$, θ_1 changes from $3\pi/2$ to $7\pi/4$, r_2 changes from 1 to $\sqrt{2}$ and θ_2 changes from $\pi/2$ to $\pi/4$. Then

$$w = -\sqrt{(\sqrt{2})(\sqrt{2})}\; e^{i(7\pi/4 + \pi/4)/2} = \sqrt{2}$$

which is not the same as the value obtained in (a).

43. Let $\sqrt{1 - z^2} = 1$ for $z = 0$. Show that as z varies from 0 to $p > 1$ along the real axis, $\sqrt{1 - z^2}$ varies from 1 to $-i\sqrt{p^2 - 1}$.

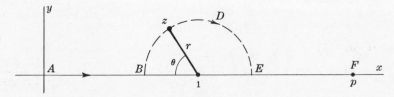

Fig. 2-29

Consider the case where z travels along path $ABDEF$, where BDE is a semi-circle as shown in Fig. 2-29. From this figure, we have

$$1 - z = 1 - x - iy = r\cos\theta - ir\sin\theta$$

so that $\sqrt{1 - z^2} = \sqrt{(1 - z)(1 + z)} = \sqrt{r}\,(\cos\theta/2 - i\sin\theta/2)\sqrt{2 - r\cos\theta + ir\sin\theta}$

Along AB: $z = x$, $r = 1 - x$, $\theta = 0$ and $\sqrt{1 - z^2} = \sqrt{1 - x}\sqrt{1 + x} = \sqrt{1 - x^2}$.

Along EF: $z = x$, $r = x - 1$, $\theta = \pi$ and $\sqrt{1 - z^2} = -i\sqrt{x - 1}\sqrt{x + 1} = -i\sqrt{x^2 - 1}$.

Hence as z varies from 0 [where $x = 0$] to p [where $x = p$], $\sqrt{1 - z^2}$ varies from 1 to $-i\sqrt{p^2 - 1}$.

44. Find a mapping function which maps the points $z = 0, \pm i, \pm 2i, \pm 3i, \ldots$ of the z plane into the point $w = 1$ of the w plane [see Figures 2-30 and 2-31].

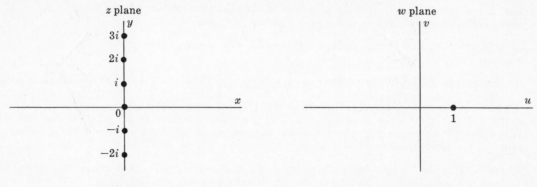

Fig. 2-30   Fig. 2-31

Since the points in the z plane are equally spaced, we are led, because of Problem 15, to consider a logarithmic function of the type $z = \ln w$.

Now if $w = 1 = e^{2k\pi i}$, $k = 0, \pm 1, \pm 2, \ldots$, then $z = \ln w = 2k\pi i$ so that the point $w = 1$ is mapped into the points $0, \pm 2\pi i, \pm 4\pi i, \ldots$.

If, however, we consider $z = (\ln w)/2\pi$, the point $w = 1$ is mapped into $z = 0, \pm i, \pm 2i, \ldots$ as required. Conversely, by means of this mapping function the points $z = 0, \pm i, \pm 2i, \ldots$ are mapped into the point $w = 1$.

Then a suitable mapping function is $z = (\ln w)/2\pi$ or $w = e^{2\pi z}$.

45. If $\lim_{n \to \infty} z_n = l$, prove that $\lim_{n \to \infty} \text{Re}\,\{z_n\} = \text{Re}\,\{l\}$ and $\lim_{n \to \infty} \text{Im}\,\{z_n\} = \text{Im}\,\{l\}$.

Let $z_n = x_n + iy_n$ and $l = l_1 + il_2$, where x_n, y_n and l_1, l_2 are the real and imaginary parts of z_n and l respectively.

By hypothesis, given any $\epsilon > 0$ we can find N such that $|z_n - l| < \epsilon$ for $n > N$, i.e.,

$$|x_n + iy_n - (l_1 + il_2)| < \epsilon \qquad \text{for} \qquad n > N$$

or $$\sqrt{(x_n - l_1)^2 + (y_n - l_2)^2} < \epsilon \qquad \text{for} \qquad n > N$$

From this it necessarily follows that

$$|x_n - l_1| < \epsilon \quad \text{and} \quad |y_n - l_2| < \epsilon \qquad \text{for} \qquad n > N$$

i.e. $\lim_{n \to \infty} x_n = l_1$ and $\lim_{n \to \infty} y_n = l_2$, as required.

46. Prove that if $|a| < 1$,

$$(a) \quad 1 + a\cos\theta + a^2\cos 2\theta + a^3\cos 3\theta + \cdots = \frac{1 - a\cos\theta}{1 - 2a\cos\theta + a^2}$$

$$(b) \quad a\sin\theta + a^2\sin 2\theta + a^3\sin 3\theta + \cdots = \frac{a\sin\theta}{1 - 2a\cos\theta + a^2}$$

Let $z = ae^{i\theta}$ in Problem 41. We can do this since $|z| = |a| < 1$. Then

$$1 + ae^{i\theta} + a^2e^{2i\theta} + a^3e^{3i\theta} + \cdots = \frac{1}{1 - ae^{i\theta}}$$

or $$(1 + a\cos\theta + a^2\cos 2\theta + \cdots) + i(a\sin\theta + a^2\sin 2\theta + \cdots) = \frac{1}{1 - ae^{i\theta}} \cdot \frac{1 - ae^{-i\theta}}{1 - ae^{-i\theta}}$$

$$= \frac{1 - a\cos\theta + ia\sin\theta}{1 - 2a\cos\theta + a^2}$$

The required results follow on equating real and imaginary parts.

Supplementary Problems

FUNCTIONS AND TRANSFORMATIONS

47. Let $w = f(z) = z(2 - z)$. Find the values of w corresponding to (a) $z = 1 + i$, (b) $z = 2 - 2i$ and graph corresponding values in the w and z planes.   *Ans.* (a) 2, (b) $4 + 4i$

48. If $w = f(z) = (1 + z)/(1 - z)$, find (a) $f(i)$, (b) $f(1 - i)$ and represent graphically.
    Ans. (a) i, (b) $-1 - 2i$

49. If $f(z) = (2z + 1)/(3z - 2)$, $z \neq 2/3$, find (a) $f(1/z)$, (b) $f\{f(z)\}$.   *Ans.* (a) $(2 + z)/(3 - 2z)$, (b) z

50. (a) If $w = f(z) = (z + 2)/(2z - 1)$, find $f(0)$, $f(i)$, $f(1 + i)$. (b) Find the values of z such that $f(z) = i$, $f(z) = 2 - 3i$. (c) Show that z is a single-valued function of w. (d) Find the values of z such that $f(z) = z$ and explain geometrically why we would call such values the *fixed* or *invariant* *points* of the transformation.   *Ans.* (a) -2, $-i$, $1 - i$, (b) $-i$, $(2 + i)/3$

51. A square S in the z plane has vertices at $(0, 0)$, $(1, 0)$, $(1, 1)$, $(0, 1)$. Determine the region in the w plane into which S is mapped under the transformations (a) $w = z^2$, (b) $w = 1/(z + 1)$.

52. Discuss Problem 51 if the square has vertices at $(1, 1)$, $(-1, 1)$, $(-1, -1)$, $(1, -1)$.

53. Separate each of the following into real and imaginary parts, i.e. find $u(x, y)$ and $v(x, y)$ such that $f(z) = u + iv$: (a) $f(z) = 2z^2 - 3iz$, (b) $f(z) = z + 1/z$, (c) $f(z) = (1 - z)/(1 + z)$, (d) $f(z) = z^{1/2}$.

    Ans. (a) $u = 2x^2 - 2y^2 + 3y$, $v = 4xy - 3x$ (c) $u = \dfrac{1 - x^2 - y^2}{(1 + x)^2 + y^2}$, $v = \dfrac{-2y}{(1 + x)^2 + y^2}$

    (b) $u = x + x/(x^2 + y^2)$,   (d) $u = r^{1/2} \cos \theta/2$, $v = r^{1/2} \sin \theta/2$
    $\qquad v = y - y/(x^2 + y^2)$ $\qquad$ where $x = r \cos \theta$, $y = r \sin \theta$

54. If $f(z) = 1/z = u + iv$, construct several members of the families $u(x, y) = \alpha$, $v(x, y) = \beta$ where α and β are constants, showing that they are families of circles.

MULTIPLE-VALUED FUNCTIONS

55. Let $w^3 = z$ and suppose that corresponding to $z = 1$ we have $w = 1$. (a) If we start at $z = 1$ in the z plane and make one complete circuit counterclockwise around the origin, find the value of w on returning to $z = 1$ for the first time. (b) What are the values of w on returning to $z = 1$ after $2, 3, 4, \ldots$ complete circuits about the origin? Discuss (a) and (b) if the paths do not enclose the origin.
    Ans. (a) $e^{2\pi i/3}$, (b) $e^{4\pi i/3}$, 1, $e^{2\pi i/3}$

56. Let $w = (1 - z^2)^{1/2}$ and suppose that corresponding to $z = 0$ we have $w = 1$. (a) If we start at $z = 0$ in the z plane and make one complete circuit counterclockwise so as to include $z = 1$ but not to include $z = -1$, find the value of w on returning to $z = 0$ for the first time. (b) What are the values of w if the circuit in (a) is repeated over and over again? (c) Work parts (a) and (b) if the circuit includes $z = -1$ but does not include $z = 1$. (d) Work parts (a) and (b) if the circuit includes both $z = 1$ and $z = -1$. (e) Work parts (a) and (b) if the circuit excludes both $z = 1$ and $z = -1$. (f) Explain why $z = 1$ and $z = -1$ are branch points. (g) What lines can be taken as branch lines?

57. Find branch points and construct branch lines for the functions (a) $f(z) = \{z/(1 - z)\}^{1/2}$, (b) $f(z) = (z^2 - 4)^{1/3}$, (c) $f(z) = \ln(z - z^2)$.

THE ELEMENTARY FUNCTIONS

58. Prove that (a) $e^{z_1}/e^{z_2} = e^{z_1 - z_2}$, (b) $|e^{iz}| = e^{-y}$.

59. Prove that there cannot be any finite values of z such that $e^z = 0$.

60. Prove that 2π is a period of e^{iz}. Are there any other periods?

61. Find all values of z for which (a) $e^{3z} = 1$, (b) $e^{4z} = i$.
    Ans. (a) $2k\pi i/3$, (b) $\frac{1}{8}\pi i + \frac{1}{2}k\pi i$, where $k = 0, \pm 1, \pm 2, \ldots$.

62. Prove (a) $\sin 2z = 2 \sin z \cos z$, (b) $\cos 2z = \cos^2 z - \sin^2 z$, (c) $\sin^2(z/2) = \frac{1}{2}(1 - \cos z)$, (d) $\cos^2(z/2) = \frac{1}{2}(1 + \cos z)$.

63. Prove (a) $1 + \tan^2 z = \sec^2 z$, (b) $1 + \cot^2 z = \csc^2 z$.

64. If $\cos z = 2$, find (a) $\cos 2z$, (b) $\cos 3z$.   *Ans.* (a) 7, (b) 26

65. Prove that all the roots of (a) $\sin z = a$, (b) $\cos z = a$, where $-1 \leqq a \leqq 1$, are real.

66. Prove that if $|\sin z| \leqq 1$ for all z, then $|\text{Im}\{z\}| \leqq \ln(\sqrt{2}+1)$.

67. Show that   (a) $\overline{\sin z} = \sin \bar{z}$,   (b) $\overline{\cos z} = \cos \bar{z}$,   (c) $\overline{\tan z} = \tan \bar{z}$.

68. For each of the following functions find $u(x, y)$ and $v(x, y)$ such that $f(z) = u + iv$, i.e. separate into real and imaginary parts: (a) $f(z) = e^{3iz}$,   (b) $f(z) = \cos z$,   (c) $f(z) = \sin 2z$,   (d) $f(z) = z^2 e^{2z}$.
Ans. (a) $u = e^{-3y}\cos 3x$, $v = e^{-3y}\sin 3x$, (b) $u = \cos x \cosh y$, $v = -\sin x \sinh y$, (c) $u = \sin 2x \cosh 2y$, $v = \cos 2x \sinh 2y$,   (d) $u = e^{2x}\{(x^2-y^2)\cos 2y - 2xy \sin 2y\}$, $v = e^{2x}\{2xy \cos 2y + (x^2-y^2)\sin 2y\}$

69. Prove that   (a) $\sinh(-z) = -\sinh z$,   (b) $\cosh(-z) = \cosh z$,   (c) $\tanh(-z) = -\tanh z$.

70. Prove that   (a) $\sinh(z_1 + z_2) = \sinh z_1 \cosh z_2 + \cosh z_1 \sinh z_2$,   (b) $\cosh 2z = \cosh^2 z + \sinh^2 z$,   (c) $1 - \tanh^2 z = \text{sech}^2 z$.

71. Prove that   (a) $\sinh^2(z/2) = \frac{1}{2}(\cosh z - 1)$,   (b) $\cosh^2(z/2) = \frac{1}{2}(\cosh z + 1)$.

72. Find $u(x, y)$ and $v(x, y)$ such that   (a) $\sinh 2z = u + iv$,   (b) $z \cosh z = u + iv$.
Ans. (a) $u = \sinh 2x \cos 2y$, $v = \cosh 2x \sin 2y$
        (b) $u = x \cosh x \cos y - y \sinh x \sin y$, $v = y \cosh x \cos y + x \sinh x \sin y$

73. Find the value of   (a) $4\sinh(\pi i/3)$,   (b) $\cosh(2k+1)\pi i/2$, $k = 0, \pm 1, \pm 2, \ldots$,   (c) $\coth 3\pi i/4$.
Ans. (a) $2i\sqrt{3}$, (b) 0, (c) i

74. (a) Show that $\ln\left(-\dfrac{1}{2} - \dfrac{\sqrt{3}}{2}i\right) = \left(\dfrac{4\pi}{3} + 2k\pi\right)i$, $k = 0, \pm 1, \pm 2, \ldots$.   (b) What is the principal value?
Ans. (b) $4\pi i/3$

75. Obtain all the values of   (a) $\ln(-4)$,   (b) $\ln(3i)$,   (c) $\ln(\sqrt{3}-i)$ and find the principal value in each case.
Ans. (a) $2\ln 2 + (\pi + 2k\pi)i$, $2\ln 2 + \pi i$. (b) $\ln 3 + (\pi/2 + 2k\pi)i$, $\ln 3 + \pi i/2$. (c) $\ln 2 + (11\pi/6 + 2k\pi)i$, $\ln 2 + 11\pi i/6$

76. Show that $\ln(z - 1) = \frac{1}{2}\ln\{(x-1)^2 + y^2\} + i\tan^{-1} y/(x-1)$,   giving restrictions if any.

77. Prove that   (a) $\cos^{-1} z = \dfrac{1}{i}\ln(z + \sqrt{z^2-1})$,   (b) $\cot^{-1} z = \dfrac{1}{2i}\ln\left(\dfrac{z+i}{z-i}\right)$ indicating any restrictions.

78. Prove that   (a) $\sinh^{-1} z = \ln(z + \sqrt{z^2+1})$,   (b) $\coth^{-1} z = \dfrac{1}{2}\ln\left(\dfrac{z+1}{z-1}\right)$.

79. Find all the values of   (a) $\sin^{-1} 2$,   (b) $\cos^{-1} i$.
Ans. (a) $\pm i\ln(2 + \sqrt{3}) + \pi/2 + 2k\pi$ (b) $-i\ln(\sqrt{2}+1) + \pi/2 + 2k\pi$, $-i\ln(\sqrt{2}-1) + 3\pi/2 + 2k\pi$

80. Find all the values of   (a) $\cosh^{-1} i$,   (b) $\sinh^{-1}\{\ln(-1)\}$.
Ans. (a) $\ln(\sqrt{2}+1) + \pi i/2 + 2k\pi i$, $\ln(\sqrt{2}-1) + 3\pi i/2 + 2k\pi i$
        (b) $\ln[(2k+1)\pi + \sqrt{(2k+1)^2\pi^2 - 1}] + \pi i/2 + 2m\pi i$,
            $\ln[\sqrt{(2k+1)^2\pi^2 - 1} - (2k+1)\pi] + 3\pi i/2 + 2m\pi i$, $k, m = 0, \pm 1, \pm 2, \ldots$

81. Determine all the values of   (a) $(1+i)^i$,   (b) $1^{\sqrt{2}}$.
Ans. (a) $e^{-\pi/4 + 2k\pi}\{\cos(\frac{1}{2}\ln 2) + i\sin(\frac{1}{2}\ln 2)\}$,   (b) $\cos(2\sqrt{2}\,k\pi) + i\sin(2\sqrt{2}\,k\pi)$

82. Find   (a) $\text{Re}\{(1-i)^{1+i}\}$,   (b) $|(-i)^{-i}|$.
Ans. (a) $e^{\frac{1}{2}\ln 2 - 7\pi/4 - 2k\pi}\cos(7\pi/4 + \frac{1}{2}\ln 2)$,   (b) $e^{3\pi/2 + 2k\pi}$

83. Find the real and imaginary parts of z^z where $z = x + iy$.

84. Show that   (a) $f(z) = (z^2-1)^{1/3}$,   (b) $f(z) = z^{1/2} + z^{1/3}$ are algebraic functions of z.

BRANCH POINTS, BRANCH LINES AND RIEMANN SURFACES

85. Prove that $z = \pm i$ are branch points of $(z^2+1)^{1/3}$.

86. Construct a Riemann surface for the functions   (a) $z^{1/3}$,   (b) $z^{1/2}(z-1)^{1/2}$,   (c) $\left(\dfrac{z+2}{z-2}\right)^{1/3}$.

87. Show that the Riemann surface for the function $z^{1/2} + z^{1/3}$ has 6 sheets.

88. Construct Riemann surfaces for the functions   (a) $\ln(z+2)$,   (b) $\sin^{-1} z$,   (c) $\tan^{-1} z$.

LIMITS

89. (a) If $f(z) = z^2 + 2z$, prove that $\lim\limits_{z \to i} f(z) = 2i - 1$.

(b) If $f(z) = \begin{cases} z^2 + 2z & z \neq i \\ 3 + 2i & z = i \end{cases}$, find $\lim\limits_{z \to i} f(z)$ and justify your answer.

90. Prove that $\lim\limits_{z \to 1+i} \dfrac{z^2 - z + 1 - i}{z^2 - 2z + 2} = 1 - \tfrac{1}{2}i$.

91. Guess at a possible value for   (a) $\lim\limits_{z \to 2+i} \dfrac{1-z}{1+z}$,   (b) $\lim\limits_{z \to 2+i} \dfrac{z^2 - 2iz}{z^2 + 4}$ and investigate the correctness of your guess.

92. If $\lim\limits_{z \to z_0} f(z) = A$ and $\lim\limits_{z \to z_0} g(z) = B$,   prove that   (a) $\lim\limits_{z \to z_0} \{2 f(z) - 3i\, g(z)\} = 2A - 3iB$,
(b) $\lim\limits_{z \to z_0} \{p\, f(z) + q\, g(z)\} = pA + qB$ where p and q are any constants.

93. If $\lim\limits_{z \to z_0} f(z) = A$,   prove that   (a) $\lim\limits_{z \to z_0} \{f(z)\}^2 = A^2$,   (b) $\lim\limits_{z \to z_0} \{f(z)\}^3 = A^3$.   Can you make a similar statement for $\lim\limits_{z \to z_0} \{f(z)\}^n$?   What restrictions, if any, must be imposed?

94. Evaluate using theorems on limits. In each case state precisely which theorems are used.

(a) $\lim\limits_{z \to 2i} (iz^4 + 3z^2 - 10i)$ (c) $\lim\limits_{z \to i/2} \dfrac{(2z - 3)(4z + i)}{(iz - 1)^2}$

(b) $\lim\limits_{z \to e^{\pi i/4}} \dfrac{z^2}{z^4 + z + 1}$ (d) $\lim\limits_{z \to i} \dfrac{z^2 + 1}{z^6 + 1}$ (e) $\lim\limits_{z \to 1+i} \left\{ \dfrac{z - 1 - i}{z^2 - 2z + 2} \right\}^2$

Ans. (a) $-12 + 6i$, (b) $\sqrt{2}\,(1 + i)/2$, (c) $-4/3 - 4i$, (d) $1/3$, (e) $-1/4$

95. Find $\lim\limits_{z \to e^{\pi i/3}} (z - e^{\pi i/3}) \left(\dfrac{z}{z^3 + 1} \right)$ *Ans.* $1/6 - i\sqrt{3}/6$

96. Prove that if $f(z) = 3z^2 + 2z$, then $\lim\limits_{z \to z_0} \dfrac{f(z) - f(z_0)}{z - z_0} = 6z_0 + 2$.

97. If $f(z) = \dfrac{2z - 1}{3z + 2}$, prove that $\lim\limits_{h \to 0} \dfrac{f(z_0 + h) - f(z_0)}{h} = \dfrac{7}{(3z_0 + 2)^2}$ provided $z_0 \neq -2/3$.

98. If we restrict ourselves to that branch of $f(z) = \sqrt{z^2 + 3}$ for which $f(0) = \sqrt{3}$, prove that

$$\lim_{z \to 1} \frac{\sqrt{z^2 + 3} - 2}{z - 1} = \frac{1}{2}$$

99. Explain exactly what is meant by the statements   (a) $\lim\limits_{z \to i} 1/(z - i)^2 = \infty$,   (b) $\lim\limits_{z \to \infty} \dfrac{2z^4 + 1}{z^4 + 1} = 2$.

100. Show that   (a) $\lim\limits_{z \to \pi/2} (\sin z)/z = 2/\pi$,   (b) $\lim\limits_{z \to \pi i/2} z^2 \cosh 4z/3 = \pi^2/8$.

101. Show that if we restrict ourselves to that branch of $f(z) = \tanh^{-1} z$ such that $f(0) = 0$, then $\lim\limits_{z \to -i} f(z) = 3\pi i/4$.

CONTINUITY

102. Let $f(z) = \dfrac{z^2 + 4}{z - 2i}$ if $z \neq 2i$, while $f(2i) = 3 + 4i$.   (a) Prove that $\lim\limits_{z \to i} f(z)$ exists and determine its value. (b) Is $f(z)$ continuous at $z = 2i$? Explain. (c) Is $f(z)$ continuous at points $z \neq 2i$? Explain.

103. Answer Problem 102 if $f(2i)$ is redefined as equal to $4i$ and explain why any differences should occur.

104. Prove that $f(z) = z/(z^4 + 1)$ is continuous at all points inside and on the unit circle $|z| = 1$ except at four points, and determine these points.   *Ans.* $e^{(2k+1)\pi i/4}$, $k = 0, 1, 2, 3$

105. If $f(z)$ and $g(z)$ are continuous at $z = z_0$, prove that $3 f(z) - 4i\, g(z)$ is also continuous at $z = z_0$.

106. If $f(z)$ is continuous at $z = z_0$, prove that (a) $\{f(z)\}^2$ and (b) $\{f(z)\}^3$ are also continuous at $z = z_0$. Can you extend the result to $\{f(z)\}^n$ where n is any positive integer?

107. Find all points of discontinuity for the following functions.

(a) $f(z) = \dfrac{2z - 3}{z^2 + 2z + 2}$, (b) $f(z) = \dfrac{3z^2 + 4}{z^4 - 16}$, (c) $f(z) = \cot z$, (d) $f(z) = \dfrac{1}{z} - \sec z$, (e) $f(z) = \dfrac{\tanh z}{z^2 + 1}$.

Ans. (a) $-1 \pm i$ (c) $k\pi$, $k = 0, \pm 1, \pm 2, \ldots$
    (b) $\pm 2, \pm 2i$ (d) 0, $(k + \frac{1}{2})\pi$, $k = 0, \pm 1, \pm 2, \ldots$ (e) $\pm i$, $(k + \frac{1}{2})\pi i$, $k = 0, \pm 1, \pm 2, \ldots$

108. Prove that $f(z) = z^2 - 2z + 3$ is continuous everywhere in the finite plane.

109. Prove that $f(z) = \dfrac{z^2 + 1}{z^3 + 9}$ is (a) continuous and (b) bounded in the region $|z| \leq 2$.

110. Prove that if $f(z)$ is continuous in a closed region, it is bounded in the region.

111. Prove that $f(z) = 1/z$ is continuous for all z such that $|z| > 0$, but that it is not bounded.

112. Prove that a polynomial is continuous everywhere in the finite plane.

113. Show that $f(z) = \dfrac{z^2 + 1}{z^2 - 3z + 2}$ is continuous for all z outside $|z| = 2$.

UNIFORM CONTINUITY

114. Prove that $f(z) = 3z - 2$ is uniformly continuous in the region $|z| \leq 10$.

115. Prove that $f(z) = 1/z^2$ (a) is not uniformly continuous in the region $|z| \leq 1$ but (b) is uniformly continuous in the region $\frac{1}{2} \leq |z| \leq 1$.

116. Prove that if $f(z)$ is continuous in a closed region $\mathcal{R}$ it is uniformly continuous in $\mathcal{R}$.

SEQUENCES AND SERIES

117. Prove that (a) $\lim\limits_{n \to \infty} \dfrac{n^2 i^n}{n^3 + 1} = 0$, (b) $\lim\limits_{n \to \infty} \left(\dfrac{n}{n + 3i} - \dfrac{in}{n + 1} \right) = 1 - i$.

118. Prove that for any complex number z, $\lim\limits_{n \to \infty} (1 + 3z/n^2) = 1$.

119. Prove that $\lim\limits_{n \to \infty} n \left(\dfrac{1 + i}{2} \right)^n = 0$.

120. Prove that $\lim\limits_{n \to \infty} n i^n$ does not exist.

121. If $\lim\limits_{n \to \infty} |u_n| = 0$, prove that $\lim\limits_{n \to \infty} u_n = 0$. Is the converse true? Justify your conclusion.

122. If $\lim\limits_{n \to \infty} a_n = A$ and $\lim\limits_{n \to \infty} b_n = B$, prove that (a) $\lim\limits_{n \to \infty} (a_n + b_n) = A + B$, (b) $\lim\limits_{n \to \infty} (a_n - b_n) = A - B$, (c) $\lim\limits_{n \to \infty} a_n b_n = AB$, (d) $\lim\limits_{n \to \infty} a_n/b_n = A/B$ if $B \neq 0$.

123. Use theorems on limits to evaluate each of the following.

(a) $\lim\limits_{n \to \infty} \dfrac{in^2 - in + 1 - 3i}{(2n + 4i - 3)(n - i)}$ (c) $\lim\limits_{n \to \infty} \sqrt{n + 2i} - \sqrt{n + i}$

(b) $\lim\limits_{n \to \infty} \left| \dfrac{(n^2 + 3i)(n - i)}{in^3 - 3n + 4 - i} \right|$ (d) $\lim\limits_{n \to \infty} \sqrt{n} \{ \sqrt{n + 2i} - \sqrt{n + i} \}$

Ans. (a) $\frac{1}{2}i$, (b) 1, (c) 0, (d) $\frac{1}{2}i$

124. If $\lim\limits_{n \to \infty} u_n = l$, prove that $\lim\limits_{n \to \infty} \dfrac{u_1 + u_2 + \cdots + u_n}{n} = l$.

125. Prove that the series $1 + i/3 + (i/3)^2 + \cdots = \sum\limits_{n=1}^{\infty} (i/3)^{n-1}$ converges and find its sum.
Ans. $(9 + 3i)/10$

126. Prove that the series $i - 2i + 3i - 4i + \cdots$ diverges.

127. If the series $\sum\limits_{n=1}^{\infty} a_n$ converges to A, and $\sum\limits_{n=1}^{\infty} b_n$ converges to B, prove that $\sum\limits_{n=1}^{\infty} (a_n + ib_n)$ converges to $A + iB$. Is the converse true?

128. Investigate the convergence of $\sum\limits_{n=1}^{\infty} \dfrac{\omega^n}{5^{n/2}}$ where $\omega = \sqrt{3} + i$. *Ans.* conv.

MISCELLANEOUS PROBLEMS

129. Let $w = \{(4 - z)(z^2 + 4)\}^{1/2}$. If $w = 4$ when $z = 0$, show that if z describes the curve C of Fig. 2-32, then the value of w at $z = 6$ is $-4i\sqrt{5}$.

130. Prove that a necessary and sufficient condition for $f(z) = u(x, y) + i\,v(x, y)$ to be continuous at $z = z_0 = x_0 + iy_0$ is that $u(x, y)$ and $v(x, y)$ be continuous at (x_0, y_0).

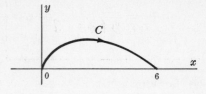

Fig. 2-32

131. Prove that the equation $\tan z = z$ has only real roots.

132. A student remarked that 1 raised to any power is equal to 1. Was he correct? Explain.

133. Show that $\quad \dfrac{\sin\theta}{2} + \dfrac{\sin 2\theta}{2^2} + \dfrac{\sin 3\theta}{2^3} + \cdots = \dfrac{2\sin\theta}{5 - 4\cos\theta}$.

134. Show that the relation $|f(x + iy)| = |f(x) + f(iy)|$ is satisfied by $f(z) = \sin z$. Can you find any other functions for which it is true?

135. Prove that $\quad \lim\limits_{z \to \infty} \dfrac{z^3 - 3z + 2}{z^4 + z^2 - 3z + 5} = 0$.

136. Prove that $|\csc z| \leqq 2e/(e^2 - 1)$ if $|y| \geqq 1$.

137. Show that $\quad \text{Re}\,\{\sin^{-1} z\} = \tfrac{1}{2}\{\sqrt{x^2 + y^2 + 2x + 1} - \sqrt{x^2 + y^2 - 2x + 1}\}$.

138. If $f(z)$ is continuous in a bounded closed region $\mathcal{R}$, prove that (a) there exists a positive number M such that for all z in $\mathcal{R}$, $|f(z)| \leqq M$, (b) $|f(z)|$ has a *least upper bound* μ in $\mathcal{R}$ and there exists at least one value z_0 in $\mathcal{R}$ such that $|f(z_0)| = \mu$.

139. Show that $|\tanh \pi(1 + i)/4| = 1$.

140. Prove that all the values of $(1 - i)^{\sqrt{2}\,i}$ lie on a straight line.

141. Evaluate (a) $\cosh \pi i/2$, (b) $\tanh^{-1} \infty$. *Ans.* (a) 0, (b) $(2k + 1)\pi i/2$, $k = 0, \pm 1, \pm 2, \dots$.

142. If $\tan z = u + iv$, show that
$$u = \frac{\sin 2x}{\cos 2x + \cosh 2y}, \qquad v = \frac{\sinh 2y}{\cos 2x + \cosh 2y}$$

143. Evaluate to 3 decimal place accuracy: (a) $e^{3 - 2i}$, (b) $\sin(5 - 4i)$.

144. Prove $\quad \text{Re}\left\{\dfrac{1 + i\tan(\theta/2)}{1 - i\tan(\theta/2)}\right\} = \cos\theta$, indicating any exceptional values.

145. If $\lim\limits_{z \to z_0} f(z) = A$ and $\lim\limits_{z \to z_0} g(z) = B \neq 0$, prove that $\lim\limits_{z \to z_0} f(z)/g(z) = A/B$ without first proving that $\lim\limits_{z \to z_0} 1/g(z) = 1/B$.

146. Let $\quad f(z) = \begin{cases} 1 & \text{if } |z| \text{ is rational} \\ 0 & \text{if } |z| \text{ is irrational} \end{cases}$. Prove that $f(z)$ is discontinuous at all values of z.

147. Prove that if $f(z) = u(x, y) + i\,v(x, y)$ is continuous in a region, then (a) $\text{Re}\,\{f(z)\} = u(x, y)$ and (b) $\text{Im}\,\{f(z)\} = v(x, y)$ are continuous in the region.

148. Prove that all the roots of $z \tan z = k$, where $k > 0$, are real.

149. Prove that if the limit of a sequence exists it must be unique.

150. (a) Prove that $\lim\limits_{n \to \infty} (\sqrt{n + 1} - \sqrt{n}) = 0$.

(b) Prove that the series $\sum\limits_{n=1}^{\infty} (\sqrt{n + 1} - \sqrt{n})$ diverges, thus showing that a series whose nth term approaches zero need not converge.

151. If $z_{n+1} = \tfrac{1}{2}(z_n + 1/z_n)$, $n = 0, 1, 2, \dots$ and $-\pi/2 < \arg z_0 < \pi/2$, prove that $\lim\limits_{n \to \infty} z_n = 1$.

Chapter 3

Complex Differentiation and The Cauchy-Riemann Equations

DERIVATIVES

If $f(z)$ is single-valued in some region $\mathcal{R}$ of the z plane, the *derivative* of $f(z)$ is defined as

$$f'(z) \;=\; \lim_{\Delta z \to 0} \frac{f(z + \Delta z) - f(z)}{\Delta z} \tag{1}$$

provided that the limit exists independent of the manner in which $\Delta z \to 0$. In such case we say that $f(z)$ is *differentiable* at z. In the definition (1) we sometimes use h instead of Δz. Although differentiability implies continuity, the reverse is not true (see Problem 4).

ANALYTIC FUNCTIONS

If the derivative $f'(z)$ exists at all points z of a region $\mathcal{R}$, then $f(z)$ is said to be *analytic in* $\mathcal{R}$ and is referred to as an *analytic function in* $\mathcal{R}$ or a function *analytic in* $\mathcal{R}$. The terms *regular* and *holomorphic* are sometimes used as synonyms for analytic.

A function $f(z)$ is said to be *analytic at a point* z_0 if there exists a neighborhood $|z - z_0| < \delta$ at all points of which $f'(z)$ exists.

CAUCHY-RIEMANN EQUATIONS

A necessary condition that $w = f(z) = u(x,y) + iv(x,y)$ be analytic in a region $\mathcal{R}$ is that, in $\mathcal{R}$, u and v satisfy the *Cauchy-Riemann equations*

$$\frac{\partial u}{\partial x} = \frac{\partial v}{\partial y}, \qquad \frac{\partial u}{\partial y} = -\frac{\partial v}{\partial x} \tag{2}$$

If the partial derivatives in (2) are continuous in $\mathcal{R}$, then the Cauchy-Riemann equations are sufficient conditions that $f(z)$ be analytic in $\mathcal{R}$. See Problem 5.

The functions $u(x,y)$ and $v(x,y)$ are sometimes called *conjugate functions*. Given one we can find the other (within an arbitrary additive constant) so that $u + iv = f(z)$ is analytic (see Problems 7 and 8).

HARMONIC FUNCTIONS

If the second partial derivatives of u and v with respect to x and y exist and are continuous in a region $\mathcal{R}$, then we find from (2) that (see Problem 6)

$$\frac{\partial^2 u}{\partial x^2} + \frac{\partial^2 u}{\partial y^2} \;=\; 0, \qquad \frac{\partial^2 v}{\partial x^2} + \frac{\partial^2 v}{\partial y^2} \;=\; 0 \tag{3}$$

It follows that under these conditions the real and imaginary parts of an analytic function satisfy *Laplace's equation* denoted by

$$\frac{\partial^2 \Psi}{\partial^2 x} + \frac{\partial^2 \Psi}{\partial^2 y} \;=\; 0 \qquad \text{or} \qquad \nabla^2 \Psi = 0 \qquad \text{where} \qquad \nabla^2 \equiv \frac{\partial^2}{\partial x^2} + \frac{\partial^2}{\partial y^2} \tag{4}$$

The operator ∇^2 is often called the *Laplacian*.

Functions such as $u(x, y)$ and $v(x, y)$ which satisfy Laplace's equation in a region $\mathcal{R}$ are called *harmonic functions* and are said to be *harmonic in* $\mathcal{R}$.

GEOMETRIC INTERPRETATION OF THE DERIVATIVE

Let z_0 [Fig. 3-1] be a point P in the z plane and let w_0 [Fig. 3-2] be its image P' in the w plane under the transformation $w = f(z)$. Since we suppose that $f(z)$ is single-valued, the point z_0 maps into only one point w_0.

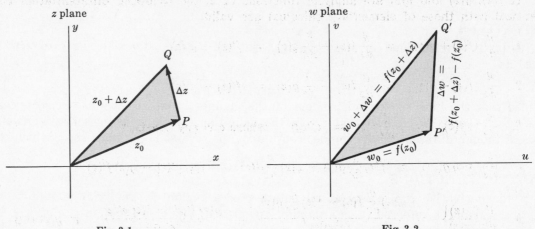

Fig. 3-1   Fig. 3-2

If we give z_0 an increment Δz we obtain the point Q of Fig. 3-1. This point has image Q' in the w plane. Thus from Fig. 3-2 we see that $P'Q'$ represents the complex number $\Delta w = f(z_0 + \Delta z) - f(z_0)$. It follows that the derivative at z_0 (if it exists) is given by

$$\lim_{\Delta z \to 0} \frac{f(z_0 + \Delta z) - f(z_0)}{\Delta z} = \lim_{Q \to P} \frac{Q'P'}{QP} \tag{5}$$

i.e. the limit of the ratio $Q'P'$ to QP as point Q approaches point P. The above interpretation clearly holds when z_0 is replaced by any point z.

DIFFERENTIALS

Let $\Delta z = dz$ be an increment given to z. Then

$$\Delta w = f(z + \Delta z) - f(z) \tag{6}$$

is called the increment in $w = f(z)$. If $f(z)$ is continuous and has a continuous first derivative in a region, then

$$\Delta w = f'(z) \Delta z + \epsilon \Delta z = f'(z) \, dz + \epsilon \, dz \tag{7}$$

where $\epsilon \to 0$ as $\Delta z \to 0$. The expression

$$dw = f'(z) \, dz \tag{8}$$

is called the *differential of w or $f(z)$*, or the *principal part* of Δw. Note that $\Delta w \neq dw$ in general. We call dz the *differential of z*.

Because of the definitions (1) and (8), we often write

$$\frac{dw}{dz} = f'(z) = \lim_{\Delta z \to 0} \frac{f(z + \Delta z) - f(z)}{\Delta z} = \lim_{\Delta z \to 0} \frac{\Delta w}{\Delta z} \tag{9}$$

It is emphasized that dz and dw are not the limits of Δz and Δw as $\Delta z \to 0$, since these limits are zero whereas dz and dw are not necessarily zero. Instead, given dz we determine dw from (8), i.e. dw is a dependent variable determined from the independent variable dz for a given z.

It is useful to think of d/dz as being an *operator* which when operating on $w = f(z)$ leads to $dw/dz = f'(z)$.

RULES FOR DIFFERENTIATION

If $f(z)$, $g(z)$ and $h(z)$ are analytic functions of z, the following differentiation rules (identical with those of elementary calculus) are valid.

1. $\dfrac{d}{dz}\{f(z) + g(z)\} \;=\; \dfrac{d}{dz}f(z) + \dfrac{d}{dz}g(z) \;=\; f'(z) + g'(z)$

2. $\dfrac{d}{dz}\{f(z) - g(z)\} \;=\; \dfrac{d}{dz}f(z) - \dfrac{d}{dz}g(z) \;=\; f'(z) - g'(z)$

3. $\dfrac{d}{dz}\{c\, f(z)\} \;=\; c\,\dfrac{d}{dz}f(z) \;=\; cf'(z)$ where c is any constant

4. $\dfrac{d}{dz}\{f(z)\,g(z)\} \;=\; f(z)\dfrac{d}{dz}g(z) + g(z)\dfrac{d}{dz}f(z) \;=\; f(z)\,g'(z) + g(z)\,f'(z)$

5. $\dfrac{d}{dz}\left\{\dfrac{f(z)}{g(z)}\right\} \;=\; \dfrac{g(z)\dfrac{d}{dz}f(z) - f(z)\dfrac{d}{dz}g(z)}{[g(z)]^2} \;=\; \dfrac{g(z)\,f'(z) - f(z)\,g'(z)}{[g(z)]^2}$ if $g(z) \neq 0$

6. If $w = f(\zeta)$ where $\zeta = g(z)$ then
$$\frac{dw}{dz} \;=\; \frac{dw}{d\zeta}\cdot\frac{d\zeta}{dz} \;=\; f'(\zeta)\frac{d\zeta}{dz} \;=\; f'\{g(z)\}\,g'(z) \tag{10}$$
Similarly, if $w = f(\zeta)$ where $\zeta = g(\eta)$ and $\eta = h(z)$, then
$$\frac{dw}{dz} \;=\; \frac{dw}{d\zeta}\cdot\frac{d\zeta}{d\eta}\cdot\frac{d\eta}{dz} \tag{11}$$

The results (10) and (11) are often called *chain rules* for differentiation of composite functions.

7. If $w = f(z)$, then $z = f^{-1}(w)$; and dw/dz and dz/dw are related by
$$\frac{dw}{dz} \;=\; \frac{1}{dz/dw} \tag{12}$$

8. If $z = f(t)$ and $w = g(t)$ where t is a parameter, then
$$\frac{dw}{dz} \;=\; \frac{dw/dt}{dz/dt} \;=\; \frac{g'(t)}{f'(t)} \tag{13}$$

Similar rules can be formulated for differentials. For example,
$$d\{f(z) + g(z)\} \;=\; df(z) + dg(z) \;=\; f'(z)\,dz + g'(z)\,dz \;=\; \{f'(z) + g'(z)\}\,dz$$
$$d\{f(z)\,g(z)\} \;=\; f(z)\,dg(z) + g(z)\,df(z) \;=\; \{f(z)\,g'(z) + g(z)\,f'(z)\}\,dz$$

DERIVATIVES OF ELEMENTARY FUNCTIONS

In the following we assume that the functions are defined as in Chapter 2. In the cases where functions have branches, i.e. are multi-valued, the branch of the function on the right is chosen so as to correspond to the branch of the function on the left. Note that the results are identical with those of elementary calculus.

1. $\dfrac{d}{dz}(c) = 0$

2. $\dfrac{d}{dz}z^n = nz^{n-1}$

3. $\dfrac{d}{dz}e^z = e^z$

4. $\dfrac{d}{dz}a^z = a^z \ln a$

5. $\dfrac{d}{dz}\sin z = \cos z$

6. $\dfrac{d}{dz}\cos z = -\sin z$

7. $\dfrac{d}{dz}\tan z = \sec^2 z$

8. $\dfrac{d}{dz}\cot z = -\csc^2 z$

9. $\dfrac{d}{dz}\sec z = \sec z \tan z$

10. $\dfrac{d}{dz}\csc z = -\csc z \cot z$

11. $\dfrac{d}{dz}\log_e z = \dfrac{d}{dz}\ln z = \dfrac{1}{z}$

12. $\dfrac{d}{dz}\log_a z = \dfrac{\log_a e}{z}$

13. $\dfrac{d}{dz}\sin^{-1} z = \dfrac{1}{\sqrt{1-z^2}}$

14. $\dfrac{d}{dz}\cos^{-1} z = \dfrac{-1}{\sqrt{1-z^2}}$

15. $\dfrac{d}{dz}\tan^{-1} z = \dfrac{1}{1+z^2}$

16. $\dfrac{d}{dz}\cot^{-1} z = \dfrac{-1}{1+z^2}$

17. $\dfrac{d}{dz}\sec^{-1} z = \dfrac{1}{z\sqrt{z^2-1}}$

18. $\dfrac{d}{dz}\csc^{-1} z = \dfrac{-1}{z\sqrt{z^2-1}}$

19. $\dfrac{d}{dz}\sinh z = \cosh z$

20. $\dfrac{d}{dz}\cosh z = \sinh z$

21. $\dfrac{d}{dz}\tanh z = \operatorname{sech}^2 z$

22. $\dfrac{d}{dz}\coth z = -\operatorname{csch}^2 z$

23. $\dfrac{d}{dz}\operatorname{sech} z = -\operatorname{sech} z \tanh z$

24. $\dfrac{d}{dz}\operatorname{csch} z = -\operatorname{csch} z \coth z$

25. $\dfrac{d}{dz}\sinh^{-1} z = \dfrac{1}{\sqrt{1+z^2}}$

26. $\dfrac{d}{dz}\cosh^{-1} z = \dfrac{1}{\sqrt{z^2-1}}$

27. $\dfrac{d}{dz}\tanh^{-1} z = \dfrac{1}{1-z^2}$

28. $\dfrac{d}{dz}\coth^{-1} z = \dfrac{1}{1-z^2}$

29. $\dfrac{d}{dz}\operatorname{sech}^{-1} z = \dfrac{-1}{z\sqrt{1-z^2}}$

30. $\dfrac{d}{dz}\operatorname{csch}^{-1} z = \dfrac{-1}{z\sqrt{z^2+1}}$

HIGHER ORDER DERIVATIVES

If $w = f(z)$ is analytic in a region, its derivative is given by $f'(z)$, w' or dw/dz. If $f'(z)$ is also analytic in the region, its derivative is denoted by $f''(z)$, w'' or $\left(\dfrac{d}{dz}\right)\left(\dfrac{dw}{dz}\right) = \dfrac{d^2w}{dz^2}$.

Similarly the nth derivative of $f(z)$, if it exists, is denoted by $f^{(n)}(z)$, $w^{(n)}$ or $\dfrac{d^nw}{dz^n}$ where n is called the *order* of the derivative. Thus derivatives of first, second, third, ... orders are given by $f'(z), f''(z), f'''(z), \ldots$. Computations of these higher order derivatives follow by repeated application of the above differentiation rules.

One of the most remarkable theorems valid for functions of a complex variable and not necessarily valid for functions of a real variable is the following

Theorem. If $f(z)$ is analytic in a region $\mathcal{R}$, so also are $f'(z), f''(z), \ldots$ analytic in $\mathcal{R}$, i.e. all higher derivatives exist in $\mathcal{R}$.

This important theorem is proved in Chapter 5.

L'HOSPITAL'S RULE

Let $f(z)$ and $g(z)$ be analytic in a region containing the point z_0 and suppose that $f(z_0) = g(z_0) = 0$ but $g'(z_0) \neq 0$. Then *L'Hospital's rule* states that

$$\lim_{z \to z_0} \frac{f(z)}{g(z)} = \frac{f'(z_0)}{g'(z_0)} \tag{14}$$

In case $f'(z_0) = g'(z_0) = 0$, the rule may be extended. See Problems 21-24.

We sometimes say that the left side of (14) has the "indeterminate form" 0/0, although such terminology is somewhat misleading since there is usually nothing indeterminate involved. Limits represented by so-called indeterminate forms ∞/∞, $0 \cdot \infty$, ∞°, 0°, 1^{∞} and $\infty - \infty$ can often be evaluated by appropriate modifications of L'Hospital's rule.

SINGULAR POINTS

A point at which $f(z)$ fails to be analytic is called a *singular point* or *singularity* of $f(z)$. Various types of singularities exist.

1. **Isolated Singularities.** The point $z = z_0$ is called an *isolated singularity* or *isolated singular point* of $f(z)$ if we can find $\delta > 0$ such that the circle $|z - z_0| = \delta$ encloses no singular point other than z_0 (i.e. there exists a deleted δ neighborhood of z_0 containing no singularity). If no such δ can be found, we call z_0 a *non-isolated singularity*.

    If z_0 is not a singular point and we can find $\delta > 0$ such that $|z - z_0| = \delta$ encloses no singular point, then we call z_0 an *ordinary point* of $f(z)$.

2. **Poles.** If we can find a positive integer n such that $\lim_{z \to z_0} (z - z_0)^n f(z) = A \neq 0$, then $z = z_0$ is called a *pole of order n*. If $n = 1$, z_0 is called a *simple pole*.

    Example 1: $f(z) = \dfrac{1}{(z-2)^3}$ has a pole of order 3 at $z = 2$.

    Example 2: $f(z) = \dfrac{3z - 2}{(z-1)^2(z+1)(z-4)}$ has a pole of order 2 at $z = 1$, and simple poles at $z = -1$ and $z = 4$.

    If $g(z) = (z - z_0)^n f(z)$, where $f(z_0) \neq 0$ and n is a positive integer, then $z = z_0$ is called a *zero of order n* of $g(z)$. If $n = 1$, z_0 is called a *simple zero*. In such case z_0 is a pole of order n of the function $1/g(z)$.

3. **Branch Points** of multiple-valued functions, already considered in Chapter 2, are singular points.

    Example 1: $f(z) = (z-3)^{1/2}$ has a branch point at $z = 3$.

    Example 2: $f(z) = \ln(z^2 + z - 2)$ has branch points where $z^2 + z - 2 = 0$, i.e. at $z = 1$ and $z = -2$.

4. **Removable Singularities.** The singular point z_0 is called a *removable singularity* of $f(z)$ if $\lim_{z \to z_0} f(z)$ exists.

    Example: The singular point $z = 0$ is a removable singularity of $f(z) = \dfrac{\sin z}{z}$ since $\lim_{z \to 0} \dfrac{\sin z}{z} = 1$.

5. **Essential Singularities.** A singularity which is not a pole, branch point or removable singularity is called an *essential singularity*.

    Example: $f(z) = e^{1/(z-2)}$ has an essential singularity at $z = 2$.

    If a function is single-valued and has a singularity, then the singularity is either a pole or an essential singularity. For this reason a pole is sometimes called a *non-essential singularity*. Equivalently, $z = z_0$ is an essential singularity if we cannot find any positive integer n such that $\lim_{z \to z_0} (z - z_0)^n f(z) = A \neq 0$.

6. Singularities at Infinity. The type of singularity of $f(z)$ at $z = \infty$ [the point at infinity; see Pages 6 and 38] is the same as that of $f(1/w)$ at $w = 0$.

> **Example:**   The function $f(z) = z^3$ has a pole of order 3 at $z = \infty$, since $f(1/w) = 1/w^3$ has a pole of order 3 at $w = 0$.

For methods of classifying singularities using infinite series, see Chapter 6.

ORTHOGONAL FAMILIES

If $w = f(z) = u(x, y) + i\,v(x, y)$ is analytic, then the one-parameter families of curves

$$u(x, y) = \alpha, \qquad v(x, y) = \beta \tag{15}$$

where α and β are constants, are *orthogonal*, i.e. each member of one family [shown heavy in Fig. 3-3] is perpendicular to each member of the other family [shown dashed in Fig. 3-3] at the point of intersection. The corresponding image curves in the w plane consisting of lines parallel to the u and v axes also form orthogonal families [see Fig. 3-4].

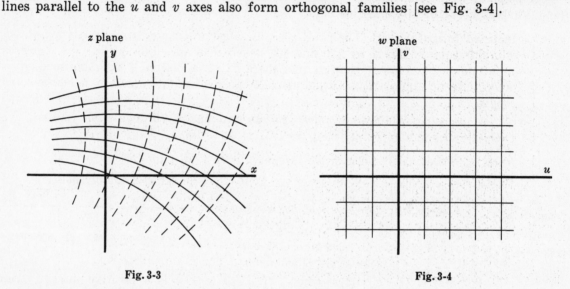

Fig. 3-3   Fig. 3-4

In view of this, one might conjecture that when the mapping function $f(z)$ is analytic the angle between any two intersecting curves C_1 and C_2 in the z plane would equal (both in magnitude and sense) the angle between corresponding intersecting image curves C_1' and C_2' in the w plane. This conjecture is in fact correct and leads to the subject of *conformal mapping* which is of such great importance in both theory and application that two chapters (8 and 9) will be devoted to it.

CURVES

If $\phi(t)$ and $\psi(t)$ are real functions of the real variable t assumed continuous in $t_1 \leq t \leq t_2$, the parametric equations

$$z = x + iy = \phi(t) + i\,\psi(t) = z(t), \qquad t_1 \leq t \leq t_2 \tag{16}$$

define a *continuous curve* or *arc* in the z plane joining points $a = z(t_1)$ and $b = z(t_2)$ [see Fig. 3-5 below].

If $t_1 \neq t_2$ while $z(t_1) = z(t_2)$, i.e. $a = b$, the endpoints coincide and the curve is said to be *closed*. A closed curve which does not intersect itself anywhere is called a *simple closed curve*. For example the curve of Fig. 3-6 is a simple closed curve while that of Fig. 3-7 is not.

If $\phi(t)$ and $\psi(t)$ [and thus $z(t)$] have continuous derivatives in $t_1 \leq t \leq t_2$, the curve is often called a *smooth curve* or *arc*. A curve which is composed of a finite number of

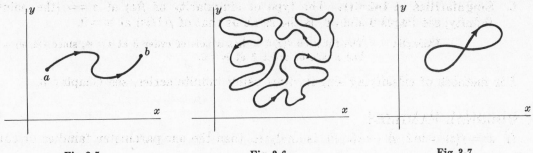

Fig. 3-5   Fig. 3-6   Fig. 3-7

smooth arcs is called a *piecewise* or *sectionally smooth* curve or sometimes a *contour*. For example, the boundary of a square is a piecewise smooth curve or contour.

Unless otherwise specified, whenever we refer to a curve or simple closed curve we shall assume it to be piecewise smooth.

APPLICATIONS TO GEOMETRY AND MECHANICS

We can consider $z(t)$ as a position vector whose terminal point describes a curve C in a definite *sense* or *direction* as t varies from t_1 to t_2. If $z(t)$ and $z(t+\Delta t)$ represent position vectors of points P and Q respectively, then

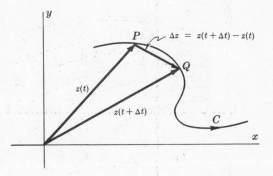

$$\frac{\Delta z}{\Delta t} = \frac{z(t+\Delta t) - z(t)}{\Delta t}$$

is a vector in the direction of Δz [Fig. 3-8].
If $\lim\limits_{\Delta t \to 0} \dfrac{\Delta z}{\Delta t} = \dfrac{dz}{dt}$ exists, the limit is a vector in
the direction of the *tangent* to C at point P and
is given by

$$\frac{dz}{dt} = \frac{dx}{dt} + i\frac{dy}{dt}$$

Fig. 3-8

If t is the time, dz/dt represents the *velocity* with which the terminal point describes the curve. Similarly, d^2z/dt^2 represents its *acceleration* along the curve.

COMPLEX DIFFERENTIAL OPERATORS

Let us define the operators ∇ (*del*) and $\overline{\nabla}$ (*del bar*) by

$$\nabla \equiv \frac{\partial}{\partial x} + i\frac{\partial}{\partial y} = 2\frac{\partial}{\partial \bar{z}}, \qquad \overline{\nabla} \equiv \frac{\partial}{\partial x} - i\frac{\partial}{\partial y} = 2\frac{\partial}{\partial z} \tag{17}$$

where the equivalence in terms of the conjugate coordinates z and $\bar{z}$ (Page 7) follows from Problem 32.

GRADIENT, DIVERGENCE, CURL AND LAPLACIAN

The operator ∇ enables us to define the following operations. In all cases we consider $F(x, y)$ as a real continuously differentiable function of x and y (scalar), while $A(x, y) = P(x, y) + i\, Q(x, y)$ is a complex continuously differentiable function of x and y (vector).

In terms of conjugate coordinates, $F(x, y) = F\left(\dfrac{z + \bar{z}}{2}, \dfrac{z - \bar{z}}{2i}\right) = G(z, \bar{z})$ and $A(x, y) = B(z, \bar{z})$.

1. Gradient. We define the *gradient* of a real function F (scalar) by

$$\operatorname{grad} F = \nabla F = \frac{\partial F}{\partial x} + i\frac{\partial F}{\partial y} = 2\frac{\partial G}{\partial \bar{z}} \tag{18}$$

Geometrically, this represents a vector normal to the curve $F(x, y) = c$ where c is a constant (see Problem 33).

Similarly, the gradient of a complex function $A = P + iQ$ (vector) is defined by

$$\text{grad } A \;=\; \nabla A \;=\; \left(\frac{\partial}{\partial x} + i\frac{\partial}{\partial y}\right)(P + iQ)$$

$$=\; \frac{\partial P}{\partial x} - \frac{\partial Q}{\partial y} + i\left(\frac{\partial P}{\partial y} + \frac{\partial Q}{\partial x}\right) \;=\; 2\frac{\partial B}{\partial \bar{z}} \tag{19}$$

In particular if B is an analytic function of z then $\partial B/\partial \bar{z} = 0$ and so the gradient is zero, i.e. $\dfrac{\partial P}{\partial x} = \dfrac{\partial Q}{\partial y}$, $\dfrac{\partial P}{\partial y} = -\dfrac{\partial Q}{\partial x}$, which shows that the Cauchy-Riemann equations are satisfied in this case.

2. **Divergence.** By using the definition of dot product of two complex numbers (Page 6) extended to the case of operators, we define the *divergence* of a complex function (vector) by

$$\text{div } A \;=\; \nabla \circ A \;=\; \text{Re}\,\{\bar{\nabla}\,A\} \;=\; \text{Re}\left\{\left(\frac{\partial}{\partial x} - i\frac{\partial}{\partial y}\right)(P + iQ)\right\}$$

$$=\; \frac{\partial P}{\partial x} + \frac{\partial Q}{\partial y} \;=\; 2\,\text{Re}\left\{\frac{\partial B}{\partial z}\right\} \tag{20}$$

Similarly we can define the divergence of a real function. It should be noted that the divergence of a complex or real function (vector or scalar) is always a real function (scalar).

3. **Curl.** By using the definition of cross product of two complex numbers (Page 6), we define the *curl* of a complex function by

$$\text{curl } A \;=\; \nabla \times A \;=\; \text{Im}\,\{\bar{\nabla}\,A\} \;=\; \text{Im}\left\{\left(\frac{\partial}{\partial x} - i\frac{\partial}{\partial y}\right)(P + iQ)\right\}$$

$$=\; \frac{\partial Q}{\partial x} - \frac{\partial P}{\partial y} \;=\; 2\,\text{Im}\left\{\frac{\partial B}{\partial z}\right\} \tag{21}$$

Similarly we can define the curl of a real function.

4. **Laplacian.** The *Laplacian operator* is defined as the dot or scalar product of ∇ with itself, i.e.,

$$\nabla \circ \nabla \;\equiv\; \nabla^2 \;\equiv\; \text{Re}\,\{\bar{\nabla}\,\nabla\} \;=\; \text{Re}\left\{\left(\frac{\partial}{\partial x} - i\frac{\partial}{\partial y}\right)\left(\frac{\partial}{\partial x} + i\frac{\partial}{\partial y}\right)\right\}$$

$$=\; \frac{\partial^2}{\partial x^2} + \frac{\partial^2}{\partial y^2} \;=\; 4\frac{\partial^2}{\partial z\,\partial \bar{z}} \tag{22}$$

Note that if A is analytic, $\nabla^2 A = 0$ so that $\nabla^2 P = 0$ and $\nabla^2 Q = 0$, i.e. P and Q are harmonic.

SOME IDENTITIES INVOLVING GRADIENT, DIVERGENCE AND CURL

The following identities hold if A_1, A_2 and A are differentiable functions.

1. grad $(A_1 + A_2)$ = grad A_1 + grad A_2
2. div $(A_1 + A_2)$ = div A_1 + div A_2
3. curl $(A_1 + A_2)$ = curl A_1 + curl A_2
4. grad $(A_1 A_2)$ = $(A_1)(\text{grad } A_2)$ + $(\text{grad } A_1)(A_2)$
5. curl grad A = 0   if A is real or, more generally, if Im $\{A\}$ is harmonic.
6. div grad A = 0   if A is imaginary or, more generally, if Re $\{A\}$ is harmonic.

Solved Problems

DERIVATIVES

1. Using the definition, find the derivative of $w = f(z) = z^3 - 2z$ at the point where
   (a) $z = z_0$, (b) $z = -1$.

   (a) By definition, the derivative at $z = z_0$ is

   $$f'(z_0) = \lim_{\Delta z \to 0} \frac{f(z_0 + \Delta z) - f(z_0)}{\Delta z} = \lim_{\Delta z \to 0} \frac{(z_0 + \Delta z)^3 - 2(z_0 + \Delta z) - \{z_0^3 - 2z_0\}}{\Delta z}$$

   $$= \lim_{\Delta z \to 0} \frac{z_0^3 + 3z_0^2 \Delta z + 3z_0(\Delta z)^2 + (\Delta z)^3 - 2z_0 - 2\Delta z - z_0^3 + 2z_0}{\Delta z}$$

   $$= \lim_{\Delta z \to 0} 3z_0^2 + 3z_0 \Delta z + (\Delta z)^2 - 2 = 3z_0^2 - 2$$

   In general, $f'(z) = 3z^2 - 2$ for all z.

   (b) From (a), or directly, we find that if $z_0 = -1$ then $f'(-1) = 3(-1)^2 - 2 = 1$.

2. Show that $\dfrac{d}{dz}\bar{z}$ does not exist anywhere, i.e. $f(z) = \bar{z}$ is non-analytic anywhere.

   By definition, $\qquad \dfrac{d}{dz}f(z) = \lim_{\Delta z \to 0} \dfrac{f(z + \Delta z) - f(z)}{\Delta z}$

   if this limit exists independent of the manner in which $\Delta z = \Delta x + i\Delta y$ approaches zero.

   Then $\qquad \dfrac{d}{dz}\bar{z} = \lim_{\Delta z \to 0} \dfrac{\overline{z + \Delta z} - \bar{z}}{\Delta z} = \lim_{\substack{\Delta x \to 0 \\ \Delta y \to 0}} \dfrac{\overline{x + iy + \Delta x + i\Delta y} - \overline{x + iy}}{\Delta x + i\Delta y}$

   $$= \lim_{\substack{\Delta x \to 0 \\ \Delta y \to 0}} \frac{x - iy + \Delta x - i\Delta y - (x - iy)}{\Delta x + i\Delta y} = \lim_{\substack{\Delta x \to 0 \\ \Delta y \to 0}} \frac{\Delta x - i\Delta y}{\Delta x + i\Delta y}$$

   If $\Delta y = 0$, the required limit is $\lim_{\Delta x \to 0} \dfrac{\Delta x}{\Delta x} = 1$.

   If $\Delta x = 0$, the required limit is $\lim_{\Delta y \to 0} \dfrac{-i\Delta y}{i\Delta y} = -1$.

   Then since the limit depends on the manner in which $\Delta z \to 0$, the derivative does not exist, i.e. $f(z) = \bar{z}$ is *non-analytic* anywhere.

3. If $w = f(z) = \dfrac{1 + z}{1 - z}$, find (a) $\dfrac{dw}{dz}$ and (b) determine where $f(z)$ is non-analytic.

   (a) *Method 1*, using the definition.

   $$\frac{dw}{dz} = \lim_{\Delta z \to 0} \frac{f(z + \Delta z) - f(z)}{\Delta z} = \lim_{\Delta z \to 0} \frac{\dfrac{1 + (z + \Delta z)}{1 - (z + \Delta z)} - \dfrac{1 + z}{1 - z}}{\Delta z}$$

   $$= \lim_{\Delta z \to 0} \frac{2}{(1 - z - \Delta z)(1 - z)} = \frac{2}{(1 - z)^2}$$

   independent of the manner in which $\Delta z \to 0$, provided $z \neq 1$.

   Method 2, using differentiation rules.

   By the quotient rule [see Problem 10(c)] we have if $z \neq 1$,

   $$\frac{d}{dz}\left(\frac{1 + z}{1 - z}\right) = \frac{(1 - z)\dfrac{d}{dz}(1 + z) - (1 + z)\dfrac{d}{dz}(1 - z)}{(1 - z)^2} = \frac{(1 - z)(1) - (1 + z)(-1)}{(1 - z)^2} = \frac{2}{(1 - z)^2}$$

   (b) The function $f(z)$ is analytic for all finite values of z except $z = 1$ where the derivative does not exist and the function is non-analytic. The point $z = 1$ is a *singular point* of $f(z)$.

4. (a) If $f(z)$ is analytic at z_0, prove that it must be continuous at z_0.

   (b) Give an example to show that the converse of (a) is not necessarily true.

(a) Since $\quad f(z_0 + h) - f(z_0) = \dfrac{f(z_0 + h) - f(z_0)}{h} \cdot h \quad$ where $\ h = \Delta z \neq 0,\ $ we have

$$\lim_{h \to 0} f(z_0 + h) - f(z_0) = \lim_{h \to 0} \frac{f(z_0 + h) - f(z_0)}{h} \cdot \lim_{h \to 0} h = f'(z_0) \cdot 0 = 0$$

because $f'(z_0)$ exists by hypothesis. Thus

$$\lim_{h \to 0} f(z_0 + h) - f(z_0) = 0 \qquad \text{or} \qquad \lim_{h \to 0} f(z_0 + h) = f(z_0)$$

showing that $f(z)$ is continuous at z_0.

(b) The function $f(z) = \bar{z}$ is continuous at z_0. However, by Problem 2, $f(z)$ is not analytic anywhere. This shows that a function which is continuous need not have a derivative, i.e. need not be analytic.

CAUCHY-RIEMANN EQUATIONS

5. Prove that a (a) necessary and (b) sufficient condition that $w = f(z) = u(x, y) + i\, v(x, y)$ be analytic in a region $\mathcal{R}$ is that the Cauchy-Riemann equations $\dfrac{\partial u}{\partial x} = \dfrac{\partial v}{\partial y},\ \dfrac{\partial u}{\partial y} = -\dfrac{\partial v}{\partial x}$ are satisfied in $\mathcal{R}$ where it is supposed that these partial derivatives are continuous in $\mathcal{R}$.

(a) *Necessity.* In order for $f(z)$ to be analytic, the limit

$$\lim_{\Delta z \to 0} \frac{f(z + \Delta z) - f(z)}{\Delta z}$$

$$= f'(z) = \lim_{\substack{\Delta x \to 0 \\ \Delta y \to 0}} \frac{\{u(x + \Delta x,\, y + \Delta y) + i\, v(x + \Delta x,\, y + \Delta y)\} - \{u(x, y) + i\, v(x, y)\}}{\Delta x + i\, \Delta y} \tag{1}$$

must exist independent of the manner in which Δz (or Δx and Δy) approaches zero. We consider two possible approaches.

Case 1. $\Delta y = 0,\ \Delta x \to 0$. In this case (1) becomes

$$\lim_{\Delta x \to 0} \left\{ \frac{u(x + \Delta x,\, y) - u(x, y)}{\Delta x} + i\left[\frac{v(x + \Delta x,\, y) - v(x, y)}{\Delta x} \right] \right\} = \frac{\partial u}{\partial x} + i\frac{\partial v}{\partial x}$$

provided the partial derivatives exist.

Case 2. $\Delta x = 0,\ \Delta y \to 0$. In this case (1) becomes

$$\lim_{\Delta y \to 0} \left\{ \frac{u(x,\, y + \Delta y) - u(x, y)}{i\, \Delta y} + \frac{v(x,\, y + \Delta y) - v(x, y)}{\Delta y} \right\} = \frac{1}{i}\frac{\partial u}{\partial y} + \frac{\partial v}{\partial y} = -i\frac{\partial u}{\partial y} + \frac{\partial v}{\partial y}$$

Now $f(z)$ cannot possibly be analytic unless these two limits are identical. Thus a necessary condition that $f(z)$ be analytic is

$$\frac{\partial u}{\partial x} + i\frac{\partial v}{\partial x} = -i\frac{\partial u}{\partial y} + \frac{\partial v}{\partial y} \qquad \text{or} \qquad \frac{\partial u}{\partial x} = \frac{\partial v}{\partial y},\quad \frac{\partial v}{\partial x} = -\frac{\partial u}{\partial y}$$

(b) *Sufficiency.* Since $\partial u/\partial x$ and $\partial u/\partial y$ are supposed continuous, we have

$$\begin{aligned}
\Delta u &= u(x + \Delta x,\, y + \Delta y) - u(x, y) \\
&= \{u(x + \Delta x,\, y + \Delta y) - u(x,\, y + \Delta y)\} + \{u(x,\, y + \Delta y) - u(x, y)\} \\
&= \left(\frac{\partial u}{\partial x} + \epsilon_1 \right) \Delta x + \left(\frac{\partial u}{\partial y} + \eta_1 \right) \Delta y = \frac{\partial u}{\partial x}\Delta x + \frac{\partial u}{\partial y}\Delta y + \epsilon_1 \Delta x + \eta_1 \Delta y
\end{aligned}$$

where $\epsilon_1 \to 0$ and $\eta_1 \to 0$ as $\Delta x \to 0$ and $\Delta y \to 0$.

Similarly, since $\partial v/\partial x$ and $\partial v/\partial y$ are supposed continuous, we have

$$\Delta v = \left(\frac{\partial v}{\partial x} + \epsilon_2 \right) \Delta x + \left(\frac{\partial v}{\partial y} + \eta_2 \right) \Delta y = \frac{\partial v}{\partial x}\Delta x + \frac{\partial v}{\partial y}\Delta y + \epsilon_2 \Delta x + \eta_2 \Delta y$$

where $\epsilon_2 \to 0$ and $\eta_2 \to 0$ as $\Delta x \to 0$ and $\Delta y \to 0$. Then

$$\Delta w = \Delta u + i\,\Delta v = \left(\frac{\partial u}{\partial x} + i\frac{\partial v}{\partial x} \right) \Delta x + \left(\frac{\partial u}{\partial y} + i\frac{\partial v}{\partial y} \right) \Delta y + \epsilon\,\Delta x + \eta\,\Delta y \tag{2}$$

where $\epsilon = \epsilon_1 + i\epsilon_2 \to 0$ and $\eta = \eta_1 + i\eta_2 \to 0$ as $\Delta x \to 0$ and $\Delta y \to 0$.

By the Cauchy-Riemann equations, (2) can be written

$$\Delta w \;=\; \left(\frac{\partial u}{\partial x} + i\frac{\partial v}{\partial x}\right)\Delta x \;+\; \left(-\frac{\partial v}{\partial x} + i\frac{\partial u}{\partial x}\right)\Delta y \;+\; \epsilon\,\Delta x \;+\; \eta\,\Delta y$$

$$=\; \left(\frac{\partial u}{\partial x} + i\frac{\partial v}{\partial x}\right)(\Delta x + i\,\Delta y) \;+\; \epsilon\,\Delta x \;+\; \eta\,\Delta y$$

Then on dividing by $\Delta z = \Delta x + i\,\Delta y$ and taking the limit as $\Delta z \to 0$, we see that

$$\frac{dw}{dz} \;=\; f'(z) \;=\; \lim_{\Delta z \to 0} \frac{\Delta w}{\Delta z} \;=\; \frac{\partial u}{\partial x} + i\frac{\partial v}{\partial x}$$

so that the derivative exists and is unique, i.e. $f(z)$ is analytic in $\mathcal{R}$.

6. If $f(z) = u + iv$ is analytic in a region $\mathcal{R}$, prove that u and v are harmonic in $\mathcal{R}$ if they have continuous second partial derivatives in $\mathcal{R}$.

If $f(z)$ is analytic in $\mathcal{R}$ then the Cauchy-Riemann equations $(1)\ \dfrac{\partial u}{\partial x} = \dfrac{\partial v}{\partial y}$ and $(2)\ \dfrac{\partial v}{\partial x} = -\dfrac{\partial u}{\partial y}$ are satisfied in $\mathcal{R}$. Assuming u and v have continuous second partial derivatives, we can differentiate both sides of (1) with respect to x and (2) with respect to y to obtain $(3)\ \dfrac{\partial^2 u}{\partial x^2} = \dfrac{\partial^2 v}{\partial x\,\partial y}$ and $(4)\ \dfrac{\partial^2 v}{\partial y\,\partial x} = -\dfrac{\partial^2 u}{\partial y^2}$ from which $\dfrac{\partial^2 u}{\partial x^2} = -\dfrac{\partial^2 u}{\partial y^2}$ or $\dfrac{\partial^2 u}{\partial x^2} + \dfrac{\partial^2 u}{\partial y^2} = 0$, i.e. u is harmonic.

Similarly, by differentiating both sides of (1) with respect to y and (2) with respect to x, we find $\dfrac{\partial^2 v}{\partial x^2} + \dfrac{\partial^2 v}{\partial y^2} = 0$ and v is harmonic.

It will be shown later (Chapter 5) that if $f(z)$ is analytic in $\mathcal{R}$, all its derivatives exist and are continuous in $\mathcal{R}$. Hence the above assumptions will not be necessary.

7. (a) Prove that $u = e^{-x}(x\sin y - y\cos y)$ is harmonic.

(b) Find v such that $f(z) = u + iv$ is analytic.

(a) $\dfrac{\partial u}{\partial x} = (e^{-x})(\sin y) + (-e^{-x})(x\sin y - y\cos y) = e^{-x}\sin y - xe^{-x}\sin y + ye^{-x}\cos y$

$$\frac{\partial^2 u}{\partial x^2} = \frac{\partial}{\partial x}(e^{-x}\sin y - xe^{-x}\sin y + ye^{-x}\cos y) = -2e^{-x}\sin y + xe^{-x}\sin y - ye^{-x}\cos y \qquad (1)$$

$\dfrac{\partial u}{\partial y} = e^{-x}(x\cos y + y\sin y - \cos y) = xe^{-x}\cos y + ye^{-x}\sin y - e^{-x}\cos y$

$$\frac{\partial^2 u}{\partial y^2} = \frac{\partial}{\partial y}(xe^{-x}\cos y + ye^{-x}\sin y - e^{-x}\cos y) = -xe^{-x}\sin y + 2e^{-x}\sin y + ye^{-x}\cos y \qquad (2)$$

Adding (1) and (2) yields $\dfrac{\partial^2 u}{\partial x^2} + \dfrac{\partial^2 u}{\partial y^2} = 0$ and u is harmonic.

(b) From the Cauchy-Riemann equations,

$$\frac{\partial v}{\partial y} = \frac{\partial u}{\partial x} = e^{-x}\sin y - xe^{-x}\sin y + ye^{-x}\cos y \qquad (3)$$

$$\frac{\partial v}{\partial x} = -\frac{\partial u}{\partial y} = e^{-x}\cos y - xe^{-x}\cos y - ye^{-x}\sin y \qquad (4)$$

Integrate (3) with respect to y, keeping x constant. Then

$$v = -e^{-x}\cos y + xe^{-x}\cos y + e^{-x}(y\sin y + \cos y) + F(x)$$
$$= ye^{-x}\sin y + xe^{-x}\cos y + F(x) \qquad (5)$$

where $F(x)$ is an arbitrary real function of x.

Substitute (5) into (4) and obtain

$$-ye^{-x}\sin y - xe^{-x}\cos y + e^{-x}\cos y + F'(x) = -ye^{-x}\sin y - xe^{-x}\cos y - ye^{-x}\sin y$$

or $F'(x) = 0$ and $F(x) = c$, a constant. Then from (5),

$$v = e^{-x}(y\sin y + x\cos y) + c$$

For another method, see Problem 40.

8. Find $f(z)$ in Problem 7.

Method 1.

We have $f(z) = f(x + iy) = u(x, y) + i v(x, y)$.

Putting $y = 0$, $f(x) = u(x, 0) + i v(x, 0)$.

Replacing x by z, $f(z) = u(z, 0) + i v(z, 0)$.

Then from Problem 7, $u(z, 0) = 0$, $v(z, 0) = ze^{-z}$ and so $f(z) = u(z, 0) + i v(z, 0) = i ze^{-z}$, apart from an arbitrary additive constant.

Method 2.

Apart from an arbitrary additive constant, we have from the results of Problem 7,

$$f(z) = u + iv = e^{-x}(x \sin y - y \cos y) + ie^{-x}(y \sin y + x \cos y)$$

$$= e^{-x}\left\{x\left(\frac{e^{iy} - e^{-iy}}{2i}\right) - y\left(\frac{e^{iy} + e^{-iy}}{2}\right)\right\} + ie^{-x}\left\{y\left(\frac{e^{iy} - e^{-iy}}{2i}\right) + x\left(\frac{e^{iy} + e^{-iy}}{2}\right)\right\}$$

$$= i(x + iy) e^{-(x+iy)} = i ze^{-z}$$

Method 3.

We have $x = \dfrac{z + \bar{z}}{2}$, $y = \dfrac{z - \bar{z}}{2i}$. Then substituting into $u(x, y) + i v(x, y)$, we find after much tedious labor that $\bar{z}$ disappears and we are left with the result $i ze^{-z}$.

In general method 1 is preferable over methods 2 and 3 when both u and v are known. If only u (or v) is known another procedure is given in Problem 101.

DIFFERENTIALS

9. If $w = f(z) = z^3 - 2z^2$, find (a) Δw, (b) dw, (c) $\Delta w - dw$.

(a) $\Delta w = f(z + \Delta z) - f(z) = \{(z + \Delta z)^3 - 2(z + \Delta z)^2\} - \{z^3 - 2z^2\}$

$\qquad = z^3 + 3z^2 \Delta z + 3z(\Delta z)^2 + (\Delta z)^3 - 2z^2 - 4z \Delta z - 2(\Delta z)^2 - z^3 + 2z^2$

$\qquad = (3z^2 - 4z) \Delta z + (3z - 2)(\Delta z)^2 + (\Delta z)^3$

(b) $dw = $ principal part of $\Delta w = (3z^2 - 4z)\Delta z = (3z^2 - 4z)dz$, since by definition $\Delta z = dz$.

Note that $f'(z) = 3z^2 - 4z$ and $dw = (3z^2 - 4z)dz$, i.e. $dw/dz = 3z^2 - 4z$.

(c) From (a) and (b), $\Delta w - dw = (3z - 2)(\Delta z)^2 + (\Delta z)^3 = \epsilon \Delta z$ where $\epsilon = (3z - 2)\Delta z + (\Delta z)^2$.

Note that $\epsilon \to 0$ as $\Delta z \to 0$, i.e. $\dfrac{\Delta w - dw}{\Delta z} \to 0$ as $\Delta z \to 0$. It follows that $\Delta w - dw$ is an infinitesimal of higher order than Δz.

DIFFERENTIATION RULES. DERIVATIVES OF ELEMENTARY FUNCTIONS

10. Prove the following assuming that $f(z)$ and $g(z)$ are analytic in a region $\mathcal{R}$.

(a) $\dfrac{d}{dz}\{f(z) + g(z)\} = \dfrac{d}{dz}f(z) + \dfrac{d}{dz}g(z)$

(b) $\dfrac{d}{dz}\{f(z) g(z)\} = f(z)\dfrac{d}{dz}g(z) + g(z)\dfrac{d}{dz}f(z)$

(c) $\dfrac{d}{dz}\left\{\dfrac{f(z)}{g(z)}\right\} = \dfrac{g(z)\dfrac{d}{dz}f(z) - f(z)\dfrac{d}{dz}g(z)}{[g(z)]^2}$ if $g(z) \neq 0$

(a) $\dfrac{d}{dz}\{f(z) + g(z)\} = \lim\limits_{\Delta z \to 0} \dfrac{f(z + \Delta z) + g(z + \Delta z) - \{f(z) + g(z)\}}{\Delta z}$

$\qquad = \lim\limits_{\Delta z \to 0} \dfrac{f(z + \Delta z) - f(z)}{\Delta z} + \lim\limits_{\Delta z \to 0} \dfrac{g(z + \Delta z) - g(z)}{\Delta z} = \dfrac{d}{dz}f(z) + \dfrac{d}{dz}g(z)$

(b) $\dfrac{d}{dz}\{f(z)\,g(z)\}$ $=$ $\displaystyle\lim_{\Delta z \to 0} \dfrac{f(z+\Delta z)\,g(z+\Delta z) - f(z)\,g(z)}{\Delta z}$

$=$ $\displaystyle\lim_{\Delta z \to 0} \dfrac{f(z+\Delta z)\{g(z+\Delta z)-g(z)\} + g(z)\{f(z+\Delta z)-f(z)\}}{\Delta z}$

$=$ $\displaystyle\lim_{\Delta z \to 0} f(z+\Delta z)\left\{\dfrac{g(z+\Delta z)-g(z)}{\Delta z}\right\} + \lim_{\Delta z \to 0} g(z)\left\{\dfrac{f(z+\Delta z)-f(z)}{\Delta z}\right\}$

$=$ $f(z)\dfrac{d}{dz}g(z) + g(z)\dfrac{d}{dz}f(z)$

Note that we have used the fact that $\displaystyle\lim_{\Delta z \to 0} f(z+\Delta z) = f(z)$ which follows since $f(z)$ is analytic and thus continuous (see Problem 4).

Another method.

Let $U = f(z)$, $V = g(z)$. Then $\Delta U = f(z+\Delta z) - f(z)$ and $\Delta V = g(z+\Delta z) - g(z)$, i.e. $f(z+\Delta z) = U + \Delta U$, $g(z+\Delta z) = V + \Delta V$. Thus

$$\dfrac{d}{dz}UV = \lim_{\Delta z \to 0} \dfrac{(U+\Delta U)(V+\Delta V) - UV}{\Delta z} = \lim_{\Delta z \to 0} \dfrac{U\,\Delta V + V\,\Delta U + \Delta U\,\Delta V}{\Delta z}$$

$$= \lim_{\Delta z \to 0}\left(U\dfrac{\Delta V}{\Delta z} + V\dfrac{\Delta U}{\Delta z} + \dfrac{\Delta U}{\Delta z}\Delta V\right) = U\dfrac{dV}{dz} + V\dfrac{dU}{dz}$$

where it is noted that $\Delta V \to 0$ as $\Delta z \to 0$, since V is supposed analytic and thus continuous.

A similar procedure can be used to prove (a).

(c) We use the second method in (b). Then

$$\dfrac{d}{dz}\left(\dfrac{U}{V}\right) = \lim_{\Delta z \to 0} \dfrac{1}{\Delta z}\left\{\dfrac{U+\Delta U}{V+\Delta V} - \dfrac{U}{V}\right\} = \lim_{\Delta z \to 0} \dfrac{V\,\Delta U - U\,\Delta V}{\Delta z(V+\Delta V)V}$$

$$= \lim_{\Delta z \to 0} \dfrac{1}{(V+\Delta V)V}\left\{V\dfrac{\Delta U}{\Delta z} - U\dfrac{\Delta V}{\Delta z}\right\} = \dfrac{V(dU/dz) - U(dV/dz)}{V^2}$$

The first method of (b) can also be used.

11. Prove that (a) $\dfrac{d}{dz}e^z = e^z$, (b) $\dfrac{d}{dz}e^{az} = ae^{az}$ where a is any constant.

(a) By definition, $w = e^z = e^{x+iy} = e^x(\cos y + i \sin y) = u + iv$ or $u = e^x \cos y$, $v = e^x \sin y$.

Since $\dfrac{\partial u}{\partial x} = e^x \cos y = \dfrac{\partial v}{\partial y}$ and $\dfrac{\partial v}{\partial x} = e^x \sin y = -\dfrac{\partial u}{\partial y}$, the Cauchy-Riemann equations are satisfied. Then by Problem 5 the required derivative exists and is equal to

$$\dfrac{\partial u}{\partial x} + i\dfrac{\partial v}{\partial x} = -i\dfrac{\partial u}{\partial y} + \dfrac{\partial v}{\partial y} = e^x \cos y + ie^x \sin y = e^z$$

(b) Let $w = e^\zeta$ where $\zeta = az$. Then by part (a) and Problem 39,

$$\dfrac{d}{dz}e^{az} = \dfrac{d}{dz}e^\zeta = \dfrac{d}{d\zeta}e^\zeta \cdot \dfrac{d\zeta}{dz} = e^\zeta \cdot a = ae^{az}$$

We can also proceed as in part (a).

12. Prove that (a) $\dfrac{d}{dz}\sin z = \cos z$, (b) $\dfrac{d}{dz}\cos z = -\sin z$, (c) $\dfrac{d}{dz}\tan z = \sec^2 z$.

(a) We have $w = \sin z = \sin(x+iy) = \sin x \cosh y + i \cos x \sinh y$. Then

$$u = \sin x \cosh y, \qquad v = \cos x \sinh y$$

Now $\dfrac{\partial u}{\partial x} = \cos x \cosh y = \dfrac{\partial v}{\partial y}$ and $\dfrac{\partial v}{\partial x} = -\sin x \sinh y = -\dfrac{\partial u}{\partial y}$ so that the Cauchy-Riemann equations are satisfied. Hence by Problem 5 the required derivative is equal to

$$\dfrac{\partial u}{\partial x} + i\dfrac{\partial v}{\partial x} = -i\dfrac{\partial u}{\partial y} + \dfrac{\partial v}{\partial y} = \cos x \cosh y - i \sin x \sinh y = \cos(x+iy) = \cos z$$

Another method.

Since $\sin z = \dfrac{e^{iz} - e^{-iz}}{2i}$, we have, using Problem 11(b),

$$\frac{d}{dz} \sin z = \frac{d}{dz}\left(\frac{e^{iz} - e^{-iz}}{2i}\right) = \frac{1}{2i}\frac{d}{dz}e^{iz} - \frac{1}{2i}\frac{d}{dz}e^{-iz} = \frac{1}{2}e^{iz} + \frac{1}{2}e^{-iz} = \cos z$$

(b)
$$\frac{d}{dz}\cos z = \frac{d}{dz}\left(\frac{e^{iz} + e^{-iz}}{2}\right) = \frac{1}{2}\frac{d}{dz}e^{iz} + \frac{1}{2}\frac{d}{dz}e^{-iz}$$

$$= \frac{i}{2}e^{iz} - \frac{i}{2}e^{-iz} = -\frac{e^{iz} - e^{-iz}}{2i} = -\sin z$$

The first method of part (a) can also be used.

(c) By the quotient rule of Problem 10(c) we have

$$\frac{d}{dz}\tan z = \frac{d}{dz}\left(\frac{\sin z}{\cos z}\right) = \frac{\cos z \dfrac{d}{dz}\sin z - \sin z \dfrac{d}{dz}\cos z}{\cos^2 z}$$

$$= \frac{(\cos z)(\cos z) - (\sin z)(-\sin z)}{\cos^2 z} = \frac{\cos^2 z + \sin^2 z}{\cos^2 z} = \frac{1}{\cos^2 z} = \sec^2 z$$

13. Prove that $\dfrac{d}{dz}z^{1/2} = \dfrac{1}{2z^{1/2}}$, realizing that $z^{1/2}$ is a multiple-valued function.

A function must be single-valued in order to have a derivative. Thus since $z^{1/2}$ is multiple-valued (in this case two-valued) we must restrict ourselves to one branch of this function at a time.

Case 1.

Let us first consider that branch of $w = z^{1/2}$ for which $w = 1$ where $z = 1$. In this case, $w^2 = z$ so that

$$\frac{dz}{dw} = 2w \quad \text{and so} \quad \frac{dw}{dz} = \frac{1}{2w} \quad \text{or} \quad \frac{d}{dz}z^{1/2} = \frac{1}{2z^{1/2}}$$

Case 2.

Next we consider that branch of $w = z^{1/2}$ for which $w = -1$ where $z = 1$. In this case too, we have $w^2 = z$ so that

$$\frac{dz}{dw} = 2w \quad \text{and} \quad \frac{dw}{dz} = \frac{1}{2w} \quad \text{or} \quad \frac{d}{dz}z^{1/2} = \frac{1}{2z^{1/2}}$$

In both cases we have $\dfrac{d}{dz}z^{1/2} = \dfrac{1}{2z^{1/2}}$. Note that the derivative does not exist at the branch point $z = 0$. In general a function does not have a derivative, i.e. is not analytic, at a branch point. Thus branch points are singular points.

14. Prove that $\dfrac{d}{dz}\ln z = \dfrac{1}{z}$.

Let $w = \ln z$. Then $z = e^w$ and $dz/dw = e^w = z$. Hence

$$\frac{d}{dz}\ln z = \frac{dw}{dz} = \frac{1}{dz/dw} = \frac{1}{z}$$

Note that the result is valid regardless of the particular branch of $\ln z$. Also observe that the derivative does not exist at the branch point $z = 0$, illustrating further the remark at the end of Problem 13.

15. Prove that $\dfrac{d}{dz}\ln f(z) = \dfrac{f'(z)}{f(z)}$.

Let $w = \ln \zeta$ where $\zeta = f(z)$. Then

$$\frac{dw}{dz} = \frac{dw}{d\zeta}\cdot\frac{d\zeta}{dz} = \frac{1}{\zeta}\cdot\frac{d\zeta}{dz} = \frac{f'(z)}{f(z)}$$

16. Prove that (a) $\dfrac{d}{dz}\sin^{-1}z = \dfrac{1}{\sqrt{1 - z^2}}$, (b) $\dfrac{d}{dz}\tanh^{-1}z = \dfrac{1}{1 - z^2}$.

(a) If we consider the principal branch of $\sin^{-1}z$, we have by Problem 22 of Chapter 2 and by Problem 15,

$$\frac{d}{dz}\sin^{-1}z = \frac{d}{dz}\left\{\frac{1}{i}\ln\left(iz + \sqrt{1-z^2}\right)\right\}$$

$$= \frac{1}{i}\frac{d}{dz}(iz + \sqrt{1-z^2})\Big/(iz + \sqrt{1-z^2})$$

$$= \frac{1}{i}\{i + \tfrac{1}{2}(1-z^2)^{-1/2}(-2z)\}\Big/(iz + \sqrt{1-z^2})$$

$$= \left(1 + \frac{iz}{\sqrt{1-z^2}}\right)\Big/(iz + \sqrt{1-z^2}) = \frac{1}{\sqrt{1-z^2}}$$

The result is also true if we consider other branches.

(b) We have, on considering the principal branch,

$$\tanh^{-1}z = \frac{1}{2}\ln\left(\frac{1+z}{1-z}\right) = \frac{1}{2}\ln(1+z) - \frac{1}{2}\ln(1-z)$$

Then

$$\frac{d}{dz}\tanh^{-1}z = \frac{1}{2}\frac{d}{dz}\ln(1+z) - \frac{1}{2}\frac{d}{dz}\ln(1-z) = \frac{1}{2}\left(\frac{1}{1+z}\right) + \frac{1}{2}\left(\frac{1}{1-z}\right) = \frac{1}{1-z^2}$$

Note that in both parts (a) and (b) the derivatives do not exist at the branch points $z = \pm 1$.

17. Using rules of differentiation, find the derivatives of each of the following:

    (a) $\cos^2(2z+3i)$, (b) $z\tan^{-1}(\ln z)$, (c) $\{\tanh^{-1}(iz+2)\}^{-1}$, (d) $(z-3i)^{4z+2}$.

(a) Let $\eta = 2z + 3i$, $\zeta = \cos\eta$, $w = \zeta^2$ from which $w = \cos^2(2z+3i)$. Then using the chain rule, we have

$$\frac{dw}{dz} = \frac{dw}{d\zeta}\cdot\frac{d\zeta}{d\eta}\cdot\frac{d\eta}{dz} = (2\zeta)(-\sin\eta)(2)$$

$$= (2\cos\eta)(-\sin\eta)(2) = -4\cos(2z+3i)\sin(2z+3i)$$

Another method.

$$\frac{d}{dz}\{\cos(2z+3i)\}^2 = 2\{\cos(2z+3i)\}\left\{\frac{d}{dz}\cos(2z+3i)\right\}$$

$$= 2\{\cos(2z+3i)\}\{-\sin(2z+3i)\}\left\{\frac{d}{dz}(2z+3i)\right\}$$

$$= -4\cos(2z+3i)\sin(2z+3i)$$

(b)
$$\frac{d}{dz}\{(z)[\tan^{-1}(\ln z)]\} = z\frac{d}{dz}[\tan^{-1}(\ln z)] + [\tan^{-1}(\ln z)]\frac{d}{dz}(z)$$

$$= z\left\{\frac{1}{1+(\ln z)^2}\right\}\frac{d}{dz}(\ln z) + \tan^{-1}(\ln z)$$

$$= \frac{1}{1+(\ln z)^2} + \tan^{-1}(\ln z)$$

(c)
$$\frac{d}{dz}\{\tanh^{-1}(iz+2)\}^{-1} = -1\{\tanh^{-1}(iz+2)\}^{-2}\frac{d}{dz}\{\tanh^{-1}(iz+2)\}$$

$$= -\{\tanh^{-1}(iz+2)\}^{-2}\left\{\frac{1}{1-(iz+2)^2}\right\}\frac{d}{dz}(iz+2)$$

$$= \frac{-i\{\tanh^{-1}(iz+2)\}^{-2}}{1-(iz+2)^2}$$

(d)
$$\frac{d}{dz}\{(z-3i)^{4z+2}\} = \frac{d}{dz}\{e^{(4z+2)\ln(z-3i)}\} = e^{(4z+2)\ln(z-3i)}\frac{d}{dz}\{(4z+2)\ln(z-3i)\}$$

$$= e^{(4z+2)\ln(z-3i)}\left\{(4z+2)\frac{d}{dz}[\ln(z-3i)] + \ln(z-3i)\frac{d}{dz}(4z+2)\right\}$$

$$= e^{(4z+2)\ln(z-3i)}\left\{\frac{4z+2}{z-3i} + 4\ln(z-3i)\right\}$$

$$= (z-3i)^{4z+1}(4z+2) + 4(z-3i)^{4z+2}\ln(z-3i)$$

18. If $w^3 - 3z^2w + 4 \ln z = 0$, find dw/dz.

Differentiating with respect to z, considering w as an implicit function of z, we have

$$\frac{d}{dz}(w^3) - 3\frac{d}{dz}(z^2w) + 4\frac{d}{dz}(\ln z) = 0 \qquad \text{or} \qquad 3w^2\frac{dw}{dz} - 3z^2\frac{dw}{dz} - 6zw + \frac{4}{z} = 0$$

Then solving for dw/dz, we obtain $\quad \dfrac{dw}{dz} = \dfrac{6zw - 4/z}{3w^2 - 3z^2}$.

19. If $w = \sin^{-1}(t-3)$ and $z = \cos(\ln t)$, find dw/dz.

$$\frac{dw}{dz} = \frac{dw/dt}{dz/dt} = \frac{1/\sqrt{1-(t-3)^2}}{-\sin(\ln t)[1/t]} = -\frac{t}{\sin(\ln t)\sqrt{1-(t-3)^2}}$$

20. In Problem 18, find d^2w/dz^2.

$$\frac{d^2w}{dz^2} = \frac{d}{dz}\left(\frac{dw}{dz}\right) = \frac{d}{dz}\left(\frac{6zw - 4/z}{3w^2 - 3z^2}\right)$$

$$= \frac{(3w^2 - 3z^2)(6z\,dw/dz + 6w + 4/z^2) - (6zw - 4/z)(6w\,dw/dz - 6z)}{(3w^2 - 3z^2)^2}$$

The required result follows on substituting the value of dw/dz from Problem 18 and simplifying.

L'HOSPITAL'S RULE

21. Prove that if $f(z)$ is analytic in a region $\mathcal{R}$ including the point z_0, then

$$f(z) = f(z_0) + f'(z_0)(z - z_0) + \eta(z - z_0)$$

where $\eta \to 0$ as $z \to z_0$.

Let $\quad \dfrac{f(z) - f(z_0)}{z - z_0} - f'(z_0) = \eta \quad$ so that

$$f(z) = f(z_0) + f'(z_0)(z - z_0) + \eta(z - z_0)$$

Then since $f(z)$ is analytic at z_0 we have as required

$$\lim_{z \to z_0} \eta = \lim_{z \to z_0}\left\{\frac{f(z) - f(z_0)}{z - z_0} - f'(z_0)\right\} = f'(z_0) - f'(z_0) = 0$$

22. Prove that if $f(z)$ and $g(z)$ are analytic at z_0, and $f(z_0) = g(z_0) = 0$ but $g'(z_0) \neq 0$, then

$$\lim_{z \to z_0} \frac{f(z)}{g(z)} = \frac{f'(z_0)}{g'(z_0)}$$

By Problem 21 we have, using the fact that $f(z_0) = g(z_0) = 0$,

$$f(z) = f(z_0) + f'(z_0)(z - z_0) + \eta_1(z - z_0) = f'(z_0)(z - z_0) + \eta_1(z - z_0)$$
$$g(z) = g(z_0) + g'(z_0)(z - z_0) + \eta_2(z - z_0) = g'(z_0)(z - z_0) + \eta_2(z - z_0)$$

where $\lim\limits_{z \to z_0} \eta_1 = \lim\limits_{z \to z_0} \eta_2 = 0$. Then, as required,

$$\lim_{z \to z_0} \frac{f(z)}{g(z)} = \lim_{z \to z_0} \frac{\{f'(z_0) + \eta_1\}(z - z_0)}{\{g'(z_0) + \eta_2\}(z - z_0)} = \frac{f'(z_0)}{g'(z_0)}$$

Another method.

$$\lim_{z \to z_0} \frac{f(z)}{g(z)} = \lim_{z \to z_0} \frac{f(z) - f(z_0)}{z - z_0} \bigg/ \frac{g(z) - g(z_0)}{z - z_0}$$

$$= \left(\lim_{z \to z_0} \frac{f(z) - f(z_0)}{z - z_0}\right) \bigg/ \left(\lim_{z \to z_0} \frac{g(z) - g(z_0)}{z - z_0}\right) = \frac{f'(z_0)}{g'(z_0)}$$

23. Evaluate $\quad$ (a) $\lim\limits_{z \to i} \dfrac{z^{10} + 1}{z^6 + 1}$, $\quad$ (b) $\lim\limits_{z \to 0} \dfrac{1 - \cos z}{z^2}$, $\quad$ (c) $\lim\limits_{z \to 0} \dfrac{1 - \cos z}{\sin z^2}$.

(a) If $f(z) = z^{10} + 1$ and $g(z) = z^6 + 1$, then $f(i) = g(i) = 0$. Also, $f(z)$ and $g(z)$ are analytic at $z = i$. Hence by L'Hospital's rule,

$$\lim_{z \to i} \frac{z^{10}+1}{z^6+1} = \lim_{z \to i} \frac{10z^9}{6z^5} = \lim_{z \to i} \frac{5}{3}z^4 = \frac{5}{3}$$

(b) If $f(z) = 1 - \cos z$ and $g(z) = z^2$, then $f(0) = g(0) = 0$. Also, $f(z)$ and $g(z)$ are analytic at $z = 0$. Hence by L'Hospital's rule,

$$\lim_{z \to 0} \frac{1 - \cos z}{z^2} = \lim_{z \to 0} \frac{\sin z}{2z}$$

Since $f_1(z) = \sin z$ and $g_1(z) = 2z$ are analytic and equal to zero when $z = 0$, we can apply L'Hospital's rule again to obtain the required limit.

$$\lim_{z \to 0} \frac{\sin z}{2z} = \lim_{z \to 0} \frac{\cos z}{2} = \frac{1}{2}$$

(c) **Method 1.** By repeated application of L'Hospital's rule, we have

$$\lim_{z \to 0} \frac{1 - \cos z}{\sin z^2} = \lim_{z \to 0} \frac{\sin z}{2z \cos z^2} = \lim_{z \to 0} \frac{\cos z}{2 \cos z^2 - 4z^2 \sin z^2} = \frac{1}{2}$$

Method 2. Since $\lim_{z \to 0} \frac{\sin z}{z} = 1$, we have by one application of L'Hospital's rule,

$$\lim_{z \to 0} \frac{1 - \cos z}{\sin z^2} = \lim_{z \to 0} \frac{\sin z}{2z \cos z^2} = \lim_{z \to 0} \left(\frac{\sin z}{z} \right) \left(\frac{1}{2 \cos z^2} \right)$$

$$= \lim_{z \to 0} \left(\frac{\sin z}{z} \right) \lim_{z \to 0} \left(\frac{1}{2 \cos z^2} \right) = (1) \left(\frac{1}{2} \right) = \frac{1}{2}$$

Method 3. Since $\lim_{z \to 0} \frac{\sin z^2}{z^2} = 1$ or, equivalently, $\lim_{z \to 0} \frac{z^2}{\sin z^2} = 1$, we can write

$$\lim_{z \to 0} \frac{1 - \cos z}{\sin z^2} = \lim_{z \to 0} \left(\frac{1 - \cos z}{z^2} \right) \left(\frac{z^2}{\sin z^2} \right) = \lim_{z \to 0} \frac{1 - \cos z}{z^2} = \frac{1}{2}$$

using part (b).

24. Evaluate $\lim_{z \to 0} (\cos z)^{1/z^2}$.

Let $w = (\cos z)^{1/z^2}$. Then $\ln w = \frac{\ln \cos z}{z^2}$ where we consider the principal branch of the logarithm. By L'Hospital's rule,

$$\lim_{z \to 0} \ln w = \lim_{z \to 0} \frac{\ln \cos z}{z^2} = \lim_{z \to 0} \frac{(-\sin z)/\cos z}{2z}$$

$$= \lim_{z \to 0} \left(\frac{\sin z}{z} \right) \left(-\frac{1}{2 \cos z} \right) = (1) \left(-\frac{1}{2} \right) = -\frac{1}{2}$$

But since the logarithm is a continuous function, we have

$$\lim_{z \to 0} \ln w = \ln \left(\lim_{z \to 0} w \right) = -\frac{1}{2}$$

or $\lim_{z \to 0} w = e^{-1/2}$ which is the required value.

Note that since $\lim_{z \to 0} \cos z = 1$ and $\lim_{z \to 0} 1/z^2 = \infty$, the required limit has the "indeterminate form" 1^{∞}.

SINGULAR POINTS

25. For each of the following functions locate and name the singularities in the finite z plane and determine whether they are isolated singularities or not.

(a) $f(z) = \dfrac{z}{(z^2+4)^2} = \dfrac{z}{\{(z+2i)(z-2i)\}^2} = \dfrac{z}{(z+2i)^2(z-2i)^2}$.

Since $\lim_{z \to 2i} (z - 2i)^2 f(z) = \lim_{z \to 2i} \dfrac{z}{(z+2i)^2} = \dfrac{1}{8i} \neq 0$, $z = 2i$ is a pole of order 2. Similarly $z = -2i$ is a pole of order 2.

Since we can find δ such that no singularity other than $z = 2i$ lies inside the circle $|z - 2i| = \delta$ (e.g. choose $\delta = 1$), it follows that $z = 2i$ is an isolated singularity. Similarly $z = -2i$ is an isolated singularity.

(b) $f(z) = \sec(1/z)$.

Since $\sec(1/z) = \dfrac{1}{\cos(1/z)}$, the singularities occur where $\cos(1/z) = 0$, i.e. $1/z = (2n+1)\pi/2$ or $z = 2/(2n+1)\pi$, where $n = 0, \pm1, \pm2, \pm3, \ldots$. Also, since $f(z)$ is not defined at $z = 0$, it follows that $z = 0$ is also a singularity.

Now by L'Hospital's rule,

$$\lim_{z \to 2/(2n+1)\pi} \left\{ z - \frac{2}{(2n+1)\pi} \right\} f(z) = \lim_{z \to 2/(2n+1)\pi} \frac{z - 2/(2n+1)\pi}{\cos(1/z)}$$

$$= \lim_{z \to 2/(2n+1)\pi} \frac{1}{-\sin(1/z)\{-1/z^2\}}$$

$$= \frac{\{2/(2n+1)\pi\}^2}{\sin(2n+1)\pi/2} = \frac{4(-1)^n}{(2n+1)^2\pi^2} \neq 0$$

Thus the singularities $z = 2/(2n+1)/\pi$, $n = 0, \pm1, \pm2, \ldots$ are *poles of order one*, i.e. *simple poles*. Note that these poles are located on the real axis at $z = \pm2/\pi, \pm2/3\pi, \pm2/5\pi, \ldots$ and that there are infinitely many in a finite interval which includes 0 (see Fig. 3-9).

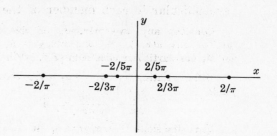

Fig. 3-9

Since we can surround each of these by a circle of radius δ which contains no other singularity, it follows that they are isolated singularities. It should be noted that the δ required is smaller the closer the singularity is to the origin.

Since we cannot find any positive integer n such that $\lim_{z \to 0} (z-0)^n f(z) = A \neq 0$, it follows that $z = 0$ is an *essential singularity*. Also since every circle of radius δ with center at $z = 0$ contains singular points other than $z = 0$, no matter how small we take δ, we see that $z = 0$ is a *non-isolated singularity*.

(c) $f(z) = \dfrac{\ln(z-2)}{(z^2+2z+2)^4}$.

The point $z = 2$ is a *branch point* and is an *isolated singularity*. Also since $z^2+2z+2 = 0$ where $z = -1 \pm i$, it follows that $z^2+2z+2 = (z+1+i)(z+1-i)$ and that $z = -1 \pm i$ are *poles of order 4* which are *isolated singularities*.

(d) $f(z) = \dfrac{\sin\sqrt{z}}{\sqrt{z}}$.

At first sight it appears as if $z = 0$ is a branch point. To test this let $z = re^{i\theta} = re^{i(\theta+2\pi)}$ where $0 \leq \theta < 2\pi$.

If $z = re^{i\theta}$, we have

$$f(z) = \frac{\sin(\sqrt{r}\,e^{i\theta/2})}{\sqrt{r}\,e^{i\theta/2}}$$

If $z = re^{i(\theta+2\pi)}$, we have

$$f(z) = \frac{\sin(\sqrt{r}\,e^{i\theta/2}\,e^{\pi i})}{\sqrt{r}\,e^{i\theta/2}\,e^{\pi i}} = \frac{\sin(-\sqrt{r}\,e^{i\theta/2})}{-\sqrt{r}\,e^{i\theta/2}} = \frac{\sin(\sqrt{r}\,e^{i\theta/2})}{\sqrt{r}\,e^{i\theta/2}}$$

Thus there is actually only one branch to the function, and so $z = 0$ cannot be a branch point.

Since $\lim_{z \to 0} \dfrac{\sin\sqrt{z}}{\sqrt{z}} = 1$, it follows in fact that $z = 0$ is a *removable singularity*.

26. (a) Locate and name all the singularities of $f(z) = \dfrac{z^8+z^4+2}{(z-1)^3(3z+2)^2}$.
    (b) Determine where $f(z)$ is analytic.

(a) The singularities in the finite z plane are located at $z = 1$ and $z = -2/3$; $z = 1$ is a *pole of order 3* and $z = -2/3$ is a *pole of order 2*.

To determine whether there is a singularity at $z = \infty$ (the point at infinity), let $z = 1/w$. Then

$$f(1/w) = \frac{(1/w)^8+(1/w)^4+2}{(1/w-1)^3(3/w+2)^2} = \frac{1+w^4+2w^8}{w^3(1-w)^3(3+2w)^2}$$

Thus since $w = 0$ is a pole of order 3 for the function $f(1/w)$, it follows that $z = \infty$ is a pole of order 3 for the function $f(z)$.

Then the given function has three singularities: a pole of order 3 at $z = 1$, a pole of order 2 at $z = -2/3$, and a pole of order 3 at $z = \infty$.

(b) From (a) it follows that $f(z)$ is analytic everywhere in the finite z plane except at the points $z = 1$ and $-2/3$.

ORTHOGONAL FAMILIES

27. Let $u(x, y) = \alpha$ and $v(x, y) = \beta$, where u and v are the real and imaginary parts of an analytic function $f(z)$ and α and β are any constants, represent two families of curves. Prove that the families are orthogonal (i.e. each member of one family is perpendicular to each member of the other family at their point of intersection).

Consider any two members of the respective families, say $u(x, y) = \alpha_1$ and $v(x, y) = \beta_1$ where α_1 and β_1 are particular constants [Fig. 3-10].

Differentiating $u(x, y) = \alpha_1$ with respect to x yields

$$\frac{\partial u}{\partial x} + \frac{\partial u}{\partial y}\frac{dy}{dx} = 0$$

Then the slope of $u(x, y) = \alpha_1$ is

$$\frac{dy}{dx} = -\frac{\partial u}{\partial x}\Big/\frac{\partial u}{\partial y}$$

Similarly the slope of $v(x, y) = \beta_1$ is

$$\frac{dy}{dx} = -\frac{\partial v}{\partial x}\Big/\frac{\partial v}{\partial y}$$

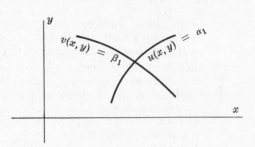

Fig. 3-10

The product of the slopes is, using the Cauchy-Riemann equations,

$$\frac{\partial u}{\partial x}\frac{\partial v}{\partial x}\Big/\frac{\partial u}{\partial y}\frac{\partial v}{\partial y} = -\frac{\partial v}{\partial y}\frac{\partial u}{\partial y}\Big/\frac{\partial u}{\partial y}\frac{\partial v}{\partial y} = -1$$

Thus the curves are orthogonal.

28. Find the orthogonal trajectories of the family of curves in the xy plane defined by $e^{-x}(x \sin y - y \cos y) = \alpha$ where α is a real constant.

By Problems 7 and 27, it follows that $e^{-x}(y \sin y + x \cos y) = \beta$, where β is a real constant, is the required equation of the orthogonal trajectories.

APPLICATIONS TO GEOMETRY AND MECHANICS

29. An ellipse C has the equation $z = a \cos \omega t + bi \sin \omega t$ where a, b, ω are positive constants, $a > b$, and t is a real variable. (a) Graph the ellipse and show that as t increases from $t = 0$ the ellipse is traversed in a counterclockwise direction. (b) Find a unit tangent vector to C at any point.

(a) As t increases from 0 to $\pi/2\omega$, $\pi/2\omega$ to π/ω, π/ω to $3\pi/2\omega$ and $3\pi/2\omega$ to $2\pi/\omega$, point z on C moves from A to B, B to D, D to E and E to A respectively, i.e. it moves in a counterclockwise direction as shown in Fig. 3-11.

(b) A tangent vector to C at any point t is

$$\frac{dz}{dt} = -a\omega \sin \omega t + b\omega i \cos \omega t$$

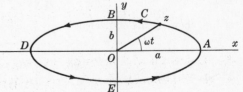

Fig. 3-11

Then a unit tangent vector to C at any point t is

$$\frac{dz/dt}{|dz/dt|} = \frac{-a\omega \sin \omega t + b\omega i \cos \omega t}{|-a\omega \sin \omega t + b\omega i \cos \omega t|} = \frac{-a \sin \omega t + bi \cos \omega t}{\sqrt{a^2 \sin^2 \omega t + b^2 \cos^2 \omega t}}$$

30. In Problem 29 suppose that z is the position vector of a particle moving on C and that t is the time.

(a) Determine the velocity and speed of the particle at any time.

(b) Determine the acceleration both in magnitude and direction at any time.

(c) Prove that $d^2z/dt^2 = -\omega^2 z$ and give a physical interpretation.

(d) Determine where the velocity and acceleration have the greatest and least magnitudes.

(a) Velocity $= dz/dt = -a\omega \sin \omega t + b\omega i \cos \omega t$

Speed $=$ magnitude of velocity $= |dz/dt| = \omega\sqrt{a^2 \sin^2 \omega t + b^2 \cos^2 \omega t}$

(b) Acceleration $= d^2z/dt^2 = -a\omega^2 \cos \omega t - b\omega^2 i \sin \omega t$

Magnitude of acceleration $= |d^2z/dt^2| = \omega^2\sqrt{a^2 \cos^2 \omega t + b^2 \sin^2 \omega t}$

(c) From (b) we see that

$$d^2z/dt^2 = -a\omega^2 \cos \omega t - b\omega^2 i \sin \omega t = -\omega^2(a \cos \omega t + bi \sin \omega t) = -\omega^2 z$$

Physically this states that the acceleration at any time is always directed toward point O and has magnitude proportional to the instantaneous distance from O. As the particle moves, its projection on the x and y axes describes what is sometimes called *simple harmonic motion* of period $2\pi/\omega$. The acceleration is sometimes known as the *centripetal acceleration*.

(d) From (a) and (b) we have

Magnitude of velocity $= \omega\sqrt{a^2 \sin^2 \omega t + b^2(1 - \sin^2 \omega t)} = \omega\sqrt{(a^2 - b^2) \sin^2 \omega t + b^2}$

Magnitude of acceleration $= \omega^2\sqrt{a^2 \cos^2 \omega t + b^2(1 - \cos^2 \omega t)} = \omega^2\sqrt{(a^2 - b^2) \cos^2 \omega t + b^2}$

Then the velocity has the greatest magnitude [given by ωa] where $\sin \omega t = \pm 1$, i.e. at points B and E [Fig. 3-11], and the least magnitude [given by ωb] where $\sin \omega t = 0$, i.e. at points A and D.

Similarly the acceleration has the greatest magnitude [given by $\omega^2 a$] where $\cos \omega t = \pm 1$, i.e. at points A and D, and the least magnitude [given by $\omega^2 b$] where $\cos \omega t = 0$, i.e. at points B and E.

Theoretically the planets of our solar system move in elliptical paths with the sun at one focus. In practice there is some deviation from an exact elliptical path.

GRADIENT, DIVERGENCE, CURL AND LAPLACIAN

31. Prove the equivalence of the operators (a) $\dfrac{\partial}{\partial x} = \dfrac{\partial}{\partial z} + \dfrac{\partial}{\partial \bar{z}}$, (b) $\dfrac{\partial}{\partial y} = i\left(\dfrac{\partial}{\partial z} - \dfrac{\partial}{\partial \bar{z}}\right)$ where $z = x + iy$, $\bar{z} = x - iy$.

If F is any continuously differentiable function, then

(a)
$$\frac{\partial F}{\partial x} = \frac{\partial F}{\partial z}\frac{\partial z}{\partial x} + \frac{\partial F}{\partial \bar{z}}\frac{\partial \bar{z}}{\partial x} = \frac{\partial F}{\partial z} + \frac{\partial F}{\partial \bar{z}}$$

showing the equivalence $\dfrac{\partial}{\partial x} = \dfrac{\partial}{\partial z} + \dfrac{\partial}{\partial \bar{z}}$.

(b)
$$\frac{\partial F}{\partial y} = \frac{\partial F}{\partial z}\frac{\partial z}{\partial y} + \frac{\partial F}{\partial \bar{z}}\frac{\partial \bar{z}}{\partial y} = \frac{\partial F}{\partial z}(i) + \frac{\partial F}{\partial \bar{z}}(-i) = i\left(\frac{\partial F}{\partial z} - \frac{\partial F}{\partial \bar{z}}\right)$$

showing the equivalence $\dfrac{\partial}{\partial y} = i\left(\dfrac{\partial}{\partial z} - \dfrac{\partial}{\partial \bar{z}}\right)$.

32. Show that (a) $\nabla \equiv \dfrac{\partial}{\partial x} + i\dfrac{\partial}{\partial y} = 2\dfrac{\partial}{\partial \bar{z}}$, (b) $\overline{\nabla} \equiv \dfrac{\partial}{\partial x} - i\dfrac{\partial}{\partial y} = 2\dfrac{\partial}{\partial z}$.

From the equivalences established in Problem 31, we have

(a)
$$\nabla \equiv \frac{\partial}{\partial x} + i\frac{\partial}{\partial y} = \frac{\partial}{\partial z} + \frac{\partial}{\partial \bar{z}} + i^2\left(\frac{\partial}{\partial z} - \frac{\partial}{\partial \bar{z}}\right) = 2\frac{\partial}{\partial \bar{z}}$$

(b)
$$\overline{\nabla} \equiv \frac{\partial}{\partial x} - i\frac{\partial}{\partial y} = \frac{\partial}{\partial z} + \frac{\partial}{\partial \bar{z}} - i^2\left(\frac{\partial}{\partial z} - \frac{\partial}{\partial \bar{z}}\right) = 2\frac{\partial}{\partial z}$$

33. If $F(x, y) = c$ [where c is a constant and F is continuously differentiable] is a curve in the xy plane, show that grad $F = \nabla F = \dfrac{\partial F}{\partial x} + i\dfrac{\partial F}{\partial y}$, is a vector normal to the curve.

We have $dF = \dfrac{\partial F}{\partial x}dx + \dfrac{\partial F}{\partial y}dy = 0$. In terms of dot product [see Page 6] this can be written

$$\left(\frac{\partial F}{\partial x} + i\frac{\partial F}{\partial y}\right) \circ (dx + i\,dy) = 0$$

But $dx + i\,dy$ is a vector tangent to C. Hence $\nabla F = \dfrac{\partial F}{\partial x} + i\dfrac{\partial F}{\partial y}$ must be perpendicular to C.

34. Show that $\dfrac{\partial P}{\partial x} - \dfrac{\partial Q}{\partial y} + i\left(\dfrac{\partial Q}{\partial x} + \dfrac{\partial P}{\partial y}\right) = 2\dfrac{\partial B}{\partial \bar{z}}$ where $B(z, \bar{z}) = P(x, y) + iQ(x, y)$.

From Problem 32, $\nabla B = 2\dfrac{\partial B}{\partial \bar{z}}$. Hence

$$\nabla B = \left(\frac{\partial}{\partial x} + i\frac{\partial}{\partial y}\right)(P + iQ) = \frac{\partial P}{\partial x} - \frac{\partial Q}{\partial y} + i\left(\frac{\partial Q}{\partial x} + \frac{\partial P}{\partial y}\right) = 2\frac{\partial B}{\partial \bar{z}}$$

35. Let C be the curve in the xy plane defined by $3x^2y - 2y^3 = 5x^4y^2 - 6x^2$. Find a unit vector normal to C at $(1, -1)$.

Let $F(x, y) = 3x^2y - 2y^3 - 5x^4y^2 + 6x^2 = 0$. By Problem 33, a vector normal to C is

$$\nabla F = \frac{\partial F}{\partial x} + i\frac{\partial F}{\partial y} = (6xy - 20x^3y^2 + 12x) + i(3x^2 - 6y^2 - 10x^4y) = -14 + 7i \quad \text{at } (1, -1)$$

Then a unit vector normal to C at $(1, -1)$ is $\dfrac{-14 + 7i}{|-14 + 7i|} = \dfrac{-2 + i}{\sqrt{5}}$. Another such unit vector is $\dfrac{2 - i}{\sqrt{5}}$.

36. If $A(x, y) = 2xy - ix^2y^3$, find (a) grad A, (b) div A, (c) curl A, (d) Laplacian of A.

(a)   grad $A = \nabla A = \left(\dfrac{\partial}{\partial x} + i\dfrac{\partial}{\partial y}\right)(2xy - ix^2y^3) = \dfrac{\partial}{\partial x}(2xy - ix^2y^3) + i\dfrac{\partial}{\partial y}(2xy - ix^2y^3)$

$$= 2y - 2ixy^3 + i(2x - 3ix^2y^2) = 2y + 3x^2y^2 + i(2x - 2xy^3)$$

(b)   div $A = \nabla \circ A = \text{Re}\{\overline{\nabla}A\} = \text{Re}\left\{\left(\dfrac{\partial}{\partial x} - i\dfrac{\partial}{\partial y}\right)(2xy - ix^2y^3)\right\}$

$$= \frac{\partial}{\partial x}(2xy) - \frac{\partial}{\partial y}(x^2y^3) = 2y - 3x^2y^2$$

(c)   curl $A = \nabla \times A = \text{Im}\{\overline{\nabla}A\} = \text{Im}\left\{\left(\dfrac{\partial}{\partial x} - i\dfrac{\partial}{\partial y}\right)(2xy - ix^2y^3)\right\}$

$$= \frac{\partial}{\partial x}(-x^2y^3) - \frac{\partial}{\partial y}(2xy) = -2xy^3 - 2x$$

(d)   Laplacian $A = \nabla^2 A = \text{Re}\{\overline{\nabla}\,\nabla A\} = \dfrac{\partial^2 A}{\partial x^2} + \dfrac{\partial^2 A}{\partial y^2} = \dfrac{\partial^2}{\partial x^2}(2xy - ix^2y^3) + \dfrac{\partial^2}{\partial y^2}(2xy - ix^2y^3)$

$$= \frac{\partial}{\partial x}(2y - 2ixy^3) + \frac{\partial}{\partial y}(2x - 3ix^2y^2) = -2iy^3 - 6ix^2y$$

MISCELLANEOUS PROBLEMS

37. Prove that in polar form the Cauchy-Riemann equations can be written

$$\frac{\partial u}{\partial r} = \frac{1}{r}\frac{\partial v}{\partial \theta}, \qquad \frac{\partial v}{\partial r} = -\frac{1}{r}\frac{\partial u}{\partial \theta}$$

We have $x = r\cos\theta$, $y = r\sin\theta$ or $r = \sqrt{x^2 + y^2}$, $\theta = \tan^{-1}(y/x)$. Then

$$\frac{\partial u}{\partial x} = \frac{\partial u}{\partial r}\frac{\partial r}{\partial x} + \frac{\partial u}{\partial \theta}\frac{\partial \theta}{\partial x} = \frac{\partial u}{\partial r}\left(\frac{x}{\sqrt{x^2 + y^2}}\right) + \frac{\partial u}{\partial \theta}\left(\frac{-y}{x^2 + y^2}\right) = \frac{\partial u}{\partial r}\cos\theta - \frac{1}{r}\frac{\partial u}{\partial \theta}\sin\theta \qquad (1)$$

$$\frac{\partial u}{\partial y} = \frac{\partial u}{\partial r}\frac{\partial r}{\partial y} + \frac{\partial u}{\partial \theta}\frac{\partial \theta}{\partial y} = \frac{\partial u}{\partial r}\left(\frac{y}{\sqrt{x^2 + y^2}}\right) + \frac{\partial u}{\partial \theta}\left(\frac{x}{x^2 + y^2}\right) = \frac{\partial u}{\partial r}\sin\theta + \frac{1}{r}\frac{\partial u}{\partial \theta}\cos\theta \qquad (2)$$

Similarly,

$$\frac{\partial v}{\partial x} = \frac{\partial v}{\partial r}\frac{\partial r}{\partial x} + \frac{\partial v}{\partial \theta}\frac{\partial \theta}{\partial x} = \frac{\partial v}{\partial r}\cos\theta - \frac{1}{r}\frac{\partial v}{\partial \theta}\sin\theta \tag{3}$$

$$\frac{\partial v}{\partial y} = \frac{\partial v}{\partial r}\frac{\partial r}{\partial y} + \frac{\partial v}{\partial \theta}\frac{\partial \theta}{\partial y} = \frac{\partial v}{\partial r}\sin\theta + \frac{1}{r}\frac{\partial v}{\partial \theta}\cos\theta \tag{4}$$

From the Cauchy-Riemann equation $\dfrac{\partial u}{\partial x} = \dfrac{\partial v}{\partial y}$ we have, using (1) and (4),

$$\left(\frac{\partial u}{\partial r} - \frac{1}{r}\frac{\partial v}{\partial \theta}\right)\cos\theta - \left(\frac{\partial v}{\partial r} + \frac{1}{r}\frac{\partial u}{\partial \theta}\right)\sin\theta = 0 \tag{5}$$

From the Cauchy-Riemann equation $\dfrac{\partial u}{\partial y} = -\dfrac{\partial v}{\partial x}$ we have, using (2) and (3),

$$\left(\frac{\partial u}{\partial r} - \frac{1}{r}\frac{\partial v}{\partial \theta}\right)\sin\theta + \left(\frac{\partial v}{\partial r} + \frac{1}{r}\frac{\partial u}{\partial \theta}\right)\cos\theta = 0 \tag{6}$$

Multiplying (5) by $\cos\theta$, (6) by $\sin\theta$ and adding yields $\quad \dfrac{\partial u}{\partial r} - \dfrac{1}{r}\dfrac{\partial v}{\partial \theta} = 0 \quad$ or $\quad \dfrac{\partial u}{\partial r} = \dfrac{1}{r}\dfrac{\partial v}{\partial \theta}$.

Multiplying (5) by $-\sin\theta$, (6) by $\cos\theta$ and adding yields $\quad \dfrac{\partial v}{\partial r} + \dfrac{1}{r}\dfrac{\partial u}{\partial \theta} = 0 \quad$ or $\quad \dfrac{\partial v}{\partial r} = -\dfrac{1}{r}\dfrac{\partial u}{\partial \theta}$.

38. Prove that the real and imaginary parts of an analytic function of a complex variable when expressed in polar form satisfy the equation [Laplace's equation in polar form]

$$\frac{\partial^2 \Psi}{\partial r^2} + \frac{1}{r}\frac{\partial \Psi}{\partial r} + \frac{1}{r^2}\frac{\partial^2 \Psi}{\partial \theta^2} = 0$$

From Problem 37, $\qquad (1) \quad \dfrac{\partial v}{\partial \theta} = r\dfrac{\partial u}{\partial r}, \qquad (2) \quad \dfrac{\partial v}{\partial r} = -\dfrac{1}{r}\dfrac{\partial u}{\partial \theta}$

To eliminate v differentiate (1) partially with respect to r and (2) with respect to θ. Then

$$(3) \quad \frac{\partial^2 v}{\partial r\,\partial \theta} = \frac{\partial}{\partial r}\left(\frac{\partial v}{\partial \theta}\right) = \frac{\partial}{\partial r}\left(r\frac{\partial u}{\partial r}\right) = r\frac{\partial^2 u}{\partial r^2} + \frac{\partial u}{\partial r}$$

$$(4) \quad \frac{\partial^2 v}{\partial \theta\,\partial r} = \frac{\partial}{\partial \theta}\left(\frac{\partial v}{\partial r}\right) = \frac{\partial}{\partial \theta}\left(-\frac{1}{r}\frac{\partial u}{\partial \theta}\right) = -\frac{1}{r}\frac{\partial^2 u}{\partial \theta^2}$$

But $\dfrac{\partial^2 v}{\partial r\,\partial \theta} = \dfrac{\partial^2 v}{\partial \theta\,\partial r}$ assuming the second partial derivatives are continuous. Hence from (3) and (4),

$$r\frac{\partial^2 u}{\partial r^2} + \frac{\partial u}{\partial r} = -\frac{1}{r}\frac{\partial^2 u}{\partial \theta^2} \qquad \text{or} \qquad \frac{\partial^2 u}{\partial r^2} + \frac{1}{r}\frac{\partial u}{\partial r} + \frac{1}{r^2}\frac{\partial^2 u}{\partial \theta^2} = 0$$

Similarly by elimination of u we find $\dfrac{\partial^2 v}{\partial r^2} + \dfrac{1}{r}\dfrac{\partial v}{\partial r} + \dfrac{1}{r^2}\dfrac{\partial^2 v}{\partial \theta^2} = 0$ so that the required result is proved.

39. If $w = f(\zeta)$ where $\zeta = g(z)$, prove that $\dfrac{dw}{dz} = \dfrac{dw}{d\zeta}\cdot\dfrac{d\zeta}{dz}$ assuming f and g are analytic in a region $\mathcal{R}$.

Let z be given an increment $\Delta z \neq 0$ so that $z + \Delta z$ is in $\mathcal{R}$. Then as a consequence ζ and w take on increments $\Delta\zeta$ and Δw respectively, where

$$\Delta w = f(\zeta + \Delta\zeta) - f(\zeta), \qquad \Delta\zeta = g(z + \Delta z) - g(z) \tag{1}$$

Note that as $\Delta z \to 0$, $\Delta w \to 0$ and $\Delta\zeta \to 0$.

If $\Delta\zeta \neq 0$, let us write $\quad \epsilon = \dfrac{\Delta w}{\Delta\zeta} - \dfrac{dw}{d\zeta} \quad$ so that $\epsilon \to 0$ as $\Delta\zeta \to 0$ and

$$\Delta w = \frac{dw}{d\zeta}\Delta\zeta + \epsilon\,\Delta\zeta \tag{2}$$

If $\Delta\zeta = 0$ for values of Δz, then (1) shows that $\Delta w = 0$ for these values of Δz. For such cases, we define $\epsilon = 0$.

It follows that in both cases, $\Delta\zeta \neq 0$ or $\Delta\zeta = 0$, (2) holds. Then dividing (2) by $\Delta z \neq 0$ and taking the limit as $\Delta z \to 0$, we have

$$
\begin{aligned}
\frac{dw}{dz} &= \lim_{\Delta z \to 0} \frac{\Delta w}{\Delta z} = \lim_{\Delta z \to 0} \left(\frac{dw}{d\zeta}\frac{\Delta\zeta}{\Delta z} + \epsilon\frac{\Delta w}{\Delta z} \right) \\
&= \frac{dw}{d\zeta} \cdot \lim_{\Delta z \to 0} \frac{\Delta\zeta}{\Delta z} + \lim_{\Delta z \to 0} \epsilon \cdot \lim_{\Delta z \to 0} \frac{\Delta w}{\Delta z} \\
&= \frac{dw}{d\zeta} \cdot \frac{d\zeta}{dz} + 0 \cdot \frac{d\zeta}{dz} = \frac{dw}{d\zeta} \cdot \frac{d\zeta}{dz}
\end{aligned}
$$

40. (a) If $u_1(x,y) = \partial u/\partial x$ and $u_2(x,y) = \partial u/\partial y$, prove that $f'(z) = u_1(z,0) - i\,u_2(z,0)$.

(b) Show how the result in (a) can be used to solve Problems 7 and 8.

(a) From Problem 5, we have $f'(z) = \dfrac{\partial u}{\partial x} - i\dfrac{\partial u}{\partial y} = u_1(x,y) - i\,u_2(x,y)$.

Putting $y = 0$, this becomes $f'(x) = u_1(x,0) - i\,u_2(x,0)$.

Then replacing x by z, we have as required $f'(z) = u_1(z,0) - i\,u_2(z,0)$.

(b) Since we are given $u = e^{-x}(x \sin y - y \cos y)$, we have

$$
u_1(x,y) = \frac{\partial u}{\partial x} = e^{-x}\sin y - xe^{-x}\sin y + ye^{-x}\cos y
$$

$$
u_2(x,y) = \frac{\partial u}{\partial y} = xe^{-x}\cos y + ye^{-x}\sin y - e^{-x}\cos y
$$

so that from part (a),

$$
f'(z) = u_1(z,0) - i\,u_2(z,0) = 0 - i(ze^{-z} - e^{-z}) = -i(ze^{-z} - e^{-z})
$$

Integrating with respect to z we have, apart from a constant, $f(z) = ize^{-z}$. By separating this into real and imaginary parts, $v = e^{-x}(y \sin y + x \cos y)$ apart from a constant.

41. Prove that curl grad $A = 0$ if A is real or, more generally, if Im A is harmonic.

If $A = P + Qi$, grad $A = \left(\dfrac{\partial}{\partial x} + i\dfrac{\partial}{\partial y}\right)(P + iQ) = \dfrac{\partial P}{\partial x} - \dfrac{\partial Q}{\partial y} + i\left(\dfrac{\partial P}{\partial y} + \dfrac{\partial Q}{\partial x}\right)$. Then

$$
\begin{aligned}
\text{curl grad } A &= \text{Im}\left[\left(\frac{\partial}{\partial x} - i\frac{\partial}{\partial y}\right)\left\{\frac{\partial P}{\partial x} - \frac{\partial Q}{\partial y} + i\left(\frac{\partial P}{\partial y} + \frac{\partial Q}{\partial x}\right)\right\}\right] \\
&= \text{Im}\left[\frac{\partial^2 P}{\partial x^2} - \frac{\partial^2 Q}{\partial x\,\partial y} + i\left(\frac{\partial^2 P}{\partial x\,\partial y} + \frac{\partial^2 Q}{\partial x^2}\right) - i\left(\frac{\partial^2 P}{\partial y\,\partial x} - \frac{\partial^2 Q}{\partial y^2}\right) + \left(\frac{\partial^2 P}{\partial y^2} + \frac{\partial^2 Q}{\partial y\,\partial x}\right)\right] \\
&= \frac{\partial^2 Q}{\partial x^2} + \frac{\partial^2 Q}{\partial y^2}
\end{aligned}
$$

Hence if $Q = 0$, i.e. A is real, or if Q is harmonic, curl grad $A = 0$.

42. Solve the partial differential equation $\dfrac{\partial^2 U}{\partial x^2} + \dfrac{\partial^2 U}{\partial y^2} = x^2 - y^2$.

Let $z = x + iy$, $\bar{z} = x - iy$ so that $x = \dfrac{z + \bar{z}}{2}$, $y = \dfrac{z - \bar{z}}{2i}$. Then

$$
x^2 - y^2 = \tfrac{1}{2}(z^2 + \bar{z}^2) \quad \text{and} \quad \frac{\partial^2 U}{\partial x^2} + \frac{\partial^2 U}{\partial y^2} = \nabla^2 U = 4\frac{\partial^2 U}{\partial z\,\partial\bar{z}}
$$

Thus the given partial differential equation becomes $4\dfrac{\partial^2 U}{\partial z\,\partial\bar{z}} = \dfrac{1}{2}(z^2 + \bar{z}^2)$ or

$$
\frac{\partial}{\partial z}\left(\frac{\partial U}{\partial\bar{z}}\right) = \frac{1}{8}(z^2 + \bar{z}^2) \tag{1}
$$

Integrating (1) with respect to z (treating $\bar{z}$ as constant),

$$\frac{\partial U}{\partial \bar{z}} = \frac{z^3}{24} + \frac{z\bar{z}^2}{8} + F_1(\bar{z}) \tag{2}$$

where $F_1(\bar{z})$ is an arbitrary function of $\bar{z}$. Integrating (2) with respect to $\bar{z}$,

$$U = \frac{z^3\bar{z}}{24} + \frac{z\bar{z}^3}{24} + F(\bar{z}) + G(z) \tag{3}$$

where $F(\bar{z})$ is the function obtained by integrating $F_1(\bar{z})$, and $G(z)$ is an arbitrary function of z. Replacing z and $\bar{z}$ by $x + iy$ and $x - iy$ respectively, we obtain

$$U = \tfrac{1}{12}(x^4 - y^4) + F(x - iy) + G(x + iy)$$

Supplementary Problems

DERIVATIVES

43. Using the definition, find the derivative of each function at the indicated points.

(a) $f(z) = 3z^2 + 4iz - 5 + i$; $z = 2$. (b) $f(z) = \dfrac{2z - i}{z + 2i}$; $z = -i$. (c) $f(z) = 3z^{-2}$; $z = 1 + i$.
 Ans. (a) $12 + 4i$ (b) $-5i$ (c) $3/2 + 3i/2$

44. Prove that $\dfrac{d}{dz}(z^2\bar{z})$ does not exist anywhere.

45. Determine whether $|z|^2$ has a derivative anywhere.

46. For each of the following functions determine the singular points, i.e. points at which the function is not analytic. Determine the derivatives at all other points. (a) $\dfrac{z}{z + i}$, (b) $\dfrac{3z - 2}{z^2 + 2z + 5}$.
 Ans. (a) $-i$, $i/(z + i)^2$; (b) $-1 \pm 2i$, $(19 + 4z - 3z^2)/(z^2 + 2z + 5)^2$

CAUCHY-RIEMANN EQUATIONS

47. Verify that the real and imaginary parts of the following functions satisfy the Cauchy-Riemann equations and thus deduce the analyticity of each function:

(a) $f(z) = z^2 + 5iz + 3 - i$,   (b) $f(z) = ze^{-z}$,   (c) $f(z) = \sin 2z$.

48. Show that the function $x^2 + iy^3$ is not analytic anywhere. Reconcile this with the fact that the Cauchy-Riemann equations are satisfied at $x = 0$, $y = 0$.

49. Prove that if $w = f(z) = u + iv$ is analytic in a region $\mathcal{R}$, then $\dfrac{dw}{dz} = \dfrac{\partial w}{\partial x} = -i\dfrac{\partial w}{\partial y}$.

50. (a) Prove that the function $u = 2x(1 - y)$ is harmonic. (b) Find a function v such that $f(z) = u + iv$ is analytic [i.e. find the conjugate function of u]. (c) Express $f(z)$ in terms of z.
 Ans. (b) $2y + x^2 - y^2$, (c) $iz^2 + 2z$

51. Answer Problem 50 for the function $u = x^2 - y^2 - 2xy - 2x + 3y$.   Ans. (b) $x^2 - y^2 + 2xy - 3x - 2y$

52. Verify that the Cauchy-Riemann equations are satisfied for the functions (a) e^{z^2}, (b) $\cos 2z$, (c) $\sinh 4z$.

53. Determine which of the following functions u are harmonic. For each harmonic function find the conjugate harmonic function v and express $u + iv$ as an analytic function of z.
(a) $3x^2y + 2x^2 - y^3 - 2y^2$, (b) $2xy + 3xy^2 - 2y^3$, (c) $xe^x \cos y - ye^x \sin y$, (d) $e^{-2xy} \sin(x^2 - y^2)$.
 Ans. (a) $v = 4xy - x^3 + 3xy^2 + c$, $f(z) = 2z^2 - iz^3 + ic$ (c) $ye^x \cos y + xe^x \sin y + c$, $ze^z + ic$
       (b) Not harmonic   (d) $-e^{-2xy} \cos(x^2 - y^2) + c$, $-ie^{iz^2} + ic$

54. (a) Prove that $\psi = \ln{[(x-1)^2 + (y-2)^2]}$ is harmonic in every region which does not include the point $(1, 2)$. (b) Find a function ϕ such that $\phi + i\psi$ is analytic. (c) Express $\phi + i\psi$ as a function of z.
Ans. (b) $-2 \tan^{-1}{\{(y-2)/(x-1)\}}$ (c) $2i \ln{(z-1-2i)}$

55. If $\mathrm{Im}\,\{f'(z)\} = 6x(2y-1)$ and $f(0) = 3-2i$, $f(1) = 6-5i$, find $f(1+i)$.   Ans. $6 + 3i$

DIFFERENTIALS

56. If $w = iz^2 - 4z + 3i$, find (a) Δw, (b) dw, (c) $\Delta w - dw$ at the point $z = 2i$.
Ans. (a) $-8\,\Delta z + i(\Delta z)^2 = -8\,dz + i(dz)^2$,   (b) $-8\,dz$,   (c) $i(dz)^2$

57. Find (a) Δw and (b) dw if $w = (2z+1)^3$, $z = -i$, $\Delta z = 1+i$.   Ans. (a) $38 - 2i$, (b) $6 - 42i$

58. If $w = 3iz^2 + 2z + 1 - 3i$, find (a) Δw, (b) dw, (c) $\Delta w/\Delta z$, (d) dw/dz where $z = i$.
Ans. (a) $-4\Delta z + 3i(\Delta z)^2$,   (b) $-4\,dz$,   (c) $-4 + 3i\,\Delta z$,   (d) -4

59. (a) If $w = \sin z$, show that $\dfrac{\Delta w}{\Delta z} = (\cos z)\left(\dfrac{\sin \Delta z}{\Delta z}\right) - 2 \sin z \left\{\dfrac{\sin^2(\Delta z/2)}{\Delta z}\right\}$.

   (b) Assuming $\displaystyle\lim_{\Delta z \to 0} \dfrac{\sin \Delta z}{\Delta z} = 1$, prove that $\dfrac{dw}{dz} = \cos z$.

   (c) Show that $dw = (\cos z)\,dz$.

60. (a) If $w = \ln z$, show that if $\Delta z/z = \zeta$, $\dfrac{\Delta w}{\Delta z} = \dfrac{1}{z} \ln{\{(1+\zeta)^{1/\zeta}\}}$.

   (b) Assuming $\displaystyle\lim_{\zeta \to 0} (1+\zeta)^{1/\zeta} = e = 2.71828\ldots$, prove that $\dfrac{dw}{dz} = \dfrac{1}{z}$.

   (c) Show that $d(\ln z) = dz/z$.

61. Prove that   (a) $d\{f(z)\,g(z)\} = \{f(z)\,g'(z) + g(z)\,f'(z)\}\,dz$
         (b) $d\{f(z)/g(z)\} = \{g(z)\,f'(z) - f(z)\,g'(z)\}\,dz/\{g(z)\}^2$
giving restrictions on $f(z)$ and $g(z)$.

DIFFERENTIATION RULES. DERIVATIVES OF ELEMENTARY FUNCTIONS.

62. Prove that if $f(z)$ and $g(z)$ are analytic in a region $\mathcal{R}$, then

   (a) $\dfrac{d}{dz}\{2i\,f(z) - (1+i)\,g(z)\} = 2i\,f'(z) - (1+i)\,g'(z)$, (b) $\dfrac{d}{dz}\{f(z)\}^2 = 2\,f(z)\,f'(z)$,   (c) $\dfrac{d}{dz}\{f(z)\}^{-1} = -\{f(z)\}^{-2}\,f'(z)$.

63. Using differentiation rules, find the derivatives of each of the following functions: (a) $(1+4i)z^2 - 3z - 2$,
   (b) $(2z+3i)(z-i)$, (c) $(2z-i)/(z+2i)$, (d) $(2iz+1)^2$, (e) $(iz-1)^{-3}$.
   Ans. (a) $(2+8i)z - 3$, (b) $4z + i$, (c) $5i/(z+2i)^2$, (d) $4i - 8z$, (e) $-3i(iz-1)^{-4}$

64. Find the derivatives of each of the following at the indicated points:
   (a) $(z+2i)(i-z)/(2z-1)$, $z = i$. (b) $\{z + (z^2+1)^2\}^2$, $z = 1+i$.
   Ans. (a) $-6/5 + 3i/5$, (b) $-108 - 78i$

65. Prove that (a) $\dfrac{d}{dz} \sec z = \sec z \tan z$, (b) $\dfrac{d}{dz} \cot z = -\csc^2 z$.

66. Prove that (a) $\dfrac{d}{dz}(z^2+1)^{1/2} = \dfrac{z}{(z^2+1)^{1/2}}$, (b) $\dfrac{d}{dz} \ln{(z^2+2z+2)} = \dfrac{2z+2}{z^2+2z+2}$ indicating restrictions if any.

67. Find the derivatives of each of the following, indicating restrictions if any.

   (a) $3 \sin^2(z/2)$,   (b) $\tan^3(z^2 - 3z + 4i)$,   (c) $\ln(\sec z + \tan z)$,   (d) $\csc\{(z^2+1)^{1/2}\}$,   (e) $(z^2-1)\cos(z+2i)$.

   Ans.   (a) $3 \sin(z/2)\cos(z/2)$

   (b) $3(2z-3)\tan^2(z^2-3z+4i)\sec^2(z^2-3z+4i)$

   (c) $\sec z$

   (d) $\dfrac{-z \csc\{(z^2+1)^{1/2}\}\cot\{(z^2+1)^{1/2}\}}{(z^2+1)^{1/2}}$

   (e) $(1-z^2)\sin(z+2i) + 2z\cos(z+2i)$

68. Prove that   (a) $\dfrac{d}{dz}(1+z^2)^{3/2} = 3z(1+z^2)^{1/2}$,   (b) $\dfrac{d}{dz}(z+2\sqrt{z})^{1/3} = \dfrac{1}{3}z^{-1/2}(z+2\sqrt{z})^{-2/3}(\sqrt{z}+1)$.

69. Prove that   (a) $\dfrac{d}{dz}(\tan^{-1}z) = \dfrac{1}{z^2+1}$,   (b) $\dfrac{d}{dz}(\sec^{-1}z) = \dfrac{1}{z\sqrt{z^2-1}}$.

70. Prove that   (a) $\dfrac{d}{dz}\sinh^{-1}z = \dfrac{1}{\sqrt{1+z^2}}$,   (b) $\dfrac{d}{dz}\operatorname{csch}^{-1}z = \dfrac{-1}{z\sqrt{z^2+1}}$.

71. Find the derivatives of each of the following:

   (a) $\{\sin^{-1}(2z-1)\}^2$ (c) $\cos^{-1}(\sin z - \cos z)$ (e) $\coth^{-1}(z \csc 2z)$

   (b) $\ln\{\cot^{-1}z^2\}$ (d) $\tan^{-1}(z+3i)^{-1/2}$ (f) $\ln(z - \tfrac{3}{2} + \sqrt{z^2-3z+2i})$

   Ans.   (a) $2\sin^{-1}(2z-1)/(z-z^2)^{1/2}$ (d) $-1/2(z+1+3i)(z+3i)^{1/2}$

   (b) $-2z/(1+z^4)\cot^{-1}z^2$ (e) $(\csc 2z)(1-2z\cot 2z)/(1-z^2\csc^2 2z)$

   (c) $-(\sin z + \cos z)/(\sin 2z)^{1/2}$ (f) $1/\sqrt{z^2-3z+2i}$

72. If $w = \cos^{-1}(z-1)$, $z = \sinh(3\zeta + 2i)$ and $\zeta = \sqrt{t}$, find dw/dt.

   Ans. $-3[\cosh(3\zeta + 2i)]/2(2z-z^2)^{1/2}t^{1/2}$

73. If $w = t\sec(t-3i)$ and $z = \sin^{-1}(2t-1)$, find dw/dz.

   Ans. $\sec(t-3i)\{1 + t\tan(t-3i)\}(t-t^2)^{1/2}$

74. If $w^2 - 2w + \sin 2z = 0$, find (a) dw/dz, (b) d^2w/dz^2.

   Ans. (a) $(\cos 2z)/(1-w)$,   (b) $\{\cos^2 2z - 2(1-w)^2\sin 2z\}/(1-w)^3$

75. Find d^2w/dz^2 at $\zeta = 0$ if $w = \cos\zeta$, $z = \tan(\zeta + \pi i)$.   *Ans.* $-\cosh^4\pi$

76. Find (a) $\dfrac{d}{dz}\{z^{\ln z}\}$,   (b) $\dfrac{d}{dz}\{[\sin(iz-2)]^{\tan^{-1}(z+3i)}\}$

   Ans. (a) $2z^{\ln z - 1}\ln z$

   (b) $\{[\sin(iz-2)]^{\tan^{-1}(z+3i)}\}\{i\tan^{-1}(z+3i)\cot(iz-2) + [\ln\sin(iz-2)]/[z^2+6iz-8]\}$

77. Find the second derivatives of each of the following:

   (a) $3\sin^2(2z-1+i)$,   (b) $\ln\tan z^2$,   (c) $\sinh(z+1)^2$,   (d) $\cos^{-1}(\ln z)$,   (e) $\operatorname{sech}^{-1}\sqrt{1+z}$.

   Ans. (a) $24\cos(4z-2+2i)$ (d) $(1 - \ln z - \ln^2 z)/z^2(1-\ln^2 z)^{3/2}$

   (b) $4\csc 2z^2 - 16z^2\csc 2z^2\cot 2z^2$ (e) $-i(1+3z)/4(1+z)^2z^{3/2}$

   (c) $2\cosh(z+1)^2 + 4(z+1)^2\sinh(z+1)^2$

L'HOSPITAL'S RULE

78. Evaluate (a) $\lim\limits_{z\to 2i}\dfrac{z^2+4}{2z^2+(3-4i)z-6i}$,   (b) $\lim\limits_{z\to e^{\pi i/3}}(z - e^{\pi i/3})\left(\dfrac{z}{z^3+1}\right)$,   (c) $\lim\limits_{z\to i}\dfrac{z^2-2iz-1}{z^4+2z^2+1}$.

   Ans. (a) $(16+12i)/25$, (b) $(1-i\sqrt{3})/6$, (c) $-1/4$

79. Evaluate (a) $\lim\limits_{z\to 0}\dfrac{z - \sin z}{z^3}$,   (b) $\lim\limits_{z\to m\pi i}(z - m\pi i)\left(\dfrac{e^z}{\sin z}\right)$.   *Ans.* (a) $1/6$, (b) $e^{m\pi i}/(\cosh m\pi)$

80. Find $\lim\limits_{z\to i}\dfrac{\tan^{-1}(z^2+1)^2}{\sin^2(z^2+1)}$ where the branch of the inverse tangent is chosen such that $\tan^{-1}0 = 0$.

   Ans. 1

81. Evaluate $\lim\limits_{z\to 0}\left(\dfrac{\sin z}{z}\right)^{1/z^2}$.   *Ans.* $e^{-1/6}$

SINGULAR POINTS

82. For each of the following functions locate and name the singularities in the finite z plane.

   (a) $\dfrac{z^2 - 3z}{z^2 + 2z + 2}$,   (b) $\dfrac{\ln(z + 3i)}{z^2}$,   (c) $\sin^{-1}(1/z)$,   (d) $\sqrt{z(z^2 + 1)}$,   (e) $\dfrac{\cos z}{(z + i)^3}$

   Ans. (a) $z = -1 \pm i$; simple poles

   (b) $z = -3i$; branch point, $z = 0$; pole of order 2

   (c) $z = 0$; essential singularity

   (d) $z = 0, \pm i$; branch points

   (e) $z = -i$; pole of order 3

83. Show that $f(z) = \dfrac{(z + 3i)^5}{(z^2 - 2z + 5)^2}$ has double poles at $z = 1 \pm 2i$ and a simple pole at infinity.

84. Show that e^{z^2} has an essential singularity at infinity.

85. Locate and name all the singularities of each of the following functions.

   (a) $(z + 3)/(z^2 - 1)$,   (b) $\csc(1/z^2)$,   (c) $(z^2 + 1)/z^{3/2}$.

   Ans. (a) $z = \pm 1$; simple poles, $z = \infty$; simple pole. (b) $z = 1/\sqrt{m\pi}$, $m = \pm 1, \pm 2, \pm 3, \ldots$; simple poles, $z = 0$; essential singularity, $z = \infty$; pole of order 2. (c) $z = 0$; branch point, $z = \infty$; branch point.

ORTHOGONAL FAMILIES

86. Find the orthogonal trajectories of the following families of curves:

   (a) $x^3 y - x y^3 = \alpha$,   (b) $e^{-x} \cos y + xy = \alpha$.

   Ans. (a) $x^4 - 6x^2 y^2 + y^4 = \beta$,   (b) $2e^{-x} \sin y + x^2 - y^2 = \beta$

87. Find the orthogonal trajectories of the family of curves $r^2 \cos 2\theta = \alpha$.   *Ans.* $r^2 \sin 2\theta = \beta$

88. By separating $f(z) = z + 1/z$ into real and imaginary parts, show that the families $(r^2 + 1) \cos \theta = \alpha r$ and $(r^2 - 1) \sin \theta = \beta r$ are orthogonal trajectories and verify this by another method.

89. If n is any real constant, prove that $r^n = \alpha \sec n\theta$ and $r^n = \beta \csc n\theta$ are orthogonal trajectories.

APPLICATIONS TO GEOMETRY AND MECHANICS

90. A particle moves along a curve $z = e^{-t}(2 \sin t + i \cos t)$.

   (a) Find a unit tangent vector to the curve at the point where $t = \pi/4$.

   (b) Determine the magnitudes of velocity and acceleration of the particle at $t = 0$ and $\pi/2$.

   Ans. (a) $\pm i$. (b) Velocity: $\sqrt{5}$, $\sqrt{5}\, e^{-\pi/2}$. Acceleration: 4, $2e^{-\pi/2}$

91. A particle moves along the curve $z = ae^{i\omega t}$. (a) Show that its speed is always constant and equal to ωa. (b) Show that the magnitude of its acceleration is always constant and equal to $\omega^2 a$. (c) Show that the acceleration is always directed toward $z = 0$. (d) Explain the relationship of this problem to the problem of a stone being twirled at the end of a string in a horizontal plane.

92. The position at time t of a particle moving in the z plane is given by $z = 3te^{-4it}$. Find the magnitudes of (a) the velocity, (b) the acceleration of the particle at $t = 0$ and $t = \pi$.

   Ans. (a) 3, $3\sqrt{1 + 16\pi^2}$. (b) 24, $24\sqrt{1 + 4\pi^2}$

93. A particle P moves along the line $x + y = 2$ in the z plane with a uniform speed of $3\sqrt{2}$ ft/sec from the point $z = -5 + 7i$ to $z = 10 - 8i$. If $w = 2z^2 - 3$ and P' is the image of P in the w plane, find the magnitudes of (a) the velocity and (b) the acceleration of P' after 3 seconds.

   Ans. (a) $24\sqrt{10}$, (b) 72

GRADIENT, DIVERGENCE, CURL AND LAPLACIAN

94. If $F = x^2y - xy^2$, find (a) ∇F, (b) $\nabla^2 F$.   Ans. (a) $(2xy - y^2) + i(x^2 - 2xy)$, (b) $2y - 2x$

95. Let $B = 3z^2 + 4\bar{z}$. Find (a) $\operatorname{grad} B$, (b) $\operatorname{div} B$, (c) $\operatorname{curl} B$, (d) Laplacian B.
    Ans. (a) 8, (b) $12x$, (c) $12y$, (d) 0

96. Let C be the curve in the xy plane defined by $x^2 - xy + y^2 = 7$. Find a unit vector normal to C at
    (a) the point $(-1, 2)$, (b) any point.
    Ans. (a) $(-4 + 5i)/\sqrt{41}$, (b) $\{2x - y + i(2y - x)\}/\sqrt{5x^2 - 8xy + 5y^2}$

97. Find an equation for the line normal to the curve $x^2y = 2xy + 6$ at the point $(3, 2)$.
    Ans. $x = 8t + 3$, $y = 3t + 2$

98. Show that $\nabla^2 |f(z)|^2 = 4 |f'(z)|^2$. Illustrate by choosing $f(z) = z^2 + iz$.

99. Prove $\nabla^2\{FG\} = F \nabla^2 G + G \nabla^2 F + 2 \nabla F \circ \nabla G$

100. Prove $\operatorname{div} \operatorname{grad} A = 0$ if A is imaginary or, more generally, if $\operatorname{Re}\{A\}$ is harmonic.

MISCELLANEOUS PROBLEMS

101. If $f(z) = u(x, y) + i v(x, y)$, prove that:
    (a) $f(z) = 2u(z/2, -iz/2) + \text{constant}$, (b) $f(z) = 2i v(z/2, -iz/2) + \text{constant}$.

102. Use Problem 101 to find $f(z)$ if (a) $u(x, y) = x^4 - 6x^2y^2 + y^4$, (b) $v(x, y) = \sinh x \cos y$.

103. If V is the instantaneous speed of a particle moving along any plane curve C, prove that the normal
    component of the acceleration at any point of C is given by V^2/R where R is the radius of curvature
    at the point.

104. Find an analytic function $f(z)$ such that $\operatorname{Re}\{f'(z)\} = 3x^2 - 4y - 3y^2$ and $f(1 + i) = 0$.
    Ans. $z^3 + 2iz^2 + 6 - 2i$

105. Show that the family of curves
$$\frac{x^2}{a^2 + \lambda} + \frac{y^2}{b^2 + \lambda} = 1$$
    with $-a^2 < \lambda < -b^2$ is orthogonal to the family with $\lambda > -b^2 > -a^2$.

106. Prove that the equation $F(x, y) = \text{constant}$ can be expressed as $u(x, y) = \text{constant}$ where u is
    harmonic if and only if $\dfrac{\partial^2 F/\partial x^2 + \partial^2 F/\partial y^2}{(\partial F/\partial x)^2 + (\partial F/\partial y)^2}$ is a function of F.

107. Illustrate the result in Problem 106 by considering $(y + 2)/(x - 1) = \text{constant}$.

108. If $f'(z) = 0$ in a region $\mathcal{R}$, prove that $f(z)$ must be a constant in $\mathcal{R}$.

109. If $w = f(z)$ is analytic and expressed in polar coordinates (r, θ), prove that
$$\frac{dw}{dz} = e^{-i\theta} \frac{\partial w}{\partial r}$$

110. If u and v are conjugate harmonic functions, prove that
$$dv = \frac{\partial u}{\partial x} dy - \frac{\partial u}{\partial y} dx$$

111. If u and v are harmonic in a region $\mathcal{R}$, prove that

$$\left(\frac{\partial u}{\partial y} - \frac{\partial v}{\partial x}\right) + i\left(\frac{\partial u}{\partial x} + \frac{\partial v}{\partial y}\right)$$

is analytic in $\mathcal{R}$.

112. Prove that $f(z) = |z|^4$ is differentiable but not analytic at $z = 0$.

113. Prove that $\psi = \ln|f(z)|$ is harmonic in a region $\mathcal{R}$ if $f(z)$ is analytic in $\mathcal{R}$ and $f(z) f'(z) \neq 0$ in $\mathcal{R}$.

114. Express the Cauchy-Riemann equations in terms of the curvilinear coordinates (ξ, η) where $x = e^\xi \cosh \eta$, $y = e^\xi \sinh \eta$.

115. Show that a solution of the differential equation

$$L\frac{d^2Q}{dt^2} + R\frac{dQ}{dt} + \frac{Q}{C} = E_0 \cos \omega t$$

where L, R, C, E_0 and ω are constants, is given by

$$Q = \text{Re}\left\{\frac{E_0 e^{i\omega t}}{i\omega[R + i(\omega L - 1/\omega C)]}\right\}$$

The equation arises in the *theory of alternating currents* of electricity.

[*Hint.* Rewrite the right hand side as $E_0 e^{i\omega t}$ and then assume a solution of the form $A e^{i\omega t}$ where A is to be determined.]

116. Show that $\nabla^2 \{f(z)\}^n = n^2 |f(z)|^{n-2} |f'(z)|^2$, stating restrictions on $f(z)$.

117. Solve the partial differential equation $\dfrac{\partial^2 U}{\partial x^2} + \dfrac{\partial^2 U}{\partial y^2} = \dfrac{8}{x^2 + y^2}$.

 Ans. $U = \frac{1}{2}\{\ln(x^2 + y^2)\}^2 + 2\{\tan^{-1}(y/x)\}^2 + F(x + iy) + G(x - iy)$

118. Prove that $\nabla^4 U = \nabla^2(\nabla^2 U) = \dfrac{\partial^4 U}{\partial x^4} + 2\dfrac{\partial^4 U}{\partial x^2 \partial y^2} + \dfrac{\partial^4 U}{\partial y^4} = 16\dfrac{\partial^4 U}{\partial z^2 \partial \bar{z}^2}$.

119. Solve the partial differential equation $\dfrac{\partial^4 U}{\partial x^4} + 2\dfrac{\partial^4 U}{\partial x^2 \partial y^2} + \dfrac{\partial^4 U}{\partial y^4} = 36(x^2 + y^2)$.

 Ans. $U = \frac{1}{16}(x^2 + y^2)^3 + (x + iy) F_1(x - iy) + G_1(x - iy) + (x - iy) F_2(x + iy) + G_2(x + iy)$

Chapter 4

Complex Integration and Cauchy's Theorem

COMPLEX LINE INTEGRALS

Let $f(z)$ be continuous at all points of a curve C [Fig. 4-1] which we shall assume has a finite length, i.e. C is a *rectifiable curve*.

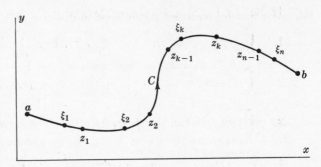

Fig. 4-1

Subdivide C into n parts by means of points $z_1, z_2, \ldots, z_{n-1}$, chosen arbitrarily, and call $a = z_0$, $b = z_n$. On each arc joining z_{k-1} to z_k [where k goes from 1 to n] choose a point ξ_k. Form the sum

$$S_n = f(\xi_1)(z_1 - a) + f(\xi_2)(z_2 - z_1) + \cdots + f(\xi_n)(b - z_{n-1}) \tag{1}$$

On writing $z_k - z_{k-1} = \Delta z_k$, this becomes

$$S_n = \sum_{k=1}^{n} f(\xi_k)(z_k - z_{k-1}) = \sum_{k=1}^{n} f(\xi_k)\,\Delta z_k \tag{2}$$

Let the number of subdivisions n increase in such a way that the largest of the chord lengths $|\Delta z_k|$ approaches zero. Then the sum S_n approaches a limit which does not depend on the mode of subdivision and we denote this limit by

$$\int_a^b f(z)\,dz \quad \text{or} \quad \int_C f(z)\,dz \tag{3}$$

called the *complex line integral* or briefly *line integral* of $f(z)$ along curve C, or the *definite integral* of $f(z)$ from a to b along curve C. In such case $f(z)$ is said to be *integrable* along C. Note that if $f(z)$ is analytic at all points of a region $\mathcal{R}$ and if C is a curve lying in $\mathcal{R}$, then $f(z)$ is certainly integrable along C.

REAL LINE INTEGRALS

If $P(x,y)$ and $Q(x,y)$ are real functions of x and y continuous at all points of curve C, the *real line integral* of $P\,dx + Q\,dy$ along curve C can be defined in a manner similar to that given above and is denoted by

$$\int_C [P(x,y)\,dx + Q(x,y)\,dy] \quad \text{or} \quad \int_C P\,dx + Q\,dy \tag{4}$$

the second notation being used for brevity. If C is smooth and has parametric equations $x = \phi(t)$, $y = \psi(t)$ where $t_1 \leqq t \leqq t_2$, the value of (4) is given by

$$\int_{t_1}^{t_2} [P\{\phi(t), \psi(t)\}\phi'(t)\,dt + Q\{\phi(t), \psi(t)\}\psi'(t)\,dt]$$

Suitable modifications can be made if C is piecewise smooth (see Problem 1).

92

CONNECTION BETWEEN REAL AND COMPLEX LINE INTEGRALS

If $f(z) = u(x,y) + i v(x,y) = u + iv$ the complex line integral (3) can be expressed in terms of real line integrals as

$$\int_C f(z)\,dz = \int_C (u+iv)(dx + i\,dy)$$

$$= \int_C u\,dx - v\,dy + i\int_C v\,dx + u\,dy \tag{5}$$

For this reason (5) is sometimes taken as a definition of a complex line integral.

PROPERTIES OF INTEGRALS

If $f(z)$ and $g(z)$ are integrable along C, then

1. $\displaystyle\int_C \{f(z) + g(z)\}\,dz = \int_C f(z)\,dz + \int_C g(z)\,dz$

2. $\displaystyle\int_C A\,f(z)\,dz = A\int_C f(z)\,dz$ where A = any constant

3. $\displaystyle\int_a^b f(z)\,dz = -\int_b^a f(z)\,dz$

4. $\displaystyle\int_a^b f(z)\,dz = \int_a^m f(z)\,dz + \int_m^b f(z)\,dz$ where points a, b, m are on C.

5. $\displaystyle\left|\int_C f(z)\,dz\right| \leq ML$

where $|f(z)| \leq M$, i.e. M is an *upper bound* of $|f(z)|$ on C, and L is the *length* of C.

There are various other ways in which the above properties can be described. For example if T, U and V are successive points on a curve, property 3 can be written $\displaystyle\int_{TUV} f(z)\,dz = -\int_{VUT} f(z)\,dz$.

Similarly if C, C_1 and C_2 represent curves from a to b, a to m and m to b respectively, it is natural for us to consider $C = C_1 + C_2$ and to write property 4 as

$$\int_{C_1 + C_2} f(z)\,dz = \int_{C_1} f(z)\,dz + \int_{C_2} f(z)\,dz$$

CHANGE OF VARIABLES

Let $z = g(\zeta)$ be a continuous function of a complex variable $\zeta = u + iv$. Suppose that curve C in the z plane corresponds to curve C' in the ζ plane and that the derivative $g'(\zeta)$ is continuous on C'. Then

$$\int_C f(z)\,dz = \int_{C'} f\{g(\zeta)\}\,g'(\zeta)\,d\zeta \tag{6}$$

These conditions are certainly satisfied if g is analytic in a region containing curve C'.

SIMPLY AND MULTIPLY CONNECTED REGIONS

A region $\mathcal{R}$ is called *simply-connected* if any simple closed curve [Page 68] which lies in $\mathcal{R}$ can be shrunk to a point without leaving $\mathcal{R}$. A region $\mathcal{R}$ which is not simply-connected is called *multiply-connected*.

For example, suppose $\mathcal{R}$ is the region defined by $|z| < 2$ shown shaded in Fig. 4-2. If Γ is any simple closed curve lying in $\mathcal{R}$ [i.e. whose points are in $\mathcal{R}$], we see that it can be shrunk to a point which lies in $\mathcal{R}$, and thus does not leave $\mathcal{R}$, so that $\mathcal{R}$ is simply-connected. On the other hand if $\mathcal{R}$ is the region defined by $1 < |z| < 2$, shown shaded in Fig. 4-3, then there is a simple closed curve Γ lying in $\mathcal{R}$ which cannot possibly be shrunk to a point without leaving $\mathcal{R}$, so that $\mathcal{R}$ is multiply-connected.

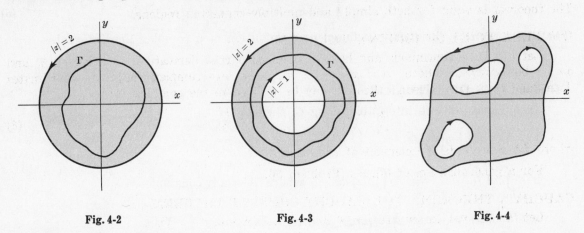

Fig. 4-2   Fig. 4-3   Fig. 4-4

Intuitively, a simply-connected region is one which does not have any "holes" in it, while a multiply-connected region is one which does. Thus the multiply-connected regions of Figures 4-3 and 4-4 have respectively one and three holes in them.

JORDAN CURVE THEOREM

Any continuous, closed curve which does not intersect itself and which may or may not have a finite length is called a *Jordan curve* [see Problem 30]. An important theorem which, although very difficult to prove, seems intuitively obvious is the following.

Jordan Curve Theorem. A Jordan curve divides the plane into two regions having the curve as common boundary. That region which is bounded [i.e. is such that all points of it satisfy $|z| < M$, where M is some positive constant] is called the *interior* or *inside* of the curve, while the other region is called the *exterior* or *outside* of the curve.

It follows from this that the region inside a simple closed curve is a simply-connected region whose boundary is the simple closed curve.

CONVENTION REGARDING TRAVERSAL OF A CLOSED PATH

The boundary C of a region is said to be traversed in the *positive sense* or *direction* if an observer travelling in this direction [and perpendicular to the plane] has the region to the left. This convention leads to the directions indicated by the arrows in Figures 4-2, 4-3 and 4-4. We use the special symbol

$$\oint_C f(z)\, dz$$

to denote integration of $f(z)$ around the boundary C in the positive sense. Note that in the case of a circle [Fig. 4-2] the positive direction is the *counterclockwise direction*. The integral around C is often called a *contour integral*.

GREEN'S THEOREM IN THE PLANE

Let $P(x, y)$ and $Q(x, y)$ be continuous and have continuous partial derivatives in a region $\mathcal{R}$ and on its boundary C. *Green's theorem* states that

$$\oint_C P\,dx \ + \ Q\,dy \ = \ \iint_{\mathcal{R}} \left(\frac{\partial Q}{\partial x} - \frac{\partial P}{\partial y} \right) dx\,dy \tag{7}$$

The theorem is valid for both simply and multiply-connected regions.

COMPLEX FORM OF GREEN'S THEOREM

Let $F(z, \bar{z})$ be continuous and have continuous partial derivatives in a region $\mathcal{R}$ and on its boundary C, where $z = x + iy$, $\bar{z} = x - iy$ are complex conjugate coordinates [see Page 7]. Then Green's theorem can be written in the complex form

$$\oint_C F(z, \bar{z})\,dz \ = \ 2i \iint_{\mathcal{R}} \frac{\partial F}{\partial \bar{z}}\,dA \tag{8}$$

where dA represents the element of area $dx\,dy$.

For a generalization of (8), see Problem 56.

CAUCHY'S THEOREM.   THE CAUCHY-GOURSAT THEOREM

Let $f(z)$ be analytic in a region $\mathcal{R}$ and on its boundary C. Then

$$\oint_C f(z)\,dz \ = \ 0 \tag{9}$$

This fundamental theorem, often called *Cauchy's integral theorem* or briefly *Cauchy's theorem,* is valid for both simply and multiply-connected regions. It was first proved by use of Green's theorem with the added restriction that $f'(z)$ be continuous in $\mathcal{R}$ [see Problem 11]. However, *Goursat* gave a proof which removed this restriction. For this reason the theorem is sometimes called the *Cauchy-Goursat theorem* [see Problems 13-16] when one desires to emphasize the removal of this restriction.

MORERA'S THEOREM

Let $f(z)$ be continuous in a simply-connected region $\mathcal{R}$ and suppose that

$$\oint_C f(z)\,dz \ = \ 0 \tag{10}$$

around *every* simple closed curve C in $\mathcal{R}$. Then $f(z)$ is analytic in $\mathcal{R}$.

This theorem, due to *Morera,* is often called the *converse of Cauchy's theorem.* It can be extended to multiply-connected regions. For a proof which assumes that $f'(z)$ is continuous in $\mathcal{R}$, see Problem 22. For a proof which eliminates this restriction, see Problem 7, Chapter 5.

INDEFINITE INTEGRALS

If $f(z)$ and $F(z)$ are analytic in a region $\mathcal{R}$ and such that $F'(z) = f(z)$, then $F(z)$ is called an *indefinite integral* or *anti-derivative* of $f(z)$ denoted by

$$F(z) \ = \ \int f(z)\,dz \tag{11}$$

Since the derivative of any constant is zero, it follows that any two indefinite integrals can differ by a constant. For this reason an arbitrary constant c is often added to the right of (11).

Example: Since $\frac{d}{dz}(3z^2 - 4\sin z) = 6z - 4\cos z$, we can write

$$\int (6z - 4\cos z)\,dz \ = \ 3z^2 - 4\sin z + c$$

INTEGRALS OF SPECIAL FUNCTIONS

Using results on Page 66 [or by direct differentiation], we can arrive at the following results (omitting a constant of integration).

1. $\int z^n\, dz = \dfrac{z^{n+1}}{n+1} \quad n \neq -1$

2. $\int \dfrac{dz}{z} = \ln z$

3. $\int e^z\, dz = e^z$

4. $\int a^z\, dz = \dfrac{a^z}{\ln a}$

5. $\int \sin z\, dz = -\cos z$

6. $\int \cos z\, dz = \sin z$

7. $\int \tan z\, dz = \ln \sec z$
   $= -\ln \cos z$

8. $\int \cot z\, dz = \ln \sin z$

9. $\int \sec z\, dz = \ln(\sec z + \tan z)$
   $= \ln \tan(z/2 + \pi/4)$

10. $\int \csc z\, dz = \ln(\csc z - \cot z)$
    $= \ln \tan(z/2)$

11. $\int \sec^2 z\, dz = \tan z$

12. $\int \csc^2 z\, dz = -\cot z$

13. $\int \sec z \tan z\, dz = \sec z$

14. $\int \csc z \cot z\, dz = -\csc z$

15. $\int \sinh z\, dz = \cosh z$

16. $\int \cosh z\, dz = \sinh z$

17. $\int \tanh z\, dz = \ln \cosh z$

18. $\int \coth z\, dz = \ln \sinh z$

19. $\int \operatorname{sech} z\, dz = \tan^{-1}(\sinh z)$

20. $\int \operatorname{csch} z\, dz = -\coth^{-1}(\cosh z)$

21. $\int \operatorname{sech}^2 z\, dz = \tanh z$

22. $\int \operatorname{csch}^2 z\, dz = -\coth z$

23. $\int \operatorname{sech} z \tanh z\, dz = -\operatorname{sech} z$

24. $\int \operatorname{csch} z \coth z\, dz = -\operatorname{csch} z$

25. $\int \dfrac{dz}{\sqrt{z^2 \pm a^2}} = \ln(z + \sqrt{z^2 \pm a^2})$

26. $\int \dfrac{dz}{z^2 + a^2} = \dfrac{1}{a}\tan^{-1}\dfrac{z}{a}$ or $-\dfrac{1}{a}\cot^{-1}\dfrac{z}{a}$

27. $\int \dfrac{dz}{z^2 - a^2} = \dfrac{1}{2a}\ln\left(\dfrac{z-a}{z+a}\right)$

28. $\int \dfrac{dz}{\sqrt{a^2 - z^2}} = \sin^{-1}\dfrac{z}{a}$ or $-\cos^{-1}\dfrac{z}{a}$

29. $\int \dfrac{dz}{z\sqrt{a^2 \pm z^2}} = \dfrac{1}{a}\ln\left(\dfrac{z}{a + \sqrt{a^2 \pm z^2}}\right)$

30. $\int \dfrac{dz}{z\sqrt{z^2 - a^2}} = \dfrac{1}{a}\cos^{-1}\dfrac{a}{z}$ or $\dfrac{1}{a}\sec^{-1}\dfrac{z}{a}$

31. $\int \sqrt{z^2 \pm a^2}\, dz = \dfrac{z}{2}\sqrt{z^2 \pm a^2}$
    $\pm \dfrac{a^2}{2}\ln(z + \sqrt{z^2 \pm a^2})$

32. $\int \sqrt{a^2 - z^2}\, dz = \dfrac{z}{2}\sqrt{a^2 - z^2} + \dfrac{a^2}{2}\sin^{-1}\dfrac{z}{a}$

33. $\int e^{az}\sin bz\, dz = \dfrac{e^{az}(a \sin bz - b \cos bz)}{a^2 + b^2}$

34. $\int e^{az}\cos bz\, dz = \dfrac{e^{az}(a \cos bz + b \sin bz)}{a^2 + b^2}$

SOME CONSEQUENCES OF CAUCHY'S THEOREM

Let $f(z)$ be analytic in a simply-connected region $\mathcal{R}$. Then the following theorems hold.

Theorem 1. If a and z are any two points in $\mathcal{R}$, then

$$\int_a^z f(z)\, dz$$

is *independent of the path* in $\mathcal{R}$ joining a and z.

Theorem 2. If a and z are any two points in $\mathcal{R}$ and

$$G(z) \;=\; \int_a^z f(z)\,dz \tag{12}$$

then $G(z)$ is analytic in $\mathcal{R}$ and $G'(z) = f(z)$.

Occasionally, confusion may arise because the variable of integration z in (12) is the same as the upper limit of integration. Since a definite integral depends only on the curve and limits of integration, any symbol can be used for the variable of integration, and for this reason we call it a *dummy variable* or *dummy symbol*. Thus (12) can be equivalently written

$$G(z) \;=\; \int_a^z f(\zeta)\,d\zeta \tag{13}$$

Theorem 3. If a and b are any two points in $\mathcal{R}$ and $F'(z) = f(z)$, then

$$\int_a^b f(z)\,dz \;=\; F(b) - F(a) \tag{14}$$

This can also be written in the form, familiar from elementary calculus,

$$\int_a^b F'(z)\,dz \;=\; F(z)\Big|_a^b \;=\; F(b) - F(a) \tag{15}$$

Example: $\displaystyle\int_{3i}^{1-i} 4z\,dz \;=\; 2z^2\Big|_{3i}^{1-i} \;=\; 2(1-i)^2 - 2(3i)^2 \;=\; 18 - 4i$

Theorem 4. Let $f(z)$ be analytic in a region bounded by two simple closed curves C and C_1 [where C_1 lies inside C as in Fig. 4-5 below] and on these curves. Then

$$\oint_C f(z)\,dz \;=\; \oint_{C_1} f(z)\,dz \tag{16}$$

where C and C_1 are both traversed in the positive sense relative to their interiors [counterclockwise in Fig. 4-5].

The result shows that if we wish to integrate $f(z)$ along curve C we can equivalently replace C by any curve C_1 so long as $f(z)$ is analytic in the region between C and C_1.

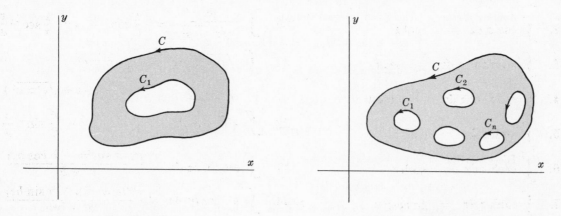

Fig. 4-5   Fig. 4-6

Theorem 5. Let $f(z)$ be analytic in a region bounded by the non-overlapping simple closed curves $C, C_1, C_2, C_3, \ldots, C_n$ [where $C_1, C_2, \ldots, C_n$ are inside C as in Fig. 4-6 above] and on these curves. Then

$$\oint_C f(z)\,dz \;=\; \oint_{C_1} f(z)\,dz + \oint_{C_2} f(z)\,dz + \cdots + \oint_{C_n} f(z)\,dz \tag{17}$$

This is a generalization of Theorem 4.

Solved Problems

LINE INTEGRALS

1. Evaluate $\displaystyle\int_{(0,3)}^{(2,4)} (2y + x^2)\, dx + (3x - y)\, dy$ along: (a) the parabola $x = 2t$, $y = t^2 + 3$;   (b) straight lines from $(0, 3)$ to $(2, 3)$ and then from $(2, 3)$ to $(2, 4)$;   (c) a straight line from $(0, 3)$ to $(2, 4)$.

(a) The points $(0, 3)$ and $(2, 4)$ on the parabola correspond to $t = 0$ and $t = 1$ respectively. Then the given integral equals

$$\int_{t=0}^{1} \{2(t^2 + 3) + (2t)^2\}\, 2\, dt + \{3(2t) - (t^2 + 3)\}\, 2t\, dt \;=\; \int_{0}^{1} (24t^2 + 12 - 2t^3 - 6t)\, dt \;=\; 33/2$$

(b) Along the straight line from $(0, 3)$ to $(2, 3)$, $y = 3$, $dy = 0$ and the line integral equals

$$\int_{x=0}^{2} (6 + x^2)\, dx + (3x - 3)0 \;=\; \int_{x=0}^{2} (6 + x^2)\, dx \;=\; 44/3$$

Along the straight line from $(2, 3)$ to $(2, 4)$, $x = 2$, $dx = 0$ and the line integral equals

$$\int_{y=3}^{4} (2y + 4)0 + (6 - y)\, dy \;=\; \int_{y=3}^{4} (6 - y)\, dy \;=\; 5/2$$

Then the required value $= 44/3 + 5/2 = 103/6$.

(c) An equation for the line joining $(0, 3)$ and $(2, 4)$ is $2y - x = 6$. Solving for x, we have $x = 2y - 6$. Then the line integral equals

$$\int_{y=3}^{4} \{2y + (2y - 6)^2\}\, 2\, dy + \{3(2y - 6) - y\}\, dy \;=\; \int_{3}^{4} (8y^2 - 39y + 54)\, dy \;=\; 97/6$$

The result can also be obtained by using $y = \tfrac{1}{2}(x + 6)$.

2. Evaluate $\displaystyle\int_{C} \bar{z}\, dz$ from $z = 0$ to $z = 4 + 2i$ along the curve C given by   (a) $z = t^2 + it$, (b) the line from $z = 0$ to $z = 2i$ and then the line from $z = 2i$ to $z = 4 + 2i$.

(a) The points $z = 0$ and $z = 4 + 2i$ on C correspond to $t = 0$ and $t = 2$ respectively. Then the line integral equals

$$\int_{t=0}^{2} (\overline{t^2 + it})\, d(t^2 + it) \;=\; \int_{0}^{2} (t^2 - it)(2t + i)\, dt \;=\; \int_{0}^{2} (2t^3 - it^2 + t)\, dt \;=\; 10 - 8i/3$$

Another Method. The given integral equals

$$\int_{C} (x - iy)(dx + i\, dy) \;=\; \int_{C} x\, dx + y\, dy \;+\; i\int_{C} x\, dy - y\, dx$$

The parametric equations of C are $x = t^2$, $y = t$ from $t = 0$ to $t = 2$. Then the line integral equals

$$\int_{t=0}^{2} (t^2)(2t\, dt) + (t)(dt) \;+\; i\int_{t=0}^{2} (t^2)(dt) - (t)(2t\, dt)$$

$$=\; \int_{0}^{2} (2t^3 + t)\, dt \;+\; i\int_{0}^{2} (-t^2)\, dt \;=\; 10 - 8i/3$$

(b) The given line integral equals

$$\int_{C} (x - iy)(dx + i\, dy) \;=\; \int_{C} x\, dx + y\, dy \;+\; i\int_{C} x\, dy - y\, dx$$

The line from $z = 0$ to $z = 2i$ is the same as the line from $(0, 0)$ to $(0, 2)$ for which $x = 0$, $dx = 0$ and the line integral equals

$$\int_{y=0}^{2} (0)(0) + y\, dy \;+\; i\int_{y=0}^{2} (0)(dy) - y(0) \;=\; \int_{y=0}^{2} y\, dy \;=\; 2$$

The line from $z = 2i$ to $z = 4 + 2i$ is the same as the line from $(0, 2)$ to $(4, 2)$ for which $y = 2$, $dy = 0$ and the line integral equals

$$\int_{x=0}^{4} x\, dx + 2 \cdot 0 \;+\; i\int_{x=0}^{4} x \cdot 0 - 2\, dx \;=\; \int_{0}^{4} x\, dx \;+\; i\int_{0}^{4} -2\, dx \;=\; 8 - 8i$$

Then the required value $= 2 + (8 - 8i) = 10 - 8i$.

3. Prove that if $f(z)$ is integrable along a curve C having finite length L and if there exists a positive number M such that $|f(z)| \leqq M$ on C, then

$$\left| \int_C f(z)\, dz \right| \leqq ML$$

By definition we have on using the notation of Page 92,

$$\int_C f(z)\, dz = \lim_{n \to \infty} \sum_{k=1}^{n} f(\xi_k)\, \Delta z_k \tag{1}$$

Now

$$\left. \begin{aligned} \left| \sum_{k=1}^{n} f(\xi_k)\, \Delta z_k \right| &\leqq \sum_{k=1}^{n} |f(\xi_k)|\, |\Delta z_k| \\[2mm] &\leqq M \sum_{k=1}^{n} |\Delta z_k| \\[2mm] &\leqq ML \end{aligned} \right\} \tag{2}$$

where we have used the facts that $|f(z)| \leqq M$ for all points z on C and that $\sum_{k=1}^{n} |\Delta z_k|$ represents the sum of all the chord lengths joining points z_{k-1} and z_k, where $k = 1, 2, \ldots, n$, and that this sum is not greater than the length of C.

Taking the limit of both sides of (2), using (1), the required result follows.

It is possible to show, more generally, that

$$\left| \int_C f(z)\, dz \right| \leqq \int_C |f(z)|\, |dz|$$

GREEN'S THEOREM IN THE PLANE

4. Prove Green's theorem in the plane if C is a simple closed curve which has the property that any straight line parallel to the coordinate axes cuts C in at most two points.

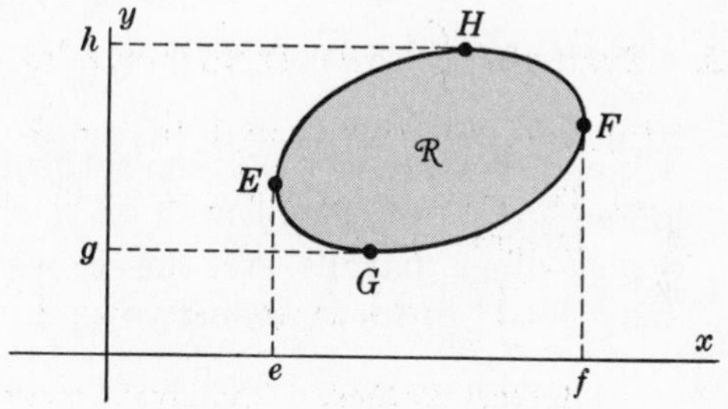

Let the equations of the curves EGF and EHF (see Fig. 4-7) be $y = Y_1(x)$ and $y = Y_2(x)$ respectively. If $\mathcal{R}$ is the region bounded by C, we have

Fig. 4-7

$$\iint_{\mathcal{R}} \frac{\partial P}{\partial y}\, dx\, dy = \int_{x=e}^{f} \left[\int_{y=Y_1(x)}^{Y_2(x)} \frac{\partial P}{\partial y}\, dy \right] dx$$

$$= \int_{x=e}^{f} P(x, y)\Big|_{y=Y_1(x)}^{Y_2(x)} dx = \int_e^f [P(x, Y_2) - P(x, Y_1)]\, dx$$

$$= -\int_e^f P(x, Y_1)\, dx - \int_f^e P(x, Y_2)\, dx = -\oint_C P\, dx$$

Then

$$\oint_C P\, dx = -\iint_{\mathcal{R}} \frac{\partial P}{\partial y}\, dx\, dy \tag{1}$$

Similarly let the equations of curves GEH and GFH be $x = X_1(y)$ and $x = X_2(y)$ respectively. Then

$$\iint_{\mathcal{R}} \frac{\partial Q}{\partial x}\, dx\, dy = \int_{y=g}^{h} \left[\int_{x=X_1(y)}^{X_2(y)} \frac{\partial Q}{\partial x}\, dx \right] dy = \int_g^h [Q(X_2, y) - Q(X_1, y)]\, dy$$

$$= \int_h^g Q(X_1, y)\, dy + \int_g^h Q(X_2, y)\, dy = \oint_C Q\, dy$$

Then

$$\oint_C Q\, dy = \iint_{\mathcal{R}} \frac{\partial Q}{\partial x}\, dx\, dy \tag{2}$$

Adding (1) and (2),

$$\oint_C P\, dx + Q\, dy = \iint_{\mathcal{R}} \left(\frac{\partial Q}{\partial x} - \frac{\partial P}{\partial y} \right) dx\, dy$$

5. Verify Green's theorem in the plane for

$$\oint_C (2xy - x^2)\,dx + (x+y^2)\,dy$$

where C is the closed curve of the region bounded by $y = x^2$ and $y^2 = x$.

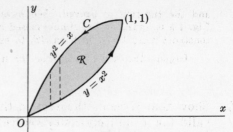

Fig. 4-8

The plane curves $y = x^2$ and $y^2 = x$ intersect at $(0,0)$ and $(1,1)$. The positive direction in traversing C is as shown in Fig. 4-8.

Along $y = x^2$, the line integral equals

$$\int_{x=0}^{1} \{(2x)(x^2) - x^2\}\,dx + \{x + (x^2)^2\}\,d(x^2) \;=\; \int_0^1 (2x^3 + x^2 + 2x^5)\,dx \;=\; 7/6$$

Along $y^2 = x$, the line integral equals

$$\int_{y=1}^{0} \{2(y^2)(y) - (y^2)^2\}\,d(y^2) + \{y^2 + y^2\}\,dy \;=\; \int_1^0 (4y^4 - 2y^5 + 2y^2)\,dy \;=\; -17/15$$

Then the required integral $= 7/6 - 17/15 = 1/30$.

$$\iint_R \left(\frac{\partial Q}{\partial x} - \frac{\partial P}{\partial y}\right) dx\,dy \;=\; \iint_R \left\{\frac{\partial}{\partial x}(x+y^2) - \frac{\partial}{\partial y}(2xy - x^2)\right\} dx\,dy$$

$$=\; \iint_R (1-2x)\,dx\,dy \;=\; \int_{x=0}^1 \int_{y=x^2}^{\sqrt{x}} (1-2x)\,dy\,dx$$

$$=\; \int_{x=0}^1 (y - 2xy)\Big|_{y=x^2}^{\sqrt{x}} dx \;=\; \int_0^1 (x^{1/2} - 2x^{3/2} - x^2 + 2x^3)\,dx \;=\; 1/30$$

Hence Green's theorem is verified.

6. Extend the proof of Green's theorem in the plane given in Problem 4 to curves C for which lines parallel to the coordinate axes may cut C in more than two points.

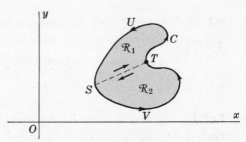

Fig. 4-9

Consider a simple closed curve C such as shown in Fig. 4-9 in which lines parallel to the axes may meet C in more than two points. By constructing line ST the region is divided into two regions R_1 and R_2 which are of the type considered in Problem 4 and for which Green's theorem applies, i.e.,

$$(1)\quad \int_{STUS} P\,dx + Q\,dy \;=\; \iint_{R_1} \left(\frac{\partial Q}{\partial x} - \frac{\partial P}{\partial y}\right) dx\,dy, \qquad (2)\quad \int_{SVTS} P\,dx + Q\,dy \;=\; \iint_R \left(\frac{\partial Q}{\partial x} - \frac{\partial P}{\partial y}\right) dx\,dy$$

Adding the left hand sides of (1) and (2), we have, omitting the integrand $P\,dx + Q\,dy$ in each case,

$$\int_{STUS} + \int_{SVTS} \;=\; \int_{ST} + \int_{TUS} + \int_{SVT} + \int_{TS} \;=\; \int_{TUS} + \int_{SVT} \;=\; \int_{TUSVT}$$

using the fact that $\displaystyle\int_{ST} = -\int_{TS}$.

Adding the right hand sides of (1) and (2), omitting the integrand,

$$\iint_{R_1} + \iint_{R_2} \;=\; \iint_R$$

Then

$$\int_{TUSVT} P\,dx + Q\,dy \;=\; \iint_R \left(\frac{\partial Q}{\partial x} - \frac{\partial P}{\partial y}\right) dx\,dy$$

and the theorem is proved. We have proved Green's theorem for the simply-connected region of Fig. 4-9 bounded by the simple closed curve C. For more complicated regions it may be necessary to construct more lines, such as ST, to establish the theorem.

Green's theorem is also true for multiply-connected regions, as shown in Problem 7.

7. Show that Green's theorem in the plane is also valid for a multiply-connected region $\mathcal{R}$ such as shown shaded in Fig. 4-10.

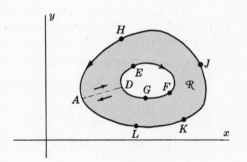

Fig. 4-10

The boundary of $\mathcal{R}$, which consists of the exterior boundary $AHJKLA$ and the interior boundary $DEFGD$, is to be traversed in the positive direction so that a person traveling in this direction always has the region on his left. It is seen that the positive directions are as indicated in the figure.

In order to establish the theorem construct a line, such as AD, called a *cross-cut*, connecting the exterior and interior boundaries. The region bounded by $ADEFGDALKJHA$ is simply-connected, and so Green's theorem is valid. Then

$$\oint_{ADEFGDALKJHA} P\,dx + Q\,dy = \iint_{\mathcal{R}} \left(\frac{\partial Q}{\partial x} - \frac{\partial P}{\partial y}\right) dx\,dy$$

But the integral on the left, leaving out the integrand, is equal to

$$\int_{AD} + \int_{DEFGD} + \int_{DA} + \int_{ALKJHA} = \int_{DEFGD} + \int_{ALKJHA}$$

since $\displaystyle\int_{AD} = -\int_{DA}$. Thus if C_1 is the curve $ALKJHA$, C_2 is the curve $DEFGD$ and C is the boundary of $\mathcal{R}$ consisting of C_1 and C_2 (traversed in the positive directions with respect to $\mathcal{R}$), then $\displaystyle\int_{C_1} + \int_{C_2} = \oint_{C}$ and so

$$\oint_{C} P\,dx + Q\,dy = \iint_{\mathcal{R}} \left(\frac{\partial Q}{\partial x} - \frac{\partial P}{\partial y}\right) dx\,dy$$

8. Let $P(x, y)$ and $Q(x, y)$ be continuous and have continuous first partial derivatives at each point of a simply-connected region $\mathcal{R}$. Prove that a necessary and sufficient condition that $\oint_{C} P\,dx + Q\,dy = 0$ around every closed path C in $\mathcal{R}$ is that $\partial P/\partial y = \partial Q/\partial x$ identically in $\mathcal{R}$.

Sufficiency. Suppose $\partial P/\partial y = \partial Q/\partial x$. Then by Green's theorem,

$$\oint_{C} P\,dx + Q\,dy = \iint_{\mathcal{R}} \left(\frac{\partial Q}{\partial x} - \frac{\partial P}{\partial y}\right) dx\,dy = 0$$

where $\mathcal{R}$ is the region bounded by C.

Necessity.

Suppose $\oint_{C} P\,dx + Q\,dy = 0$ around every closed path C in $\mathcal{R}$ and that $\partial P/\partial y \neq \partial Q/\partial x$ at some point of $\mathcal{R}$. In particular suppose $\partial P/\partial y - \partial Q/\partial x > 0$ at the point (x_0, y_0).

By hypothesis $\partial P/\partial y$ and $\partial Q/\partial x$ are continuous in $\mathcal{R}$ so that there must be some region τ containing (x_0, y_0) as an interior point for which $\partial P/\partial y - \partial Q/\partial x > 0$. If Γ is the boundary of τ, then by Green's theorem

$$\oint_{\Gamma} P\,dx + Q\,dy = \iint_{\tau} \left(\frac{\partial Q}{\partial x} - \frac{\partial P}{\partial y}\right) dx\,dy > 0$$

contradicting the hypothesis that $\oint_C P\,dx + Q\,dy = 0$ for *all* closed curves in $\mathcal{R}$. Thus $\partial Q/\partial x - \partial P/\partial y$ cannot be positive.

Similarly we can show that $\partial Q/\partial x - \partial P/\partial y$ cannot be negative and it follows that it must be identically zero, i.e. $\partial P/\partial y = \partial Q/\partial x$ identically in $\mathcal{R}$.

The results can be extended to multiply-connected regions.

9. Let P and Q be defined as in Problem 8. Prove that a necessary and sufficient condition that $\int_A^B P\,dx + Q\,dy$ be independent of the path in $\mathcal{R}$ joining points A and B is that $\partial P/\partial y = \partial Q/\partial x$ identically in $\mathcal{R}$.

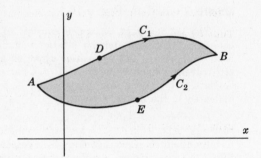

Fig. 4-11

Sufficiency. If $\partial P/\partial y = \partial Q/\partial x$, then by Problem 8

$$\int_{ADBEA} P\,dx + Q\,dy \;=\; 0$$

[see Fig. 4-11]. From this, omitting for brevity the integrand $P\,dx + Q\,dy$, we have

$$\int_{ADB} + \int_{BEA} = 0, \qquad \int_{ADB} = -\int_{BEA} = \int_{AEB} \qquad \text{and so} \qquad \int_{C_1} = \int_{C_2}$$

i.e. the integral is independent of the path.

Necessity.

If the integral is independent of the path, then for all paths C_1 and C_2 in $\mathcal{R}$ we have

$$\int_{C_1} = \int_{C_2}, \qquad \int_{ADB} = \int_{AEB} \qquad \text{and} \qquad \int_{ADBEA} = 0$$

From this it follows that the line integral around any closed path in $\mathcal{R}$ is zero, and hence by Problem 8 that $\partial P/\partial y = \partial Q/\partial x$.

The results can be extended to multiply-connected regions.

COMPLEX FORM OF GREEN'S THEOREM

10. If $B(z, \bar{z})$ is continuous and has continuous partial derivatives in a region $\mathcal{R}$ and on its boundary C, where $z = x + iy$ and $\bar{z} = x - iy$, prove that Green's theorem can be written in complex form as

$$\oint_C B(z, \bar{z})\,dz \;=\; 2i \iint_{\mathcal{R}} \frac{\partial B}{\partial \bar{z}}\,dx\,dy$$

Let $B(z, \bar{z}) = P(x, y) + iQ(x, y)$. Then using Green's theorem, we have

$$\oint_C B(z, \bar{z})\,dz \;=\; \oint_C (P + iQ)(dx + i\,dy) \;=\; \oint_C P\,dx - Q\,dy \;+\; i\oint_C Q\,dx + P\,dy$$

$$=\; -\iint_{\mathcal{R}} \left(\frac{\partial Q}{\partial x} + \frac{\partial P}{\partial y} \right) dx\,dy \;+\; i\iint_{\mathcal{R}} \left(\frac{\partial P}{\partial x} - \frac{\partial Q}{\partial y} \right) dx\,dy$$

$$=\; i\iint_{\mathcal{R}} \left[\left(\frac{\partial P}{\partial x} - \frac{\partial Q}{\partial y} \right) + i\left(\frac{\partial P}{\partial y} + \frac{\partial Q}{\partial x} \right) \right] dx\,dy$$

$$=\; 2i \iint_{\mathcal{R}} \frac{\partial B}{\partial \bar{z}}\,dx\,dy$$

from Problem 34, Page 83. The result can also be written in terms of curl B [see Page 70].

CAUCHY'S THEOREM AND THE CAUCHY-GOURSAT THEOREM

11. Prove Cauchy's theorem $\oint_C f(z)\,dz = 0$ if $f(z)$ is analytic with derivative $f'(z)$ which is continuous at all points inside and on a simple closed curve C.

Since $f(z) = u + iv$ is analytic and has a continuous derivative

$$f'(z) \;=\; \frac{\partial u}{\partial x} + i\frac{\partial v}{\partial x} \;=\; \frac{\partial v}{\partial y} - i\frac{\partial u}{\partial y}$$

it follows that the partial derivatives $(1)\ \dfrac{\partial u}{\partial x} = \dfrac{\partial v}{\partial y}$, $(2)\ \dfrac{\partial v}{\partial x} = -\dfrac{\partial u}{\partial y}$ are continuous inside and on C. Thus Green's theorem can be applied and we have

$$\oint_C f(z)\,dz \;=\; \oint_C (u+iv)(dx+i\,dy) \;=\; \oint_C u\,dx - v\,dy \;+\; i\oint_C v\,dx + u\,dy$$

$$=\; \iint_{\mathcal{R}} \left(-\frac{\partial v}{\partial x} - \frac{\partial u}{\partial y}\right) dx\,dy \;+\; i\iint_{\mathcal{R}} \left(\frac{\partial u}{\partial x} - \frac{\partial v}{\partial y}\right) dx\,dy \;=\; 0$$

using the Cauchy-Riemann equations (1) and (2).

By using the fact that Green's theorem is applicable to multiply-connected regions, we can extend the result to multiply-connected regions under the given conditions on $f(z)$.

The *Cauchy-Goursat theorem* [see Problems 13-16] removes the restriction that $f'(z)$ be continuous.

Another method.

The result can be obtained from the complex form of Green's theorem [Problem 10] by noting that if $B(z,\bar{z}) = f(z)$ is independent of $\bar{z}$, then $\partial B/\partial \bar{z} = 0$ and so $\oint_C f(z)\,dz = 0$.

12. Prove $(a)\ \oint_C dz = 0,\quad (b)\ \oint_C z\,dz = 0,\quad (c)\ \oint_C (z-z_0)\,dz = 0$ where C is any simple closed curve and z_0 is a constant.

These follow at once from Cauchy's theorem since the functions 1, z and $z - z_0$ are analytic inside C and have continuous derivatives.

The results can also be established directly from the definition of an integral (see Problem 90).

13. Prove the *Cauchy-Goursat* theorem for the case of a triangle.

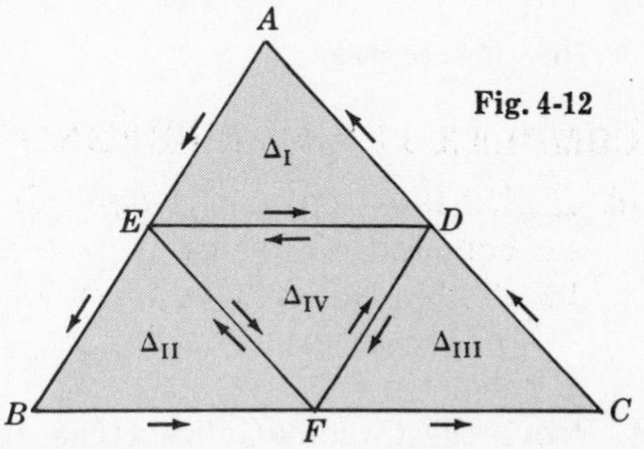

Consider any triangle in the z plane such as ABC, denoted briefly by Δ, in Fig. 4-12. Join the midpoints D, E and F of sides AB, AC and BC respectively to form four triangles indicated briefly by Δ_I, Δ_II, Δ_III and Δ_IV.

If $f(z)$ is analytic inside and on triangle ABC we have, omitting the integrand on the right,

Fig. 4-12

$$\oint_{ABCA} f(z)\,dz \;=\; \int_{DAE} + \int_{EBF} + \int_{FCD}$$

$$=\; \left\{\int_{DAE} + \int_{ED}\right\} + \left\{\int_{EBF} + \int_{FE}\right\} + \left\{\int_{FCD} + \int_{DF}\right\} + \left\{\int_{DE} + \int_{EF} + \int_{FD}\right\}$$

$$=\; \int_{DAED} + \int_{EBFE} + \int_{FCDF} + \int_{DEFD}$$

$$=\; \oint_{\Delta_\mathrm{I}} f(z)\,dz \;+\; \oint_{\Delta_\mathrm{II}} f(z)\,dz \;+\; \oint_{\Delta_\mathrm{III}} f(z)\,dz \;+\; \oint_{\Delta_\mathrm{IV}} f(z)\,dz$$

where in the second line we have made use of the fact that

$$\int_{ED} = -\int_{DE}\,, \qquad \int_{FE} = -\int_{EF}\,, \qquad \int_{DF} = -\int_{FD}$$

Then

$$\left| \oint_{\Delta} f(z)\, dz \right| \;\leqq\; \left| \oint_{\Delta_{\mathrm{I}}} f(z)\, dz \right| + \left| \oint_{\Delta_{\mathrm{II}}} f(z)\, dz \right| + \left| \oint_{\Delta_{\mathrm{III}}} f(z)\, dz \right| + \left| \oint_{\Delta_{\mathrm{IV}}} f(z)\, dz \right| \tag{1}$$

Let Δ_1 be the triangle corresponding to that term on the right of (1) having largest value (if there are two or more such terms then Δ_1 is any of the associated triangles). Then

$$\left| \oint_{\Delta} f(z)\, dz \right| \;\leqq\; 4 \left| \oint_{\Delta_1} f(z)\, dz \right| \tag{2}$$

By joining midpoints of the sides of triangle Δ_1, we obtain similarly a triangle Δ_2 such that

$$\left| \oint_{\Delta_1} f(z)\, dz \right| \;\leqq\; 4 \left| \oint_{\Delta_2} f(z)\, dz \right| \tag{3}$$

so that

$$\left| \oint_{\Delta} f(z)\, dz \right| \;\leqq\; 4^2 \left| \oint_{\Delta_2} f(z)\, dz \right| \tag{4}$$

After n steps we obtain a triangle Δ_n such that

$$\left| \oint_{\Delta} f(z)\, dz \right| \;\leqq\; 4^n \left| \oint_{\Delta_n} f(z)\, dz \right| \tag{5}$$

Now $\Delta, \Delta_1, \Delta_2, \Delta_3, \ldots$ is a sequence of triangles each of which is contained in the preceding (i.e. a sequence of *nested triangles*) and there exists a point z_0 which lies in every triangle of the sequence.

Since z_0 lies inside or on the boundary of Δ, it follows that $f(z)$ is analytic at z_0. Then by Problem 21, Page 78,

$$f(z) \;=\; f(z_0) + f'(z_0)(z - z_0) + \eta(z - z_0) \tag{6}$$

where for any $\epsilon > 0$ we can find δ such that $|\eta| < \epsilon$ whenever $|z - z_0| < \delta$.

Thus by integration of both sides of (6) and using Problem 12,

$$\oint_{\Delta_n} f(z)\, dz \;=\; \oint_{\Delta_n} \eta(z - z_0)\, dz \tag{7}$$

Now if P is the perimeter of Δ, then the perimeter of Δ_n is $P_n = P/2^n$. If z is any point on Δ_n, then as seen from Fig. 4-13 we must have $|z - z_0| < P/2^n < \delta$. Hence from ($7$) and Property 5, Page 93 we have

$$\left| \oint_{\Delta_n} f(z)\, dz \right| \;=\; \left| \oint_{\Delta_n} \eta(z - z_0)\, dz \right| \;\leqq\; \epsilon \cdot \frac{P}{2^n} \cdot \frac{P}{2^n} \;=\; \frac{\epsilon P^2}{4^n}$$

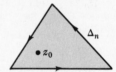

Then (5) becomes

$$\left| \oint_{\Delta} f(z)\, dz \right| \;\leqq\; 4^n \cdot \frac{\epsilon P^2}{4^n} \;=\; \epsilon P^2$$

Fig. 4-13

Since ϵ can be made arbitrarily small it follows that, as required,

$$\oint_{\Delta} f(z)\, dz \;=\; 0$$

14. **Prove the Cauchy-Goursat theorem for any closed polygon.**

Consider for example a closed polygon $ABCDEFA$ such as indicated in Fig. 4-14. By constructing the lines BF, CF and DF the polygon is subdivided into triangles. Then by Cauchy's theorem for triangles [Problem 13] and the fact that the integrals along BF and FB, CF and FC, DF and FD cancel, we find as required

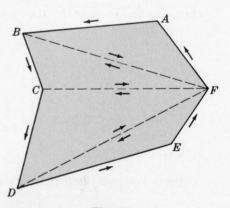

$$\int_{ABCDEFA} f(z)\, dz \;=\; \int_{ABFA} f(z)\, dz \;+\; \int_{BCFB} f(z)\, dz$$

$$+\; \int_{CDFC} f(z)\, dz \;+\; \int_{DEFD} f(z)\, dz$$

$$=\; 0$$

Fig. 4-14

where we suppose that $f(z)$ is analytic inside and on the polygon.

It should be noted that we have proved the result for simple polygons whose sides do not cross. A proof can also be given for any polygon which intersects itself (see Problem 66).

15. Prove the Cauchy-Goursat theorem for any simple closed curve.

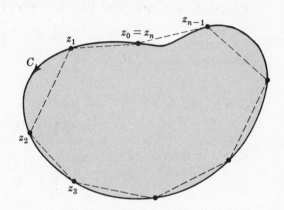

Fig. 4-15

Let us assume that C is contained in a region $\mathcal{R}$ in which $f(z)$ is analytic.

Choose n points of subdivision $z_1, z_2, \ldots, z_n$ on curve C [Fig. 4-15] where for convenience of notation we consider $z_0 = z_n$. Construct polygon P by joining these points.

Let us define the sum

$$S_n = \sum_{k=1}^{n} f(z_k)\, \Delta z_k$$

where $\Delta z_k = z_k - z_{k-1}$. Since

$$\lim S_n = \oint_C f(z)\, dz$$

[where the limit on the left means that $n \to \infty$ in such a way that the largest of $|\Delta z_k| \to 0$], it follows that given any $\epsilon > 0$ we can choose N so that for $n > N$

$$\left| \oint_C f(z)\, dz - S_n \right| < \frac{\epsilon}{2} \tag{1}$$

Consider now the integral along polygon P. Since this is zero by Problem 14, we have

$$\oint_P f(z)\, dz = 0 = \int_{z_0}^{z_1} f(z)\, dz + \int_{z_1}^{z_2} f(z)\, dz + \cdots + \int_{z_{n-1}}^{z_n} f(z)\, dz$$

$$= \int_{z_0}^{z_1} \{f(z) - f(z_1) + f(z_1)\}\, dz + \cdots + \int_{z_{n-1}}^{z_n} \{f(z) - f(z_n) + f(z_n)\}\, dz$$

$$= \int_{z_0}^{z_1} \{f(z) - f(z_1)\}\, dz + \cdots + \int_{z_{n-1}}^{z_n} \{f(z) - f(z_n)\}\, dz + S_n$$

so that

$$S_n = \int_{z_0}^{z_1} \{f(z_1) - f(z)\}\, dz + \cdots + \int_{z_{n-1}}^{z_n} \{f(z_n) - f(z)\}\, dz \tag{2}$$

Let us now choose N so large that on the lines joining z_0 and z_1, z_1 and z_2, $\ldots$, z_{n-1} and z_n,

$$|f(z_1) - f(z)| < \frac{\epsilon}{2L}, \quad |f(z_2) - f(z)| < \frac{\epsilon}{2L}, \quad \ldots, \quad |f(z_n) - f(z)| < \frac{\epsilon}{2L} \tag{3}$$

where L is the length of C. Then from (2) and (3) we have

$$|S_n| \leq \left| \int_{z_0}^{z_1} \{f(z_1) - f(z)\}\, dz \right| + \left| \int_{z_1}^{z_2} \{f(z_2) - f(z)\}\, dz \right| + \cdots + \left| \int_{z_{n-1}}^{z_n} \{f(z_n) - f(z)\}\, dz \right|$$

or

$$|S_n| \leq \frac{\epsilon}{2L} \{|z_1 - z_0| + |z_2 - z_1| + \cdots + |z_n - z_{n-1}|\} = \frac{\epsilon}{2} \tag{4}$$

From

$$\oint_C f(z)\, dz = \oint_C f(z)\, dz - S_n + S_n$$

we have, using (1) and (4),

$$\left| \oint_C f(z)\, dz \right| \leq \left| \oint_C f(z)\, dz - S_n \right| + |S_n| < \frac{\epsilon}{2} + \frac{\epsilon}{2} = \epsilon$$

Thus since ϵ is arbitrary, it follows that $\oint_C f(z)\, dz = 0$ as required.

16. Prove the Cauchy-Goursat theorem for multiply-connected regions.

We shall present a proof for the multiply-connected region $\mathcal{R}$ bounded by the simple closed curves C_1 and C_2 as indicated in Fig. 4-16. Extensions to other multiply-connected regions are easily made (see Problem 67).

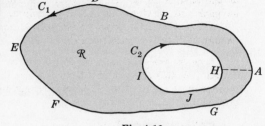

Fig. 4-16

Construct cross-cut AH. Then the region bounded by $ABDEFGAHJIHA$ is simply-connected so that by Problem 15,

$$\oint_{ABDEFGAHJIHA} f(z)\,dz = 0$$

Hence

$$\int_{ABDEFGA} f(z)\,dz + \int_{AH} f(z)\,dz + \int_{HJIH} f(z)\,dz + \int_{HA} f(z)\,dz = 0$$

Since $\displaystyle\int_{AH} f(z)\,dz = -\int_{HA} f(z)\,dz$, this becomes

$$\int_{ABDEFGA} f(z)\,dz + \int_{HJIH} f(z)\,dz = 0$$

This however amounts to saying that

$$\oint_C f(z)\,dz = 0$$

where C is the complete boundary of $\mathcal{R}$ (consisting of $ABDEFGA$ and $HJIH$) traversed in the sense that an observer walking on the boundary always has the region $\mathcal{R}$ on his left.

CONSEQUENCES OF CAUCHY'S THEOREM

17. If $f(z)$ is analytic in a simply-connected region $\mathcal{R}$, prove that $\displaystyle\int_a^b f(z)\,dz$ is independent of the path in $\mathcal{R}$ joining any two points a and b in $\mathcal{R}$.

By Cauchy's theorem,

$$\int_{ADBEA} f(z)\,dz = 0$$

or

$$\int_{ADB} f(z)\,dz + \int_{BEA} f(z)\,dz = 0$$

Hence

$$\int_{ADB} f(z)\,dz = -\int_{BEA} f(z)\,dz = \int_{AEB} f(z)\,dz$$

Thus

$$\int_{C_1} f(z)\,dz = \int_{C_2} f(z)\,dz = \int_a^b f(z)\,dz$$

which yields the required result.

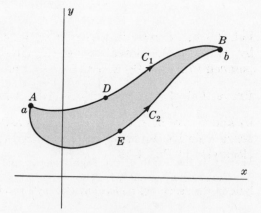

Fig. 4-17

18. Let $f(z)$ be analytic in a simply-connected region $\mathcal{R}$ and let a and z be points in $\mathcal{R}$. Prove that (a) $F(z) = \displaystyle\int_a^z f(u)\,du$ is analytic in $\mathcal{R}$ and (b) $F'(z) = f(z)$.

We have

$$\frac{F(z+\Delta z) - F(z)}{\Delta z} - f(z) = \frac{1}{\Delta z}\left\{\int_a^{z+\Delta z} f(u)\,du - \int_a^z f(u)\,du\right\} - f(z)$$

$$= \frac{1}{\Delta z}\int_z^{z+\Delta z} [f(u) - f(z)]\,du \tag{1}$$

By Cauchy's theorem, the last integral is independent of the path joining z and $z + \Delta z$ so long as the path is in $\mathcal{R}$. In particular we can choose as path the straight line segment joining z and $z + \Delta z$ (see Fig. 4-18) provided we choose $|\Delta z|$ small enough so that this path lies in $\mathcal{R}$.

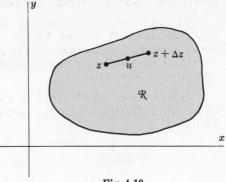

Now by the continuity of $f(z)$ we have for all points u on this straight line path $|f(u) - f(z)| < \epsilon$ whenever $|u - z| < \delta$, which will certainly be true if $|\Delta z| < \delta$.

Furthermore, we have

$$\left| \int_z^{z+\Delta z} [f(u) - f(z)]\, du \right| < \epsilon |\Delta z| \qquad (2)$$

Fig. 4-18

so that from (1)

$$\left| \frac{F(z + \Delta z) - F(z)}{\Delta z} - f(z) \right| = \frac{1}{|\Delta z|} \left| \int_z^{z+\Delta z} [f(u) - f(z)]\, du \right| < \epsilon$$

for $|\Delta z| < \delta$. This, however, amounts to saying that $\displaystyle\lim_{\Delta z \to 0} \frac{F(z + \Delta z) - F(z)}{\Delta z} = f(z)$, i.e. $F(z)$ is analytic and $F'(z) = f(z)$.

19. A function $F(z)$ such that $F'(z) = f(z)$ is called an *indefinite integral* of $f(z)$ and is denoted by $\displaystyle\int f(z)\, dz$. Show that (a) $\displaystyle\int \sin z\, dz = -\cos z + c$, (b) $\displaystyle\int \frac{dz}{z} = \ln z + c$ where c is an arbitrary constant.

(a) Since $\dfrac{d}{dz}(-\cos z + c) = \sin z$, we have $\displaystyle\int \sin z\, dz = -\cos z + c$.

(b) Since $\dfrac{d}{dz}(\ln z + c) = \dfrac{1}{z}$, we have $\displaystyle\int \frac{dz}{z} = \ln z + c$.

20. Let $f(z)$ be analytic in a region $\mathcal{R}$ bounded by two simple closed curves C_1 and C_2 [shaded in Fig. 4-19] and also on C_1 and C_2.

Prove that $\displaystyle\oint_{C_1} f(z)\, dz = \oint_{C_2} f(z)\, dz$, where C_1 and C_2 are both traversed in the positive sense relative to their interiors [counterclockwise in Fig. 4-19].

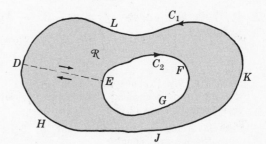

Construct cross-cut DE. Then since $f(z)$ is analytic in the region $\mathcal{R}$, we have by Cauchy's theorem

Fig. 4-19

$$\int_{DEFGEDHJKLD} f(z)\, dz = 0$$

or

$$\int_{DE} f(z)\, dz + \int_{EFGE} f(z)\, dz + \int_{ED} f(z)\, dz + \int_{DHJKLD} f(z)\, dz = 0$$

Hence since $\displaystyle\int_{DE} f(z)\, dz = -\int_{ED} f(z)\, dz$,

$$\int_{DHJKLD} f(z)\, dz = -\int_{EFGE} f(z)\, dz = \int_{EGFE} f(z)\, dz \qquad \text{or} \qquad \oint_{C_1} f(z)\, dz = \oint_{C_2} f(z)\, dz$$

21. Evaluate $\displaystyle\oint_C \frac{dz}{z-a}$ where C is any simple closed curve C and $z=a$ is   (a) outside C,

(b) inside C.

(a) If a is outside C, then $f(z) = 1/(z-a)$ is analytic everywhere inside and on C. Hence by Cauchy's theorem,

$$\oint_C \frac{dz}{z-a} = 0.$$

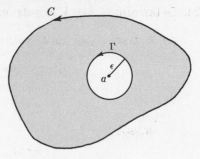

Fig. 4-20

(b) Suppose a is inside C and let Γ be a circle of radius ϵ with center at $z=a$ so that Γ is inside C [this can be done since $z=a$ is an interior point].

By Problem 20,

$$\oint_C \frac{dz}{z-a} = \oint_\Gamma \frac{dz}{z-a} \qquad (1)$$

Now on Γ, $|z-a| = \epsilon$ or $z-a = \epsilon e^{i\theta}$, i.e. $z = a + \epsilon e^{i\theta}$, $0 \leqq \theta < 2\pi$. Thus since $dz = i\epsilon e^{i\theta}\, d\theta$, the right side of (1) becomes

$$\int_{\theta=0}^{2\pi} \frac{i\epsilon\, e^{i\theta}\, d\theta}{\epsilon e^{i\theta}} = i\int_0^{2\pi} d\theta = 2\pi i$$

which is the required value.

22. Evaluate $\displaystyle\oint_C \frac{dz}{(z-a)^n}$, $n = 2, 3, 4, \ldots$ where $z=a$ is inside the simple closed curve C.

As in Problem 21,

$$\oint_C \frac{dz}{(z-a)^n} = \oint_\Gamma \frac{dz}{(z-a)^n}$$

$$= \int_0^{2\pi} \frac{i\epsilon\, e^{i\theta}\, d\theta}{\epsilon^n\, e^{in\theta}} = \frac{i}{\epsilon^{n-1}} \int_0^{2\pi} e^{(1-n)i\theta}\, d\theta$$

$$= \frac{i}{\epsilon^{n-1}} \frac{e^{(1-n)i\theta}}{(1-n)i}\bigg|_0^{2\pi} = \frac{1}{(1-n)\,\epsilon^{n-1}}\big[e^{2(1-n)\pi i} - 1\big] = 0$$

where $n \neq 1$.

23. If C is the curve $y = x^3 - 3x^2 + 4x - 1$ joining points $(1,1)$ and $(2,3)$, find the value of

$$\int_C (12z^2 - 4iz)\, dz$$

Method 1. By Problem 17, the integral is independent of the path joining $(1,1)$ and $(2,3)$. Hence any path can be chosen. In particular let us choose the straight line paths from $(1,1)$ to $(2,1)$ and then from $(2,1)$ to $(2,3)$.

Case 1. Along the path from $(1,1)$ to $(2,1)$, $y=1$, $dy=0$ so that $z = x+iy = x+i$, $dz = dx$. Then the integral equals

$$\int_{x=1}^2 \{12(x+i)^2 - 4i(x+i)\}\, dx = \{4(x+i)^3 - 2i(x+i)^2\}\bigg|_1^2 = 20 + 30i$$

Case 2. Along the path from $(2,1)$ to $(2,3)$, $x=2$, $dx=0$ so that $z = x+iy = 2+iy$, $dz = i\, dy$. Then the integral equals

$$\int_{y=1}^3 \{12(2+iy)^2 - 4i(2+iy)\}i\, dy = \{4(2+iy)^3 - 2i(2+iy)^2\}\bigg|_1^3 = -176 + 8i$$

Then adding, the required value $= (20 + 30i) + (-176 + 8i) = -156 + 38i$.

Method 2. The given integral equals

$$\int_{1+i}^{2+3i} (12z^2 - 4iz)\, dz = (4z^3 - 2iz^2)\bigg|_{1+i}^{2+3i} = -156 + 38i$$

It is clear that Method 2 is easier.

INTEGRALS OF SPECIAL FUNCTIONS

24. Determine *(a)* $\int \sin 3z \, \cos 3z \, dz$,   *(b)* $\int \cot (2z + 5) \, dz$.

   (a) ***Method 1.*** Let $\sin 3z = u$. Then $du = 3 \cos 3z \, dz$ or $\cos 3z \, dz = du/3$. Then

$$\int \sin 3z \, \cos 3z \, dz \;=\; \int u \frac{du}{3} \;=\; \frac{1}{3} \int u \, du \;=\; \frac{1}{3} \frac{u^2}{2} + c$$

$$=\; \frac{1}{6} u^2 + c \;=\; \frac{1}{6} \sin^2 3z + c$$

   Method 2.

$$\int \sin 3z \, \cos 3z \, dz \;=\; \frac{1}{3} \int \sin 3z \; d(\sin 3z) \;=\; \frac{1}{6} \sin^2 3z + c$$

   Method 3. Let $\cos 3z = u$. Then $du = -3 \sin 3z \, dz$ or $\sin 3z \, dz = -du/3$. Then

$$\int \sin 3z \, \cos 3z \, dz \;=\; -\frac{1}{3} \int u \, du \;=\; -\frac{1}{6} u^2 + c_1 \;=\; -\frac{1}{6} \cos^2 3z + c_1$$

   Note that the results of Methods 1 and 3 differ by a constant.

   (b) ***Method 1.***

$$\int \cot (2z + 5) \, dz \;=\; \int \frac{\cos (2z + 5)}{\sin (2z + 5)} \, dz$$

   Let $u = \sin (2z + 5)$. Then $du = 2 \cos (2z + 5) \, dz$ and $\cos (2z + 5) \, dz = du/2$. Thus

$$\int \frac{\cos (2z + 5) \, dz}{\sin (2z + 5)} \;=\; \frac{1}{2} \int \frac{du}{u} \;=\; \frac{1}{2} \ln u + c \;=\; \frac{1}{2} \ln \sin (2z + 5) + c$$

   Method 2.

$$\int \cot (2z + 5) \, dz \;=\; \int \frac{\cos (2z + 5)}{\sin (2z + 5)} \, dz \;=\; \frac{1}{2} \int \frac{d\{\sin (2z + 5)\}}{\sin (2z + 5)}$$

$$=\; \frac{1}{2} \ln \sin (2z + 5) + c$$

25. *(a)* Prove that $\int F(z) \, G'(z) \, dz \;=\; F(z) \, G(z) \;-\; \int F'(z) \, G(z) \, dz$.

   (b) Find $\int z \, e^{2z} \, dz$ and $\int_0^1 z \, e^{2z} \, dz$.

   (c) Find $\int z^2 \sin 4z \, dz$ and $\int_0^{2\pi} z^2 \sin 4z \, dz$.

   (d) Evaluate $\int_C (z + 2) e^{iz} \, dz$ along the parabola C defined by $\pi^2 y = x^2$ from $(0, 0)$ to $(\pi, 1)$.

   (a) We have

$$d \{F(z) \, G(z)\} \;=\; F(z) \, G'(z) \, dz \;+\; F'(z) \, G(z) \, dz$$

Integrating both sides yields

$$\int d \{F(z) \, G(z)\} \;=\; F(z) \, G(z) \;=\; \int F(z) \, G'(z) \, dz \;+\; \int F'(z) \, G(z) \, dz$$

Then

$$\int F(z) \, G'(z) \, dz \;=\; F(z) \, G(z) \;-\; \int F'(z) \, G(z) \, dz$$

   The method is often called *integration by parts*.

   (b) Let $F(z) = z$, $G'(z) = e^{2z}$. Then $F'(z) = 1$ and $G(z) = \frac{1}{2}e^{2z}$, omitting the constant of integration. Thus by part *(a)*,

$$\int z e^{2z} \, dz \;=\; \int F(z) \, G'(z) \, dz \;=\; F(z) \, G(z) \;-\; \int F'(z) \, G(z) \, dz$$

$$=\; (z)(\tfrac{1}{2}e^{2z}) \;-\; \int 1 \cdot \tfrac{1}{2}e^{2z} \, dz \;=\; \tfrac{1}{2}z e^{2z} - \tfrac{1}{4}e^{2z} + c$$

Hence $\displaystyle\int_0^1 ze^{2z}\,dz = (\tfrac{1}{2}ze^{2z} - \tfrac{1}{4}e^{2z} + c)\Big|_0^1 = \tfrac{1}{2}e^2 - \tfrac{1}{4}e^2 + \tfrac{1}{4} = \tfrac{1}{4}(e^2 + 1)$

(c) Integrating by parts choosing $F(z) = z^2$, $G'(z) = \sin 4z$ we have

$$\int z^2 \sin 4z\,dz = (z^2)(-\tfrac{1}{4}\cos 4z) - \int (2z)(-\tfrac{1}{4}\cos 4z)\,dz$$

$$= -\tfrac{1}{4}z^2\cos 4z + \tfrac{1}{2}\int z\cos 4z\,dz$$

Integrating this last integral by parts, this time choosing $F(z) = z$ and $G'(z) = \cos 4z$, we find

$$\int z\cos 4z\,dz = (z)(\tfrac{1}{4}\sin 4z) - \int (1)(\tfrac{1}{4}\sin 4z)\,dz = \tfrac{1}{4}z\sin 4z + \tfrac{1}{16}\cos 4z$$

Hence $\displaystyle\int z^2\sin 4z\,dz = -\tfrac{1}{4}z^2\cos 4z + \tfrac{1}{8}z\sin 4z + \tfrac{1}{32}\cos 4z + c$

and $\displaystyle\int_0^{2\pi} z^2\sin 4z\,dz = -\pi^2 + \tfrac{1}{32} - \tfrac{1}{32} = -\pi^2$

The double integration by parts can be indicated in a suggestive manner by writing

$$\int z^2\sin 4z\,dz = (z^2)(-\tfrac{1}{4}\cos 4z) - (2z)(-\tfrac{1}{16}\sin 4z) + (2)(\tfrac{1}{64}\cos 4z) + c$$

$$= -\tfrac{1}{4}z^2\cos 4z + \tfrac{1}{8}z\sin 4z + \tfrac{1}{32}\cos 4z$$

where the first parentheses in each term [after the first] is obtained by differentiating z^2 successively, the second parentheses is obtained by integrating $\sin 4z$ successively, and the terms alternate in sign.

(d) The points $(0,0)$ and $(\pi, 1)$ correspond to $z = 0$ and $z = \pi + i$. Since $(z + 2)e^{iz}$ is analytic, we see by Problem 17 that the integral is independent of the path and is equal to

$$\int_0^{1+i} (z+2)\,e^{iz}\,dz = \left\{(z+2)\left(\frac{e^{iz}}{i}\right) - (1)(-e^{iz})\right\}\Big|_0^{\pi+i}$$

$$= (\pi + i + 2)\left(\frac{e^{i(\pi+i)}}{i}\right) + e^{i(\pi+i)} - \frac{2}{i} - 1$$

$$= -2e^{-1} - 1 + i(2 + \pi e^{-1} + 2e^{-1})$$

26. Show that $\displaystyle\int \frac{dz}{z^2 + a^2} = \frac{1}{a}\tan^{-1}\frac{z}{a} + c_1 = \frac{1}{2ai}\ln\left(\frac{z - ai}{z + ai}\right) + c_2.$

Let $z = a\tan u$. Then

$$\int \frac{dz}{z^2 + a^2} = \int \frac{a\sec^2 u\,du}{a^2(\tan^2 u + 1)} = \frac{1}{a}\int du = \frac{1}{a}\tan^{-1}\frac{z}{a} + c_1$$

Also, $\displaystyle\frac{1}{z^2 + a^2} = \frac{1}{(z - ai)(z + ai)} = \frac{1}{2ai}\left(\frac{1}{z - ai} - \frac{1}{z + ai}\right)$

and so $\displaystyle\int \frac{dz}{z^2 + a^2} = \frac{1}{2ai}\int \frac{dz}{z - ai} - \frac{1}{2ai}\int \frac{dz}{z + ai}$

$$= \frac{1}{2ai}\ln(z - ai) - \frac{1}{2ai}\ln(z + ai) + c_2 = \frac{1}{2ai}\ln\left(\frac{z - ai}{z + ai}\right) + c_2$$

MISCELLANEOUS PROBLEMS

27. Prove Morera's theorem [Page 95] under the assumption that $f(z)$ has a continuous derivative in $\mathcal{R}$.

If $f(z)$ has a continuous derivative in $\mathcal{R}$, then we can apply Green's theorem to obtain

$$\oint_C f(z)\,dz = \oint_C u\,dx - v\,dy + i\oint_C v\,dx + u\,dy$$

$$= \iint_{\mathcal{R}} \left(-\frac{\partial v}{\partial x} - \frac{\partial u}{\partial y}\right)dx\,dy + i\iint_{\mathcal{R}} \left(\frac{\partial u}{\partial x} - \frac{\partial v}{\partial y}\right)dx\,dy$$

Then if $\oint_C f(z)\,dz = 0$ around every closed path C in $\mathcal{R}$, we must have

$$\oint_C u\,dx - v\,dy = 0, \qquad \oint_C v\,dx + u\,dy = 0$$

around every closed path C in $\mathcal{R}$. Hence from Problem 8, the Cauchy-Riemann equations

$$\frac{\partial u}{\partial x} = \frac{\partial v}{\partial y}, \qquad \frac{\partial v}{\partial x} = -\frac{\partial u}{\partial y}$$

are satisfied and thus [since these partial derivatives are continuous] it follows [Problem 5, Chapter 3] that $u + iv = f(z)$ is analytic.

28. A force field is given by $F = 3z + 5$. Find the work done in moving an object in this force field along the parabola $z = t^2 + it$ from $z = 0$ to $z = 4 + 2i$.

$$\text{Total work done} = \int_C F \circ dz = \text{Re}\int_C \bar{F}\,dz = \text{Re}\left\{\int_C (3\bar{z}+5)\,dz\right\}$$

$$= \text{Re}\left\{3\int_C \bar{z}\,dz + 5\int_C dz\right\} = \text{Re}\left\{3(10 - \tfrac{8}{3}i) + 5(4+2i)\right\} = 50$$

using the result of Problem 2.

29. Find   (a) $\displaystyle\int e^{ax}\sin bx\,dx$,   (b) $\displaystyle\int e^{ax}\cos bx\,dx$.

Omitting the constant of integration, we have

$$\int e^{(a+ib)x}\,dx = \frac{e^{(a+ib)x}}{a+ib}$$

which can be written

$$\int e^{ax}(\cos bx + i\sin bx)\,dx = \frac{e^{ax}(\cos bx + i\sin bx)}{a+ib} = \frac{e^{ax}(\cos bx + i\sin bx)(a-ib)}{a^2 + b^2}$$

Then equating real and imaginary parts,

$$\int e^{ax}\cos bx\,dx = \frac{e^{ax}(a\cos bx + b\sin bx)}{a^2 + b^2}$$

$$\int e^{ax}\sin bx\,dx = \frac{e^{ax}(a\sin bx - b\cos bx)}{a^2 + b^2}$$

30. Give an example of a continuous, closed, non-intersecting curve which lies in a bounded region $\mathcal{R}$ but which has an infinite length.

Consider equilateral triangle ABC [Fig. 4-21] with sides of unit length. By trisecting each side, construct equilateral triangles DEF, GHJ and KLM. Then omitting sides DF, GJ and KM, we obtain the closed non-intersecting curve $ADEFBGHJCKLMA$ of Fig. 4-22.

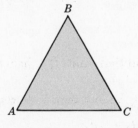

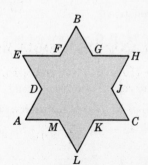

Fig. 4-21   Fig. 4-22

The process can now be continued by trisecting sides DE, EF, FB, BG, GH, etc., and constructing equilateral triangles as before. By repeating the process indefinitely [see Fig. 4-23] we obtain a continuous closed non-intersecting curve which is the boundary of a region with finite area equal to

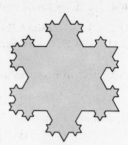

$$\tfrac{1}{4}\sqrt{3} \;+\; (3)(\tfrac{1}{3})^2 \frac{\sqrt{3}}{4} \;+\; (9)(\tfrac{1}{9})^2 \frac{\sqrt{3}}{4} \;+\; (27)(\tfrac{1}{27})^2 \frac{\sqrt{3}}{4} \;+\; \cdots$$

$$= \;\frac{\sqrt{3}}{4}(1 + \tfrac{1}{3} + \tfrac{1}{9} + \cdots) \;=\; \frac{\sqrt{3}}{4}\frac{1}{1-1/3} \;=\; \frac{3\sqrt{3}}{8}$$

or 1.5 times the area of triangle ABC, and which has infinite length (see Problem 91).

Fig. 4-23

31. Let $F(x, y)$ and $G(x, y)$ be continuous and have continuous first and second partial derivatives in a simply-connected region $\mathcal{R}$ bounded by a simple closed curve C. Prove that

$$\oint_C F\left(\frac{\partial G}{\partial y} dx - \frac{\partial G}{\partial x} dy\right) \;=\; -\iint_{\mathcal{R}} \left[F\left(\frac{\partial^2 G}{\partial x^2} + \frac{\partial^2 G}{\partial y^2}\right) + \left(\frac{\partial F}{\partial x}\frac{\partial G}{\partial x} + \frac{\partial F}{\partial y}\frac{\partial G}{\partial y}\right)\right] dx\, dy$$

Let $\;P = F\dfrac{\partial G}{\partial y},\; Q = -F\dfrac{\partial G}{\partial x}\;$ in Green's theorem

$$\oint_C P\, dx + Q\, dy \;=\; \iint_{\mathcal{R}} \left(\frac{\partial Q}{\partial x} - \frac{\partial P}{\partial y}\right) dx\, dy$$

Then as required

$$\oint_C F\left(\frac{\partial G}{\partial y} dx - \frac{\partial G}{\partial x} dy\right) \;=\; \iint_{\mathcal{R}} \left(\frac{\partial}{\partial x}\left\{-F\frac{\partial G}{\partial x}\right\} - \frac{\partial}{\partial y}\left\{F\frac{\partial G}{\partial y}\right\}\right) dx\, dy$$

$$=\; -\iint_{\mathcal{R}} \left[F\left(\frac{\partial^2 G}{\partial x^2} + \frac{\partial^2 G}{\partial y^2}\right) + \left(\frac{\partial F}{\partial x}\frac{\partial G}{\partial x} + \frac{\partial F}{\partial y}\frac{\partial G}{\partial y}\right)\right] dx\, dy$$

Supplementary Problems

LINE INTEGRALS

32. Evaluate $\displaystyle\int_{(0,1)}^{(2,5)} (3x + y)\, dx + (2y - x)\, dy$ along (a) the curve $y = x^2 + 1$, (b) the straight line joining $(0, 1)$ and $(2, 5)$, (c) the straight lines from $(0, 1)$ to $(0, 5)$ and then from $(0, 5)$ to $(2, 5)$, (d) the straight lines from $(0, 1)$ to $(2, 1)$ and then from $(2, 1)$ to $(2, 5)$.
 Ans. (a) 88/3, (b) 32, (c) 40, (d) 24

33. (a) Evaluate $\displaystyle\oint_C (x + 2y)\, dx + (y - 2x)\, dy$ around the ellipse C defined by $x = 4\cos\theta$, $y = 3\sin\theta$, $0 \le \theta < 2\pi$ if C is described in a counterclockwise direction. (b) What is the answer to (a) if C is described in a clockwise direction? *Ans.* (a) -48π, (b) 48π

34. Evaluate $\displaystyle\int_C (x^2 - iy^2)\, dz$ along (a) the parabola $y = 2x^2$ from $(1, 2)$ to $(2, 8)$, (b) the straight lines from $(1, 1)$ to $(1, 8)$ and then from $(1, 8)$ to $(2, 8)$, (c) the straight line from $(1, 1)$ to $(2, 8)$.
 Ans. (a) $\frac{511}{3} - \frac{49}{5}i$, (b) $\frac{518}{3} - 57i$, (c) $\frac{518}{3} - 8i$

35. Evaluate $\displaystyle\oint_C |z|^2\, dz$ around the square with vertices at $(0, 0), (1, 0), (1, 1), (0, 1)$. *Ans.* $-1 + i$

36. Evaluate $\int_C (z^2 + 3z)\, dz$ along (a) the circle $|z| = 2$ from $(2, 0)$ to $(0, 2)$ in a counterclockwise direction, (b) the straight line from $(2, 0)$ to $(0, 2)$, (c) the straight lines from $(2, 0)$ to $(2, 2)$ and then from $(2, 2)$ to $(0, 2)$.   *Ans.* $-\frac{44}{3} - \frac{8}{3}i$ for all cases

37. If $f(z)$ and $g(z)$ are integrable, prove that

$$(a) \quad \int_a^b f(z)\, dz = -\int_b^a f(z)\, dz$$

$$(b) \quad \int_C \{2\, f(z) - 3i\, g(z)\}\, dz = 2\int_C f(z)\, dz - 3i\int_C g(z)\, dz.$$

38. Evaluate $\int_i^{2-i} (3xy + iy^2)\, dz$ (a) along the straight line joining $z = i$ and $z = 2 - i$, (b) along the curve $x = 2t - 2$, $y = 1 + t - t^2$.   *Ans.* (a) $-\frac{4}{3} + \frac{8}{3}i$, (b) $-\frac{1}{3} + \frac{79}{30}i$

39. Evaluate $\oint_C \bar{z}^2\, dz$ around the circles (a) $|z| = 1$, (b) $|z - 1| = 1$.   *Ans.* (a) 0, (b) $4\pi i$

40. Evaluate $\oint_C (5z^4 - z^3 + 2)\, dz$ around (a) the circle $|z| = 1$, (b) the square with vertices at $(0, 0)$, $(1, 0)$, $(1, 1)$ and $(0, 1)$, (c) the curve consisting of the parabolas $y = x^2$ from $(0, 0)$ to $(1, 1)$ and $y^2 = x$ from $(1, 1)$ to $(0, 0)$.   *Ans.* 0 in all cases

41. Evaluate $\int_C (z^2 + 1)^2\, dz$ along the arc of the cycloid $x = a(\theta - \sin \theta)$, $y = a(1 - \cos \theta)$ from the point where $\theta = 0$ to the point where $\theta = 2\pi$.   *Ans.* $(96\pi^5 a^5 + 80\pi^3 a^3 + 30\pi a)/15$

42. Evaluate $\int_C \bar{z}^2\, dz + z^2\, d\bar{z}$ along the curve C defined by $z^2 + 2z\bar{z} + \bar{z}^2 = (2 - 2i)z + (2 + 2i)\bar{z}$ from the point $z = 1$ to $z = 2 + 2i$.   *Ans.* 248/15

43. Evaluate $\oint_C \frac{dz}{z - 2}$ around (a) the circle $|z - 2| = 4$, (b) the circle $|z - 1| = 5$, (c) the square with vertices at $3 \pm 3i$, $-3 \pm 3i$.   *Ans.* $2\pi i$ in all cases

44. Evaluate $\oint_C (x^2 + iy^2)\, ds$ around the circle $|z| = 2$ where s is the arc length.   *Ans.* $8\pi(1 + i)$

GREEN'S THEOREM IN THE PLANE

45. Verify Green's theorem in the plane for $\oint_C (x^2 - 2xy)\, dx + (y^2 - x^3 y)\, dy$ where C is a square with vertices at $(0, 0)$, $(2, 0)$, $(2, 2)$, $(0, 2)$.   *Ans.* common value $= -8$

46. Evaluate $\oint_C (5x + 6y - 3)\, dx + (3x - 4y + 2)\, dy$ around a triangle in the xy plane with vertices at $(0, 0)$, $(4, 0)$ and $(4, 3)$.   *Ans.* -18

47. Let C be any simple closed curve bounding a region having area A. Prove that

$$A = \frac{1}{2} \oint_C x\, dy - y\, dx$$

48. Use the result of Problem 47 to find the area bounded by the ellipse $x = a \cos \theta$, $y = b \sin \theta$, $0 \leq \theta < 2\pi$.   *Ans.* πab

49. Find the area bounded by the hypocycloid $x^{2/3} + y^{2/3} = a^{2/3}$ shown shaded in Fig. 4-24. [*Hint.* Parametric equations are $x = a \cos^3 \theta$, $y = a \sin^3 \theta$, $0 \leq \theta < 2\pi$.]   *Ans.* $3\pi a^2/8$

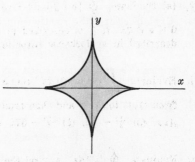

Fig. 4-24

50. Verify Green's theorem in the plane for $\oint_C x^2y\,dx + (y^3 - xy^2)\,dy$ where C is the boundary of the region enclosed by the circles $x^2 + y^2 = 4$, $x^2 + y^2 = 16$.   *Ans.* common value $= 120\pi$

51. (a) Prove that $\oint_C (y^2 \cos x - 2e^y)\,dx + (2y \sin x - 2xe^y)\,dy = 0$ around any simple closed curve C.

(b) Evaluate the integral in (a) along the parabola $y = x^2$ from $(0, 0)$ to (π, π^2).   *Ans.* (b) $-2\pi e^{\pi^2}$

52. (a) Show that $\int_{(2,1)}^{(3,2)} (2xy^3 - 2y^2 - 6y)\,dx + (3x^2y^2 - 4xy - 6x)\,dy$ is independent of the path joining points $(2, 1)$ and $(3, 2)$. (b) Evaluate the integral in (a).   *Ans.* (b) 24

COMPLEX FORM OF GREEN'S THEOREM

53. If C is a simple closed curve enclosing a region of area A, prove that $A = \dfrac{1}{2i} \oint_C \bar{z}\,dz$.

54. Evaluate $\oint_C \bar{z}\,dz$ around (a) the circle $|z - 2| = 3$, (b) the square with vertices at $z = 0, 2, 2i$ and $2 + 2i$, (c) the ellipse $|z - 3| + |z + 3| = 10$.   *Ans.* (a) $18\pi i$, (b) $8i$, (c) $40\pi i$

55. Evaluate $\oint_C (8\bar{z} + 3z)\,dz$ around the hypocycloid $x^{2/3} + y^{2/3} = a^{2/3}$.   *Ans.* $6\pi i a^2$

56. Let $P(z, \bar{z})$ and $Q(z, \bar{z})$ be continuous and have continuous partial derivatives in a region $\mathcal{R}$ and on its boundary C. Prove that

$$\oint_C P(z, \bar{z})\,dz + Q(z, \bar{z})\,d\bar{z} = 2i \iint_{\mathcal{R}} \left(\frac{\partial P}{\partial \bar{z}} - \frac{\partial Q}{\partial z} \right) dA$$

57. Show that the area in Problem 53 can be written in the form $A = \dfrac{1}{4i} \oint_C \bar{z}\,dz - z\,d\bar{z}$.

58. Show that the centroid of the region of Problem 53 is given in conjugate coordinates by $(\hat{z}, \hat{\bar{z}})$ where

$$\hat{z} = -\frac{1}{4Ai} \oint_C z^2\,d\bar{z}, \qquad \hat{\bar{z}} = \frac{1}{4Ai} \oint_C \bar{z}^2\,dz$$

59. Find the centroid of the region bounded above by $|z| = a > 0$ and below by $\operatorname{Im} z = 0$.
Ans. $\hat{z} = 2ai/\pi$, $\hat{\bar{z}} = -2ai/\pi$

CAUCHY'S THEOREM AND THE CAUCHY-GOURSAT THEOREM

60. Verify Cauchy's theorem for the functions (a) $3z^2 + iz - 4$, (b) $5 \sin 2z$, (c) $3 \cosh(z + 2)$ if C is the square with vertices at $1 \pm i, -1 \pm i$.

61. Verify Cauchy's theorem for the function $z^3 - iz^2 - 5z + 2i$ if C is (a) the circle $|z| = 1$, (b) the circle $|z - 1| = 2$, (c) the ellipse $|z - 3i| + |z + 3i| = 20$.

62. If C is the circle $|z - 2| = 5$, determine whether $\oint_C \dfrac{dz}{z - 3} = 0$. (b) Does your answer to (a) contradict Cauchy's theorem?

63. Explain clearly the relationship between the observations

$$\oint_C (x^2 - y^2 + 2y)\,dx + (2x - 2xy)\,dy = 0 \quad \text{and} \quad \oint_C (z^2 - 2iz)\,dz = 0$$

where C is any simple closed curve.

64. By evaluating $\oint_C e^z\,dz$ around the circle $|z| = 1$, show that

$$\int_0^{2\pi} e^{\cos \theta} \cos(\theta + \sin \theta)\,d\theta = \int_0^{2\pi} e^{\cos \theta} \sin(\theta + \sin \theta)\,d\theta = 0$$

65. State and prove Cauchy's theorem for multiply-connected regions.

66. Prove the Cauchy-Goursat theorem for a polygon, such as *ABCDEFGA* shown in Fig. 4-25, which may intersect itself.

67. Prove the Cauchy-Goursat theorem for the multiply-connected region $\mathcal{R}$ shown shaded in Fig. 4-26.

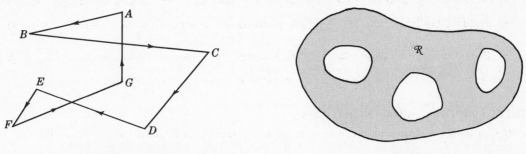

Fig. 4-25   Fig. 4-26

68. (a) Prove the Cauchy-Goursat theorem for a rectangle and (b) show how the result of (a) can be used to prove the theorem for any simple closed curve C.

69. Let P and Q be continuous and have continuous first partial derivatives in a region $\mathcal{R}$. Let C be any simple closed curve in $\mathcal{R}$ and suppose that for any such curve

$$\oint_C P\,dx + Q\,dy = 0$$

(a) Prove that there exists an analytic function $f(z)$ such that Re $\{f(z)\,dz\} = P\,dx + Q\,dy$ is an exact differential.

(b) Determine p and q in terms of P and Q such that Im $\{f(z)\,dz\} = p\,dx + q\,dy$ and verify that

$$\oint_C p\,dx + q\,dy = 0.$$

(c) Discuss the connection between (a) and (b) and Cauchy's theorem.

70. Illustrate the results of Problem 69 if $P = 2x + y - 2xy$, $Q = x - 2y - x^2 + y^2$ by finding p, q and $f(z)$.   *Ans.* One possibility is $p = x^2 - y^2 + 2y - x$, $q = 2x + y - 2xy$, $f(z) = iz^2 + (2-i)z$.

71. Let P and Q be continuous and have continuous partial derivatives in a region $\mathcal{R}$. Suppose that for any simple closed curve C in $\mathcal{R}$ we have $\oint_C P\,dx + Q\,dy = 0$.   (a) Prove that $\oint_C Q\,dx - P\,dy = 0$.
(b) Discuss the relationship of (a) with Cauchy's theorem.

CONSEQUENCES OF CAUCHY'S THEOREM

72. Show directly that $\displaystyle\int_{3+4i}^{4-3i} (6z^2 + 8iz)\,dz$ has the same value along the following paths C joining the points $3 + 4i$ and $4 - 3i$:   (a) a straight line,   (b) the straight lines from $3 + 4i$ to $4 + 4i$ and then from $4 + 4i$ to $4 - 3i$,   (c) the circle $|z| = 5$.   Determine this value.   *Ans.* $338 - 266i$

73. Show that $\displaystyle\int_C e^{-2z}\,dz$ is independent of the path C joining the points $1 - \pi i$ and $2 + 3\pi i$ and determine its value.   *Ans.* $\frac{1}{2}e^{-2}(1 - e^{-2})$

74. Given $G(z) = \displaystyle\int_{\pi - \pi i}^{z} \cos 3\zeta\,d\zeta$.   (a) Prove that $G(z)$ is independent of the path joining $\pi - \pi i$ and the arbitrary point z.   (b) Determine $G(\pi i)$.   (c) Prove that $G'(z) = \cos 3z$.   *Ans.* (b) 0

75. Given $G(z) = \displaystyle\int_{1+i}^{z} \sin \zeta^2\,d\zeta$.   (a) Prove that $G(z)$ is an analytic function of z.   (b) Prove that $G'(z) = \sin z^2$.

76. State and prove a theorem corresponding to (a) Problem 17, (b) Problem 18, (c) Problem 20 for the real line integral $\displaystyle\int_C P\,dx + Q\,dy$.

77. Prove Theorem 5, Page 97 for the region of Fig. 4-26.

78. (a) If C is the circle $|z| = R$, show that

$$\lim_{R \to \infty} \oint_C \frac{z^2 + 2z - 5}{(z^2 + 4)(z^2 + 2z + 2)} \, dz \;=\; 0$$

(b) Use the result of (a) to deduce that if C_1 is the circle $|z - 2| = 5$, then

$$\oint_{C_1} \frac{z^2 + 2z - 5}{(z^2 + 4)(z^2 + 2z + 2)} \, dz \;=\; 0$$

(c) Is the result in (b) true if C_1 is the circle $|z + 1| = 2$? Explain.

INTEGRALS OF SPECIAL FUNCTIONS

79. Find each of the following integrals:

(a) $\displaystyle\int e^{-2z} \, dz$,   (b) $\displaystyle\int z \sin z^2 \, dz$,   (c) $\displaystyle\int \frac{z^2 + 1}{z^3 + 3z + 2} \, dz$,   (d) $\displaystyle\int \sin^4 2z \cos 2z \, dz$

(e) $\displaystyle\int z^2 \tanh(4z^3) \, dz$

     Ans. (a) $-\tfrac{1}{2} e^{-2z} + c$ (c) $\tfrac{1}{3} \ln(z^3 + 3z + 2) + c$

            (b) $-\tfrac{1}{2} \cos z^2 + c$ (d) $\tfrac{1}{10} \sin^5 2z + c$ (e) $\tfrac{1}{12} \ln \cosh(4z^3) + c$

80. Find each of the following integrals:

(a) $\displaystyle\int z \cos 2z \, dz$,   (b) $\displaystyle\int z^2 e^{-z} \, dz$,   (c) $\displaystyle\int z \ln z \, dz$,   (d) $\displaystyle\int z^3 \sinh z \, dz$.

     Ans. (a) $\tfrac{1}{2} z \sin 2z + \tfrac{1}{4} \cos 2z + c$ (c) $\tfrac{1}{2} z^2 \ln z - \tfrac{1}{4} + c$

            (b) $-e^{-z}(z^2 + 2z + 2) + c$ (d) $(z^3 + 6z) \cosh z - 3(z^2 + 2) \sinh z + c$

81. Evaluate each of the following:

(a) $\displaystyle\int_{\pi i}^{2\pi i} e^{3z} \, dz$,   (b) $\displaystyle\int_0^{\pi i} \sinh 5z \, dz$,   (c) $\displaystyle\int_0^{\pi + i} z \cos 2z \, dz$.

     Ans. (a) $2/3$,   (b) $-2/5$,   (c) $\tfrac{1}{4} \cosh 2 - \tfrac{1}{2} \sinh 2 + \tfrac{1}{2} \pi i \sinh 2$

82. Show that $\displaystyle\int_0^{\pi/2} \sin^2 z \, dz = \int_0^{\pi/2} \cos^2 z \, dz = \pi/4$.

83. Show that $\displaystyle\int \frac{dz}{z^2 - a^2} = \frac{1}{2a} \ln\left(\frac{z - a}{z + a}\right) + c_1 = \frac{1}{a} \coth^{-1} \frac{z}{a} + c_2$.

84. Show that if we restrict ourselves to the same branch of the square root,

$$\int z\sqrt{2z + 5} \, dz \;=\; \frac{1}{20}(2z + 5)^{5/2} - \frac{5}{6}(2z + 5)^{3/2} + c$$

85. Evaluate $\displaystyle\int \sqrt{1 + \sqrt{z + 1}} \, dz$, stating conditions under which your result is valid.

     Ans. $\tfrac{4}{5}(1 + \sqrt{z + 1})^{5/2} - \tfrac{4}{3}(1 + \sqrt{z + 1})^{3/2} + c$

MISCELLANEOUS PROBLEMS

86. Use the definition of an integral to prove that along any arbitrary path joining points a and b,

(a) $\displaystyle\int_a^b dz = b - a$,   (b) $\displaystyle\int_a^b z \, dz = \tfrac{1}{2}(b^2 - a^2)$.

87. Prove the theorem concerning change of variables on Page 93.
     [*Hint.* Express each side as two real line integrals and use the Cauchy-Riemann equations.]

88. Let $u(x, y)$ be harmonic and have continuous derivatives, of order two at least, in a region $\mathcal{R}$.
     (a) Show that

$$v(x, y) \;=\; \int_{(a,b)}^{(x,y)} -\frac{\partial u}{\partial y} \, dx + \frac{\partial u}{\partial x} \, dy$$

     is independent of the path in $\mathcal{R}$ joining (a, b) to (x, y).

     (b) Prove that $u + iv$ is an analytic function of $z = x + iy$ in $\mathcal{R}$.

     (c) Prove that v is harmonic in $\mathcal{R}$.

89. Work Problem 88 for the special cases   (a) $u = 3x^2y + 2x^2 - y^3 - 2y^2$,   (b) $u = xe^x \cos y - ye^x \sin y$. [See Problems 53(a) and (c), Page 86.]

90. Using the definition of an integral, verify directly that

$$(a) \oint_C dz = 0, \qquad (b) \oint_C z\, dz = 0, \qquad (c) \oint_C (z - z_0)\, dz = 0$$

where C is a simple closed curve and z_0 is any constant.

91. Find the length of the closed curve of Problem 30 after n steps and verify that as $n \to \infty$ the length of the curve becomes infinite.

92. Evaluate $\displaystyle\int_C \frac{dz}{z^2 + 4}$ along the line $x + y = 1$ in the direction of increasing x.   *Ans.* $\pi/2$

93. Show that $\displaystyle\int_0^\infty xe^{-x} \sin x\, dx = \tfrac{1}{2}$.

94. Evaluate $\displaystyle\int_{-2 - 2\sqrt{3}\,i}^{-2 + 2\sqrt{3}\,i} z^{1/2}\, dz$ along a straight line path if we choose that branch of $z^{1/2}$ such that $z^{1/2} = 1$ for $z = 1$.   *Ans.* 32/3

95. Does Cauchy's theorem hold for the function $f(z) = z^{1/2}$ where C is the circle $|z| = 1$?   Explain.

96. Does Cauchy's theorem hold for a curve, such as *EFGHFJE* in Fig. 4-27, which intersects itself? Justify your answer.

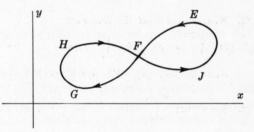

97. If n is the direction of the outward drawn normal to a simple closed curve C, s is the arc length parameter and U is any continuously differentiable function, prove that

$$\frac{\partial U}{\partial n} = \frac{\partial U}{\partial x}\frac{dx}{ds} + \frac{\partial U}{\partial y}\frac{dy}{ds}$$

Fig. 4-27

98. Prove *Green's first identity*,

$$\iint_{\mathcal{R}} U\, \nabla^2 V\, dx\, dy + \iint_{\mathcal{R}} \left(\frac{\partial U}{\partial x}\frac{\partial V}{\partial x} + \frac{\partial U}{\partial y}\frac{\partial V}{\partial y}\right) dx\, dy = \oint_C U\frac{\partial V}{\partial n}\, ds$$

where $\mathcal{R}$ is the region bounded by the simple closed curve C, $\nabla^2 = \dfrac{\partial^2}{\partial x^2} + \dfrac{\partial^2}{\partial y^2}$,   while n and s are as in Problem 97.

99. Use Problem 98 to prove *Green's second identity*

$$\iint_{\mathcal{R}} (U\, \nabla^2 V - V\, \nabla^2 U)\, dA = \oint_C \left(U\frac{\partial V}{\partial n} - V\frac{\partial U}{\partial n}\right) ds$$

where dA is an element of area of $\mathcal{R}$.

100. Write the result of Problem 31 in terms of the operator ∇.

101. Evaluate $\displaystyle\oint_C \frac{dz}{\sqrt{z^2 + 2z + 2}}$ around the unit circle $|z| = 1$ starting with $z = 1$, assuming the integrand positive for this value.

102. If n is a positive integer, show that

$$\int_0^{2\pi} e^{\sin n\theta} \cos(\theta - \cos n\theta)\, d\theta = \int_0^{2\pi} e^{\sin n\theta} \sin(\theta - \cos n\theta)\, d\theta = 0$$

Chapter 5

Cauchy's Integral Formulas and Related Theorems

CAUCHY'S INTEGRAL FORMULAS

If $f(z)$ is analytic inside and on a simple closed curve C and a is any point inside C [Fig. 5-1], then

$$f(a) = \frac{1}{2\pi i} \oint_C \frac{f(z)}{z-a} \, dz \qquad (1)$$

where C is traversed in the positive (counterclockwise) sense.

Also the nth derivative of $f(z)$ at $z=a$ is given by

$$f^{(n)}(a) = \frac{n!}{2\pi i} \oint_C \frac{f(z)}{(z-a)^{n+1}} \, dz \qquad n = 1, 2, 3, \ldots \qquad (2)$$

The result (1) can be considered a special case of (2) with $n=0$ if we define $0! = 1$.

The results (1) and (2) are called *Cauchy's integral formulas* and are quite remarkable because they show that if a function $f(z)$ is known *on* the simple closed curve C then the values of the function and all its derivatives can be found at all points *inside C*. Thus if a function of a complex variable has a first derivative, i.e. is analytic, in a simply-connected region $\mathcal{R}$, all its higher derivatives exist in $\mathcal{R}$. This is not necessarily true for functions of real variables.

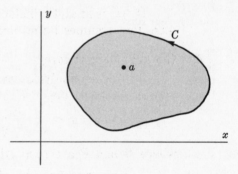

Fig. 5-1

SOME IMPORTANT THEOREMS

The following is a list of some important theorems which are consequences of Cauchy's integral formulas.

1. **Morera's theorem** (converse of Cauchy's theorem).

    If $f(z)$ is continuous in a simply-connected region $\mathcal{R}$ and if $\oint_C f(z) \, dz = 0$ around every simple closed curve C in $\mathcal{R}$, then $f(z)$ is analytic in $\mathcal{R}$.

2. **Cauchy's inequality.**

    If $f(z)$ is analytic inside and on a circle C of radius r and center at $z=a$, then

$$|f^{(n)}(a)| \leq \frac{M \cdot n!}{r^n} \qquad n = 0, 1, 2, \ldots \qquad (3)$$

    where M is a constant such that $|f(z)| < M$ on C, i.e. M is an upper bound of $|f(z)|$ on C.

118

3. **Liouville's theorem.**

Suppose that for all z in the entire complex plane, (i) $f(z)$ is analytic and (ii) $f(z)$ is bounded, i.e. $|f(z)| < M$ for some constant M. Then $f(z)$ must be a constant.

4. **Fundamental theorem of algebra.**

Every polynomial equation $P(z) = a_0 + a_1 z + a_2 z^2 + \cdots + a_n z^n = 0$ with degree $n \geqq 1$ and $a_n \neq 0$ has at least one root.

From this it follows that $P(z) = 0$ has exactly n roots, due attention being paid to multiplicities of roots.

5. **Gauss' mean value theorem.**

If $f(z)$ is analytic inside and on a circle C with center at a and radius r, then $f(a)$ is the mean of the values of $f(z)$ on C, i.e.,

$$f(a) = \frac{1}{2\pi} \int_0^{2\pi} f(a + re^{i\theta}) \, d\theta \qquad (4)$$

6. **Maximum modulus theorem.**

If $f(z)$ is analytic inside and on a simple closed curve C and is not identically equal to a constant, then the maximum value of $|f(z)|$ occurs on C.

7. **Minimum modulus theorem.**

If $f(z)$ is analytic inside and on a simple closed curve C and $f(z) \neq 0$ inside C, then $|f(z)|$ assumes its minimum value on C.

8. **The argument theorem.**

Let $f(z)$ be analytic inside and on a simple closed curve C except for a finite number of poles inside C. Then

$$\frac{1}{2\pi i} \oint_C \frac{f'(z)}{f(z)} \, dz = N - P \qquad (5)$$

where N and P are respectively the number of zeros and poles of $f(z)$ inside C.

For a generalization of this theorem see Problem 90.

9. **Rouché's theorem.**

If $f(z)$ and $g(z)$ are analytic inside and on a simple closed curve C and if $|g(z)| < |f(z)|$ on C, then $f(z) + g(z)$ and $f(z)$ have the same number of zeros inside C.

10. **Poisson's integral formulas for a circle.**

Let $f(z)$ be analytic inside and on the circle C defined by $|z| = R$. Then if $z = re^{i\theta}$ is any point inside C, we have

$$f(re^{i\theta}) = \frac{1}{2\pi} \int_0^{2\pi} \frac{(R^2 - r^2) f(Re^{i\phi})}{R^2 - 2Rr \cos(\theta - \phi) + r^2} \, d\phi \qquad (6)$$

If $u(r, \theta)$ and $v(r, \theta)$ are the real and imaginary parts of $f(re^{i\theta})$ while $u(R, \phi)$ and $v(R, \phi)$ are the real and imaginary parts of $f(Re^{i\phi})$, then

$$u(r, \theta) = \frac{1}{2\pi} \int_0^{2\pi} \frac{(R^2 - r^2) u(R, \phi)}{R^2 - 2Rr \cos(\theta - \phi) + r^2} \, d\phi \qquad (7)$$

$$v(r, \theta) = \frac{1}{2\pi} \int_0^{2\pi} \frac{(R^2 - r^2) v(R, \phi)}{R^2 - 2Rr \cos(\theta - \phi) + r^2} \, d\phi \qquad (8)$$

These results are called *Poisson's integral formulas for a circle*. They express the values of a harmonic function inside a circle in terms of its values on the boundary.

11. Poisson's integral formulas for a half plane.

Let $f(z)$ be analytic in the upper half $y \geqq 0$ of the z plane and let $\zeta = \xi + i\eta$ be any point in this upper half plane. Then

$$f(\zeta) = \frac{1}{\pi} \int_{-\infty}^{\infty} \frac{\eta \, f(x)}{(x - \xi)^2 + \eta^2} \, dx \qquad (9)$$

In terms of the real and imaginary parts of $f(\zeta)$ this can be written

$$u(\xi, \eta) = \frac{1}{\pi} \int_{-\infty}^{\infty} \frac{\eta \, u(x, 0)}{(x - \xi)^2 + \eta^2} \, dx \qquad (10)$$

$$v(\xi, \eta) = \frac{1}{\pi} \int_{-\infty}^{\infty} \frac{\eta \, v(x, 0)}{(x - \xi)^2 + \eta^2} \, dx \qquad (11)$$

These are called *Poisson's integral formulas for a half plane*. They express the values of a harmonic function in the upper half plane in terms of the values on the x axis [the boundary] of the half plane.

Solved Problems

CAUCHY'S INTEGRAL FORMULAS

1. If $f(z)$ is analytic inside and on the boundary C of a simply-connected region $\mathcal{R}$, prove *Cauchy's integral formula*

$$f(a) = \frac{1}{2\pi i} \oint_C \frac{f(z)}{z - a} \, dz$$

Method 1.

The function $f(z)/(z - a)$ is analytic inside and on C except at the point $z = a$ (see Fig. 5-2). By Theorem 4, Page 97, we have

$$\oint_C \frac{f(z)}{z - a} \, dz = \oint_\Gamma \frac{f(z)}{z - a} \, dz \qquad (1)$$

where we can choose Γ as a circle of radius ϵ with center at a. Then an equation for Γ is $|z - a| = \epsilon$ or $z - a = \epsilon e^{i\theta}$ where $0 \leqq \theta < 2\pi$. Substituting $z = a + \epsilon e^{i\theta}$, $dz = i\epsilon e^{i\theta}$, the integral on the right of (1) becomes

$$\oint_\Gamma \frac{f(z)}{z - a} \, dz = \int_0^{2\pi} \frac{f(a + \epsilon e^{i\theta}) \, i\epsilon e^{i\theta}}{\epsilon e^{i\theta}} \, d\theta$$

$$= i \int_0^{2\pi} f(a + \epsilon e^{i\theta}) \, d\theta$$

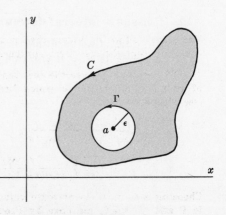

Fig. 5-2

Thus we have from (1),

$$\oint_C \frac{f(z)}{z-a}\, dz \;=\; i \int_0^{2\pi} f(a + \epsilon e^{i\theta})\, d\theta \qquad (2)$$

Taking the limit of both sides of (2) and making use of the continuity of $f(z)$, we have

$$\oint_C \frac{f(z)}{z-a}\, dz \;=\; \lim_{\epsilon \to 0} i \int_0^{2\pi} f(a + \epsilon e^{i\theta})\, d\theta$$

$$=\; i \int_0^{2\pi} \lim_{\epsilon \to 0} f(a + \epsilon e^{i\theta})\, d\theta \;=\; i \int_0^{2\pi} f(a)\, d\theta \;=\; 2\pi i\, f(a) \qquad (3)$$

so that we have, as required,

$$f(a) \;=\; \frac{1}{2\pi i} \oint_C \frac{f(z)}{z-a}\, dz$$

Method 2. The right side of equation (1) of Method 1 can be written as

$$\oint_\Gamma \frac{f(z)}{z-a}\, dz \;=\; \oint_\Gamma \frac{f(z) - f(a)}{z-a}\, dz \;+\; \oint_\Gamma \frac{f(a)}{z-a}\, dz$$

$$=\; \oint_\Gamma \frac{f(z) - f(a)}{z-a}\, dz \;+\; 2\pi i\, f(a)$$

using Problem 21, Chapter 4. The required result will follow if we can show that

$$\oint_\Gamma \frac{f(z) - f(a)}{z-a}\, dz \;=\; 0$$

But by Problem 21, Chapter 3,

$$\oint_\Gamma \frac{f(z) - f(a)}{z-a}\, dz \;=\; \oint_\Gamma f'(a)\, dz \;+\; \oint_\Gamma \eta\, dz \;=\; \oint_\Gamma \eta\, dz$$

Then choosing Γ so small that for all points on Γ we have $|\eta| < \delta/2\pi$, we find

$$\left| \oint_\Gamma \eta\, dz \right| \;<\; \left(\frac{\delta}{2\pi} \right)(2\pi\epsilon) \;=\; \epsilon$$

Thus $\displaystyle\oint_\Gamma \eta\, dz = 0$ and the proof is complete.

2. If $f(z)$ is analytic inside and on the boundary C of a simply-connected region $\mathcal{R}$, prove that

$$f'(a) \;=\; \frac{1}{2\pi i} \oint_C \frac{f(z)}{(z-a)^2}\, dz$$

From Problem 1 if a and $a + h$ lie in $\mathcal{R}$, we have

$$\frac{f(a+h) - f(a)}{h} \;=\; \frac{1}{2\pi i} \oint_C \frac{1}{h}\left\{ \frac{1}{z-(a+h)} - \frac{1}{z-a} \right\} f(z)\, dz \;=\; \frac{1}{2\pi i} \oint_C \frac{f(z)\, dz}{(z-a-h)(z-a)}$$

$$=\; \frac{1}{2\pi i} \oint_C \frac{f(z)\, dz}{(z-a)^2} \;+\; \frac{h}{2\pi i} \oint_C \frac{f(z)\, dz}{(z-a-h)(z-a)^2}$$

The result follows on taking the limit as $h \to 0$ if we can show that the last term approaches zero.

To show this we use the fact that if Γ is a circle of radius ϵ and center a which lies entirely in $\mathcal{R}$ (see Fig. 5-3), then

$$\frac{h}{2\pi i} \oint_C \frac{f(z)\, dz}{(z-a-h)(z-a)^2}$$

$$=\; \frac{h}{2\pi i} \oint_\Gamma \frac{f(z)\, dz}{(z-a-h)(z-a)^2}$$

Choosing h so small in absolute value that $a + h$ lies in Γ and $|h| < \epsilon/2$, we have by Problem 7(c), Chapter 1, and the fact that Γ has equation $|z - a| = \epsilon$,

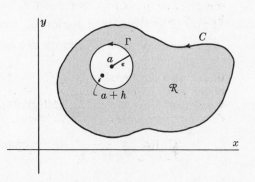

Fig. 5-3

$$|z - a - h| \;\geqq\; |z - a| - |h| \;>\; \epsilon - \epsilon/2 = \epsilon/2$$

Also since $f(z)$ is analytic in $\mathcal{R}$, we can find a positive number M such that $|f(z)| < M$.

Then since the length of Γ is $2\pi\epsilon$, we have

$$\left| \frac{h}{2\pi i} \oint_\Gamma \frac{f(z)\,dz}{(z - a - h)(z - a)^2} \right| \;\leqq\; \frac{|h|}{2\pi} \frac{M(2\pi\epsilon)}{(\epsilon/2)(\epsilon^2)} = \frac{2\,|h|\,M}{\epsilon^2}$$

and it follows that the left side approaches zero as $h \to 0$, thus completing the proof.

It is of interest to observe that the result is equivalent to

$$\frac{d}{da} f(a) \;=\; \frac{d}{da} \left\{ \frac{1}{2\pi i} \oint_C \frac{f(z)}{z - a} dz \right\} \;=\; \frac{1}{2\pi i} \oint_C \frac{\partial}{\partial a} \left\{ \frac{f(z)}{z - a} \right\} dz$$

which is an extension to contour integrals of *Leibnitz's rule* for differentiating under the integral sign.

3. Prove that under the conditions of Problem 2,

$$f^{(n)}(a) \;=\; \frac{n!}{2\pi i} \oint_C \frac{f(z)}{(z - a)^{n+1}} dz \qquad n = 0, 1, 2, 3, \ldots$$

The cases where $n = 0$ and 1 follow from Problems 1 and 2 respectively provided we define $f^{(0)}(a) = f(a)$ and $0! = 1$.

To establish the case where $n = 2$, we use Problem 2 where a and $a + h$ lie in $\mathcal{R}$ to obtain

$$\frac{f'(a + h) - f'(a)}{h} \;=\; \frac{1}{2\pi i} \oint_C \frac{1}{h} \left\{ \frac{1}{(z - a - h)^2} - \frac{1}{(z - a)^2} \right\} f(z)\,dz$$

$$= \frac{2!}{2\pi i} \oint_C \frac{f(z)}{(z - a)^3} dz \;+\; \frac{h}{2\pi i} \oint_C \frac{3(z - a) - 2h}{(z - a - h)^2 (z - a)^3} f(z)\,dz$$

The result follows on taking the limit as $h \to 0$ if we can show that the last term approaches zero. The proof is similar to that of Problem 2, for using the fact that the integral around C equals the integral around Γ, we have

$$\left| \frac{h}{2\pi i} \oint_\Gamma \frac{3(z - a) - 2h}{(z - a - h)^2 (z - a)^3} f(z)\,dz \right| \;\leqq\; \frac{|h|}{2\pi} \frac{M(2\pi\epsilon)}{(\epsilon/2)^2 (\epsilon^3)} = \frac{4\,|h|\,M}{\epsilon^4}$$

Since M exists such that $|\{3(z - a) - 2h\} f(z)| < M$.

In a similar manner we can establish the result for $n = 3, 4, \ldots$ (see Problems 36 and 37).

The result is equivalent to (see last paragraph of Problem 2)

$$\frac{d^n}{da^n} f(a) \;=\; \frac{d^n}{da^n} \left\{ \frac{1}{2\pi i} \oint_C \frac{f(z)}{z - a} dz \right\} \;=\; \frac{1}{2\pi i} \oint_C \frac{\partial^n}{\partial a^n} \left\{ \frac{f(z)}{z - a} \right\} dz$$

4. If $f(z)$ is analytic in a region $\mathcal{R}$, prove that $f'(z), f''(z), \ldots$ are analytic in $\mathcal{R}$.

This follows from Problems 2 and 3.

5. Evaluate   (a) $\displaystyle\oint_C \frac{\sin \pi z^2 + \cos \pi z^2}{(z - 1)(z - 2)} dz$,

   (b) $\displaystyle\oint_C \frac{e^{2z}}{(z + 1)^4} dz$ where C is the circle $|z| = 3$.

(a) Since $\displaystyle\frac{1}{(z - 1)(z - 2)} = \frac{1}{z - 2} - \frac{1}{z - 1}$,   we have

$$\oint_C \frac{\sin \pi z^2 + \cos \pi z^2}{(z-1)(z-2)} \, dz = \oint_C \frac{\sin \pi z^2 + \cos \pi z^2}{z-2} \, dz - \oint_C \frac{\sin \pi z^2 + \cos \pi z^2}{z-1} \, dz$$

By Cauchy's integral formula with $a = 2$ and $a = 1$ respectively, we have

$$\oint_C \frac{\sin \pi z^2 + \cos \pi z^2}{z-2} \, dz = 2\pi i \{\sin \pi (2)^2 + \cos \pi (2)^2\} = 2\pi i$$

$$\oint_C \frac{\sin \pi z^2 + \cos \pi z^2}{z-1} \, dz = 2\pi i \{\sin \pi (1)^2 + \cos \pi (1)^2\} = -2\pi i$$

since $z = 1$ and $z = 2$ are inside C and $\sin \pi z^2 + \cos \pi z^2$ is analytic inside C. Then the required integral has the value $2\pi i - (-2\pi i) = 4\pi i$.

(b) Let $f(z) = e^{2z}$ and $a = -1$ in the Cauchy integral formula

$$f^{(n)}(a) = \frac{n!}{2\pi i} \oint_C \frac{f(z)}{(z-a)^{n+1}} \, dz \qquad\qquad (1)$$

If $n = 3$, then $f'''(z) = 8e^{2z}$ and $f'''(-1) = 8e^{-2}$. Hence (1) becomes

$$8e^{-2} = \frac{3!}{2\pi i} \oint_C \frac{e^{2z}}{(z+1)^4} \, dz$$

from which we see that the required integral has the value $8\pi i e^{-2}/3$.

6. **Prove Cauchy's integral formula for multiply-connected regions.**

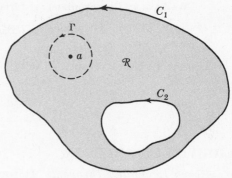

Fig. 5-4

We present a proof for the multiply-connected region $\mathcal{R}$ bounded by the simple closed curves C_1 and C_2 as indicated in Fig. 5-4. Extensions to other multiply-connected regions are easily made (see Problem 40).

Construct a circle Γ having center at any point a in $\mathcal{R}$ so that Γ lies entirely in $\mathcal{R}$. Let $\mathcal{R}'$ consist of the set of points in $\mathcal{R}$ which are exterior to Γ. Then the function $\frac{f(z)}{z-a}$ is analytic inside and on the boundary of $\mathcal{R}'$. Hence by Cauchy's theorem for multiply-connected regions (Problem 16, Chapter 4),

$$\frac{1}{2\pi i} \oint_{C_1} \frac{f(z)}{z-a} \, dz - \frac{1}{2\pi i} \oint_{C_2} \frac{f(z)}{z-a} \, dz - \frac{1}{2\pi i} \oint_{\Gamma} \frac{f(z)}{z-a} \, dz = 0 \qquad\qquad (1)$$

But by Cauchy's integral formula for simply-connected regions, we have

$$f(a) = \frac{1}{2\pi i} \oint_{\Gamma} \frac{f(z)}{z-a} \, dz \qquad\qquad (2)$$

so that from (1),

$$f(a) = \frac{1}{2\pi i} \oint_{C_1} \frac{f(z)}{z-a} \, dz - \frac{1}{2\pi i} \oint_{C_2} \frac{f(z)}{z-a} \, dz \qquad\qquad (3)$$

Then if C represents the entire boundary of $\mathcal{R}$ (suitably traversed so that an observer moving around C always has $\mathcal{R}$ lying to his left), we can write (3) as

$$f(a) = \frac{1}{2\pi i} \oint_C \frac{f(z)}{z-a} \, dz$$

In a similar manner we can show that the other Cauchy integral formulas

$$f^{(n)}(a) = \frac{n!}{2\pi i} \oint_C \frac{f(z)}{(z-a)^{n+1}} \, dz \qquad n = 1, 2, 3, \ldots$$

hold for multiply-connected regions (see Problem 40).

MORERA'S THEOREM

7. Prove *Morera's theorem* (the converse of Cauchy's theorem): If $f(z)$ is continuous in a simply-connected region $\mathcal{R}$ and if

$$\oint_C f(z)\,dz = 0$$

around every simple closed curve C in $\mathcal{R}$, then $f(z)$ is analytic in $\mathcal{R}$.

If $\oint_C f(z)\,dz = 0$ independent of C, it follows by Problem 17, Chapter 4, that $F(z) = \int_a^z f(z)\,dz$ is independent of the path joining a and z, so long as this path is in $\mathcal{R}$.

Then by reasoning identical with that used in Problem 18, Chapter 4, it follows that $F(z)$ is analytic in $\mathcal{R}$ and $F'(z) = f(z)$. However, by Problem 2, it follows that $F'(z)$ is also analytic if $F(z)$ is. Hence $f(z)$ is analytic in $\mathcal{R}$.

CAUCHY'S INEQUALITY

8. If $f(z)$ is analytic inside and on a circle C of radius r and center at $z = a$, prove *Cauchy's inequality*

$$|f^{(n)}(a)| \leq \frac{M \cdot n!}{r^n} \qquad n = 0, 1, 2, 3, \ldots$$

where M is a constant such that $|f(z)| < M$.

We have by Cauchy's integral formulas,

$$f^{(n)}(a) = \frac{n!}{2\pi i} \oint_C \frac{f(z)}{(z-a)^{n+1}}\,dz \qquad n = 0, 1, 2, 3, \ldots$$

Then by Problem 3, Chapter 4, since $|z - a| = r$ on C and the length of C is $2\pi r$,

$$|f^{(n)}(a)| = \frac{n!}{2\pi} \left| \oint_C \frac{f(z)}{(z-a)^{n+1}}\,dz \right| \leq \frac{n!}{2\pi} \cdot \frac{M}{r^{n+1}} \cdot 2\pi r = \frac{M \cdot n!}{r^n}$$

LIOUVILLE'S THEOREM

9. Prove *Liouville's theorem*: If for all z in the entire complex plane, (i) $f(z)$ is analytic and (ii) $f(z)$ is bounded [i.e. we can find a constant M such that $|f(z)| < M$], then $f(z)$ must be a constant.

Let a and b be any two points in the z plane. Suppose that C is a circle of radius r having center at a and enclosing point b (see Fig. 5-5).

From Cauchy's integral formula, we have

$$f(b) - f(a) = \frac{1}{2\pi i} \oint_C \frac{f(z)}{z-b}\,dz - \frac{1}{2\pi i} \oint_C \frac{f(z)}{z-a}\,dz$$

$$= \frac{b-a}{2\pi i} \oint_C \frac{f(z)\,dz}{(z-b)(z-a)}$$

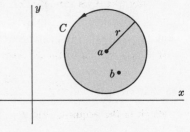

Fig. 5-5

Now we have

$$|z-a| = r, \qquad |z-b| = |z-a+a-b| \geq |z-a| - |a-b| = r - |a-b| \geq r/2$$

if we choose r so large that $|a - b| < r/2$. Then since $|f(z)| < M$ and the length of C is $2\pi r$, we have by Problem 3, Chapter 4,

$$|f(b) - f(a)| = \frac{|b-a|}{2\pi} \left| \oint_C \frac{f(z)\,dz}{(z-b)(z-a)} \right| \leq \frac{|b-a|\,M(2\pi r)}{2\pi(r/2)r} = \frac{2\,|b-a|\,M}{r}$$

Letting $r \to \infty$ we see that $|f(b) - f(a)| = 0$ or $f(b) = f(a)$, which shows that $f(z)$ must be a constant.

Another method. Letting $n = 1$ in Problem 8 and replacing a by z we have,

$$|f'(z)| \leq M/r$$

Letting $r \to \infty$, we deduce that $|f'(z)| = 0$ and so $f'(z) = 0$. Hence $f(z) = \text{constant}$, as required.

FUNDAMENTAL THEOREM OF ALGEBRA

10. Prove the *fundamental theorem of algebra*: Every polynomial equation $P(z) = a_0 + a_1 z + a_2 z^2 + \cdots + a_n z^n = 0$, where the degree $n \geq 1$ and $a_n \neq 0$, has at least one root.

If $P(z) = 0$ has no root, then $f(z) = \dfrac{1}{P(z)}$ is analytic for all z. Also $|f(z)| = \dfrac{1}{|P(z)|}$ is bounded (and in fact approaches zero) as $|z| \to \infty$.

Then by Liouville's theorem (Problem 9) it follows that $f(z)$ and thus $P(z)$ must be a constant. Thus we are led to a contradiction and conclude that $P(z) = 0$ must have at least one root or, as is sometimes said, $P(z)$ has at least one *zero*.

11. Prove that every polynomial equation $P(z) = a_0 + a_1 z + a_2 z^2 + \cdots + a_n z^n = 0$, where the degree $n \geq 1$ and $a_n \neq 0$, has exactly n roots.

By the fundamental theorem of algebra (Problem 10), $P(z)$ has at least one root. Denote this root by α. Then $P(\alpha) = 0$. Hence

$$
\begin{aligned}
P(z) - P(\alpha) &= a_0 + a_1 z + a_2 z^2 + \cdots + a_n z^n - (a_0 + a_1 \alpha + a_2 \alpha^2 + \cdots + a_n \alpha^n) \\
&= a_1(z - \alpha) + a_2(z^2 - \alpha^2) + \cdots + a_n(z^n - \alpha^n) \\
&= (z - \alpha)\, Q(z)
\end{aligned}
$$

where $Q(z)$ is a polynomial of degree $(n - 1)$.

Applying the fundamental theorem of algebra again, we see that $Q(z)$ has at least one zero which we can denote by β [which may equal α] and so $P(z) = (z - \alpha)(z - \beta)\, R(z)$. Continuing in this manner we see that $P(z)$ has exactly n zeros.

GAUSS' MEAN VALUE THEOREM

12. Let $f(z)$ be analytic inside and on a circle C with center at a. Prove *Gauss' mean value theorem* that the mean of the values of $f(z)$ on C is $f(a)$.

By Cauchy's integral formula,

$$f(a) = \frac{1}{2\pi i} \oint_C \frac{f(z)}{z - a}\, dz \tag{1}$$

If C has radius r, the equation of C is $|z - a| = r$ or $z = a + re^{i\theta}$. Thus (1) becomes

$$f(a) = \frac{1}{2\pi i} \int_0^{2\pi} \frac{f(a + re^{i\theta})\, ire^{i\theta}}{re^{i\theta}}\, d\theta = \frac{1}{2\pi} \int_0^{2\pi} f(a + re^{i\theta})\, d\theta$$

which is the required result.

MAXIMUM MODULUS THEOREM

13. Prove the *maximum modulus theorem*: If $f(z)$ is analytic inside and on a simple closed curve C, then the maximum value of $|f(z)|$ occurs on C, unless $f(z)$ is a constant.

Method 1.

Since $f(z)$ is analytic and hence continuous inside and on C, it follows that $|f(z)|$ does have a maximum value M for at least one value of z inside or on C. Suppose this maximum value is not attained on the boundary of C but is attained at an interior point a, i.e. $|f(a)| = M$. Let C_1 be a circle inside C with center at a (see Fig. 5-6). If we exclude $f(z)$ from being a constant inside C_1, then there must be a point inside C_1, say b, such that $|f(b)| < M$ or, what is the same thing, $|f(b)| = M - \epsilon$ where $\epsilon > 0$.

Now by the continuity of $|f(z)|$ at b, we see that for any $\epsilon > 0$ we can find $\delta > 0$ such that

$$||f(z)| - |f(b)|| < \tfrac{1}{2}\epsilon \quad \text{whenever} \quad |z - b| < \delta \qquad (1)$$

i.e.,

$$|f(z)| < |f(b)| + \tfrac{1}{2}\epsilon = M - \epsilon + \tfrac{1}{2}\epsilon = M - \tfrac{1}{2}\epsilon \qquad (2)$$

for all points interior to a circle C_2 with center at b and radius δ, as shown shaded in the figure.

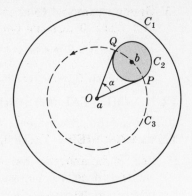

Construct a circle C_3 with center at a which passes through b (dashed in Fig. 5-6). On part of this circle [namely that part PQ included in C_2] we have from (2), $|f(z)| < M - \tfrac{1}{2}\epsilon$. On the remaining part of the circle we have $|f(z)| \leq M$.

Fig. 5-6

If we measure θ counterclockwise from OP and let $\angle POQ = \alpha$, it follows from Problem 12 that if $r = |b - a|$,

$$f(a) = \frac{1}{2\pi} \int_0^\alpha f(a + re^{i\theta})\, d\theta + \frac{1}{2\pi} \int_\alpha^{2\pi} f(a + re^{i\theta})\, d\theta$$

Then

$$|f(a)| \leq \frac{1}{2\pi} \int_0^\alpha |f(a + re^{i\theta})|\, d\theta + \frac{1}{2\pi} \int_\alpha^{2\pi} |f(a + re^{i\theta})|\, d\theta$$

$$\leq \frac{1}{2\pi} \int_0^\alpha (M - \tfrac{1}{2}\epsilon)\, d\theta + \frac{1}{2\pi} \int_\alpha^{2\pi} M\, d\theta$$

$$= \frac{\alpha}{2\pi}(M - \tfrac{1}{2}\epsilon) + \frac{M}{2\pi}(2\pi - \alpha)$$

$$= M - \frac{\alpha\epsilon}{4\pi}$$

i.e. $|f(a)| = M \leq M - \frac{\alpha\epsilon}{4\pi}$, an impossible situation. By virtue of this contradiction we conclude that $|f(z)|$ cannot attain its maximum at any interior point of C and so must attain its maximum on C.

Method 2.

From Problem 12, we have

$$|f(a)| \leq \frac{1}{2\pi} \int_0^{2\pi} |f(a + re^{i\theta})|\, d\theta \qquad (3)$$

Let us suppose that $|f(a)|$ is a maximum so that $|f(a + re^{i\theta})| \leq |f(a)|$. If $|f(a + re^{i\theta})| < |f(a)|$ for one value of θ then, by continuity of f, it would hold for a finite arc, say $\theta_1 < \theta < \theta_2$. But in such case the mean value of $|f(a + re^{i\theta})|$ is less than $|f(a)|$, which would contradict (3). It follows therefore that in any δ neighborhood of a, i.e. for $|z - a| < \delta$, $f(z)$ must be a constant. If $f(z)$ is not a constant, the maximum value of $|f(z)|$ must occur on C.

For another method, see Problem 57.

MINIMUM MODULUS THEOREM

14. Prove the *minimum modulus theorem*: Let $f(z)$ be analytic inside and on a simple closed curve C. Prove that if $f(z) \neq 0$ inside C, then $|f(z)|$ must assume its minimum value on C.

Since $f(z)$ is analytic inside and on C and since $f(z) \neq 0$ inside C, it follows that $1/f(z)$ is analytic inside C. By the maximum modulus theorem it follows that $1/|f(z)|$ cannot assume its maximum value inside C and so $|f(z)|$ cannot assume its minimum value inside C. Then since $|f(z)|$ has a minimum, this minimum must be attained on C.

15. Give an example to show that if $f(z)$ is analytic inside and on a simple closed curve C and $f(z) = 0$ at some point inside C, then $|f(z)|$ need not assume its minimum value on C.

Let $f(z) = z$ for $|z| \leq 1$, so that C is a circle with center at the origin and radius one. We have $f(z) = 0$ at $z = 0$. If $z = re^{i\theta}$, then $|f(z)| = r$ and it is clear that the minimum value of $|f(z)|$ does not occur on C but occurs inside C where $r = 0$, i.e. at $z = 0$.

THE ARGUMENT THEOREM

16. Let $f(z)$ be analytic inside and on a simple closed curve C except for a pole $z = \alpha$ of order (multiplicity) p inside C. Suppose also that inside C $f(z)$ has only one zero $z = \beta$ of order (multiplicity) n and no zeros on C. Prove that

$$\frac{1}{2\pi i} \oint_C \frac{f'(z)}{f(z)} \, dz \;=\; n - p$$

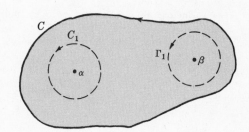

Fig. 5-7

Let C_1 and Γ_1 be non-overlapping circles lying inside C and enclosing $z = \alpha$ and $z = \beta$ respectively. Then

$$\frac{1}{2\pi i} \oint_C \frac{f'(z)}{f(z)} \, dz \;=\; \frac{1}{2\pi i} \oint_{C_1} \frac{f'(z)}{f(z)} \, dz \;+\; \frac{1}{2\pi i} \oint_{\Gamma_1} \frac{f'(z)}{f(z)} \, dz \tag{1}$$

Since $f(z)$ has a pole of order p at $z = \alpha$, we have

$$f(z) \;=\; \frac{F(z)}{(z - \alpha)^p} \tag{2}$$

where $F(z)$ is analytic and different from zero inside and on C_1. Then taking logarithms in (2) and differentiating, we find

$$\frac{f'(z)}{f(z)} \;=\; \frac{F'(z)}{F(z)} - \frac{p}{z - \alpha} \tag{3}$$

so that

$$\frac{1}{2\pi i} \oint_{C_1} \frac{f'(z)}{f(z)} \, dz \;=\; \frac{1}{2\pi i} \oint_{C_1} \frac{F'(z)}{F(z)} \, dz \;-\; \frac{p}{2\pi i} \oint_{C_1} \frac{dz}{z - \alpha} \;=\; 0 - p \;=\; -p \tag{4}$$

Since $f(z)$ has a zero of order n at $z = \beta$, we have

$$f(z) \;=\; (z - \beta)^n \, G(z) \tag{5}$$

where $G(z)$ is analytic and different from zero inside and on Γ_1.

Then by logarithmic differentiation, we have

$$\frac{f'(z)}{f(z)} \;=\; \frac{n}{z - \beta} + \frac{G'(z)}{G(z)} \tag{6}$$

so that

$$\frac{1}{2\pi i} \oint_{\Gamma_1} \frac{f'(z)}{f(z)} \, dz \;=\; \frac{n}{2\pi i} \oint_{\Gamma_1} \frac{dz}{z - \beta} \;+\; \frac{1}{2\pi i} \oint \frac{G'(z)}{G(z)} \, dz \;=\; n \tag{7}$$

Hence from (1), (4) and (7), we have the required result

$$\frac{1}{2\pi i} \oint_C \frac{f'(z)}{f(z)} \, dz \;=\; \frac{1}{2\pi i} \oint_{C_1} \frac{f'(z)}{f(z)} \, dz \;+\; \frac{1}{2\pi i} \oint_{\Gamma_1} \frac{f'(z)}{f(z)} \, dz \;=\; n - p$$

17. Let $f(z)$ be analytic inside and on a simple closed curve C except for a finite number of poles inside C. Suppose that $f(z) \neq 0$ on C. If N and P are respectively the number of zeros and poles of $f(z)$ inside C, counting multiplicities, prove that

$$\frac{1}{2\pi i} \oint_C \frac{f'(z)}{f(z)}\, dz \;=\; N - P$$

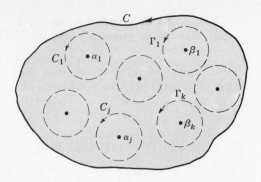

Fig. 5-8

Let $\alpha_1, \alpha_2, \ldots, \alpha_j$ and $\beta_1, \beta_2, \ldots, \beta_k$ be the respective poles and zeros of $f(z)$ lying inside C [Fig. 5-8] and suppose their multiplicities are $p_1, p_2, \ldots, p_j$ and $n_1, n_2, \ldots, n_k$.

Enclose each pole and zero by non-overlapping circles $C_1, C_2, \ldots, C_j$ and $\Gamma_1, \Gamma_2, \ldots, \Gamma_k$. This can always be done since the poles and zeros are isolated.

Then we have, using the results of Problem 16,

$$\frac{1}{2\pi i} \oint_C \frac{f'(z)}{f(z)}\, dz \;=\; \sum_{r=1}^{j} \frac{1}{2\pi i} \oint_{\Gamma_r} \frac{f'(z)}{f(z)}\, dz \;+\; \sum_{r=1}^{k} \frac{1}{2\pi i} \oint_{C_r} \frac{f'(z)}{f(z)}\, dz$$

$$=\; \sum_{r=1}^{j} n_r \;-\; \sum_{r=1}^{k} p_r$$

$$=\; N - P$$

ROUCHÉ'S THEOREM

18. Prove *Rouché's theorem*: If $f(z)$ and $g(z)$ are analytic inside and on a simple closed curve C and if $|g(z)| < |f(z)|$ on C, then $f(z) + g(z)$ and $f(z)$ have the same number of zeros inside C.

Let $F(z) = g(z)/f(z)$ so that $g(z) = f(z) F(z)$ or briefly $g = fF$. Then if N_1 and N_2 are the number of zeros inside C of $f + g$ and f respectively, we have by Problem 17, using the fact that these functions have no poles inside C,

$$N_1 \;=\; \frac{1}{2\pi i} \oint_C \frac{f' + g'}{f + g}\, dz, \qquad N_2 \;=\; \frac{1}{2\pi i} \oint_C \frac{f'}{f}\, dz$$

Then

$$N_1 - N_2 \;=\; \frac{1}{2\pi i} \oint_C \frac{f' + f'F + fF'}{f + fF}\, dz \;-\; \frac{1}{2\pi i} \oint_C \frac{f'}{f}\, dz$$

$$=\; \frac{1}{2\pi i} \oint_C \frac{f'(1 + F) + fF'}{f(1 + F)}\, dz \;-\; \frac{1}{2\pi i} \oint_C \frac{f'}{f}\, dz$$

$$=\; \frac{1}{2\pi i} \oint_C \left\{ \frac{f'}{f} + \frac{F'}{1 + F} \right\} dz \;-\; \frac{1}{2\pi i} \oint_C \frac{f'}{f}\, dz$$

$$=\; \frac{1}{2\pi i} \oint_C \frac{F'}{1 + F}\, dz \;=\; \frac{1}{2\pi i} \int_C F'(1 - F + F^2 - F^3 + \cdots)\, dz$$

$$=\; 0$$

using the given fact that $|F| < 1$ on C so that the series is uniformly convergent on C and term by term integration yields the value zero. Thus $N_1 = N_2$ as required.

19. Use Rouché's theorem (Problem 18) to prove that *every polynomial of degree n has exactly n zeros* (fundamental theorem of algebra).

Suppose the polynomial to be $a_0 + a_1 z + a_2 z^2 + \cdots + a_n z^n$, where $a_n \neq 0$. Choose $f(z) = a_n z^n$ and $g(z) = a_0 + a_1 z + a_2 z^2 + \cdots + a_{n-1} z^{n-1}$.

If C is a circle having center at the origin and radius $r > 1$, then on C we have

$$\left|\frac{g(z)}{f(z)}\right| \;=\; \frac{|a_0 + a_1 z + a_2 z^2 + \cdots + a_{n-1} z^{n-1}|}{|a_n z^n|}$$

$$\leqq \; \frac{|a_0| + |a_1|\,r + |a_2|\,r^2 + \cdots + |a_{n-1}|\,r^{n-1}}{|a_n|\,r^n}$$

$$\leqq \; \frac{|a_0|\,r^{n-1} + |a_1|\,r^{n-1} + |a_2|\,r^{n-1} + \cdots + |a_{n-1}|\,r^{n-1}}{|a_n|\,r^n}$$

$$= \; \frac{|a_0| + |a_1| + |a_2| + \cdots + |a_{n-1}|}{|a_n|\,r}$$

Then by choosing r large enough we can make $\left|\dfrac{g(z)}{f(z)}\right| < 1$, i.e. $|g(z)| < |f(z)|$. Hence by Rouché's theorem the given polynomial $f(z) + g(z)$ has the same number of zeros as $f(z) = a_n z^n$. But since this last function has n zeros all located at $z = 0$, $f(z) + g(z)$ also has n zeros and the proof is complete.

20. Prove that all the roots of $z^7 - 5z^3 + 12 = 0$ lie between the circles $|z| = 1$ and $|z| = 2$.

Consider the circle C_1: $|z| = 1$. Let $f(z) = 12$, $g(z) = z^7 - 5z^3$. On C_1 we have

$$|g(z)| \;=\; |z^7 - 5z^3| \;\leqq\; |z^7| + |5z^3| \;\leqq\; 6 \;<\; 12 \;=\; |f(z)|$$

Hence by Rouché's theorem $f(z) + g(z) = z^7 - 5z^3 + 12$ has the same number of zeros inside $|z| = 1$ as $f(z) = 12$, i.e. there are no zeros inside C_1.

Consider the circle C_2: $|z| = 2$. Let $f(z) = z^7$, $g(z) = 12 - 5z^3$. On C_2 we have

$$|g(z)| \;=\; |12 - 5z^3| \;\leqq\; |12| + |5z^3| \;\leqq\; 60 \;<\; 2^7 \;=\; |f(z)|$$

Hence by Rouché's theorem $f(z) + g(z) = z^7 - 5z^3 + 12$ has the same number of zeros inside $|z| = 2$ as $f(z) = z^7$, i.e. all the zeros are inside C_2.

Hence all the roots lie inside $|z| = 2$ but outside $|z| = 1$, as required.

POISSON'S INTEGRAL FORMULAS FOR A CIRCLE

21. (a) Let $f(z)$ be analytic inside and on the circle C defined by $|z| = R$, and let $z = re^{i\theta}$ be any point inside C. Prove that

$$f(re^{i\theta}) \;=\; \frac{1}{2\pi} \int_0^{2\pi} \frac{R^2 - r^2}{R^2 - 2Rr\cos(\theta - \phi) + r^2}\, f(Re^{i\phi})\, d\phi$$

(b) If $u(r, \theta)$ and $v(r, \theta)$ are the real and imaginary parts of $f(re^{i\theta})$, prove that

$$u(r, \theta) \;=\; \frac{1}{2\pi} \int_0^{2\pi} \frac{(R^2 - r^2)\, u(R, \phi)\, d\phi}{R^2 - 2Rr\cos(\theta - \phi) + r^2}$$

$$v(r, \theta) \;=\; \frac{1}{2\pi} \int_0^{2\pi} \frac{(R^2 - r^2)\, v(R, \phi)\, d\phi}{R^2 - 2Rr\cos(\theta - \phi) + r^2}$$

The results are called *Poisson's integral formulas for the circle*.

(a) Since $z = re^{i\theta}$ is any point inside C, we have by Cauchy's integral formula

$$f(z) \;=\; f(re^{i\theta}) \;=\; \frac{1}{2\pi i} \oint_C \frac{f(w)}{w - z}\, dw \qquad (1)$$

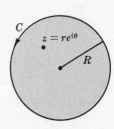

Fig. 5-9

The *inverse of the point z* with respect to C lies outside C and is given by $R^2/\bar{z}$. Hence by Cauchy's theorem,

$$0 \;=\; \frac{1}{2\pi i} \oint_C \frac{f(w)}{w - R^2/\bar{z}}\, dw \qquad (2)$$

If we subtract (2) from (1), we find

$$f(z) = \frac{1}{2\pi i} \oint_C \left\{ \frac{1}{w-z} - \frac{1}{w - R^2/\bar{z}} \right\} f(w)\, dw$$

$$= \frac{1}{2\pi i} \oint_C \frac{z - R^2/\bar{z}}{(w-z)(w - R^2/\bar{z})} f(w)\, dw \tag{3}$$

Now let $z = re^{i\theta}$ and $w = Re^{i\phi}$. Then since $\bar{z} = re^{-i\theta}$, (3) yields

$$f(re^{i\theta}) = \frac{1}{2\pi i} \int_0^{2\pi} \frac{\{re^{i\theta} - (R^2/r)e^{i\theta}\}\, f(Re^{i\phi})\, iRe^{i\phi}\, d\phi}{\{Re^{i\phi} - re^{i\theta}\}\{Re^{i\phi} - (R^2/r)e^{i\theta}\}}$$

$$= \frac{1}{2\pi} \int_0^{2\pi} \frac{(r^2 - R^2)\, e^{i(\theta + \phi)}\, f(Re^{i\phi})\, d\phi}{(Re^{i\phi} - re^{i\theta})(re^{i\phi} - Re^{i\theta})}$$

$$= \frac{1}{2\pi} \int_0^{2\pi} \frac{(R^2 - r^2)\, f(Re^{i\phi})\, d\phi}{(Re^{i\phi} - re^{i\theta})(Re^{-i\phi} - re^{-i\theta})}$$

$$= \frac{1}{2\pi} \int_0^{2\pi} \frac{(R^2 - r^2)\, f(Re^{i\phi})\, d\phi}{R^2 - 2Rr \cos(\theta - \phi) + r^2}$$

(b) Since $f(re^{i\theta}) = u(r, \theta) + i\, v(r, \theta)$ and $f(Re^{i\phi}) = u(R, \phi) + i\, v(R, \phi)$, we have from part (a),

$$u(r, \theta) + i\, v(r, \theta) = \frac{1}{2\pi} \int_0^{2\pi} \frac{(R^2 - r^2)\{u(R, \phi) + i\, v(R, \phi)\}\, d\phi}{R^2 - 2Rr \cos(\theta - \phi) + r^2}$$

$$= \frac{1}{2\pi} \int_0^{2\pi} \frac{(R^2 - r^2)\, u(R, \phi)\, d\phi}{R^2 - 2Rr \cos(\theta - \phi) + r^2} + \frac{i}{2\pi} \int_0^{2\pi} \frac{(R^2 - r^2)\, v(R, \phi)\, d\phi}{R^2 - 2Rr \cos(\theta - \phi) + r^2}$$

Then the required result follows on equating real and imaginary parts.

POISSON'S INTEGRAL FORMULAS FOR A HALF PLANE

22. Derive Poisson's formulas for the half plane [see Page 120].

Let C be the boundary of a semicircle of radius R [see Fig. 5-10] containing ζ as an interior point. Since C encloses ζ but does not enclose $\bar{\zeta}$, we have by Cauchy's integral formula,

$$f(\zeta) = \frac{1}{2\pi i} \oint_C \frac{f(z)}{z - \zeta}\, dz, \qquad 0 = \frac{1}{2\pi i} \oint_C \frac{f(z)}{z - \bar{\zeta}}\, dz$$

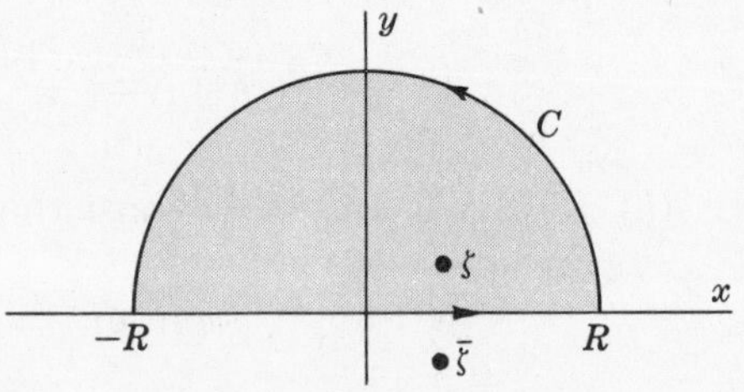

Then by subtraction,

$$f(\zeta) = \frac{1}{2\pi i} \oint_C f(z) \left\{ \frac{1}{z - \zeta} - \frac{1}{z - \bar{\zeta}} \right\} dz$$

$$= \frac{1}{2\pi i} \oint_C \frac{(\zeta - \bar{\zeta})\, f(z)\, dz}{(z - \zeta)(z - \bar{\zeta})}$$

Fig. 5-10

Letting $\zeta = \xi + i\eta$, $\bar{\zeta} = \xi - i\eta$, this can be written

$$f(\zeta) = \frac{1}{\pi} \int_{-R}^{R} \frac{\eta\, f(x)\, dx}{(x - \xi)^2 + \eta^2} + \frac{1}{\pi} \int_\Gamma \frac{\eta\, f(z)\, dz}{(z - \zeta)(z - \bar{\zeta})}$$

where Γ is the semicircular arc of C. As $R \to \infty$, this last integral approaches zero [see Problem 76] and we have

$$f(\zeta) = \frac{1}{\pi} \int_{-\infty}^{\infty} \frac{\eta\, f(x)\, dx}{(x - \xi)^2 + \eta^2}$$

Writing $f(\zeta) = f(\xi + i\eta) = u(\xi, \eta) + i\, v(\xi, \eta)$, $f(x) = u(x, 0) + i\, v(x, 0)$, we obtain as required,

$$u(\xi, \eta) = \frac{1}{\pi} \int_{-\infty}^{\infty} \frac{\eta\, u(x, 0)\, dx}{(x - \xi)^2 + \eta^2}, \qquad v(\xi, \eta) = \frac{1}{\pi} \int_{-\infty}^{\infty} \frac{\eta\, v(x, 0)\, dx}{(x - \xi)^2 + \eta^2}$$

MISCELLANEOUS PROBLEMS

23. Let $f(z)$ be analytic in a region $\mathcal{R}$ bounded by two concentric circles C_1 and C_2 and on the boundary [Fig. 5-11]. Prove that if z_0 is any point in $\mathcal{R}$, then

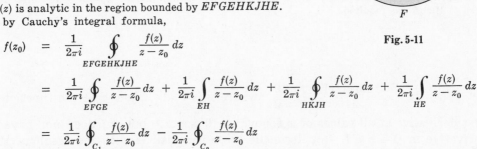

$$f(z_0) = \frac{1}{2\pi i} \oint_{C_1} \frac{f(z)}{z - z_0} dz - \frac{1}{2\pi i} \oint_{C_2} \frac{f(z)}{z - z_0} dz$$

Method 1.

Construct cross-cut EH connecting circles C_1 and C_2. Then $f(z)$ is analytic in the region bounded by $EFGEHKJHE$. Hence by Cauchy's integral formula,

$$f(z_0) = \frac{1}{2\pi i} \oint_{EFGEHKJHE} \frac{f(z)}{z - z_0} dz$$

Fig. 5-11

$$= \frac{1}{2\pi i} \oint_{EFGE} \frac{f(z)}{z - z_0} dz + \frac{1}{2\pi i} \int_{EH} \frac{f(z)}{z - z_0} dz + \frac{1}{2\pi i} \oint_{HKJH} \frac{f(z)}{z - z_0} dz + \frac{1}{2\pi i} \int_{HE} \frac{f(z)}{z - z_0} dz$$

$$= \frac{1}{2\pi i} \oint_{C_1} \frac{f(z)}{z - z_0} dz - \frac{1}{2\pi i} \oint_{C_2} \frac{f(z)}{z - z_0} dz$$

since the integrals along EH and HE cancel.

Similar results can be established for the derivatives of $f(z)$.

Method 2. The result also follows from equation (3) of Problem 6 if we replace the simple closed curves C_1 and C_2 by the circles of Fig. 5-11.

24. Prove that $\displaystyle\int_0^{2\pi} \cos^{2n} \theta \, d\theta = \frac{1 \cdot 3 \cdot 5 \cdots (2n-1)}{2 \cdot 4 \cdot 6 \cdots (2n)} 2\pi$ where $n = 1, 2, 3, \dots$.

Let $z = e^{i\theta}$. Then $dz = ie^{i\theta} d\theta = iz \, d\theta$ or $d\theta = dz/iz$ and $\cos\theta = \frac{1}{2}(e^{i\theta} + e^{-i\theta}) = \frac{1}{2}(z + 1/z)$. Hence if C is the unit circle $|z| = 1$, we have

$$\int_0^{2\pi} \cos^{2n}\theta \, d\theta = \oint_C \left\{ \frac{1}{2}\left(z + \frac{1}{z}\right)\right\}^{2n} \frac{dz}{iz}$$

$$= \frac{1}{2^{2n} i} \oint_C \frac{1}{z}\left\{ z^{2n} + \binom{2n}{1}(z^{2n-1})\left(\frac{1}{z}\right) + \cdots + \binom{2n}{k}(z^{2n-k})\left(\frac{1}{z}\right)^k + \cdots + \left(\frac{1}{z}\right)^{2n}\right\} dz$$

$$= \frac{1}{2^{2n} i} \oint_C \left\{ z^{2n-1} + \binom{2n}{1}z^{2n-3} + \cdots + \binom{2n}{k}z^{2n-2k-1} + \cdots + z^{-2n}\right\} dz$$

$$= \frac{1}{2^{2n} i} \cdot 2\pi i \binom{2n}{n} = \frac{1}{2^{2n}}\binom{2n}{n} 2\pi$$

$$= \frac{1}{2^{2n}} \frac{(2n)!}{n! \, n!} 2\pi = \frac{(2n)(2n-1)(2n-2)\cdots(n)(n-1)\cdots 1}{2^{2n} n! \, n!} 2\pi$$

$$= \frac{1 \cdot 3 \cdot 5 \cdots (2n-1)}{2 \cdot 4 \cdot 6 \cdots 2n} 2\pi$$

25. If $f(z) = u(x, y) + i \, v(x, y)$ is analytic in a region $\mathcal{R}$, prove that u and v are harmonic in $\mathcal{R}$.

In Problem 6, Chapter 3, we proved that u and v are harmonic in $\mathcal{R}$, i.e. satisfy the equation $\frac{\partial^2 \phi}{\partial x^2} + \frac{\partial^2 \phi}{\partial y^2} = 0$, under the *assumption* of existence of the second partial derivatives of u and v, i.e. the existence of $f''(z)$.

This assumption is no longer necessary since we have in fact proved in Problem 4 that if $f(z)$ is analytic in $\mathcal{R}$ then *all* the derivatives of $f(z)$ exist.

26. Prove *Schwarz's theorem*: Let $f(z)$ be analytic for $|z| \leq R$, $f(0) = 0$ and $|f(z)| \leq M$. Then

$$|f(z)| \; \leq \; \frac{M\,|z|}{R}$$

The function $f(z)/z$ is analytic in $|z| \leq R$. Hence on $|z| = R$ we have by the maximum modulus theorem,

$$\left| \frac{f(z)}{z} \right| \; \leq \; \frac{M}{R}$$

However, since this inequality must also hold for points inside $|z| = R$, we have for $|z| \leq R$, $|f(z)| \leq M|z|/R$ as required.

27. Let $f(x) = \begin{cases} x^2 \sin(1/x) & x \neq 0 \\ 0 & x = 0 \end{cases}$ where x is real. Show that the function (*a*) has a first derivative at all values of x for which $0 \leq x \leq 1$ but (*b*) does not have a second derivative in $0 \leq x \leq 1$. (*c*) Reconcile these conclusions with the result of Problem 4.

(*a*) The only place where there is any question as to existence of the first derivative is at $x = 0$. But at $x = 0$ the derivative is

$$\lim_{\Delta x \to 0} \frac{f(0 + \Delta x) - f(0)}{\Delta x} \;=\; \lim_{\Delta x \to 0} \frac{(\Delta x)^2 \sin(1/\Delta x) - 0}{\Delta x}$$

$$=\; \lim_{\Delta x \to 0} \Delta x \sin(1/\Delta x) \;=\; 0$$

and so exists.

At all other values of x in $0 \leq x \leq 1$, the derivative is given (using elementary differentiation rules) by

$$x^2 \cos(1/x) \,\{-1/x^2\} \,+\, (2x) \sin(1/x) \;=\; 2x \sin(1/x) \,-\, \cos(1/x)$$

(*b*) From part (*a*), we have

$$f'(x) \;=\; \begin{cases} 2x \sin(1/x) \,-\, \cos(1/x) & x \neq 0 \\ 0 & x = 0 \end{cases}$$

The second derivative exists for all x such that $0 < x \leq 1$. At $x = 0$ the second derivative is given by

$$\lim_{\Delta x \to 0} \frac{f'(0 + \Delta x) \,-\, f'(0)}{\Delta x} \;=\; \lim_{\Delta x \to 0} \frac{2\,\Delta x \sin(1/\Delta x) \,-\, \cos(1/\Delta x) \,-\, 0}{\Delta x}$$

$$=\; \lim_{\Delta x \to 0} \{2 \sin(1/\Delta x) \,-\, (1/\Delta x) \cos(1/\Delta x)\}$$

which does not exist.

It follows that the second derivative of $f(x)$ does not exist in $0 \leq x \leq 1$.

(*c*) According to Problem 4, if $f(z)$ is analytic in a region $\mathcal{R}$ then all higher derivatives exist and are analytic in $\mathcal{R}$. The above results do not conflict with this, since the function $f(z) = z^2 \sin(1/z)$ is not analytic in any region which includes $z = 0$.

28. (*a*) If $F(z)$ is analytic inside and on a simple closed curve C except for a pole of order m at $z = a$ inside C, prove that

$$\frac{1}{2\pi i} \oint_C F(z)\,dz \;=\; \lim_{z \to a} \frac{1}{(m-1)!} \frac{d^{m-1}}{dz^{m-1}} \{(z-a)^m\,F(z)\}$$

(*b*) How would you modify the result in (*a*) if more than one pole were inside C?

(a) If $F(z)$ has a pole of order m at $z = a$, then $F(z) = f(z)/(z-a)^m$ where $f(z)$ is analytic inside and on C, and $f(a) \neq 0$. Then by Cauchy's integral formula,

$$\frac{1}{2\pi i} \oint_C F(z)\, dz \;=\; \frac{1}{2\pi i} \oint_C \frac{f(z)}{(z-a)^m}\, dz \;=\; \frac{f^{(m-1)}(a)}{(m-1)!}$$

$$=\; \lim_{z \to a} \frac{1}{(m-1)!} \frac{d^{m-1}}{dz^{m-1}} \{(z-a)^m F(z)\}$$

(b) Suppose there are two poles at $z = a_1$ and $z = a_2$ inside C, of orders m_1 and m_2 respectively. Let Γ_1 and Γ_2 be circles inside C having radii ϵ_1 and ϵ_2 and centers at a_1 and a_2 respectively. Then

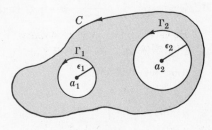

Fig. 5-12

$$\frac{1}{2\pi i} \oint_C F(z)\, dz \;=\; \frac{1}{2\pi i} \oint_{\Gamma_1} F(z)\, dz$$

$$+\; \frac{1}{2\pi i} \oint_{\Gamma_2} F(z)\, dz \qquad (1)$$

If $F(z)$ has a pole of order m_1 at $z = a_1$, then

$$F(z) \;=\; \frac{f_1(z)}{(z-a_1)^{m_1}} \qquad \text{where } f_1(z) \text{ is analytic and } f_1(a_1) \neq 0$$

If $F(z)$ has a pole of order m_2 at $z = a_2$, then

$$F(z) \;=\; \frac{f_2(z)}{(z-a_2)^{m_2}} \qquad \text{where } f_2(z) \text{ is analytic and } f_2(a_2) \neq 0$$

Then by (1) and part (a),

$$\frac{1}{2\pi i} \oint_C F(z)\, dz \;=\; \frac{1}{2\pi i} \oint_{\Gamma_1} \frac{f_1(z)}{(z-a_1)^{m_1}}\, dz \;+\; \frac{1}{2\pi i} \oint_{\Gamma_2} \frac{f_2(z)}{(z-a_2)^{m_2}}\, dz$$

$$=\; \lim_{z \to a_1} \frac{1}{(m_1-1)!} \frac{d^{m_1-1}}{dz^{m_1-1}} \{(z-a_1)^{m_1} F(z)\}$$

$$+\; \lim_{z \to a_2} \frac{1}{(m_2-1)!} \frac{d^{m_2-1}}{dz^{m_2-1}} \{(z-a_2)^{m_2} F(z)\}$$

If the limits on the right are denoted by R_1 and R_2, we can write

$$\oint_C F(z)\, dz \;=\; 2\pi i (R_1 + R_2)$$

where R_1 and R_2 are called the *residues* of $F(z)$ at the poles $z = a_1$ and $z = a_2$.

In general if $F(z)$ has a number of poles inside C with residues $R_1, R_2, \ldots$, then $\oint_C F(z)\, dz =$ $2\pi i$ times the sum of the residues. This result is called the *residue theorem*. Applications of this theorem together with generalization to singularities other than poles, are treated in Chap. 7.

29. Evaluate $\displaystyle\oint_C \frac{e^z}{(z^2+\pi^2)^2}\, dz$ where C is the circle $|z| = 4$.

The poles of $\dfrac{e^z}{(z^2+\pi^2)^2} = \dfrac{e^z}{(z-\pi i)^2 (z+\pi i)^2}$ are at $z = \pm\pi i$ inside C and are both of order two.

Residue at $z = \pi i$ is $\displaystyle\lim_{z \to \pi i} \frac{1}{1!} \frac{d}{dz} \left\{ (z-\pi i)^2 \frac{e^z}{(z-\pi i)^2(z+\pi i)^2} \right\} = \frac{\pi+i}{4\pi^3}.$

Residue at $z = -\pi i$ is $\displaystyle\lim_{z \to -\pi i} \frac{1}{1!} \frac{d}{dz} \left\{ (z+\pi i)^2 \frac{e^z}{(z-\pi i)^2(z+\pi i)^2} \right\} = \frac{\pi-i}{4\pi^3}.$

Then $\displaystyle\oint_C \frac{e^z}{(z^2+\pi^2)^2}\, dz = 2\pi i(\text{sum of residues}) = 2\pi i \left(\frac{\pi+i}{4\pi^3} + \frac{\pi-i}{4\pi^3} \right) = \frac{i}{\pi}.$

Supplementary Problems

CAUCHY'S INTEGRAL FORMULAS

30. Evaluate $\dfrac{1}{2\pi i}\displaystyle\oint_C \dfrac{e^z}{z-2}\,dz$ if C is (a) the circle $|z|=3$, (b) the circle $|z|=1$.   *Ans.* (a) e^2, (b) 0

31. Evaluate $\displaystyle\oint_C \dfrac{\sin 3z}{z+\pi/2}\,dz$ if C is the circle $|z|=5$.   *Ans.* $2\pi i$

32. Evaluate $\displaystyle\oint_C \dfrac{e^{3z}}{z-\pi i}\,dz$ if C is (a) the circle $|z-1|=4$, (b) the ellipse $|z-2|+|z+2|=6$.
Ans. (a) $-2\pi i$, (b) 0

33. Evaluate $\dfrac{1}{2\pi i}\displaystyle\oint_C \dfrac{\cos\pi z}{z^2-1}\,dz$ around a rectangle with vertices at: (a) $2\pm i,\ -2\pm i$; (b) $-i,\ 2-i,\ 2+i,\ i$.
Ans. (a) 0, (b) $-\frac{1}{2}$

34. Show that $\dfrac{1}{2\pi i}\displaystyle\oint_C \dfrac{e^{zt}}{z^2+1}\,dz = \sin t$ if $t>0$ and C is the circle $|z|=3$.

35. Evaluate $\displaystyle\oint_C \dfrac{e^{iz}}{z^3}\,dz$ where C is the circle $|z|=2$.   *Ans.* $-\pi i$

36. Prove that $f'''(a) = \dfrac{3!}{2\pi i}\displaystyle\oint_C \dfrac{f(z)\,dz}{(z-a)^4}$ if C is a simple closed curve enclosing $z=a$ and $f(z)$ is analytic inside and on C.

37. Prove Cauchy's integral formulas for all positive integral values of n. [*Hint:* Use mathematical induction.]

38. Find the value of (a) $\displaystyle\oint_C \dfrac{\sin^6 z}{z-\pi/6}\,dz$, (b) $\displaystyle\oint_C \dfrac{\sin^6 z}{(z-\pi/6)^3}\,dz$ if C is the circle $|z|=1$.
Ans. (a) $\pi i/32$, (b) $21\pi i/16$

39. Evaluate $\dfrac{1}{2\pi i}\displaystyle\oint_C \dfrac{e^{zt}}{(z^2+1)^2}\,dz$ if $t>0$ and C is the circle $|z|=3$.   *Ans.* $\frac{1}{2}(\sin t - t\cos t)$

40. Prove Cauchy's integral formulas for the multiply-connected region of Fig. 4-26, Page 115.

MORERA'S THEOREM

41. (a) Determine whether $G(z) = \displaystyle\int_1^z \dfrac{d\zeta}{\zeta}$ is independent of the path joining 1 and z.

(b) Discuss the relationship of your answer to part (a) with Morera's theorem.

42. Does Morera's theorem apply in a multiply-connected region? Justify your answer.

43. (a) If $P(x,y)$ and $Q(x,y)$ are conjugate harmonic functions and C is any simple closed curve, prove that $\displaystyle\oint_C P\,dx + Q\,dy = 0$.

(b) If for all simple closed curves C in a region $\mathcal{R}$, $\displaystyle\oint_C P\,dx + Q\,dy = 0$, is it true that P and Q are conjugate harmonic functions, i.e. is the converse of (a) true? Justify your conclusion.

CAUCHY'S INEQUALITY

44. (a) Use Cauchy's inequality to obtain estimates for the derivatives of $\sin z$ at $z=0$ and (b) determine how good these estimates are.

45. (a) Show that if $f(z) = 1/(1-z)$, then $f^{(n)}(z) = n!/(1-z)^{n+1}$.

(b) Use (a) to show that the Cauchy inequality is "best possible", i.e. the estimate of growth of the nth derivative cannot be improved for *all* functions.

46. Prove that the equality in Cauchy's inequality (3), Page 118, holds if and only if $f(z) = kMz^n/r^n$ where $|k| = 1$.

47. Discuss Cauchy's inequality for the function $f(z) = e^{-1/z^2}$ in the neighborhood of $z = 0$.

LIOUVILLE'S THEOREM

48. The function of a real variable defined by $f(x) = \sin x$ is (a) analytic everywhere and (b) bounded, i.e. $|\sin x| \leq 1$ for all x but it is certainly not a constant. Does this contradict Liouville's theorem? Explain.

49. A non-constant function $F(z)$ is such that $F(z + a) = F(z)$, $F(z + bi) = F(z)$ where $a > 0$ and $b > 0$ are given constants. Prove that $F(z)$ cannot be analytic in the rectangle $0 \leq x \leq a$, $0 \leq y \leq b$.

FUNDAMENTAL THEOREM OF ALGEBRA

50. (a) Carry out the details of proof of the fundamental theorem of algebra to show that the particular function $f(z) = z^4 - z^2 - 2z + 2$ has exactly four zeros. (b) Determine the zeros of $f(z)$.
    Ans. (b) $1, 1, -1 \pm i$

51. Determine all the roots of the equations (a) $z^3 - 3z + 4i = 0$, (b) $z^4 + z^2 + 1 = 0$.
    Ans. (a) i, $\frac{1}{2}(-i \pm \sqrt{15})$, (b) $\frac{1}{2}(-1 \pm \sqrt{3}\,i)$, $\frac{1}{2}(1 \pm \sqrt{3}\,i)$

GAUSS' MEAN VALUE THEOREM

52. Evaluate $\dfrac{1}{2\pi} \displaystyle\int_0^{2\pi} \sin^2\left(\pi/6 + 2e^{i\theta}\right) d\theta$ *Ans.* 1/4

53. Show that the mean value of any harmonic function over a circle is equal to the value of the function at the center.

54. Find the mean value of $x^2 - y^2 + 2y$ over the circle $|z - 5 + 2i| = 3$.   *Ans.* 5

55. Prove that $\displaystyle\int_0^\pi \ln \sin \theta \, d\theta = -\pi \ln 2$.   [*Hint.* Consider $f(z) = \ln(1+z)$.]

MAXIMUM MODULUS THEOREM

56. Find the maximum of $|f(z)|$ in $|z| \leq 1$ for the functions $f(z)$ given by (a) $z^2 - 3z + 2$, (b) $z^4 + z^2 + 1$, (c) $\cos 3z$, (d) $(2z + 1)/(2z - 1)$.

57. (a) If $f(z)$ is analytic inside and on the simple closed curve C enclosing $z = a$, prove that

$$\{f(a)\}^n = \frac{1}{2\pi i} \oint_C \frac{\{f(z)\}^n}{z - a}\, dz \qquad n = 0, 1, 2, \ldots$$

(b) Use (a) to prove that $|f(a)|^n \leq M^n/2\pi D$ where D is the minimum distance from a to the curve C and M is the maximum value of $|f(z)|$ on C.

(c) By taking the nth root of both sides of the inequality in (b) and letting $n \to \infty$, prove the maximum modulus theorem.

58. Let $U(x, y)$ be harmonic inside and on a simple closed curve C. Prove that the (a) maximum and (b) minimum values of $U(x, y)$ are attained on C. Are there other restrictions on $U(x, y)$?

59. Verify Problem 58 for the functions (a) $x^2 - y^2$ and (b) $x^3 - 3xy^2$ if C is the circle $|z| = 1$.

60. Is the maximum modulus theorem valid for multiply-connected regions? Justify your answer.

THE ARGUMENT THEOREM

61. If $f(z) = z^5 - 3iz^2 + 2z - 1 + i$, evaluate $\oint_C \dfrac{f'(z)}{f(z)} dz$ where C encloses all the zeros of $f(z)$.
      Ans. $10\pi i$

62. Let $f(z) = \dfrac{(z^2 + 1)^2}{(z^2 + 2z + 2)^3}$. Evaluate $\dfrac{1}{2\pi i} \oint_C \dfrac{f'(z)}{f(z)} dz$ where C is the circle $|z| = 4$.   *Ans.* -2

63. Evaluate $\oint_C \dfrac{f'(z)}{f(z)} dz$ if C is the circle $|z| = \pi$ and   (a) $f(z) = \sin \pi z$,   (b) $f(z) = \cos \pi z$,   (c) $f(z) = \tan \pi z$.

      Ans. (a) $14\pi i$,   (b) $12\pi i$,   (c) $2\pi i$

64. If $f(z) = z^4 - 2z^3 + z^2 - 12z + 20$ and C is the circle $|z| = 5$, evaluate $\oint_C \dfrac{z f'(z)}{f(z)} dz$.   *Ans.* $4\pi i$

ROUCHÉ'S THEOREM

65. If $a > e$, prove that the equation $az^n = e^z$ has n roots inside $|z| = 1$.

66. Prove that $ze^z = a$ where $a \neq 0$ is real has infinitely many roots.

67. Prove that $\tan z = az$, $a > 0$ has (a) infinitely many real roots, (b) only two pure imaginary roots if $0 < a < 1$, (c) all real roots if $a \geqq 1$.

68. Prove that $z \tan z = a$, $a > 0$ has infinitely many real roots but no imaginary roots.

POISSON'S INTEGRAL FORMULAS FOR A CIRCLE

69. Show that $\displaystyle\int_0^{2\pi} \dfrac{R^2 - r^2}{R^2 - 2Rr \cos(\theta - \phi) + r^2} d\phi = 2\pi$

      (a) with, (b) without Poisson's integral formula for a circle.

70. Show that   (a) $\displaystyle\int_0^{2\pi} \dfrac{e^{\cos\phi} \cos(\sin\phi)}{5 - 4\cos(\theta - \phi)} d\phi = \dfrac{2\pi}{3} e^{\cos\theta} \cos(\sin\theta)$

               (b) $\displaystyle\int_0^{2\pi} \dfrac{e^{\cos\phi} \sin(\sin\phi)}{5 - 4\cos(\theta - \phi)} d\phi = \dfrac{2\pi}{3} e^{\cos\theta} \sin(\sin\theta)$

71. (a) Prove that the function $U(r, \theta) = \dfrac{2}{\pi} \tan^{-1}\left(\dfrac{2r \sin\theta}{1 - r^2}\right)$, $0 < r < 1$, $0 \leqq \theta < 2\pi$ is harmonic inside the circle $|z| = 1$.

      (b) Show that $\displaystyle\lim_{r \to 1-} U(r, \theta) = \begin{cases} 1 & 0 < \theta < \pi \\ -1 & \pi < \theta < 2\pi \end{cases}$.

      (c) Can you derive the expression for $U(r, \theta)$ from Poisson's integral formula for a circle?

72. If $f(z)$ is analytic inside and on the circle C defined by $|z| = R$ and if $z = re^{i\theta}$ is any point inside C, show that

$$f'(re^{i\theta}) = \dfrac{i}{2\pi} \int_0^{2\pi} \dfrac{R(R^2 - r^2) f(Re^{i\phi}) \sin(\theta - \phi)}{[R^2 - 2Rr \cos(\theta - \phi) + r^2]^2} d\phi$$

73. Verify that the functions u and v of equations (7) and (8), Page 119, satisfy Laplace's equation.

POISSON'S INTEGRAL FORMULAS FOR A HALF PLANE

74. Find a function which is harmonic in the upper half plane $y > 0$ and which on the x axis takes the values -1 if $x < 0$ and 1 if $x > 0$.   *Ans.* $1 - (2/\pi) \tan^{-1}(y/x)$

75. Work Problem 74 if the function takes the values -1 if $x < -1$, 0 if $-1 < x < 1$, and 1 if $x > 1$.

      Ans. $1 - \dfrac{1}{\pi} \tan^{-1}\left(\dfrac{y}{x+1}\right) - \dfrac{1}{\pi} \tan^{-1}\left(\dfrac{y}{x-1}\right)$

76. Prove the statement made in Problem 22 that the integral over Γ approaches zero as $R \to \infty$.

77. Prove that under suitable restrictions on $f(x)$,

$$\lim_{\eta \to 0+} \frac{1}{\pi} \int_{-\infty}^{\infty} \frac{\eta f(x)}{(x - \xi)^2 + \eta^2} \, dx \;=\; f(\xi)$$

and state these restrictions.

78. Verify that the functions u and v of equations (10) and (11), Page 120, satisfy Laplace's equation.

MISCELLANEOUS PROBLEMS

79. Evaluate $\dfrac{1}{2\pi i} \oint_C \dfrac{z^2 \, dz}{z^2 + 4}$ where C is the square with vertices at $\pm 2, \pm 2 + 4i$.   *Ans. i*

80. Evaluate $\oint_C \dfrac{\cos^2 tz}{z^3} \, dz$ where C is the circle $|z| = 1$ and $t > 0$.   *Ans. $-2\pi i t^2$*

81. (a) Show that $\oint_C \dfrac{dz}{z + 1} = 2\pi i$ if C is the circle $|z| = 2$.

   (b) Use (a) to show that

   $$\oint_C \frac{(x+1) \, dx + y \, dy}{(x+1)^2 + y^2} \;=\; 0, \qquad \oint_C \frac{(x+1) \, dy - y \, dx}{(x+1)^2 + y^2} \;=\; 2\pi$$

   and verify these results directly.

82. Find all functions $f(z)$ which are analytic everywhere in the entire complex plane and which satisfy the conditions (a) $f(2 - i) = 4i$ and (b) $|f(z)| < e^2$ for all z.

83. If $f(z)$ is analytic inside and on a simple closed curve C, prove that

   $$(a) \quad f'(a) \;=\; \frac{1}{2\pi} \int_0^{2\pi} e^{-i\theta} f(a + e^{i\theta}) \, d\theta$$

   $$(b) \quad \frac{f^{(n)}(a)}{n!} \;=\; \frac{1}{2\pi} \int_0^{2\pi} e^{-ni\theta} f(a + e^{i\theta}) \, d\theta$$

84. Prove that $8z^4 - 6z + 5 = 0$ has one root in each quadrant.

85. Show that (a) $\displaystyle\int_0^{2\pi} e^{\cos \theta} \cos (\sin \theta) \, d\theta \;=\; 0$, (b) $\displaystyle\int_0^{2\pi} e^{\cos \theta} \sin (\sin \theta) \, d\theta \;=\; 2\pi$.

86. Extend the result of Problem 23 so as to obtain formulas for the derivatives of $f(z)$ at any point in $\mathcal{R}$.

87. Prove that $z^3 e^{1-z} = 1$ has exactly two roots inside the circle $|z| = 1$.

88. If $t > 0$ and C is any simple closed curve enclosing $z = -1$, prove that

   $$\frac{1}{2\pi i} \oint_C \frac{z e^{zt}}{(z+1)^3} \, dz \;=\; \left(t - \frac{t^2}{2} \right) e^{-t}$$

89. Find all functions $f(z)$ which are analytic in $|z| < 1$ and which satisfy the conditions (a) $f(0) = 1$ and (b) $|f(z)| \geqq 1$ for $|z| < 1$.

90. Let $f(z)$ and $g(z)$ be analytic inside and on a simple closed curve C except that $f(z)$ has zeros at $a_1, a_2, \ldots, a_m$ and poles at $b_1, b_2, \ldots, b_n$ of orders (multiplicities) $p_1, p_2, \ldots, p_m$ and $q_1, q_2, \ldots, q_n$ respectively. Prove that

   $$\frac{1}{2\pi i} \oint_C g(z) \frac{f'(z)}{f(z)} \, dz \;=\; \sum_{k=1}^{m} p_k \, g(a_k) \;-\; \sum_{k=1}^{n} q_k \, g(b_k)$$

91. If $f(z) = a_0 z^n + a_1 z^{n-1} + a_2 z^{n-2} + \cdots + a_n$ where $a_0 \neq 0$, a_1, ..., a_n are complex constants and C encloses all the zeros of $f(z)$, evaluate $\quad$ (a) $\dfrac{1}{2\pi i} \oint_C \dfrac{z f'(z)}{f(z)} \, dz$, $\quad$ (b) $\dfrac{1}{2\pi i} \oint_C \dfrac{z^2 f'(z)}{f(z)} \, dz$ $\quad$ and interpret the results. $\qquad$ Ans. (a) $-a_1/a_0$, $\quad$ (b) $(a_1^2 - 2a_0 a_2)/a_0^2$

92. Find all functions $f(z)$ which are analytic in the region $|z| \leq 1$ and are such that $\quad$ (a) $f(0) = 3$ and (b) $|f(z)| \leq 3$ for all z such that $|z| < 1$.

93. Prove that $z^6 + 192z + 640 = 0$ has one root in the first and fourth quadrants and two roots in the second and third quadrants.

94. Prove that the function $xy(x^2 - y^2)$ cannot have an absolute maximum or minimum inside the circle $|z| = 1$.

95. (a) If a function is analytic in a region $\mathcal{R}$, is it bounded in $\mathcal{R}$? $\quad$ (b) In view of your answer to (a), is it necessary to state that $f(z)$ is bounded in Liouville's theorem?

96. Find all functions $f(z)$ which are analytic everywhere, have a zero of order two at $z = 0$, satisfy the condition $|f'(z)| \leq 6|z|$ for all z, and are such that $f(i) = -2$.

97. Prove that all the roots of $z^5 + z - 16i = 0$ lie between the circles $|z| = 1$ and $|z| = 2$.

98. If U is harmonic inside and on a simple closed curve C, prove that

$$\oint_C \frac{\partial U}{\partial n} \, ds = 0$$

where n is a unit normal to C in the z plane and s is the arc length parameter.

99. A theorem of Cauchy states that all the roots of the equation $\quad z^n + a_1 z^{n-1} + a_2 z^{n-2} + \cdots + a_n = 0$, where $a_1, a_2, \ldots, a_n$ are real, lie inside the circle $\quad |z| = 1 + \max \{a_1, a_2, \ldots, a_n\}$, i.e. $|z| = 1$ plus the maximum of the values $a_1, a_2, \ldots, a_n$. Verify this theorem for the special cases (a) $z^3 - z^2 + z - 1 = 0$, $\quad$ (b) $z^4 + z^2 + 1 = 0$, $\quad$ (c) $z^4 - z^2 - 2z + 2 = 0$, $\quad$ (d) $z^4 + 3z^2 - 6z + 10 = 0$.

100. Prove the theorem of Cauchy stated in Problem 99.

101. Let $P(z)$ be any polynomial. If m is any positive integer and $\omega = e^{2\pi i/m}$, prove that

$$\frac{P(1) + P(\omega) + P(\omega^2) + \cdots + P(\omega^{m-1})}{m} = P(0)$$

and give a geometric interpretation.

102. Is the result of Problem 101 valid for any function $f(z)$? Justify your answer.

103. Prove *Jensen's theorem*: If $f(z)$ is analytic inside and on the circle $|z| = R$ except for zeros at $a_1, a_2, \ldots, a_m$ of multiplicities $p_1, p_2, \ldots, p_m$ and poles at $b_1, b_2, \ldots, b_n$ of multiplicities $q_1, q_2, \ldots, q_n$ respectively, and if $f(0)$ is finite and different from zero, then

$$\frac{1}{2\pi} \int_0^{2\pi} \ln |f(Re^{i\theta})| \, d\theta = \ln |f(0)| + \sum_{k=1}^m p_k \ln \left(\frac{R}{|a_k|} \right) - \sum_{k=1}^n q_k \ln \left(\frac{R}{|b_k|} \right)$$

[*Hint.* Consider $\displaystyle\oint_C \ln z \, \{f'(z)/f(z)\} \, dz$ where C is the circle $|z| = R$.]

Chapter 6

Infinite Series
Taylor's and Laurent's Series

SEQUENCES OF FUNCTIONS

The ideas of Chapter 2, Pages 40 and 41, for sequences and series of constants are easily extended to sequences and series of functions.

Let $u_1(z), u_2(z), \ldots, u_n(z), \ldots$, denoted briefly by $\{u_n(z)\}$, be a sequence of functions of z defined and single-valued in some region of the z plane. We call $U(z)$ the *limit* of $u_n(z)$ as $n \to \infty$, and write $\lim_{n \to \infty} u_n(z) = U(z)$, if given any positive number ϵ we can find a number N [depending in general on both ϵ and z] such that

$$|u_n(z) - U(z)| < \epsilon \quad \text{for all } n > N$$

In such case we say that the sequence *converges* or is *convergent* to $U(z)$.

If a sequence converges for all values of z (points) in a region $\mathcal{R}$, we call $\mathcal{R}$ the *region of convergence* of the sequence. A sequence which is not convergent at some value (point) z is called *divergent* at z.

The theorems on limits given on Page 40 can be extended to sequences of functions.

SERIES OF FUNCTIONS

From the sequence of functions $\{u_n(z)\}$ let us form a new sequence $\{S_n(z)\}$ defined by

$$
\begin{aligned}
S_1(z) &= u_1(z) \\
S_2(z) &= u_1(z) + u_2(z) \\
&\vdots \\
S_n(z) &= u_1(z) + u_2(z) + \cdots + u_n(z)
\end{aligned}
$$

where $S_n(z)$, called the nth *partial sum*, is the sum of the first n terms of the sequence $\{u_n(z)\}$.

The sequence $S_1(z), S_2(z), \ldots$ or $\{S_n(z)\}$ is symbolized by

$$u_1(z) + u_2(z) + \cdots = \sum_{n=1}^{\infty} u_n(z) \tag{1}$$

called an *infinite series*. If $\lim_{n \to \infty} S_n(z) = S(z)$, the series is called *convergent* and $S(z)$ is its *sum*; otherwise the series is called *divergent*. We sometimes write $\sum_{n=1}^{\infty} u_n(z)$ as $\Sigma u_n(z)$ or Σu_n for brevity.

As we have already seen, a necessary condition that the series (1) converge is $\lim_{n \to \infty} u_n(z) = 0$, but this is not sufficient. See, for example, Problem 150, Chapter 2, and also Problems $67(c)$, $67(d)$ and $111(a)$.

If a series converges for all values of z (points) in a region $\mathcal{R}$, we call $\mathcal{R}$ the *region of convergence* of the series.

ABSOLUTE CONVERGENCE

A series $\sum_{n=1}^{\infty} u_n(z)$ is called *absolutely convergent* if the series of absolute values, i.e. $\sum_{n=1}^{\infty} |u_n(z)|$, converges.

If $\sum_{n=1}^{\infty} u_n(z)$ converges but $\sum_{n=1}^{\infty} |u_n(z)|$ does not converge, we call $\sum_{n=1}^{\infty} u_n(z)$ *conditionally convergent*.

UNIFORM CONVERGENCE OF SEQUENCES AND SERIES

In the definition of limit of a sequence of functions it was pointed out that the number N depends in general on ϵ and the particular value of z. It may happen, however, that we can find a number N such that $|u_n(z) - U(z)| < \epsilon$ for all $n > N$, where the same number N holds for all z in a region $\mathcal{R}$ [i.e. N depends only on ϵ and not on the particular value of z (point) in the region]. In such case we say that $u_n(z)$ *converges uniformly*, or is *uniformly convergent*, to $U(z)$ for all z in $\mathcal{R}$.

Similarly if the sequence of partial sums $\{S_n(z)\}$ converges uniformly to $S(z)$ in a region, we say that the infinite series (1) *converges uniformly*, or is *uniformly convergent*, to $S(z)$ in the region.

If we call $R_n(z) = u_{n+1}(z) + u_{n+2}(z) + \cdots = S(z) - S_n(z)$ the *remainder* of the infinite series (1) after n terms, we can equivalently say that the series is uniformly convergent to $S(z)$ in $\mathcal{R}$ if given any $\epsilon > 0$ we can find a number N such that for all z in $\mathcal{R}$,

$$|R_n(z)| = |S(z) - S_n(z)| < \epsilon \qquad \text{for all } n > N$$

POWER SERIES

A series having the form

$$a_0 + a_1(z-a) + a_2(z-a)^2 + \cdots = \sum_{n=0}^{\infty} a_n(z-a)^n \tag{2}$$

is called a *power series* in $z - a$. We shall sometimes indicate (2) briefly by $\Sigma a_n(z-a)^n$.

Clearly the power series (2) converges for $z = a$, and this may indeed be the only point for which it converges [see Problem 13(b)]. In general, however, the series converges for other points as well. In such case we can show that there exists a positive number R such that (2) converges for $|z-a| < R$ and diverges for $|z-a| > R$, while for $|z-a| = R$ it may or may not converge.

Geometrically if Γ is a circle of radius R with center at $z = a$, then the series (2) converges at all points inside Γ and diverges at all points outside Γ, while it may or may not converge on the circle Γ. We can consider the special cases $R = 0$ and $R = \infty$ respectively to be the cases where (2) converges only at $z = a$ or converges for all (finite) values of z. Because of this geometrical interpretation, R is often called the *radius of convergence* of (2) and the corresponding circle is called the *circle of convergence*.

SOME IMPORTANT THEOREMS

For reference purposes we list here some important theorems involving sequences and series. Many of these will be familiar from their analogs for real variables.

A. General Theorems

    Theorem 1. If a sequence has a limit, the limit is unique [i.e. it is the only one].

    Theorem 2. Let $u_n = a_n + ib_n$, $n = 1, 2, 3, \ldots$, where a_n and b_n are real. Then a necessary and sufficient condition that $\{u_n\}$ converge is that $\{a_n\}$ and $\{b_n\}$ converge.

Theorem 3.   Let $\{a_n\}$ be a real sequence with the property that

$$\text{(i)} \quad a_{n+1} \geqq a_n \ \text{ or } \ a_{n+1} \leqq a_n, \qquad \text{(ii)} \quad |a_n| < M \text{ (a constant)}$$

Then $\{a_n\}$ converges.

If the first condition in Property (i) holds the sequence is called *monotonic increasing*, while if the second condition holds it is called *monotonic decreasing*. If Property (ii) holds, the sequence is said to be *bounded*. Thus the theorem states that every bounded monotonic (increasing or decreasing) sequence has a limit.

Theorem 4.   A necessary and sufficient condition that $\{u_n\}$ converges is that given any $\epsilon > 0$, we can find a number N such that $|u_p - u_q| < \epsilon$ for all $p > N, q > N$.

This result, which has the advantage that the limit itself is not present, is called *Cauchy's convergence criterion*.

Theorem 5.   A necessary condition that Σu_n converge is that $\lim\limits_{n \to \infty} u_n = 0$. However, the condition is not sufficient.

Theorem 6.   Multiplication of each term of a series by a constant different from zero does not affect the convergence or divergence. Removal (or addition) of a finite number of terms from (or to) a series does not affect the convergence or divergence.

Theorem 7.   A necessary and sufficient condition that $\sum\limits_{n=1}^{\infty} (a_n + ib_n)$ converge, where a_n and b_n are real, is that $\sum\limits_{n=1}^{\infty} a_n$ and $\sum\limits_{n=1}^{\infty} b_n$ converge.

B. Theorems on Absolute Convergence

Theorem 8.   If $\sum\limits_{n=1}^{\infty} |u_n|$ converges, then $\sum\limits_{n=1}^{\infty} u_n$ converges. In words, an absolutely convergent series is convergent.

Theorem 9.   The terms of an absolutely convergent series can be rearranged in any order and all such rearranged series converge to the same sum. Also the sum, difference and product of absolutely convergent series is absolutely convergent.

These are not so for conditionally convergent series (see Problem 127).

C. Special Tests for Convergence

Theorem 10.   (*Comparison tests.*)

(a) If $\Sigma |v_n|$ converges and $|u_n| \leqq |v_n|$, then Σu_n converges absolutely.

(b) If $\Sigma |v_n|$ diverges and $|u_n| \geqq |v_n|$, then $\Sigma |u_n|$ diverges but Σu_n may or may not converge.

Theorem 11.   (*Ratio test.*)

If $\lim\limits_{n \to \infty} \left| \dfrac{u_{n+1}}{u_n} \right| = L$, then Σu_n converges (absolutely) if $L < 1$ and diverges if $L > 1$. If $L = 1$, the test fails.

Theorem 12.   (*nth Root test.*)

If $\lim\limits_{n \to \infty} \sqrt[n]{|u_n|} = L$, then Σu_n converges (absolutely) if $L < 1$ and diverges if $L > 1$. If $L = 1$, the test fails.

Theorem 13.   (*Integral test.*)   If $f(x) \geqq 0$ for $x \geqq a$, then $\Sigma f(n)$ converges or diverges according as $\lim\limits_{M \to \infty} \displaystyle\int_a^M f(x)\,dx$ converges or diverges.

Theorem 14.   (*Raabe's test.*)

If $\lim\limits_{n \to \infty} n \left(1 - \left| \dfrac{u_{n+1}}{u_n} \right| \right) = L$, then Σu_n converges (absolutely) if $L > 1$ and diverges or converges conditionally if $L < 1$. If $L = 1$, the test fails.

Theorem 15. (*Gauss' test.*)

If $\left| \dfrac{u_{n+1}}{u_n} \right| = 1 - \dfrac{L}{n} + \dfrac{c_n}{n^2}$ where $|c_n| < M$ for all $n > N$, then Σu_n converges (absolutely) if $L > 1$ and diverges or converges conditionally if $L \leqq 1$.

Theorem 16. (*Alternating series test.*)

If $a_n \geqq 0$, $a_{n+1} \leqq a_n$ for $n = 1, 2, 3, \ldots$ and $\lim\limits_{n \to \infty} a_n = 0$, then $a_1 - a_2 + a_3 - \cdots = \Sigma(-1)^{n-1} a_n$ converges.

D. Theorems on Uniform Convergence

Theorem 17. (*Weierstrass M test.*)

If $|u_n(z)| \leqq M_n$ where M_n is independent of z in a region $\mathcal{R}$ and ΣM_n converges, then $\Sigma u_n(z)$ is uniformly convergent in $\mathcal{R}$.

Theorem 18. The sum of a uniformly convergent series of continuous functions is continuous, i.e. if $u_n(z)$ is continuous in $\mathcal{R}$ and $S(z) = \Sigma u_n(z)$ is uniformly convergent in $\mathcal{R}$, then $S(z)$ is continuous in $\mathcal{R}$.

Theorem 19. If $\{u_n(z)\}$ are continuous in $\mathcal{R}$, $S(z) = \Sigma u_n(z)$ is uniformly convergent in $\mathcal{R}$ and C is a curve in $\mathcal{R}$, then

$$\int_C S(z)\, dz \;=\; \int_C u_1(z)\, dz \;+\; \int_C u_2(z)\, dz \;+\; \cdots$$

or

$$\int_C \{\Sigma u_n(z)\}\, dz \;=\; \Sigma \int_C u_n(z)\, dz$$

In words, a uniformly convergent series of continuous functions can be integrated term by term.

Theorem 20. If $u_n'(z) = \dfrac{d}{dz} u_n(z)$ exists in $\mathcal{R}$, $\Sigma u_n'(z)$ converges uniformly in $\mathcal{R}$ and $\Sigma u_n(z)$ converges in $\mathcal{R}$, then $\dfrac{d}{dz} \Sigma u_n(z) = \Sigma u_n'(z)$.

Theorem 21. If $\{u_n(z)\}$ are analytic and $\Sigma u_n(z)$ is uniformly convergent in $\mathcal{R}$, then $S(z) = \Sigma u_n(z)$ is analytic in $\mathcal{R}$.

E. Theorems on Power Series

Theorem 22. A power series converges uniformly and absolutely in any region which lies entirely inside its circle of convergence.

Theorem 23.

(a) A power series can be differentiated term by term in any region which lies entirely inside its circle of convergence.

(b) A power series can be integrated term by term along any curve C which lies entirely inside its circle of convergence.

(c) The sum of a power series is continuous in any region which lies entirely inside its circle of convergence.

These follow from Theorems 17, 18, 19 and 21.

Theorem 24. (*Abel's theorem.*)

Let $\Sigma a_n z^n$ have radius of convergence R and suppose that z_0 is a point on the circle of convergence such that $\Sigma a_n z_0^n$ converges. Then $\lim\limits_{z \to z_0} \Sigma a_n z^n = \Sigma a_n z_0^n$ where $z \to z_0$ from within the circle of convergence.

Extensions to other power series are easily made.

Theorem 25. If $\Sigma a_n z^n$ converges to zero for all z such that $|z| < R$ where $R > 0$, then $a_n = 0$. Equivalently, if $\Sigma a_n z^n = \Sigma b_n z^n$ for all z such that $|z| < R$, then $a_n = b_n$.

TAYLOR'S THEOREM

Let $f(z)$ be analytic inside and on a simple closed curve C. Let a and $a+h$ be two points inside C. Then

$$f(a+h) \;=\; f(a) \;+\; h\,f'(a) \;+\; \frac{h^2}{2!}f''(a) \;+\; \cdots \;+\; \frac{h^n}{n!}f^{(n)}(a) \;+\; \cdots \qquad (3)$$

or writing $z = a + h$, $h = z - a$,

$$f(z) \;=\; f(a) \;+\; f'(a)\,(z-a) \;+\; \frac{f''(a)}{2!}(z-a)^2 \;+\; \cdots \;+\; \frac{f^{(n)}(a)}{n!}(z-a)^n \;+\; \cdots \qquad (4)$$

This is called *Taylor's theorem* and the series (3) or (4) is called a *Taylor series* or *expansion* for $f(a+h)$ or $f(z)$.

The region of convergence of the series (4) is given by $|z-a| < R$, where the radius of convergence R is the distance from a to the nearest singularity of the function $f(z)$. On $|z-a| = R$, the series may or may not converge. For $|z-a| > R$, the series diverges.

If the nearest singularity of $f(z)$ is at infinity, the radius of convergence is infinite, i.e. the series converges for all z.

If $a = 0$ in (3) or (4), the resulting series is often called a *Maclaurin series*.

SOME SPECIAL SERIES

The following list shows some special series together with their regions of convergence. In the case of multiple-valued functions, the principal branch is used.

1. $e^z \qquad = 1 + z + \dfrac{z^2}{2!} + \dfrac{z^3}{3!} + \cdots + \dfrac{z^n}{n!} + \cdots$ $\qquad\qquad |z| < \infty$

2. $\sin z \quad = z - \dfrac{z^3}{3!} + \dfrac{z^5}{5!} - \cdots (-1)^{n-1}\dfrac{z^{2n-1}}{(2n-1)!} + \cdots$ $\qquad |z| < \infty$

3. $\cos z \quad = 1 - \dfrac{z^2}{2!} + \dfrac{z^4}{4!} - \cdots (-1)^{n-1}\dfrac{z^{2n-2}}{(2n-2)!} + \cdots$ $\qquad |z| < \infty$

4. $\ln(1+z) = z - \dfrac{z^2}{2} + \dfrac{z^3}{3} - \cdots (-1)^{n-1}\dfrac{z^n}{n} + \cdots$ $\qquad\qquad |z| < 1$

5. $\tan^{-1} z \;= z - \dfrac{z^3}{3} + \dfrac{z^5}{5} - \cdots (-1)^{n-1}\dfrac{z^{2n-1}}{2n-1} + \cdots$ $\qquad\quad |z| < 1$

6. $(1+z)^p \;= 1 + pz + \dfrac{p(p-1)}{2!}z^2 + \cdots + \dfrac{p(p-1)\cdots(p-n+1)}{n!}z^n + \cdots \quad |z| < 1$

   This is the *binomial theorem* or *formula*. If $(1+z)^p$ is multiple-valued the result is valid for that branch of the function which has the value 1 when $z = 0$.

LAURENT'S THEOREM

Let C_1 and C_2 be concentric circles of radii R_1 and R_2 respectively and center at a [Fig. 6-1]. Suppose that $f(z)$ is single-valued and analytic on C_1 and C_2 and in the ring-shaped region $\mathcal{R}$ [also called *annulus* or *annular region*] between C_1 and C_2, shown shaded in the figure. Let $a + h$ be any point in $\mathcal{R}$. Then we have

$$f(a+h) \;=\; a_0 + a_1 h + a_2 h^2 + \cdots$$

$$+ \frac{a_{-1}}{h} + \frac{a_{-2}}{h^2} + \frac{a_{-3}}{h^3} + \cdots \qquad (5)$$

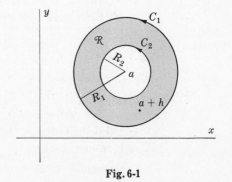

Fig. 6-1

where
$$a_n = \frac{1}{2\pi i} \oint_{C_1} \frac{f(z)}{(z-a)^{n+1}} dz \qquad n = 0, 1, 2, \ldots$$
$$a_{-n} = \frac{1}{2\pi i} \oint_{C_2} (z-a)^{n-1} f(z)\, dz \qquad n = 1, 2, 3, \ldots \qquad \Bigg\} \qquad (6)$$

C_1 and C_2 being traversed in the positive direction with respect to their interiors.

We can in the above integrations replace C_1 and C_2 by any concentric circle C between C_1 and C_2 [see Problem 100]. Then the coefficients (6) can be written in a single formula,

$$a_n = \frac{1}{2\pi i} \oint_C \frac{f(z)}{(z-a)^{n+1}} dz \qquad n = 0, \pm 1, \pm 2, \ldots \qquad (7)$$

With an appropriate change of notation, we can write the above as

$$f(z) = a_0 + a_1(z-a) + a_2(z-a)^2 + \cdots + \frac{a_{-1}}{z-a} + \frac{a_{-2}}{(z-a)^2} + \cdots \qquad (8)$$

where
$$a_n = \frac{1}{2\pi i} \oint_C \frac{f(\zeta)}{(\zeta-a)^{n+1}} d\zeta \qquad n = 0, \pm 1, \pm 2, \ldots \qquad (9)$$

This is called *Laurent's theorem* and (5) or (8) with coefficients (6), (7) or (9) is called a *Laurent series* or *expansion*.

The part $a_0 + a_1(z-a) + a_2(z-a)^2 + \cdots$ is called the *analytic part* of the Laurent series, while the remainder of the series which consists of inverse powers of $z-a$ is called the *principal part*. If the principal part is zero, the Laurent series reduces to a Taylor series.

CLASSIFICATION OF SINGULARITIES

It is possible to classify the singularities of a function $f(z)$ by examination of its Laurent series. For this purpose we assume that in Fig. 6-1, $R_2 = 0$, so that $f(z)$ is analytic inside and on C_1 except at $z = a$ which is an isolated singularity [see Page 67]. In the following, all singularities are assumed isolated unless otherwise indicated.

1. **Poles.** If $f(z)$ has the form (8) in which the principal part has only a finite number of terms given by

$$\frac{a_{-1}}{z-a} + \frac{a_{-2}}{(z-a)^2} + \cdots + \frac{a_{-n}}{(z-a)^n}$$

where $a_{-n} \neq 0$, then $z = a$ is called a *pole of order n*. If $n = 1$, it is called a *simple pole*.

If $f(z)$ has a pole at $z = a$, then $\lim\limits_{z \to a} f(z) = \infty$ [see Problem 32].

2. **Removable singularities.** If a single-valued function $f(z)$ is not defined at $z = a$ but $\lim\limits_{z \to a} f(z)$ exists, then $z = a$ is called a *removable singularity*. In such case we define $f(z)$ at $z = a$ as equal to $\lim\limits_{z \to a} f(z)$.

> **Example:** If $f(z) = \sin z / z$, then $z = 0$ is a removable singularity since $f(0)$ is not defined but $\lim\limits_{z \to 0} \sin z / z = 1$. We define $f(0) = \lim\limits_{z \to 0} \sin z / z = 1$. Note that in this case
> $$\frac{\sin z}{z} = \frac{1}{z} \left\{ z - \frac{z^3}{3!} + \frac{z^5}{5!} - \frac{z^7}{7!} + \cdots \right\} = 1 - \frac{z^2}{3!} + \frac{z^4}{5!} - \frac{z^6}{7!} + \cdots$$

3. **Essential singularities.** If $f(z)$ is single-valued, then any singularity which is not a pole or removable singularity is called an *essential singularity*. If $z = a$ is an essential singularity of $f(z)$, the principal part of the Laurent expansion has infinitely many terms.

> **Example:** Since $e^{1/z} = 1 + \frac{1}{z} + \frac{1}{2! \, z^2} + \frac{1}{3! \, z^3} + \cdots$, $z = 0$ is an essential singularity.

The following two related theorems are of interest (see Problems 153-155):

Casorati-Weierstrass theorem. In any neighborhood of an isolated essential singularity a, an otherwise analytic function $f(z)$ comes arbitrarily close to any complex number A. In symbols, given any positive numbers δ and ϵ and any complex number A, there exists a value of z inside the circle $|z - a| = \delta$ for which $|f(z) - A| < \epsilon$.

Picard's theorem. In the neighborhood of an isolated essential singularity a, an otherwise analytic function $f(z)$ can take on any value whatsoever with perhaps one exception.

4. **Branch points.** A point $z = z_0$ is called a *branch point* of the multiple-valued function $f(z)$ if the branches of $f(z)$ are interchanged when z describes a closed path about z_0 [see Page 37]. Since each of the branches of a multiple-valued function is analytic, all of the theorems for analytic functions, in particular Taylor's theorem, apply.

> **Example:** The branch of $f(z) = z^{1/2}$ which has the value 1 for $z = 1$, has a Taylor series of the form $a_0 + a_1(z - 1) + a_2(z - 1)^2 + \cdots$ with radius of convergence $R = 1$ [the distance from $z = 1$ to the nearest singularity, namely the branch point $z = 0$].

5. **Singularities at infinity.** By letting $z = 1/w$ in $f(z)$, we obtain the function $f(1/w) = F(w)$. Then the nature of the singularity at $z = \infty$ [the point at infinity] is defined to be the same as that of $F(w)$ at $w = 0$.

> **Example:** $f(z) = z^3$ has a pole of order 3 at $z = \infty$, since $F(w) = f(1/w) = 1/w^3$ has a pole of order 3 at $w = 0$. Similarly $f(z) = e^z$ has an essential singularity at $z = \infty$, since $F(w) = f(1/w) = e^{1/w}$ has an essential singularity at $w = 0$.

ENTIRE FUNCTIONS

A function which is analytic everywhere in the finite plane [i.e. everywhere except at ∞] is called an *entire function* or *integral function*. The functions e^z, $\sin z$, $\cos z$ are entire functions.

An entire function can be represented by a Taylor series which has an infinite radius of convergence. Conversely if a power series has an infinite radius of convergence, it represents an entire function.

Note that by Liouville's theorem [Chapter 5, Page 119] a function which is analytic *everywhere including* ∞ must be a constant.

MEROMORPHIC FUNCTIONS

A function which is analytic everywhere in the finite plane except at a finite number of poles is called a *meromorphic function*.

> **Example:** $\dfrac{z}{(z - 1)(z + 3)^2}$ which is analytic everywhere in the finite plane except at the poles $z = 1$ (simple pole) and $z = -3$ (pole of order two) is a meromorphic function.

LAGRANGE'S EXPANSION

Let z be that root of $z = a + \zeta\,\phi(z)$ which has the value $z = a$ when $\zeta = 0$. Then if $\phi(z)$ is analytic inside and on a circle C containing $z = a$, we have

$$z \;=\; a + \sum_{n=1}^{\infty} \frac{\zeta^n}{n!} \frac{d^{n-1}}{da^{n-1}} \{[\phi(a)]^n\} \tag{11}$$

More generally, if $F(z)$ is analytic inside and on C, then

$$F(z) \;=\; F(a) + \sum_{n=1}^{\infty} \frac{\zeta^n}{n!} \frac{d^{n-1}}{da^{n-1}} \{F'(a)\,[\phi(a)]^n\} \tag{12}$$

The expansion (*12*) and the special case (*11*) are often referred to as *Lagrange's expansions*.

ANALYTIC CONTINUATION

Suppose that we do not know the precise form of an analytic function $f(z)$ but only know that inside some circle of convergence C_1 with center at a [Fig. 6-2] $f(z)$ is represented by a Taylor series

$$a_0 + a_1(z-a) + a_2(z-a)^2 + \cdots \qquad (13)$$

Choosing a point b inside C_1, we can find the value of $f(z)$ and its derivatives at b from (13) and thus arrive at a new series

$$b_0 + b_1(z-b) + b_2(z-b)^2 + \cdots \qquad (14)$$

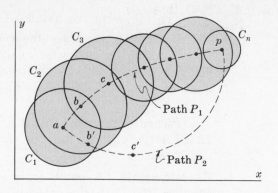

Fig. 6-2

having circle of convergence C_2. If C_2 extends beyond C_1, then the values of $f(z)$ and its derivatives can be obtained in this extended portion and so we have achieved more information concerning $f(z)$.

We say in this case that $f(z)$ has been *extended analytically* beyond C_1 and call the process *analytic continuation* or *analytic extension*.

The process can of course be repeated indefinitely. Thus choosing point c inside C_2, we arrive at a new series having circle of convergence C_3 which may extend beyond C_1 and C_2, etc.

The collection of all such power series representations, i.e. all possible analytic continuations, is defined as the analytic function $f(z)$ and each power series is sometimes called an *element* of $f(z)$.

In performing analytic continuations we must avoid singularities. For example, there cannot be any singularity in Fig. 6-2 which is both inside C_2 and on the boundary of C_1, since otherwise (14) would diverge at this point. In some cases the singularities on a circle of convergence are so numerous that analytic continuation is impossible. In these cases the boundary of the circle is called a *natural boundary* or barrier [see Prob. 30]. The function represented by a series having a natural boundary is called a *lacunary* function.

In going from circle C_1 to circle C_n [Fig. 6-2], we have chosen the path of centers $a, b, c, \ldots, p$ which we represent briefly by *path P_1*. Many other paths are also possible, e.g. $a, b', c', \ldots, p$ represented briefly by *path P_2*. A question arises as to whether one obtains the same series representation valid inside C_n when one chooses different paths. The answer is *yes* so long as the region bounded by paths P_1 and P_2 has no singularity.

For further discussion of analytic continuation, see Chapter 10.

Solved Problems

SEQUENCES AND SERIES OF FUNCTIONS

1. Using the definition, prove that $\lim\limits_{n \to \infty} \left(1 + \dfrac{z}{n}\right) = 1$ for all z.

Given any number $\epsilon > 0$, we must find N such that $|1 + z/n - 1| < \epsilon$ for $n > N$. Then $|z/n| < \epsilon$, i.e. $|z|/n < \epsilon$ if $n > |z|/\epsilon = N$.

2. (a) Prove that the series $z(1-z) + z^2(1-z) + z^3(1-z) + \cdots$ converges for $|z| < 1$, and   (b) find its sum.

The sum of the first n terms of the series is

$$\begin{aligned} S_n(z) &= z(1-z) + z^2(1-z) + \cdots + z^n(1-z) \\ &= z - z^2 + z^2 - z^3 + \cdots + z^n - z^{n+1} \\ &= z - z^{n+1} \end{aligned}$$

Now $|S_n(z) - z| = |-z^{n+1}| = |z|^{n+1} < \epsilon$ for $(n+1)\ln|z| < \ln\epsilon$,   i.e. $n+1 > \dfrac{\ln\epsilon}{\ln|z|}$ or $n > \dfrac{\ln\epsilon}{\ln|z|} - 1$ if $z \neq 0$.

If $z = 0$, $S_n(0) = 0$ and $|S_n(0) - 0| < \epsilon$ for all n.

Hence $\lim\limits_{n\to\infty} S_n(z) = z$,   the required sum for all z such that $|z| < 1$.

Another method.

Since $S_n(z) = z - z^{n+1}$,   we have [by Problem 41, Chapter 2, in which we showed that $\lim\limits_{n\to\infty} z^n = 0$ if $|z| < 1$]

$$\text{Required sum} = S(z) = \lim_{n\to\infty} S_n(z) = \lim_{n\to\infty} (z - z^{n+1}) = z$$

ABSOLUTE AND UNIFORM CONVERGENCE

3. (a) Prove that the series in Problem 2 converges uniformly to the sum z for $|z| \leqq \tfrac{1}{2}$.

(b) Does the series converge uniformly for $|z| \leqq 1$?   Explain.

(a) In Problem 2 we have shown that $|S_n(z) - z| < \epsilon$ for all $n > \dfrac{\ln\epsilon}{\ln|z|} - 1$,   i.e. the series converges to the sum z for $|z| < 1$ and thus for $|z| \leqq \tfrac{1}{2}$.

Now if $|z| \leqq \tfrac{1}{2}$, the largest value of $\dfrac{\ln\epsilon}{\ln|z|} - 1$ occurs where $|z| = \tfrac{1}{2}$ and is given by $\dfrac{\ln\epsilon}{\ln(1/2)} - 1 = N$. It follows that $|S_n(z) - z| < \epsilon$ for all $n > N$ where N depends only on ϵ and not on the particular z in $|z| \leqq \tfrac{1}{2}$. Thus the series converges uniformly to z for $|z| \leqq \tfrac{1}{2}$.

(b) The same argument given in part (a) serves to show that the series converges uniformly to sum z for $|z| \leqq .9$ or $|z| \leqq .99$ by using $N = \dfrac{\ln\epsilon}{\ln(.9)} - 1$ and $N = \dfrac{\ln\epsilon}{\ln(.99)} - 1$ respectively.

However, it is clear that we cannot extend the argument to $|z| \leqq 1$ since this would require $N = \dfrac{\ln\epsilon}{\ln 1} - 1$ which is infinite, i.e. there is no finite value of N which can be used in this case. Thus the series does not converge uniformly for $|z| \leqq 1$.

4. (a) Prove that the sequence $\left\{ \dfrac{1}{1+nz} \right\}$ is uniformly convergent to zero for all z such that $|z| \geqq 2$.   (b) Can the region of uniform convergence in (a) be extended?   Explain.

(a) We have $\left| \dfrac{1}{1+nz} - 0 \right| < \epsilon$ when $\dfrac{1}{|1+nz|} < \epsilon$ or $|1+nz| > 1/\epsilon$. Now $|1+nz| \leqq |1| + |nz| = 1 + n|z|$ and $1 + n|z| \geqq |1+nz| > 1/\epsilon$ for $n > \dfrac{1/\epsilon - 1}{|z|}$. Thus the sequence converges to zero for $|z| > 2$.

To determine whether it converges uniformly to zero, note that the largest value of $\dfrac{1/\epsilon - 1}{|z|}$ in $|z| \geqq 2$ occurs for $|z| = 2$ and is given by $\tfrac{1}{2}\{(1/\epsilon) - 1\} = N$. It follows that $\left| \dfrac{1}{1+nz} - 0 \right| < \epsilon$ for all $n > N$ where N depends only on ϵ and not on the particular z in $|z| \geqq 2$. Thus the sequence is uniformly convergent to zero in this region.

(b) If δ is any positive number, the largest value of $\dfrac{(1/\epsilon) - 1}{|z|}$ in $|z| \geqq \delta$ occurs for $|z| = \delta$ and is given by $\dfrac{(1/\epsilon) - 1}{\delta}$. As in part (a), it follows that the sequence converges uniformly to zero for all z such that $|z| \geqq \delta$, i.e. in any region which excludes all points in a neighborhood of $z = 0$.

Since δ can be chosen arbitrarily close to zero, it follows that the region of (a) can be extended considerably.

5. Show that (a) the sum function in Problem 2 is discontinuous at $z = 1$, (b) the limit in Problem 4 is discontinuous at $z = 0$.

(a) From Problem 2, $S_n(z) = z - z^{n+1}$, $S(z) = \lim\limits_{n \to \infty} S_n(z)$. If $|z| < 1$, $S(z) = \lim\limits_{n \to \infty} S_n(z) = z$. If $z = 1$, $S_n(z) = S_n(1) = 0$ and $\lim\limits_{n \to \infty} S_n(1) = 0$. Hence $S(z)$ is discontinuous at $z = 1$.

(b) From Problem 4 if we write $u_n(z) = \dfrac{1}{1 + nz}$ and $U(z) = \lim\limits_{n \to \infty} u_n(z)$ we have $U(z) = 0$ if $z \neq 0$ and 1 if $z = 0$. Thus $U(z)$ is discontinuous at $z = 0$.

These are consequences of the fact [see Problem 16] that if a series of continuous functions is uniformly convergent in a region $\mathcal{R}$, then the sum function must be continuous in $\mathcal{R}$. Hence if the sum function is not continuous, the series cannot be uniformly convergent. A similar result holds for sequences.

6. Prove that the series of Problem 2 is absolutely convergent for $|z| < 1$.

Let
$$\begin{aligned}
T_n(z) &= |z(1-z)| + |z^2(1-z)| + \cdots + |z^n(1-z)| \\
&= |1 - z| \{ |z| + |z|^2 + |z|^3 + \cdots + |z|^n \} \\
&= |1 - z| \, |z| \left\{ \frac{1 - |z|^n}{1 - |z|} \right\}
\end{aligned}$$

If $|z| < 1$, then $\lim\limits_{n \to \infty} |z|^n = 0$ and $\lim\limits_{n \to \infty} T_n(z)$ exists so that the series converges absolutely.

Note that the series of absolute values converges in this case to $\dfrac{|1 - z| \, |z|}{1 - |z|}$.

SPECIAL CONVERGENCE TESTS

7. If $\Sigma |v_n|$ converges and $|u_n| \leqq |v_n|$, $n = 1, 2, 3, \ldots$, prove that $\Sigma |u_n|$ also converges (i.e. establish the comparison test for convergence).

Let $S_n = |u_1| + |u_2| + \cdots + |u_n|$, $T_n = |v_1| + |v_2| + \cdots + |v_n|$.

Since $\Sigma |v_n|$ converges, $\lim\limits_{n \to \infty} T_n$ exists and equals T, say. Also since $|v_n| \geqq 0$, $T_n \leqq T$.

Then $S_n = |u_1| + |u_2| + \cdots + |u_n| \leqq |v_1| + |v_2| + \cdots + |v_n| \leqq T$ or $0 \leqq S_n \leqq T$.

Thus S_n is a bounded monotonic increasing sequence and must have a limit [Theorem 3, Page 141], i.e. $\Sigma |u_n|$ converges.

8. Prove that $\dfrac{1}{1^p} + \dfrac{1}{2^p} + \dfrac{1}{3^p} + \cdots = \sum\limits_{n=1}^{\infty} \dfrac{1}{n^p}$ converges for any constant $p > 1$.

We have
$$\frac{1}{1^p} = \frac{1}{1^{p-1}}$$

$$\frac{1}{2^p} + \frac{1}{3^p} \leqq \frac{1}{2^p} + \frac{1}{2^p} = \frac{1}{2^{p-1}}$$

$$\frac{1}{4^p} + \frac{1}{5^p} + \frac{1}{6^p} + \frac{1}{7^p} \leqq \frac{1}{4^p} + \frac{1}{4^p} + \frac{1}{4^p} + \frac{1}{4^p} = \frac{1}{4^{p-1}}$$

etc., where we consider $1, 2, 4, 8, \ldots$ terms of the series. It follows that the sum of any finite number of terms of the given series is less than the geometric series

$$\frac{1}{1^{p-1}} + \frac{1}{2^{p-1}} + \frac{1}{4^{p-1}} + \frac{1}{8^{p-1}} + \cdots = \frac{1}{1 - 1/2^{p-1}}$$

which converges for $p > 1$. Thus the given series, sometimes called the p *series*, converges.

By using a method analogous to that used here together with the comparison test for divergence [Theorem 10(b), Page 141], we can show that $\sum_{n=1}^{\infty} \frac{1}{n^p}$ diverges for $p \leqq 1$.

9. Prove that an absolutely convergent series is convergent.

Given that $\Sigma \, |u_n|$ converges, we must show that Σu_n converges.

Let $S_M = u_1 + u_2 + \cdots + u_M$ and $T_M = |u_1| + |u_2| + \cdots + |u_M|$. Then

$$
\begin{aligned}
S_M + T_M &= (u_1 + |u_1|) + (u_2 + |u_2|) + \cdots + (u_M + |u_M|) \\
&\leqq 2\,|u_1| + 2\,|u_2| + \cdots + 2\,|u_M|
\end{aligned}
$$

Since $\Sigma \, |u_n|$ converges and $u_n + |u_n| \geqq 0$ for $n = 1, 2, 3, \ldots$, it follows that $S_M + T_M$ is a bounded monotonic increasing sequence and so $\lim_{M \to \infty} (S_M + T_M)$ exists.

Also since $\lim_{M \to \infty} T_M$ exists [because by hypothesis the series is absolutely convergent],

$$\lim_{M \to \infty} S_M = \lim_{M \to \infty} (S_M + T_M - T_M) = \lim_{M \to \infty} (S_M + T_M) - \lim_{M \to \infty} T_M$$

must also exist and the result is proved.

10. Prove that $\sum_{n=1}^{\infty} \frac{z^n}{n(n+1)}$ converges (absolutely) for $|z| \leqq 1$.

If $|z| \leqq 1$, then $\left| \dfrac{z^n}{n(n+1)} \right| = \dfrac{|z|^n}{n(n+1)} \leqq \dfrac{1}{n(n+1)} \leqq \dfrac{1}{n^2}$.

Taking $u_n = \dfrac{z^n}{n(n+1)}$, $v_n = \dfrac{1}{n^2}$ in the comparison test and recognizing that $\Sigma \dfrac{1}{n^2}$ converges by Problem 8 with $p = 2$, we see that $\Sigma \, |u_n|$ converges, i.e. Σu_n converges absolutely.

11. Establish the ratio test for convergence.

We must show that if $\lim_{n \to \infty} \left| \dfrac{u_{n+1}}{u_n} \right| = L < 1$, then $\Sigma \, |u_n|$ converges or, by Problem 9, Σu_n is (absolutely) convergent.

By hypothesis, we can choose an integer N so large that for all $n \geqq N$, $\left| \dfrac{u_{n+1}}{u_n} \right| \leqq r$ where r is some constant such that $L < r < 1$. Then

$$
\begin{aligned}
|u_{N+1}| &\leqq r\,|u_N| \\
|u_{N+2}| &\leqq r\,|u_{N+1}| < r^2\,|u_N| \\
|u_{N+3}| &\leqq r\,|u_{N+2}| < r^3\,|u_N|
\end{aligned}
$$

etc. By addition,

$$|u_{N+1}| + |u_{N+2}| + \cdots \leqq |u_N| (r + r^2 + r^3 + \cdots)$$

and so $\Sigma \, |u_n|$ converges by the comparison test since $0 < r < 1$.

12. Find the region of convergence of the series $\sum_{n=1}^{\infty} \frac{(z+2)^{n-1}}{(n+1)^3 \, 4^n}$.

If $u_n = \dfrac{(z+2)^{n-1}}{(n+1)^3 \, 4^n}$, then $u_{n+1} = \dfrac{(z+2)^n}{(n+2)^3 \, 4^{n+1}}$. Hence, excluding $z = -2$ for which the given series converges, we have

$$\lim_{n \to \infty} \left| \frac{u_{n+1}}{u_n} \right| = \lim_{n \to \infty} \left| \frac{(z+2)}{4} \frac{(n+1)^3}{(n+2)^3} \right| = \frac{|z+2|}{4}$$

Then the series converges (absolutely) for $\dfrac{|z+2|}{4} < 1$, i.e. $|z+2| < 4$. The point $z = -2$ is included in $|z+2| < 4$.

If $\dfrac{|z+2|}{4} = 1$, i.e. $|z+2| = 4$, the ratio test fails. However it is seen that in this case

$$\left|\frac{(z+2)^{n-1}}{(n+1)^3\,4^n}\right| \;=\; \frac{1}{4(n+1)^3} \;\leqq\; \frac{1}{n^3}$$

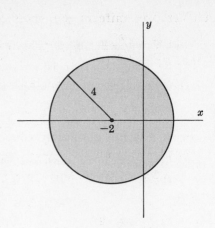

and since $\displaystyle\sum \frac{1}{n^3}$ converges [p series with $p = 3$], the given series converges (absolutely).

It follows that the given series converges (absolutely) for $|z+2| \leqq 4$. Geometrically this is the set of all points inside and on the circle of radius 4 with center at $z = -2$, called the *circle of convergence* [shown shaded in Fig. 6-3]. The *radius of convergence* is equal to 4.

Fig. 6-3

13. Find the region of convergence of the series   (a) $\displaystyle\sum_{n=1}^{\infty} \frac{(-1)^{n-1}\,z^{2n-1}}{(2n-1)!}$,   (b) $\displaystyle\sum_{n=1}^{\infty} n!\,z^n$.

(a) If $u_n = \dfrac{(-1)^{n-1}\,z^{2n-1}}{(2n-1)!}$, then $u_{n+1} = \dfrac{(-1)^n\,z^{2n+1}}{(2n+1)!}$. Hence, excluding $z = 0$ for which the given series converges, we have

$$\lim_{n\to\infty}\left|\frac{u_{n+1}}{u_n}\right| \;=\; \lim_{n\to\infty}\left|-\frac{z^2(2n-1)!}{(2n+1)!}\right| \;=\; \lim_{n\to\infty}\frac{(2n-1)!\,|z|^2}{(2n+1)(2n)(2n-1)!}$$

$$=\; \lim_{n\to\infty}\frac{|z|^2}{(2n+1)(2n)} \;=\; 0$$

for all finite z. Thus the series converges (absolutely) for all z, and we say that the series converges for $|z| < \infty$. We can equivalently say that the circle of convergence is infinite or that the radius of convergence is infinite.

(b) If $u_n = n!\,z^n$, $u_{n+1} = (n+1)!\,z^{n+1}$. Then excluding $z = 0$ for which the given series converges, we have

$$\lim_{n\to\infty}\left|\frac{u_{n+1}}{u_n}\right| \;=\; \lim_{n\to\infty}\left|\frac{(n+1)!\,z^{n+1}}{n!\,z^n}\right| \;=\; \lim_{n\to\infty}(n+1)\,|z| \;=\; \infty$$

Thus the series converges only for $z = 0$.

THEOREMS ON UNIFORM CONVERGENCE

14. Prove the Weierstrass M test, i.e. if in a region $\mathcal{R}$, $|u_n(z)| \leqq M_n$, $n = 1, 2, 3, \ldots$, where M_n are positive constants such that ΣM_n converges, then $\Sigma u_n(z)$ is uniformly (and absolutely) convergent in $\mathcal{R}$.

The remainder of the series $\Sigma u_n(z)$ after n terms is $R_n(z) = u_{n+1}(z) + u_{n+2}(z) + \cdots$. Now

$$|R_n(z)| \;=\; |u_{n+1}(z) + u_{n+2}(z) + \cdots| \;\leqq\; |u_{n+1}(z)| + |u_{n+2}(z)| + \cdots$$

$$\leqq\; M_{n+1} + M_{n+2} + \cdots$$

But $M_{n+1} + M_{n+2} + \cdots$ can be made less than ϵ by choosing $n > N$, since ΣM_n converges. Since N is clearly independent of z, we have $|R_n(z)| < \epsilon$ for $n > N$, and the series is uniformly convergent. The absolute convergence follows at once from the comparison test.

15. Test for uniform convergence in the indicated region:

(a) $\displaystyle\sum_{n=1}^{\infty} \frac{z^n}{n\sqrt{n+1}}$, $|z| \leq 1$;   (b) $\displaystyle\sum_{n=1}^{\infty} \frac{1}{n^2+z^2}$, $1 < |z| < 2$;   (c) $\displaystyle\sum_{n=1}^{\infty} \frac{\cos nz}{n^3}$, $|z| \leq 1$.

(a) If $u_n(z) = \dfrac{z^n}{n\sqrt{n+1}}$, then $|u_n(z)| = \dfrac{|z|^n}{n\sqrt{n+1}} \leq \dfrac{1}{n^{3/2}}$ if $|z| \leq 1$. Calling $M_n = \dfrac{1}{n^{3/2}}$, we see that ΣM_n converges (p series with $p = 3/2$). Hence by the Weierstrass M test the given series converges uniformly (and absolutely) for $|z| \leq 1$.

(b) The given series is $\dfrac{1}{1^2+z^2} + \dfrac{1}{2^2+z^2} + \dfrac{1}{3^2+z^2} + \cdots$. The first two terms can be omitted without affecting the uniform convergence of the series. For $n \geq 3$ and $1 < |z| < 2$, we have

$$|n^2 + z^2| \geq |n^2| - |z^2| \geq n^2 - 4 \geq \tfrac{1}{2}n^2 \quad \text{or} \quad \left| \frac{1}{n^2+z^2} \right| \leq \frac{2}{n^2}$$

Since $\displaystyle\sum_{n=3}^{\infty} \frac{2}{n^2}$ converges, it follows from the Weierstrass M test (with $M_n = 2/n^2$) that the given series converges uniformly (and absolutely) for $1 < |z| < 2$.

Note that the convergence, and thus uniform convergence, breaks down if $|z| = 1$ or $|z| = 2$ [namely at $z = \pm i$ and $z = \pm 2i$]. Hence the series cannot converge uniformly for $1 \leq |z| \leq 2$.

(c) If $z = x + iy$, we have

$$\frac{\cos nz}{n^3} = \frac{e^{inz} + e^{-inz}}{2n^3} = \frac{e^{inx-ny} + e^{-inx+ny}}{2n^3}$$

$$= \frac{e^{-ny}(\cos nx + i \sin nx)}{2n^3} + \frac{e^{ny}(\cos nx - i \sin nx)}{2n^3}$$

The series $\displaystyle\sum_{n=1}^{\infty} \frac{e^{ny}(\cos nx - i \sin nx)}{2n^3}$ and $\displaystyle\sum_{n=1}^{\infty} \frac{e^{-ny}(\cos nx + i \sin nx)}{2n^3}$ cannot converge for $y > 0$ and $y < 0$ respectively [since in these cases the nth term does not approach zero]. Hence the series does not converge for all z such that $|z| \leq 1$, and so cannot possibly be uniformly convergent in this region.

The series does converge for $y = 0$, i.e. if z is real. In this case $z = x$ and the series becomes $\displaystyle\sum_{n=1}^{\infty} \frac{\cos nx}{n^3}$. Then since $\left| \dfrac{\cos nx}{n^3} \right| \leq \dfrac{1}{n^3}$ and $\displaystyle\sum_{n=1}^{\infty} \frac{1}{n^3}$ converges, it follows from the Weierstrass M test (with $M_n = 1/n^3$) that the given series converges uniformly in any interval on the real axis.

16. Prove Theorem 18, Page 142, i.e. if $u_n(z)$, $n = 1, 2, 3, \ldots$, are continuous in $\mathcal{R}$ and $\displaystyle\sum_{n=1}^{\infty} u_n(z)$ is uniformly convergent to $S(z)$ in $\mathcal{R}$, then $S(z)$ is continuous in $\mathcal{R}$.

If $S_n(z) = u_1(z) + u_2(z) + \cdots + u_n(z)$, and $R_n(z) = u_{n+1}(z) + u_{n+2}(z) + \cdots$ is the remainder after n terms, it is clear that

$$S(z) = S_n(z) + R_n(z) \quad \text{and} \quad S(z+h) = S_n(z+h) + R_n(z+h)$$

and so

$$S(z+h) - S(z) = S_n(z+h) - S_n(z) + R_n(z+h) - R_n(z) \tag{1}$$

where z and $z+h$ are in $\mathcal{R}$.

Since $S_n(z)$ is the sum of a finite number of continuous functions, it must also be continuous. Then given $\epsilon > 0$, we can find δ so that

$$|S_n(z+h) - S_n(z)| < \epsilon/3 \quad \text{whenever} \quad |h| < \delta \tag{2}$$

Since the series, by hypothesis, is uniformly convergent, we can choose N so that for all z in $\mathcal{R}$,

$$|R_n(z)| < \epsilon/3 \quad \text{and} \quad |R_n(z+h)| < \epsilon/3 \quad \text{for } n > N \tag{3}$$

Then from (1), (2) and (3),

$$|S(z+h) - S(z)| \leq |S_n(z+h) - S_n(z)| + |R_n(z+h)| + |R_n(z)| < \epsilon$$

for $|h| < \delta$ and all z in $\mathcal{R}$, and so the continuity is established.

17. Prove Theorem 19, Page 142, i.e. if $\{u_n(z)\}$, $n = 1, 2, 3, \ldots$, are continuous in $\mathcal{R}$, $S(z) = \sum\limits_{n=1}^{\infty} u_n(z)$ is uniformly convergent in $\mathcal{R}$ and C is a curve in $\mathcal{R}$, then

$$\int_C S(z)\, dz \;=\; \int_C \left(\sum_{n=1}^{\infty} u_n(z) \right) dz \;=\; \sum_{n=1}^{\infty} \int_C u_n(z)\, dz$$

As in Problem 16, we have $S(z) = S_n(z) + R_n(z)$ and so since these are continuous in $\mathcal{R}$ [by Problem 16] their integrals exist, i.e.,

$$\int_C S(z)\, dz \;=\; \int_C S_n(z)\, dz \;+\; \int_C R_n(z)\, dz$$

$$=\; \int_C u_1(z)\, dz \;+\; \int_C u_2(z)\, dz \;+\; \cdots \;+\; \int_C u_n(z)\, dz \;+\; \int_C R_n(z)\, dz$$

By hypothesis the series is uniformly convergent, so that given any $\epsilon > 0$ we can find a number N independent of z in $\mathcal{R}$ such that $|R_n(z)| < \epsilon$ when $n > N$. Denoting by L the length of C, we have [using Property 5, Page 93]

$$\left| \int_C R_n(z)\, dz \right| \;<\; \epsilon L$$

Then $\left| \int_C S(z)\, dz - \int_C S_n(z)\, dz \right|$ can be made as small as we like by choosing n large enough, and the result is proved.

THEOREMS ON POWER SERIES

18. If a power series $\Sigma a_n z^n$ converges for $z = z_0 \neq 0$, prove that it converges (a) absolutely for $|z| < |z_0|$, (b) uniformly for $|z| \leqq |z_1|$ where $|z_1| < |z_0|$.

(a) Since $\Sigma a_n z_0^n$ converges, $\lim\limits_{n \to \infty} a_n z_0^n = 0$ and so we can make $|a_n z_0^n| < 1$ by choosing n large enough, i.e. $|a_n| < \dfrac{1}{|z_0|^n}$ for $n > N$. Then

$$\sum_{N+1}^{\infty} |a_n z^n| \;=\; \sum_{N+1}^{\infty} |a_n|\,|z|^n \;\leqq\; \sum_{N+1}^{\infty} \frac{|z|^n}{|z_0|^n} \tag{1}$$

But the last series in (1) converges for $|z| < |z_0|$ and so by the comparison test the first series converges, i.e. the given series is absolutely convergent.

(b) Let $M_n = \dfrac{|z_1|^n}{|z_0|^n}$. Then ΣM_n converges, since $|z_1| < |z_0|$. As in part (a), $|a_n z^n| < M_n$ for $|z| \leqq |z_1|$ so that, by the Weierstrass M test, $\Sigma a_n z^n$ is uniformly convergent.

It follows that a power series is uniformly convergent in any region which lies entirely inside its circle of convergence.

19. Prove that both the power series $\sum\limits_{n=0}^{\infty} a_n z^n$ and the corresponding series of derivatives $\sum\limits_{n=0}^{\infty} n a_n z^{n-1}$ have the same radius of convergence.

Let $R > 0$ be the radius of convergence of $\Sigma a_n z^n$. Let $0 < |z_0| < R$. Then as in Problem 18 we can choose N so that $|a_n| < \dfrac{1}{|z_0|^n}$ for $n > N$.

Thus the terms of the series $\Sigma |n a_n z^{n-1}| = \Sigma n |a_n|\,|z|^{n-1}$ can for $n > N$ be made less than corresponding terms of the series $\Sigma n \dfrac{|z|^{n-1}}{|z_0|^n}$ which converges, by the ratio test, for $|z| < |z_0| < R$.

Hence $\Sigma n a_n z^{n-1}$ converges absolutely for all points such that $|z| < |z_0|$ (no matter how close $|z_0|$ is to R), i.e. for $|z| < R$.

If however $|z| > R$, $\lim\limits_{n \to \infty} a_n z^n \neq 0$ and thus $\lim\limits_{n \to \infty} n a_n z^{n-1} \neq 0$, so that $\Sigma n a_n z^{n-1}$ does not converge.

Thus R is the radius of convergence of $\Sigma n a_n z^{n-1}$. This is also true if $R = 0$.

Note that the series of derivatives may or may not converge for values of z such that $|z| = R$.

20. Prove that in any region which lies entirely within its circle of convergence, a power series (a) represents a continuous function, say $f(z)$, (b) can be integrated term by term to yield the integral of $f(z)$, (c) can be differentiated term by term to yield the derivative of $f(z)$.

We consider the power series $\Sigma a_n z^n$, although analogous results hold for $\Sigma a_n(z-a)^n$.

(a) This follows from Problem 16 and the fact that each term $a_n z^n$ of the series is continuous.

(b) This follows from Problem 17 and the fact that each term $a_n z^n$ of the series is continuous and thus integrable.

(c) From Problem 19 the derivative of a power series converges within the circle of convergence of the original power series and therefore is uniformly convergent in any region entirely within the circle of convergence. Thus the required result follows from Theorem 20, Page 142.

21. Prove that the series $\displaystyle\sum_{n=1}^{\infty} \frac{z^n}{n^2}$ has a finite value at all points inside and on its circle of convergence but that this is not true for the series of derivatives.

By the ratio test the series converges for $|z| < 1$ and diverges for $|z| > 1$. If $|z| = 1$, then $|z^n/n^2| = 1/n^2$ and the series is convergent (absolutely). Thus the series converges for $|z| \leqq 1$ and so has a finite value inside and on its circle of convergence.

The series of derivatives is $\displaystyle\sum_{n=1}^{\infty} \frac{z^{n-1}}{n}$. By the ratio test the series converges for $|z| < 1$. However, the series does not converge for all z such that $|z| = 1$, for example if $z = 1$ the series diverges.

TAYLOR'S THEOREM

22. Prove Taylor's theorem: If $f(z)$ is analytic inside a circle C with center at a, then for all z inside C,

$$f(z) \;=\; f(a) \;+\; f'(a)\,(z-a) \;+\; \frac{f''(a)}{2!}\,(z-a)^2 \;+\; \frac{f'''(a)}{3!}\,(z-a)^3 \;+\; \cdots$$

Let z be any point inside C. Construct a circle C_1 with center at a and enclosing z (see Fig. 6-4). Then by Cauchy's integral formula,

$$f(z) \;=\; \frac{1}{2\pi i} \oint_{C_1} \frac{f(w)}{w-z}\, dw \qquad (1)$$

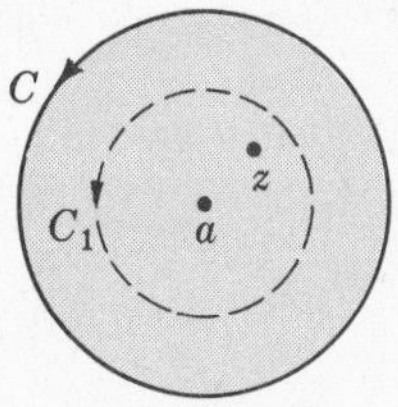

Fig. 6-4

We have

$$\frac{1}{w-z} \;=\; \frac{1}{(w-a)-(z-a)} \;=\; \frac{1}{(w-a)}\left\{\frac{1}{1-(z-a)/(w-a)}\right\}$$

$$=\; \frac{1}{(w-a)}\left\{ 1 + \left(\frac{z-a}{w-a}\right) + \left(\frac{z-a}{w-a}\right)^2 + \cdots + \left(\frac{z-a}{w-a}\right)^{n-1}\right.$$

$$\left. +\; \left(\frac{z-a}{w-a}\right)^n \frac{1}{1-(z-a)/(w-a)}\right\}$$

or
$$\frac{1}{w-z} \;=\; \frac{1}{w-a} + \frac{z-a}{(w-a)^2} + \frac{(z-a)^2}{(w-a)^3} + \cdots + \frac{(z-a)^{n-1}}{(w-a)^n} + \left(\frac{z-a}{w-a}\right)^n \frac{1}{w-z} \qquad (2)$$

Multiplying both sides of (2) by $f(w)$ and using (1), we have

$$f(z) \;=\; \frac{1}{2\pi i} \oint_{C_1} \frac{f(w)}{w-a}\, dw + \frac{z-a}{2\pi i} \oint_{C_1} \frac{f(w)}{(w-a)^2}\, dw + \cdots + \frac{(z-a)^{n-1}}{2\pi i} \oint_{C_1} \frac{f(w)}{(w-a)^n}\, dw + U_n \qquad (3)$$

where
$$U_n \;=\; \frac{1}{2\pi i} \oint_{C_1} \left(\frac{z-a}{w-a}\right)^n \frac{f(w)}{w-z}\, dw$$

Using Cauchy's integral formulas

$$f^{(n)}(a) = \frac{n!}{2\pi i} \oint_{C_1} \frac{f(w)}{(w-a)^{n+1}}\,dw \qquad n = 0, 1, 2, 3, \ldots$$

(3) becomes

$$f(z) = f(a) + f'(a)(z-a) + \frac{f''(a)}{2!}(z-a)^2 + \cdots + \frac{f^{(n-1)}(a)}{(n-1)!}(z-a)^{n-1} + U_n$$

If we can now show that $\lim\limits_{n\to\infty} U_n = 0$, we will have proved the required result. To do this we note that since w is on C_1,

$$\left|\frac{z-a}{w-a}\right| = \gamma < 1$$

where γ is a constant. Also we have $|f(w)| < M$ where M is a constant, and

$$|w - z| = |(w-a) - (z-a)| \geqq r_1 - |z-a|$$

where r_1 is the radius of C_1. Hence from Property 5, Page 93, we have

$$|U_n| = \frac{1}{2\pi}\left|\oint_{C_1}\left(\frac{z-a}{w-a}\right)^n \frac{f(w)}{w-z}\,dw\right|$$

$$\leqq \frac{1}{2\pi}\frac{\gamma^n M}{r_1 - |z-a|}\cdot 2\pi r_1 = \frac{\gamma^n M r_1}{r_1 - |z-a|}$$

and we see that $\lim\limits_{n\to\infty} U_n = 0$, completing the proof.

23. Let $f(z) = \ln(1+z)$, where we consider that branch which has the value zero when $z = 0$. (a) Expand $f(z)$ in a Taylor series about $z = 0$. (b) Determine the region of convergence for the series in (a). (c) Expand $\ln\left(\dfrac{1+z}{1-z}\right)$ in a Taylor series about $z = 0$.

(a)

$$\begin{array}{llll}
f(z) & = \ln(1+z) & f(0) & = 0 \\[4pt]
f'(z) & = \dfrac{1}{1+z} = (1+z)^{-1} & f'(0) & = 1 \\[4pt]
f''(z) & = -(1+z)^{-2} & f''(0) & = -1 \\[4pt]
f'''(z) & = (-1)(-2)(1+z)^{-3} & f'''(0) & = 2! \\[4pt]
\cdot & & \cdot & \\
\cdot & & \cdot & \\
\cdot & & \cdot & \\[4pt]
f^{(n+1)}(z) & = (-1)^n n!\,(1+z)^{-(n+1)} & f^{(n+1)}(0) & = (-1)^n n!
\end{array}$$

Then

$$f(z) = \ln(1+z) = f(0) + f'(0)z + \frac{f''(0)}{2!}z^2 + \frac{f'''(0)}{3!}z^3 + \cdots$$

$$= z - \frac{z^2}{2} + \frac{z^3}{3} - \frac{z^4}{4} + \cdots$$

Another method. If $|z| < 1$,

$$\frac{1}{1+z} = 1 - z + z^2 - z^3 + \cdots$$

Then integrating from 0 to z yields

$$\ln(1+z) = z - \frac{z^2}{2} + \frac{z^3}{3} - \frac{z^4}{4} + \cdots$$

(b) The nth term is $u_n = \dfrac{(-1)^{n-1}z^n}{n}$. Using the ratio test,

$$\lim_{n\to\infty}\left|\frac{u_{n+1}}{u_n}\right| = \lim_{n\to\infty}\left|\frac{nz}{n+1}\right| = |z|$$

and the series converges for $|z| < 1$. The series can be shown to converge for $|z| = 1$ except for $z = -1$.

This result also follows from the fact that the series converges in a circle which extends to the nearest singularity (i.e. $z = -1$) of $f(z)$.

(c) From the result in (a) we have, on replacing z by $-z$,

$$\ln(1+z) = z - \frac{z^2}{2} + \frac{z^3}{3} - \frac{z^4}{4} + \cdots$$

$$\ln(1-z) = -z - \frac{z^2}{2} - \frac{z^3}{3} - \frac{z^4}{4} - \cdots$$

both series convergent for $|z| < 1$. By subtraction, we have

$$\ln\left(\frac{1+z}{1-z}\right) = 2\left(z + \frac{z^3}{3} + \frac{z^5}{5} + \cdots\right) = \sum_{n=0}^{\infty} \frac{2z^{2n+1}}{2n+1}$$

which converges for $|z| < 1$. We can also show that this series converges for $|z| = 1$ except for $z = \pm 1$.

24. (a) Expand $f(z) = \sin z$ in a Taylor series about $z = \pi/4$ and (b) determine the region of convergence of this series.

(a) $f(z) = \sin z$, $f'(z) = \cos z$, $f''(z) = -\sin z$, $f'''(z) = -\cos z$, $f^{IV}(z) = \sin z$, $\ldots$

$f(\pi/4) = \sqrt{2}/2$, $f'(\pi/4) = \sqrt{2}/2$, $f''(\pi/4) = -\sqrt{2}/2$, $f'''(\pi/4) = -\sqrt{2}/2$, $f^{IV}(\pi/4) = \sqrt{2}/2$, $\ldots$

Then, since $a = \pi/4$,

$$\begin{aligned}
f(z) &= f(a) + f'(a)(z-a) + \frac{f''(a)(z-a)^2}{2!} + \frac{f'''(a)(z-a)^3}{3!} + \cdots \\
&= \frac{\sqrt{2}}{2} + \frac{\sqrt{2}}{2}(z - \pi/4) - \frac{\sqrt{2}}{2 \cdot 2!}(z - \pi/4)^2 - \frac{\sqrt{2}}{2 \cdot 3!}(z - \pi/4)^3 + \cdots \\
&= \frac{\sqrt{2}}{2}\left\{ 1 + (z - \pi/4) - \frac{(z - \pi/4)^2}{2!} - \frac{(z - \pi/4)^3}{3!} + \cdots \right\}
\end{aligned}$$

Another method.

Let $u = z - \pi/4$ or $z = u + \pi/4$. Then we have,

$$\begin{aligned}
\sin z &= \sin(u + \pi/4) = \sin u \cos(\pi/4) + \cos u \sin(\pi/4) \\
&= \frac{\sqrt{2}}{2}(\sin u + \cos u) \\
&= \frac{\sqrt{2}}{2}\left\{ \left(u - \frac{u^3}{3!} + \frac{u^5}{5!} - \cdots\right) + \left(1 - \frac{u^2}{2!} + \frac{u^4}{4!} - \cdots\right) \right\} \\
&= \frac{\sqrt{2}}{2}\left\{ 1 + u - \frac{u^2}{2!} - \frac{u^3}{3!} + \frac{u^4}{4!} + \cdots \right\} \\
&= \frac{\sqrt{2}}{2}\left\{ 1 + (z - \pi/4) - \frac{(z - \pi/4)^2}{2!} - \frac{(z - \pi/4)^3}{3!} + \cdots \right\}
\end{aligned}$$

(b) Since the singularity of $\sin z$ nearest to $\pi/4$ is at infinity, the series converges for all finite values of z, i.e. $|z| < \infty$. This can also be established by the ratio test.

LAURENT'S THEOREM

25. Prove *Laurent's theorem*: If $f(z)$ is analytic inside and on the boundary of the ring-shaped region $\mathcal{R}$ bounded by two concentric circles C_1 and C_2 with center at a and respective radii r_1 and $r_2 (r_1 > r_2)$, then for all z in $\mathcal{R}$,

$$f(z) = \sum_{n=0}^{\infty} a_n(z-a)^n + \sum_{n=1}^{\infty} \frac{a_{-n}}{(z-a)^n}$$

where

$$a_n = \frac{1}{2\pi i} \oint_{C_1} \frac{f(w)}{(w-a)^{n+1}}\, dw \qquad n = 0, 1, 2, \ldots$$

$$a_{-n} = \frac{1}{2\pi i} \oint_{C_2} \frac{f(w)}{(w-a)^{-n+1}}\, dw \qquad n = 1, 2, 3, \ldots$$

By Cauchy's integral formula [see Problem 23, Page 131] we have

$$f(z) = \frac{1}{2\pi i} \oint_{C_1} \frac{f(w)}{w-z} dw - \frac{1}{2\pi i} \oint_{C_2} \frac{f(w)}{w-z} dw \qquad (1)$$

Consider the first integral in (1). As in Problem 22, equation (2), we have

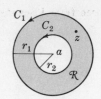

Fig. 6-5

$$\frac{1}{w-z} = \frac{1}{(w-a)\{1-(z-a)/(w-a)\}}$$

$$= \frac{1}{w-a} + \frac{z-a}{(w-a)^2} + \cdots + \frac{(z-a)^{n-1}}{(w-a)^n} + \left(\frac{z-a}{w-a}\right)^n \frac{1}{w-z} \qquad (2)$$

so that

$$\frac{1}{2\pi i} \oint_{C_1} \frac{f(w)}{w-z} dw = \frac{1}{2\pi i} \oint_{C_1} \frac{f(w)}{w-a} dw + \frac{z-a}{2\pi i} \oint_{C_1} \frac{f(w)}{(w-a)^2} dw$$

$$+ \cdots + \frac{(z-a)^{n-1}}{2\pi i} \oint_{C_1} \frac{f(w)}{(w-a)^n} dw + U_n$$

$$= a_0 + a_1(z-a) + \cdots + a_{n-1}(z-a)^{n-1} + U_n \qquad (3)$$

where

$$a_0 = \frac{1}{2\pi i} \oint_{C_1} \frac{f(w)}{w-a} dw, \quad a_1 = \frac{1}{2\pi i} \oint_{C_1} \frac{f(w)}{(w-a)^2} dw, \quad \ldots, \quad a_{n-1} = \frac{1}{2\pi i} \oint_{C_1} \frac{f(w)}{(w-a)^n} dw$$

and

$$U_n = \frac{1}{2\pi i} \oint_{C_1} \left(\frac{z-a}{w-a}\right)^n \frac{f(w)}{w-z} dw$$

Let us now consider the second integral in (1). We have on interchanging w and z in (2),

$$-\frac{1}{w-z} = \frac{1}{(z-a)\{1-(w-a)/(z-a)\}}$$

$$= \frac{1}{z-a} + \frac{w-a}{(z-a)^2} + \cdots + \frac{(w-a)^{n-1}}{(z-a)^n} + \left(\frac{w-a}{z-a}\right)^n \frac{1}{z-w}$$

so that

$$-\frac{1}{2\pi i} \oint_{C_2} \frac{f(w)}{w-z} dw = \frac{1}{2\pi i} \oint_{C_2} \frac{f(w)}{z-a} dw + \frac{1}{2\pi i} \oint_{C_2} \frac{w-a}{(z-a)^2} f(w) \, dw$$

$$+ \cdots + \frac{1}{2\pi i} \oint_{C_2} \frac{(w-a)^{n-1}}{(z-a)^n} f(w) \, dw + V_n$$

$$= \frac{a_{-1}}{z-a} + \frac{a_{-2}}{(z-a)^2} + \cdots + \frac{a_{-n}}{(z-a)^n} + V_n \qquad (4)$$

where

$$a_{-1} = \frac{1}{2\pi i} \oint_{C_2} f(w) \, dw, \quad a_{-2} = \frac{1}{2\pi i} \oint_{C_2} (w-a) f(w) \, dw, \quad \ldots, \quad a_{-n} = \frac{1}{2\pi i} \oint_{C_2} (w-a)^{n-1} f(w) \, dw$$

and

$$V_n = \frac{1}{2\pi i} \oint_{C_2} \left(\frac{w-a}{z-a}\right)^n \frac{f(w)}{z-w} dw$$

From (1), (3) and (4) we have

$$f(z) = \{a_0 + a_1(z-a) + \cdots + a_{n-1}(z-a)^{n-1}\}$$

$$+ \left\{\frac{a_{-1}}{z-a} + \frac{a_{-2}}{(z-a)^2} + \cdots + \frac{a_{-n}}{(z-a)^n}\right\} + U_n + V_n \qquad (5)$$

The required result follows if we can show that (a) $\lim_{n \to \infty} U_n = 0$ and (b) $\lim_{n \to \infty} V_n = 0$. The proof of (a) follows from Problem 22. To prove (b) we first note that since w is on C_2,

$$\left|\frac{w-a}{z-a}\right| = \kappa < 1$$

where κ is a constant. Also we have $|f(w)| < M$ where M is a constant and

$$|z-w| = |(z-a)-(w-a)| \geq |z-a| - r_2$$

Hence from Property 5, Page 93, we have

$$|V_n| \;=\; \frac{1}{2\pi}\left| \oint_{C_2} \left(\frac{w-a}{z-a}\right)^n \frac{f(w)}{z-w}\, dw \right|$$

$$\leqq\; \frac{1}{2\pi}\frac{\kappa^n M}{|z-a|-r_2}\,2\pi r_2 \;=\; \frac{\kappa^n M r_2}{|z-a|-r_2}$$

Then $\lim\limits_{n\to\infty} V_n = 0$ and the proof is complete.

26. Find Laurent series about the indicated singularity for each of the following functions. Name the singularity in each case and give the region of convergence of each series.

(a) $\dfrac{e^{2z}}{(z-1)^3};\quad z=1.$ Let $z-1=u$. Then $z=1+u$ and

$$\frac{e^{2z}}{(z-1)^3} \;=\; \frac{e^{2+2u}}{u^3} \;=\; \frac{e^2}{u^3}\cdot e^{2u} \;=\; \frac{e^2}{u^3}\left\{ 1 + 2u + \frac{(2u)^2}{2!} + \frac{(2u)^3}{3!} + \frac{(2u)^4}{4!} + \cdots \right\}$$

$$=\; \frac{e^2}{(z-1)^3} + \frac{2e^2}{(z-1)^2} + \frac{2e^2}{z-1} + \frac{4e^2}{3} + \frac{2e^2}{3}(z-1) + \cdots$$

$z=1$ is a *pole of order 3*, or *triple pole*.

The series converges for all values of $z \neq 1$.

(b) $(z-3)\sin\dfrac{1}{z+2};\quad z=-2.$ Let $z+2=u$ or $z=u-2$.   Then

$$(z-3)\sin\frac{1}{z+2} \;=\; (u-5)\sin\frac{1}{u} \;=\; (u-5)\left\{ \frac{1}{u} - \frac{1}{3!\,u^3} + \frac{1}{5!\,u^5} - \cdots \right\}$$

$$=\; 1 - \frac{5}{u} - \frac{1}{3!\,u^2} + \frac{5}{3!\,u^3} + \frac{1}{5!\,u^4} - \cdots$$

$$=\; 1 - \frac{5}{z+2} - \frac{1}{6(z+2)^2} + \frac{5}{6(z+2)^3} + \frac{1}{120(z+2)^4} - \cdots$$

$z=-2$ is an *essential singularity*.

The series converges for all values of $z \neq -2$.

(c) $\dfrac{z-\sin z}{z^3};\quad z=0.$

$$\frac{z-\sin z}{z^3} \;=\; \frac{1}{z^3}\left\{ z - \left(z - \frac{z^3}{3!} + \frac{z^5}{5!} - \frac{z^7}{7!} + \cdots \right) \right\}$$

$$=\; \frac{1}{z^3}\left\{ \frac{z^3}{3!} - \frac{z^5}{5!} + \frac{z^7}{7!} - \cdots \right\} \;=\; \frac{1}{3!} - \frac{z^2}{5!} + \frac{z^4}{7!} - \cdots$$

$z=0$ is a *removable singularity*.

The series converges for all values of z.

(d) $\dfrac{z}{(z+1)(z+2)};\quad z=-2.$ Let $z+2=u$.   Then

$$\frac{z}{(z+1)(z+2)} \;=\; \frac{u-2}{(u-1)u} \;=\; \frac{2-u}{u}\cdot\frac{1}{1-u} \;=\; \frac{2-u}{u}(1+u+u^2+u^3+\cdots)$$

$$=\; \frac{2}{u} + 1 + u + u^2 + \cdots \;=\; \frac{2}{z+2} + 1 + (z+2) + (z+2)^2 + \cdots$$

$z=-2$ is a pole of order 1, or simple pole.

The series converges for all values of z such that $0 < |z+2| < 1$.

(e) $\dfrac{1}{z^2(z-3)^2};\quad z=3.$ Let $z-3=u$.   Then by the binomial theorem,

$$\frac{1}{z^2(z-3)^2} = \frac{1}{u^2(3+u)^2} = \frac{1}{9u^2(1+u/3)^2}$$

$$= \frac{1}{9u^2}\left\{1 + (-2)\left(\frac{u}{3}\right) + \frac{(-2)(-3)}{2!}\left(\frac{u}{3}\right)^2 + \frac{(-2)(-3)(-4)}{3!}\left(\frac{u}{3}\right)^3 + \cdots\right\}$$

$$= \frac{1}{9u^2} - \frac{2}{27u} + \frac{1}{27} - \frac{4}{243}u + \cdots$$

$$= \frac{1}{9(z-3)^2} - \frac{2}{27(z-3)} + \frac{1}{27} - \frac{4(z-3)}{243} + \cdots$$

$z = 3$ is a *pole of order 2* or *double pole*.

The series converges for all values of z such that $0 < |z-3| < 3$.

27. Expand $f(z) = \dfrac{1}{(z+1)(z+3)}$ in a Laurent series valid for *(a)* $1 < |z| < 3$, *(b)* $|z| > 3$, *(c)* $0 < |z+1| < 2$, *(d)* $|z| < 1$.

(a) Resolving into partial fractions, $\dfrac{1}{(z+1)(z+3)} = \dfrac{1}{2}\left(\dfrac{1}{z+1}\right) - \dfrac{1}{2}\left(\dfrac{1}{z+3}\right)$.

If $|z| > 1$,

$$\frac{1}{2(z+1)} = \frac{1}{2z(1+1/z)} = \frac{1}{2z}\left(1 - \frac{1}{z} + \frac{1}{z^2} - \frac{1}{z^3} + \cdots\right) = \frac{1}{2z} - \frac{1}{2z^2} + \frac{1}{2z^3} - \frac{1}{2z^4} + \cdots$$

If $|z| < 3$,

$$\frac{1}{2(z+3)} = \frac{1}{6(1+z/3)} = \frac{1}{6}\left(1 - \frac{z}{3} + \frac{z^2}{9} - \frac{z^3}{27} + \cdots\right) = \frac{1}{6} - \frac{z}{18} + \frac{z^2}{54} - \frac{z^3}{162} + \cdots$$

Then the required Laurent expansion valid for both $|z| > 1$ and $|z| < 3$, i.e. $1 < |z| < 3$, is

$$\cdots - \frac{1}{2z^4} + \frac{1}{2z^3} - \frac{1}{2z^2} + \frac{1}{2z} - \frac{1}{6} + \frac{z}{18} - \frac{z^2}{54} + \frac{z^3}{162} - \cdots$$

(b) If $|z| > 1$, we have as in part *(a)*,

$$\frac{1}{2(z+1)} = \frac{1}{2z} - \frac{1}{2z^2} + \frac{1}{2z^3} - \frac{1}{2z^4} + \cdots$$

If $|z| > 3$,

$$\frac{1}{2(z+3)} = \frac{1}{2z(1+3/z)} = \frac{1}{2z}\left(1 - \frac{3}{z} + \frac{9}{z^2} - \frac{27}{z^3} + \cdots\right) = \frac{1}{2z} - \frac{3}{2z^2} + \frac{9}{2z^3} - \frac{27}{2z^4} + \cdots$$

Then the required Laurent expansion valid for both $|z| > 1$ and $|z| > 3$, i.e. $|z| > 3$, is by subtraction

$$\frac{1}{z^2} - \frac{4}{z^3} + \frac{13}{z^4} - \frac{40}{z^5} + \cdots$$

(c) Let $z + 1 = u$. Then

$$\frac{1}{(z+1)(z+3)} = \frac{1}{u(u+2)} = \frac{1}{2u(1+u/2)} = \frac{1}{2u}\left(1 - \frac{u}{2} + \frac{u^2}{4} - \frac{u^3}{8} + \cdots\right)$$

$$= \frac{1}{2(z+1)} - \frac{1}{4} + \frac{1}{8}(z+1) - \frac{1}{16}(z+1)^2 + \cdots$$

valid for $|u| < 2$, $u \neq 0$ or $0 < |z+1| < 2$.

(d) If $|z| < 1$,

$$\frac{1}{2(z+1)} = \frac{1}{2(1+z)} = \frac{1}{2}(1 - z + z^2 - z^3 + \cdots) = \frac{1}{2} - \frac{1}{2}z + \frac{1}{2}z^2 - \frac{1}{2}z^3 + \cdots$$

If $|z| < 3$, we have by part *(a)*,

$$\frac{1}{2(z+3)} = \frac{1}{6} - \frac{z}{18} + \frac{z^2}{54} - \frac{z^3}{162} + \cdots$$

Then the required Laurent expansion, valid for both $|z| < 1$ and $|z| < 3$, i.e. $|z| < 1$, is by subtraction

$$\frac{1}{3} - \frac{4}{9}z + \frac{13}{27}z^2 - \frac{40}{81}z^3 + \cdots$$

This is a *Taylor series*.

LAGRANGE'S EXPANSION

28. Prove Lagrange's expansion *(11)* on Page 145.

Let us assume that C is taken so that there is only one simple zero of $z = a + \zeta\,\phi(z)$ inside C. Then from Problem 90, Page 137, with $g(z) = z$ and $f(z) = z - a - \zeta\,\phi(z)$, we have

$$z = \frac{1}{2\pi i} \oint_C w \left\{ \frac{1 - \zeta\,\phi'(w)}{w - a - \zeta\,\phi(w)} \right\} dw$$

$$= \frac{1}{2\pi i} \oint_C \frac{w}{w - a} \{1 - \zeta\,\phi'(w)\} \left\{ \frac{1}{1 - \zeta\,\phi(w)/(w-a)} \right\} dw$$

$$= \frac{1}{2\pi i} \oint_C \frac{w}{w - a} \{1 - \zeta\,\phi'(w)\} \left\{ \sum_{n=0}^{\infty} \zeta^n\,\phi^n(w)/(w-a)^n \right\} dw$$

$$= \frac{1}{2\pi i} \oint_C \frac{w}{w - a}\,dw + \sum_{n=1}^{\infty} \frac{\zeta^n}{2\pi i} \oint_C \left\{ \frac{w\,\phi^n(w)}{(w-a)^{n+1}} - \frac{w\,\phi^{n-1}(w)\,\phi'(w)}{(w-a)^n} \right\} dw$$

$$= a - \sum_{n=1}^{\infty} \frac{\zeta^n}{2\pi i} \oint_C \frac{w}{n} \frac{d}{dw} \left\{ \frac{\phi^n(w)}{(w-a)^n} \right\} dw$$

$$= a + \sum_{n=1}^{\infty} \frac{\zeta^n}{2\pi i n} \oint_C \frac{\phi^n(w)}{(w-a)^n}\,dw$$

$$= a + \sum_{n=1}^{\infty} \frac{\zeta^n}{n!} \frac{d^{n-1}}{da^{n-1}} [\phi^n(a)]$$

ANALYTIC CONTINUATION

29. Show that the series $(a)\ \displaystyle\sum_{n=0}^{\infty} \frac{z^n}{2^{n+1}}$ and $(b)\ \displaystyle\sum_{n=0}^{\infty} \frac{(z-i)^n}{(2-i)^{n+1}}$ are analytic continuations of each other.

(a) By the ratio test, the series converges for $|z| < 2$ [shaded in Fig. 6-6]. In this circle the series [which is a geometric series with first term $\frac{1}{2}$ and ratio $z/2$] can be summed and represents the function $\dfrac{1/2}{1 - z/2} = \dfrac{1}{2-z}$.

(b) By the ratio test, the series converges for $\left| \dfrac{z-i}{2-i} \right| < 1$,

i.e. $|z - i| < \sqrt{5}$, [see Fig. 6-6]. In this circle the series [which is a geometric series with first term $1/(2-i)$ and ratio $(z-i)/(2-i)$] can be summed and represents the function $\dfrac{1/(2-i)}{1 - (z-i)/(2-i)} = \dfrac{1}{2-z}$.

Since the power series represent the same function in the regions common to the interiors of the circles $|z| = 2$ and $|z - i| = \sqrt{5}$, it follows that they are analytic continuations of each other.

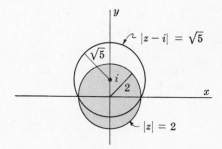

Fig. 6-6

30. Prove that the series $1 + z + z^2 + z^4 + z^8 + \cdots = 1 + \displaystyle\sum_{n=0}^{\infty} z^{2^n}$ cannot be continued analytically beyond $|z| = 1$.

Let $F(z) = 1 + z + z^2 + z^4 + z^8 + \cdots$. Then $F(z) = z + F(z^2)$, $F(z) = z + z^2 + F(z^4)$, $F(z) = z + z^2 + z^4 + F(z^8)$, $\cdots$.

From these it is clear that the values of z given by $z = 1$, $z^2 = 1$, $z^4 = 1$, $z^8 = 1$, $\ldots$ are all singularities of $F(z)$. These singularities all lie on the circle $|z| = 1$. Given any small arc of this circle, there will be infinitely many such singularities. These represent an impassable barrier and analytic continuation beyond $|z| = 1$ is therefore impossible. The circle $|z| = 1$ constitutes a *natural boundary*.

MISCELLANEOUS PROBLEMS

31. Let $\{f_k(z)\}$, $k = 1, 2, 3, \ldots$ be a sequence of functions analytic in a region $\mathcal{R}$. Suppose that

$$F(z) = \sum_{k=1}^{\infty} f_k(z)$$

is uniformly convergent in $\mathcal{R}$. Prove that $F(z)$ is analytic in $\mathcal{R}$.

Let $S_n(z) = \sum_{k=1}^{n} f_k(z)$. By definition of uniform convergence, given any $\epsilon > 0$ we can find a positive integer N depending on ϵ and not on z such that for all z in $\mathcal{R}$,

$$|F(z) - S_n(z)| < \epsilon \qquad \text{for all } n > N \tag{1}$$

Now suppose that C is any simple closed curve lying entirely in $\mathcal{R}$ and denote its length by L. Then by Problem 16, since $f_k(z)$, $k = 1, 2, 3, \ldots$ are continuous, $F(z)$ is also continuous so that $\oint_C F(z)\,dz$ exists. Also, using (1) we see that for $n > N$,

$$\left| \oint_C F(z)\,dz - \sum_{k=1}^{n} \oint_C f_k(z)\,dz \right| = \left| \oint_C \{F(z) - S_n(z)\}\,dz \right|$$

$$< \epsilon L$$

Because ϵ can can be made as small as we please, we see that

$$\oint_C F(z)\,dz = \sum_{k=1}^{\infty} \oint_C f_k(z)\,dz$$

But by Cauchy's theorem, $\oint_C f_k(z)\,dz = 0$. Hence

$$\oint_C F(z)\,dz = 0$$

and so by Morera's theorem (Page 118, Chapter 5) $F(z)$ must be analytic.

32. Prove that an analytic function cannot be bounded in the neighborhood of an isolated singularity.

Let $f(z)$ be analytic inside and on a circle C of radius r, except at the isolated singularity $z = a$ taken to be the center of C. Then by Laurent's theorem $f(z)$ has a Laurent expansion

$$f(z) = \sum_{k=-\infty}^{\infty} a_k(z-a)^k \tag{1}$$

where the coefficients a_k are given by equation (7), Page 144. In particular,

$$a_{-n} = \frac{1}{2\pi i} \oint_C \frac{f(z)}{(z-a)^{-n+1}}\,dz \qquad n = 1, 2, 3, \ldots \tag{2}$$

Now if $|f(z)| < M$ for a constant M, i.e. if $f(z)$ is bounded, then from (2),

$$|a_{-n}| = \frac{1}{2\pi} \left| \oint_C (z-a)^{n-1} f(z)\,dz \right|$$

$$\leq \frac{1}{2\pi} r^{n-1} \cdot M \cdot 2\pi r = M r^n$$

Hence since r can be made arbitrarily small, we have $a_{-n} = 0$, $n = 1, 2, 3, \ldots$, i.e. $a_{-1} = a_{-2} = a_{-3} = \cdots = 0$, and the Laurent series reduces to a Taylor series about $z = a$. This shows that $f(z)$ is analytic at $z = a$ so that $z = a$ is not a singularity, contrary to hypothesis. This contradiction shows that $f(z)$ cannot be bounded in the neighborhood of an isolated singularity.

33. Prove that if $z \neq 0$, then

$$e^{\frac{1}{2}\alpha(z - 1/z)} = \sum_{n=-\infty}^{\infty} J_n(\alpha) z^n$$

where

$$J_n(\alpha) = \frac{1}{2\pi} \int_0^{2\pi} \cos(n\theta - \alpha \sin \theta) \, d\theta \qquad n = 0, 1, 2, \ldots$$

The point $z = 0$ is the only finite singularity of the function $e^{\frac{1}{2}\alpha(z - 1/z)}$ and it follows that the function must have a Laurent series expansion of the form

$$e^{\frac{1}{2}\alpha(z - 1/z)} = \sum_{n=-\infty}^{\infty} J_n(\alpha) z^n \tag{1}$$

which holds for $|z| > 0$. By equation (7), Page 144, the coefficients $J_n(\alpha)$ are given by

$$J_n(\alpha) = \frac{1}{2\pi i} \oint_C \frac{e^{\frac{1}{2}\alpha(z - 1/z)}}{z^{n+1}} \, dz \tag{2}$$

where C is any simple closed curve having $z = 0$ inside.

Let us in particular choose C to be a circle of radius 1 having center at the origin; i.e. the equation of C is $|z| = 1$ or $z = e^{i\theta}$. Then (2) becomes

$$J_n(\alpha) = \frac{1}{2\pi i} \int_0^{2\pi} \frac{e^{\frac{1}{2}\alpha(e^{i\theta} - e^{-i\theta})}}{e^{i(n+1)\theta}} \, i \, e^{i\theta} \, d\theta$$

$$= \frac{1}{2\pi} \int_0^{2\pi} e^{i\alpha \sin \theta - in\theta} \, d\theta$$

$$= \frac{1}{2\pi} \int_0^{2\pi} \cos(\alpha \sin \theta - n\theta) \, d\theta + \frac{i}{2\pi} \int_0^{2\pi} \sin(\alpha \sin \theta - n\theta) \, d\theta$$

$$= \frac{1}{2\pi} \int_0^{2\pi} \cos(n\theta - \alpha \sin \theta) \, d\theta$$

using the fact that $I = \int_0^{2\pi} \sin(\alpha \sin \theta - n\theta) \, d\theta = 0$. This last result follows since on letting $\theta = 2\pi - \phi$, we find

$$I = \int_0^{2\pi} \sin(-\alpha \sin \phi - 2\pi n + n\phi) \, d\phi = -\int_0^{2\pi} \sin(\alpha \sin \phi - n\phi) \, d\phi = -I$$

so that $I = -I$ and $I = 0$. The required result is thus established.

The function $J_n(\alpha)$ is called a *Bessel function* of the first kind of order n.

For further discussion of Bessel functions, see Chapter 10.

34. The *Legendre polynomials* $P_n(t)$, $n = 0, 1, 2, 3, \ldots$ are defined by *Rodrigues' formula*

$$P_n(t) = \frac{1}{2^n n!} \frac{d^n}{dt^n} (t^2 - 1)^n$$

(a) Prove that if C is any simple closed curve enclosing the point $z = t$, then

$$P_n(t) = \frac{1}{2\pi i} \cdot \frac{1}{2^n} \oint_C \frac{(z^2 - 1)^n}{(z - t)^{n+1}} \, dz$$

This is called *Schlaefli's representation* for $P_n(t)$, or *Schlaefli's formula*.

(b) Prove that

$$P_n(t) = \frac{1}{2\pi} \int_0^{2\pi} (t + \sqrt{t^2 - 1} \cos \theta)^n \, d\theta$$

(a) By Cauchy's integral formulas, if C encloses point t,

$$f^{(n)}(t) = \frac{d^n}{dt^n} f(t) = \frac{n!}{2\pi i} \oint_C \frac{f(z)}{(z - t)^{n+1}} \, dz$$

Then taking $f(t) = (t^2 - 1)^n$ so that $f(z) = (z^2 - 1)^n$, we have the required result

$$P_n(t) = \frac{1}{2^n n!} \frac{d^n}{dt^n} (t^2 - 1)^n$$

$$= \frac{1}{2^n} \cdot \frac{1}{2\pi i} \oint_C \frac{(z^2 - 1)^n}{(z - t)^{n+1}} \, dz$$

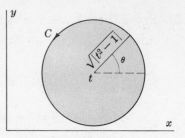

(b) Choose C as a circle with center at t and radius $\sqrt{|t^2 - 1|}$ as shown in Fig. 6-7. Then an equation for C is $|z - t| = \sqrt{|t^2 - 1|}$ or $z = t + \sqrt{t^2 - 1}\, e^{i\theta}$, $0 \le \theta < 2\pi$. Using this in part (a), we have

Fig. 6-7

$$P_n(t) = \frac{1}{2^n} \cdot \frac{1}{2\pi i} \int_0^{2\pi} \frac{\{(t + \sqrt{t^2 - 1}\, e^{i\theta})^2 - 1\}^n \sqrt{t^2 - 1}\, i e^{i\theta} \, d\theta}{(\sqrt{t^2 - 1}\, e^{i\theta})^{n+1}}$$

$$= \frac{1}{2^n} \cdot \frac{1}{2\pi} \int_0^{2\pi} \frac{\{(t^2 - 1) + 2t\sqrt{t^2 - 1}\, e^{i\theta} + (t^2 - 1)e^{2i\theta}\}^n \, e^{-in\theta} \, d\theta}{(t^2 - 1)^{n/2}}$$

$$= \frac{1}{2^n} \cdot \frac{1}{2\pi} \int_0^{2\pi} \frac{\{(t^2 - 1)e^{-i\theta} + 2t\sqrt{t^2 - 1} + (t^2 - 1)e^{i\theta}\}^n \, d\theta}{(t^2 - 1)^{n/2}}$$

$$= \frac{1}{2^n} \cdot \frac{1}{2\pi} \int_0^{2\pi} \frac{\{2t\sqrt{t^2 - 1} + 2(t^2 - 1)\cos\theta\}^n \, d\theta}{(t^2 - 1)^{n/2}}$$

$$= \frac{1}{2\pi} \int_0^{2\pi} (t + \sqrt{t^2 - 1}\, \cos\theta)^n \, d\theta$$

For further discussion of Legendre polynomials, see Chapter 10.

Supplementary Problems

SEQUENCES AND SERIES OF FUNCTIONS

35. Using the definition, prove: (a) $\lim\limits_{n \to \infty} \dfrac{3n - 2z}{n + z} = 3$,   (b) $\lim\limits_{n \to \infty} \dfrac{nz}{n^2 + z^2} = 0$.

36. If $\lim\limits_{n \to \infty} u_n(z) = U(z)$ and $\lim\limits_{n \to \infty} v_n(z) = V(z)$, prove that   (a) $\lim\limits_{n \to \infty} \{u_n(z) \pm v_n(z)\} = U(z) \pm V(z)$,
    (b) $\lim\limits_{n \to \infty} \{u_n(z)\, v_n(z)\} = U(z)\, V(z)$,   (c) $\lim\limits_{n \to \infty} u_n(z)/v_n(z) = U(z)/V(z)$ if $V(z) \ne 0$.

37. (a) Prove that the series $\dfrac{1}{2} + \dfrac{z}{2^2} + \dfrac{z^2}{2^3} + \cdots = \sum\limits_{n=1}^{\infty} \dfrac{z^{n-1}}{2^n}$ converges for $|z| < 2$ and (b) find its sum.

    Ans. (a) $S_n(z) = \{1 - (z/2)^n\}/(2 - z)$ and $\lim\limits_{n \to \infty} S_n(z)$ exists if $|z| < 2$,   (b) $S(z) = 1/(2 - z)$

38. (a) Determine the set of values of z for which the series $\sum\limits_{n=0}^{\infty} (-1)^n (z^n + z^{n+1})$ converges and
    (b) find its sum.   *Ans.* (a) $|z| < 1$,   (b) 1

39. (a) For what values of z does the series $\sum\limits_{n=1}^{\infty} \dfrac{1}{(z^2 + 1)^n}$ converge and   (b) what is its sum?

    Ans. (a) All z such that $|z^2 + 1| > 1$,   (b) $1/z^2$

40. If $\lim\limits_{n \to \infty} |u_n(z)| = 0$, prove that $\lim\limits_{n \to \infty} u_n(z) = 0$. Is the converse true? Justify your answer.

41. Prove that for all finite z, $\lim\limits_{n \to \infty} z^n/n! = 0$.

42. Let $\{a_n\}$, $n = 1, 2, 3, \ldots$ be a sequence of positive numbers having zero as a limit. Suppose that $|u_n(z)| \leqq a_n$ for $n = 1, 2, 3, \ldots$. Prove that $\lim\limits_{n \to \infty} u_n(z) = 0$.

43. Prove that the convergence or divergence of a series is not affected by adding (or removing) a finite number of terms.

44. Let $S_n = z + 2z^2 + 3z^3 + \cdots + nz^n$, $T_n = z + z^2 + z^3 + \cdots + z^n$. (a) Show that $S_n = (T_n - nz^{n+1})/(1 - z)$.
(b) Use (a) to find the sum of the series $\sum\limits_{n=1}^{\infty} nz^n$ and determine the set of values for which the series converges.   *Ans.* (b) $z/(1 - z)^2$, $|z| < 1$

45. Find the sum of the series $\sum\limits_{n=0}^{\infty} \dfrac{n+1}{2^n}$.   ***Ans.* 4**

ABSOLUTE AND UNIFORM CONVERGENCE

46. (a) Prove that $u_n(z) = 3z + 4z^2/n$, $n = 1, 2, 3, \ldots$, converges uniformly to $3z$ for all z inside or on the circle $|z| = 1$. (b) Can the circle of part (a) be enlarged? Explain.

47. (a) Determine whether the sequence $u_n(z) = nz/(n^2 + z^2)$ [Problem 35(b)] converges uniformly to zero for all z inside $|z| = 3$. (b) Does the result of (a) hold for all finite values of z?

48. Prove that the series $1 + az + a^2z^2 + \cdots$ converges uniformly to $1/(1 - az)$ inside or on the circle $|z| = R$ where $R < 1/|a|$.

49. Investigate the (a) absolute and (b) uniform convergence of the series

$$\frac{z}{3} + \frac{z(3 - z)}{3^2} + \frac{z(3 - z)^2}{3^3} + \frac{z(3 - z)^3}{3^4} + \cdots$$

Ans. (a) Converges absolutely if $|z - 3| < 3$ or $z = 0$. (b) Converges uniformly for $|z - 3| \leqq R$ where $0 < R < 3$; does not converge uniformly in any neighborhood which includes $z = 0$.

50. Investigate the (a) absolute and (b) uniform convergence of the series in Problem 38.
Ans. (a) Converges absolutely if $|z| < 1$. (b) Converges uniformly if $|z| \leqq R$ where $R < 1$.

51. Investigate the (a) absolute and (b) uniform convergence of the series in Problem 39.
Ans. (a) Converges absolutely if $|z^2 + 1| > 1$. (b) Converges uniformly if $|z^2 + 1| \geqq R$ where $R > 1$.

52. Let $\{a_n\}$ be a sequence of positive constants having limit zero; and suppose that for all z in a region $\mathcal{R}$, $|u_n(z)| \leqq a_n$, $n = 1, 2, 3, \ldots$. Prove that $\lim\limits_{n \to \infty} u_n(z) = 0$ uniformly in $\mathcal{R}$.

53. (a) Prove that the sequence $u_n(z) = nze^{-nz^2}$ converges to zero for all finite z such that $\text{Re}\,\{z^2\} > 0$, and represent this region geometrically. (b) Discuss the uniform convergence of the sequence in (a).
Ans. (b) Not uniformly convergent in any region which includes $z = 0$.

54. If $\sum\limits_{n=0}^{\infty} a_n$ and $\sum\limits_{n=0}^{\infty} b_n$ converge absolutely, prove that $\sum\limits_{n=0}^{\infty} c_n$, where $c_n = a_0b_n + a_1 b_{n-1} + \cdots + a_nb_0$, converges absolutely.

55. Prove that if each of two series is absolutely and uniformly convergent in $\mathcal{R}$, their product is absolutely and uniformly convergent in $\mathcal{R}$.

SPECIAL CONVERGENCE TESTS

56. Test for convergence:

(a) $\sum\limits_{n=1}^{\infty} \dfrac{1}{2^n + 1}$,   (b) $\sum\limits_{n=1}^{\infty} \dfrac{n}{3^n - 1}$,   (c) $\sum\limits_{n=1}^{\infty} \dfrac{n+3}{3n^2 - n + 2}$,   (d) $\sum\limits_{n=1}^{\infty} \dfrac{(-1)^n}{4n + 3}$,   (e) $\sum\limits_{n=1}^{\infty} \dfrac{2n - 1}{\sqrt{n^3 + n + 2}}$.
Ans. (a) conv., (b) conv., (c) div., (d) conv., (e) div.

57. Investigate the convergence of:

(a) $\sum_{n=1}^{\infty} \dfrac{1}{n+|z|}$,   (b) $\sum_{n=1}^{\infty} \dfrac{(-1)^n}{n+|z|}$,   (c) $\sum_{n=1}^{\infty} \dfrac{1}{n^2+|z|}$,   (d) $\sum_{n=1}^{\infty} \dfrac{1}{n^2+z}$.

Ans. (a) Diverges for all finite z. (b) Converges for all z. (c) Converges for all z. (d) Converges for all z except $z = -n^2$, $n = 1, 2, 3, \ldots$.

58. Investigate the convergence of $\sum_{n=0}^{\infty} \dfrac{ne^{n\pi i/4}}{e^n - 1}$.   Ans. Conv.

59. Find the region of convergence of   (a) $\sum_{n=0}^{\infty} \dfrac{(z+i)^n}{(n+1)(n+2)}$,   (b) $\sum_{n=1}^{\infty} \dfrac{1}{n^2 \cdot 3^n} \left(\dfrac{z+1}{z-1} \right)^n$,   (c) $\sum_{n=1}^{\infty} \dfrac{(-1)^n z^n}{n!}$.

Ans. (a) $|z+i| \leqq 1$,   (b) $|(z+1)/(z-1)| \leqq 3$,   (c) $|z| < \infty$

60. Investigate the region of absolute convergence of $\sum_{n=1}^{\infty} \dfrac{n(-1)^n (z-i)^n}{4^n (n^2+1)^{5/2}}$.

Ans. Conv. abs. for $|z-i| \leqq 4$.

61. Find the region of convergence of $\sum_{n=0}^{\infty} \dfrac{e^{2\pi i n z}}{(n+1)^{3/2}}$.

Ans. Converges if Im $z \geqq 0$.

62. Prove that the series $\sum_{n=1}^{\infty} (\sqrt{n+1} - \sqrt{n})$ diverges although the nth term approaches zero.

63. Let N be a positive integer and suppose that for all $n > N$, $|u_n| > 1/(n \ln n)$. Prove that $\sum_{n=1}^{\infty} u_n$ diverges.

64. Establish the validity of the   (a) nth root test [Theorem 12],   (b) integral test [Theorem 13], on Page 141.

65. Find the interval of convergence of $1 + 2z + z^2 + 2z^3 + z^4 + 2z^5 + \cdots$.   Ans. $|z| < 1$

66. Prove Raabe's test (Theorem 14) on Page 141.

67. Test for convergence: (a) $\dfrac{1}{2 \ln^2 2} + \dfrac{1}{3 \ln^2 3} + \dfrac{1}{4 \ln^2 4} + \cdots$,   (b) $\dfrac{1}{5} + \dfrac{1 \cdot 4}{5 \cdot 8} + \dfrac{1 \cdot 4 \cdot 7}{5 \cdot 8 \cdot 11} + \cdots$,   (c) $\dfrac{2}{5} +$

$\dfrac{2 \cdot 7}{5 \cdot 10} + \dfrac{2 \cdot 7 \cdot 12}{5 \cdot 10 \cdot 15} + \cdots$,   (d) $\dfrac{\ln 2}{2} + \dfrac{\ln 3}{3} + \dfrac{\ln 4}{4} + \cdots$.

Ans. (a) conv., (b) conv., (c) div., (d) div.

THEOREMS ON UNIFORM CONVERGENCE AND POWER SERIES

68. Determine the regions in which each of the following series is uniformly convergent:

(a) $\sum_{n=1}^{\infty} \dfrac{z^n}{3^n + 1}$,   (b) $\sum_{n=1}^{\infty} \dfrac{(z-i)^{2n}}{n^2}$,   (c) $\sum_{n=1}^{\infty} \dfrac{1}{(n+1)z^n}$,   (d) $\sum_{n=1}^{\infty} \dfrac{\sqrt{n+1}}{n^2 + |z|^2}$.

Ans. (a) $|z| \leqq R$ where $R < 3$.   (b) $|z-i| \leqq 1$.   (c) $|z| \geqq R$ where $R > 1$.   (d) All z

69. Prove Theorem 20, Page 142.

70. State and prove theorems for sequences analogous to Theorems 18, 19 and 20, Page 142, for series.

71. (a) By differentiating both sides of the identity

$$\dfrac{1}{1-z} = 1 + z + z^2 + z^3 + \cdots \qquad |z| < 1$$

find the sum of the series $\sum_{n=1}^{\infty} nz^n$ for $|z| < 1$. Justify all steps.

(b) Find the sum of the series $\sum_{n=1}^{\infty} n^2 z^n$ for $|z| < 1$.

Ans. (a) $z/(1-z)^2$ [compare Problem 44],   (b) $z(1+z)/(1-z)^3$

72. Let z be real and such that $0 \leq z \leq 1$, and let $u_n(z) = nze^{-nz^2}$.   (a) Find $\displaystyle\lim_{n \to \infty} \int_0^1 u_n(z)\, dz$.   (b) Find $\displaystyle\int_0^1 \left\{ \lim_{n \to \infty} u_n(z) \right\} dz$.   (c) Explain why the answers to (a) and (b) are not equal. [See Problem 53.]
Ans. (a) 1/2, (b) 0

73. Prove Abel's theorem [Theorem 24, Page 142].

74. (a) Prove that $\dfrac{1}{1+z^2} = 1 - z^2 + z^4 - z^6 + \cdots$ for $|z| < 1$.

(b) If we choose that branch of $f(z) = \tan^{-1} z$ such that $f(0) = 0$, use (a) to prove that
$$\tan^{-1} z = \int_0^z \frac{dz}{1+z^2} = z - \frac{z^3}{3} + \frac{z^5}{5} - \frac{z^7}{7} + \cdots$$

(c) Prove that $\dfrac{\pi}{4} = 1 - \dfrac{1}{3} + \dfrac{1}{5} - \dfrac{1}{7} + \cdots$.

75. Prove Theorem 25, Page 142.

76. (a) Determine $Y(z) = \displaystyle\sum_{n=0}^{\infty} a_n z^n$ such that for all z in $|z| \leq 1$, $Y'(z) = Y(z)$, $Y(0) = 1$. State all theorems used and verify that the result obtained is a solution.

(b) Is the result obtained in (a) valid outside of $|z| \leq 1$? Justify your answer.

(c) Show that $Y(z) = e^z$ satisfies the differential equation and conditions in (a).

(d) Can we identify the series in (a) with e^z? Explain.

Ans. (a) $Y(z) = 1 + z + \dfrac{z^2}{2!} + \dfrac{z^3}{3!} + \cdots$

77. (a) Use series methods on the differential equation $Y''(z) + Y(z) = 0$, $Y(0) = 0$, $Y'(0) = 1$ to obtain the series expansion
$$\sin z = z - \frac{z^3}{3!} + \frac{z^5}{5!} - \frac{z^7}{7!} + \cdots$$

(b) How could you obtain a corresponding series for $\cos z$?

TAYLOR'S THEOREM

78. Expand each of the following functions in a Taylor series about the indicated point and determine the region of convergence in each case.
(a) e^{-z}; $z = 0$ (b) $\cos z$; $z = \pi/2$ (c) $1/(1+z)$; $z = 1$ (d) $z^3 - 3z^2 + 4z - 2$; $z = 2$ (e) ze^{2z}; $z = -1$

79. If each of the following functions were expanded into a Taylor series about the indicated points, what would be the region of convergence? Do not perform the expansion.
(a) $\sin z/(z^2 + 4)$; $z = 0$ (c) $(z+3)/(z-1)(z-4)$; $z = 2$ (e) $e^z/z(z-1)$; $z = 4i$ (g) $\sec \pi z$; $z = 1$
(b) $z/(e^z + 1)$; $z = 0$ (d) $e^{-z^2} \sinh(z+2)$; $z = 0$ (f) $z \coth 2z$; $z = 0$
Ans. (a) $|z| < 2$, (b) $|z| < \pi$, (c) $|z-2| < 1$, (d) $|z| < \infty$, (e) $|z - 4i| < 4$, (f) $|z| < \pi/2$, (g) $|z-1| < 1/2$

80. Verify the expansions 1, 2, 3 for e^z, $\sin z$ and $\cos z$ on Page 143.

81. Show that $\sin z^2 = z^2 - \dfrac{z^6}{3!} + \dfrac{z^{10}}{5!} - \dfrac{z^{14}}{7!} + \cdots$, $|z| < \infty$.

82. Prove that $\tan^{-1} z = z - \dfrac{z^3}{3} + \dfrac{z^5}{5} - \dfrac{z^7}{7} + \cdots$, $|z| < 1$.

83. Show that   (a) $\tan z = z + \dfrac{z^3}{3} + \dfrac{2z^5}{15} + \cdots$, $|z| < \pi/2$

(b) $\sec z = 1 + \dfrac{z^2}{2} + \dfrac{5z^4}{24} + \cdots$, $|z| < \pi/2$

(c) $\csc z = \dfrac{1}{z} + \dfrac{z}{6} + \dfrac{7z^3}{360} + \cdots$, $0 < |z| < \pi$

84. By replacing z by iz in the expansion of Problem 82, obtain the result in Problem 23(c) on Page 155.

85. How would you obtain series for (a) $\tanh z$, (b) $\operatorname{sech} z$, (c) $\operatorname{csch} z$ from the series in Problem 83?

86. Prove the uniqueness of the Taylor series expansion of $f(z)$ about $z = a$.

[*Hint.* Assume $f(z) = \sum_{n=0}^{\infty} c_n(z-a)^n = \sum_{n=0}^{\infty} d_n(z-a)^n$ and show that $c_n = d_n$, $n = 0, 1, 2, 3, \ldots$.]

87. Prove the binomial Theorem 6 on Page 143.

88. If we choose that branch of $\sqrt{1 + z^3}$ having the value 1 for $z = 0$, show that

$$\frac{1}{\sqrt{1+z^3}} \;=\; 1 - \frac{1}{2}z^3 + \frac{1\cdot 3}{2\cdot 4}z^6 - \frac{1\cdot 3\cdot 5}{2\cdot 4\cdot 6}z^9 + \cdots \qquad |z| < 1$$

89. (a) Choosing that branch of $\sin^{-1} z$ having the value zero for $z = 0$, show that

$$\sin^{-1} z \;=\; z + \frac{1}{2}\frac{z^3}{3} + \frac{1\cdot 3}{2\cdot 4}\frac{z^5}{5} + \frac{1\cdot 3\cdot 5}{2\cdot 4\cdot 6}\frac{z^7}{7} + \cdots \qquad |z| < 1$$

(b) Prove that the result in (a) is valid for $z = i$.

90. (a) Expand $f(z) = \ln(3 - iz)$ in powers of $z - 2i$, choosing that branch of the logarithm for which $f(0) = \ln 3$, and (b) determine the region of convergence.

Ans. (a) $\ln 5 - \dfrac{i(z-2i)}{5} + \dfrac{(z-2i)^2}{2\cdot 5^2} + \dfrac{i(z-2i)^3}{3\cdot 5^3} - \dfrac{(z-2i)^4}{4\cdot 5^4} - \cdots$ (b) $|z - 2i| < 5$

LAURENT'S THEOREM

91. Expand $f(z) = 1/(z - 3)$ in a Laurent series valid for (a) $|z| < 3$, (b) $|z| > 3$.

Ans. (a) $-\dfrac{1}{3} - \dfrac{1}{9}z - \dfrac{1}{27}z^2 - \dfrac{1}{81}z^3 - \cdots$ (b) $z^{-1} + 3z^{-2} + 9z^{-3} + 27z^{-4} + \cdots$

92. Expand $f(z) = \dfrac{z}{(z-1)(2-z)}$ in a Laurent series valid for:

(a) $|z| < 1$, (b) $1 < |z| < 2$, (c) $|z| > 2$, (d) $|z-1| > 1$, (e) $0 < |z-2| < 1$.

Ans. (a) $-\dfrac{1}{2}z - \dfrac{3}{4}z^2 - \dfrac{7}{8}z^3 - \dfrac{15}{16}z^4 - \cdots$ (b) $\cdots + \dfrac{1}{z^2} + \dfrac{1}{z} + 1 + \dfrac{1}{2}z + \dfrac{1}{4}z^2 + \dfrac{1}{8}z^3 + \cdots$

(c) $-\dfrac{1}{2} - \dfrac{3}{z^2} - \dfrac{7}{z^3} - \dfrac{15}{z^4} - \cdots$ (d) $-(z-1)^{-1} - 2(z-1)^{-2} - 2(z-1)^{-3} - \cdots$

(e) $1 - 2(z-2)^{-1} - (z-2) + (z-2)^2 - (z-2)^3 + (z-2)^4 - \cdots$

93. Expand $f(z) = 1/z(z-2)$ in a Laurent series valid for (a) $0 < |z| < 2$, (b) $|z| > 2$.

94. Find an expansion of $f(z) = z/(z^2 + 1)$ valid for $|z - 3| > 2$.

95. Expand $f(z) = 1/(z-2)^2$ in a Laurent series valid for (a) $|z| < 2$, (b) $|z| > 2$

96. Expand each of the following functions in a Laurent series about $z = 0$, naming the type of singularity in each case.

(a) $(1 - \cos z)/z$, (b) e^{z^2}/z^3, (c) $z^{-1}\cosh z^{-1}$, (d) $z^2 e^{-z^4}$, (e) $z \sinh \sqrt{z}$.

Ans. (a) $\dfrac{z}{2!} - \dfrac{z^3}{4!} + \dfrac{z^5}{6!} - \cdots$; removable singularity   (d) $z^2 - z^6 + \dfrac{z^{10}}{2!} - \dfrac{z^{14}}{3!} + \cdots$;

ordinary point

(b) $\dfrac{1}{z^3} + \dfrac{1}{z} + \dfrac{z}{2!} + \dfrac{z^3}{3!} + \dfrac{z^5}{4!} + \dfrac{z^7}{5!} + \cdots$;

pole of order 3

(e) $z^{3/2} + \dfrac{z^{5/2}}{3!} + \dfrac{z^{7/2}}{5!} + \dfrac{z^{9/2}}{7!} + \cdots$;

(c) $\dfrac{1}{z} - \dfrac{1}{2!\,z^3} + \dfrac{1}{4!\,z^5} - \cdots$; essential singularity   branch point

97. Show that if $\tan z$ is expanded into a Laurent series about $z = \pi/2$, (a) the principal part is $-1/(z - \pi/2)$, (b) the series converges for $0 < |z - \pi/2| < \pi/2$, (c) $z = \pi/2$ is a simple pole.

98. Determine and classify all the singularities of the functions:

(a) $1/(2 \sin z - 1)^2$,   (b) $z/(e^{1/z} - 1)$,   (c) $\cos(z^2 + z^{-2})$,   (d) $\tan^{-1}(z^2 + 2z + 2)$,   (e) $z/(e^z - 1)$.

Ans. (a) $\pi/6 + 2m\pi$, $(2m+1)\pi - \pi/6$, $m = 0, \pm 1, \pm 2, \ldots$; poles of order 2

   (b) $i/2m\pi$, $m = \pm 1, \pm 2, \ldots$; simple poles, $z = 0$; essential singularity, $z = \infty$; pole of order 2

   (c) $z = 0, \infty$; essential singularities   (d) $z = -1 \pm i$; branch points

   (e) $z = 2m\pi i$, $m = \pm 1, \pm 2, \ldots$; simple poles, $z = 0$; removable singularity, $z = \infty$; essential singularity

99. (a) Expand $f(z) = e^{z/(z-2)}$ in a Laurent series about $z = 2$ and (b) determine the region of convergence of this series. (c) Classify the singularities of $f(z)$.

Ans. (a) $e\left\{ 1 + 2(z-2)^{-1} + \dfrac{2^2(z-2)^{-2}}{2!} + \dfrac{2^3(z-2)^{-3}}{3!} + \cdots \right\}$ (b) $|z-2| > 0$ (c) $z = 2$; essential

   singularity, $z = \infty$; removable singularity

100. Establish the result (7), Page 144, for the coefficients in a Laurent series.

101. Prove that the only singularities of a rational function are poles.

102. Prove the converse of Problem 101, i.e. if the only singularities of a function are poles, the function must be rational.

LAGRANGE'S EXPANSION

103. Show that the root of the equation $z = 1 + \zeta z^p$, which is equal to 1 when $\zeta = 0$, is given by

$$ z = 1 + \zeta + \frac{2p}{2!}\zeta^2 + \frac{(3p)(3p-1)}{3!}\zeta^3 + \frac{(4p)(4p-1)(4p-2)}{4!}\zeta^4 + \cdots $$

104. Calculate the root in Problem 103 if $p = 1/2$ and $\zeta = 1$, (a) by series and (b) exactly, and compare the two answers.   *Ans.* 2.62 to two decimal accuracy

105. By considering the equation $z = a + \frac{1}{2}\zeta(z^2 - 1)$, show that

$$ \frac{1}{\sqrt{1 - 2a\zeta + \zeta^2}} = 1 + \sum_{n=1}^{\infty} \frac{\zeta^n}{2^n\, n!} \frac{d^n}{da^n}(a^2 - 1)^n $$

106. Show how Lagrange's expansion can be used to solve Kepler's problem of determining that root of $z = a + \zeta \sin z$ for which $z = a$ when $\zeta = 0$.

107. Prove the Lagrange expansion (12) on Page 145.

ANALYTIC CONTINUATION

108. (a) Prove that $F_2(z) = \dfrac{1}{1+i} \displaystyle\sum_{n=0}^{\infty} \left(\dfrac{z+i}{1+i} \right)^n$ is an analytic continuation of $F_1(z) = \displaystyle\sum_{n=0}^{\infty} z^n$, showing graphically the regions of convergence of the series.

(b) Determine the function represented by all analytic continuations of $F_1(z)$.   *Ans.* (b) $1/(1-z)$

109. Let $F_1(z) = \displaystyle\sum_{n=0}^{\infty} \frac{z^{n+1}}{3^n}$.   (a) Find an analytic continuation of $F_1(z)$ which converges for $z = 3 - 4i$.

(b) Determine the value of the analytic continuation in (a) for $z = 3 - 4i$.   *Ans.* (b) $-3 - \frac{9}{4}i$

110. Prove that the series

$$ z^{1!} + z^{2!} + z^{3!} + \cdots $$

has the natural boundary $|z| = 1$.

MISCELLANEOUS PROBLEMS

111. (a) Prove that $\displaystyle\sum_{n=1}^{\infty} \frac{1}{n^p}$ diverges if the constant $p \leq 1$.

(b) Prove that if p is complex the series in (a) converges if $\operatorname{Re}\{p\} > 1$.

(c) Investigate the convergence or divergence of the series in (a) if $\operatorname{Re}\{p\} \leq 1$.

112. Test for convergence or divergence: (a) $\sum\limits_{n=1}^{\infty} \dfrac{\sqrt{n}}{n+i}$,   (b) $\sum\limits_{n=1}^{\infty} \dfrac{n+\sin^2 n}{ie^n + (2-i)n}$,   (c) $\sum\limits_{n=1}^{\infty} n \sin^{-1}(1/n^3)$,

(d) $\sum\limits_{n=2}^{\infty} \dfrac{(i)^n}{n \ln n}$,   (e) $\sum\limits_{n=1}^{\infty} \coth^{-1} n$,   (f) $\sum\limits_{n=1}^{\infty} ne^{-n^2}$.

Ans. (a) div., (b) conv., (c) conv., (d) conv., (e) div., (f) conv.

113. Euler presented the following argument to show that $\sum\limits_{-\infty}^{\infty} z^n = 0$:

$$\frac{z}{1-z} = z + z^2 + z^3 + \cdots = \sum_1^{\infty} z^n, \qquad \frac{z}{z-1} = \frac{1}{1-1/z} = 1 + \frac{1}{z} + \frac{1}{z^2} + \cdots = \sum_0^{-\infty} z^n$$

Then adding, $\sum\limits_{-\infty}^{\infty} z^n = 0$. Explain the fallacy.

114. Show that for $|z-1| < 1$, $z \ln z = (z-1) + \dfrac{(z-1)^2}{1 \cdot 2} - \dfrac{(z-1)^3}{2 \cdot 3} + \dfrac{(z-1)^4}{3 \cdot 4} - \cdots$.

115. Expand $\sin^3 z$ in a Maclaurin series.   *Ans.* $\sum\limits_{n=1}^{\infty} \dfrac{(3 - 3^{2n-1})z^{2n-1}}{4(2n-1)!}$

116. Given the series $z^2 + \dfrac{z^2}{1+z^2} + \dfrac{z^2}{(1+z^2)^2} + \dfrac{z^2}{(1+z^2)^3} + \cdots$.

  (a) Show that the sum of the first n terms is $S_n(z) = 1 + z^2 - 1/(1+z^2)^{n-1}$.

  (b) Show that the sum of the series is $1 + z^2$ for $z \neq 0$, and 0 for $z = 0$; and hence that $z = 0$ is a point of discontinuity.

  (c) Show that the series is not uniformly convergent in the region $|z| \leq \delta$ where $\delta > 0$.

117. If $F(z) = \dfrac{3z - 3}{(2z-1)(z-2)}$, find a Laurent series of $F(z)$ about $z = 1$ convergent for $\tfrac{1}{2} < |z-1| < 1$.

Ans. $\cdots - \tfrac{1}{8}(z-1)^{-4} + \tfrac{1}{4}(z-1)^{-3} - \tfrac{1}{2}(z-1)^{-2} + (z-1)^{-1} - 1 - (z-1) - (z-1)^2 - \cdots$

118. Let $G(z) = (\tan^{-1} z)/z^4$.   (a) Expand $G(z)$ in a Laurent series.   (b) Determine the region of convergence of the series in (a).   (c) Evaluate $\oint_C G(z)\,dz$ where C is a square with vertices at $2 \pm 2i$, $-2 \pm 2i$.   *Ans.* (a) $\dfrac{1}{z^3} - \dfrac{1}{3z} + \dfrac{z}{5} - \dfrac{z^3}{7} + \cdots$ (b) $|z| > 0$ (c) $-1/3$

119. For each of the functions ze^{1/z^2}, $(\sin^2 z)/z$, $1/z(4-z)$ which have singularities at $z = 0$:   (a) give a Laurent expansion about $z = 0$ and determine the region of convergence;   (b) state in each case whether $z = 0$ is a removable singularity, essential singularity or a pole;   (c) evaluate the integral of the function about the circle $|z| = 2$.

Ans. (a) $z + z^{-1} + z^{-3}/2! + z^{-5}/3! + \cdots$; $|z| > 0$, $2z - 2z^3/3 + 4z^5/45 - \cdots$; $|z| \geq 0$, $z^{-1}/4 + 1/16 + z/64 + z^2/256 + \cdots$; $0 < |z| < 4$

    (b) essential singularity, removable singularity, pole   (c) $2\pi i$, 0, $\pi i/2$

120. (a) Investigate the convergence of $\sum\limits_{n=1}^{\infty} \dfrac{1}{n^{1+1/n}}$.   (b) Does your answer to (a) contradict Problem 8, Page 148?   *Ans.* (a) diverges

121. (a) Show that the series $\dfrac{\sin z}{1^2 + 1} + \dfrac{\sin^2 z}{2^2 + 1} + \dfrac{\sin^3 z}{3^2 + 1} + \cdots$, where $z = x + iy$, converges absolutely in the region bounded by $\sin^2 x + \sinh^2 y = 1$.   (b) Graph the region of (a).

122. If $|z| > 0$, prove that
$$\cosh(z + 1/z) = c_0 + c_1(z + 1/z) + c_2(z^2 + 1/z^2) + \cdots$$
where
$$c_n = \frac{1}{2\pi} \int_0^{2\pi} \cos n\phi \cosh(2 \cos \phi)\,d\phi$$

123. If $f(z)$ has simple zeros at $1 - i$ and $1 + i$, double poles at $-1 + i$ and $-1 - i$, but no other finite singularities, prove that the function must be given by
$$f(z) = \kappa \frac{z^2 - 2z + 2}{(z^2 + 2z + 2)^2}$$
where κ is an arbitrary constant.

124. Prove that for all z, $e^z \sin z = \sum_{n=1}^{\infty} \frac{2^{n/2} \sin(n\pi/4)}{n!} z^n$.

125. Show that $\ln 2 = 1 - \frac{1}{2} + \frac{1}{3} - \frac{1}{4} + \cdots$,   justifying all steps.   [*Hint*. Use Problem 23.]

126. Investigate the uniform convergence of the series $\sum_{n=1}^{\infty} \frac{z}{[1+(n-1)z][1+nz]}$.

$\left[\right.$ *Hint*. Resolve the nth term into partial fractions and show that the nth partial sum is $S_n(z) = 1 - \frac{1}{1+nz}$.$\left.\right]$

Ans. Not uniformly convergent in any region which includes $z = 0$; uniformly convergent in a region $|z| \geqq \delta$, where δ is any positive number.

127. If $1 - \frac{1}{2} + \frac{1}{3} - \frac{1}{4} + \cdots$ converges to S, prove that the rearranged series $1 + \frac{1}{3} - \frac{1}{2} + \frac{1}{5} + \frac{1}{7} - \frac{1}{4} + \frac{1}{9} + \frac{1}{11} - \frac{1}{6} + \cdots = \frac{3}{2}S$.   Explain.

[*Hint*. Take 1/2 of the first series and write it as $0 + \frac{1}{2} + 0 - \frac{1}{4} + 0 + \frac{1}{6} + \cdots$;   then add term by term to the first series. Note that $S = \ln 2$, as shown in Problem 125.]

128. Prove that the *hypergeometric series*
$$1 + \frac{a \cdot b}{1 \cdot c} z + \frac{a(a+1)\, b(b+1)}{1 \cdot 2 \cdot c(c+1)} z^2 + \frac{a(a+1)(a+2)\, b(b+1)(b+2)}{1 \cdot 2 \cdot 3 \cdot c(c+1)(c+2)} z^3 + \cdots$$

(a) converges absolutely if $|z| < 1$,   (b) diverges for $|z| > 1$,   (c) converges absolutely for $z = 1$ if $\text{Re}\{a+b-c\} < 0$,   (d) satisfies the differential equation $z(1-z)Y'' + \{c - (a+b+1)z\}Y' - abY = 0$.

129. Prove that for $|z| < 1$,
$$(\sin^{-1} z)^2 = z^2 + \frac{2}{3} \cdot \frac{z^4}{2} + \frac{2 \cdot 4}{3 \cdot 5} \cdot \frac{z^6}{3} + \frac{2 \cdot 4 \cdot 6}{3 \cdot 5 \cdot 7} \cdot \frac{z^8}{4} + \cdots$$

130. Prove that $\sum_{n=1}^{\infty} \frac{1}{n^{1+i}}$ diverges.

131. Show that $\frac{1}{1 \cdot 2} - \frac{1}{2 \cdot 3} + \frac{1}{3 \cdot 4} - \frac{1}{4 \cdot 5} + \cdots = 2\ln 2 - 1$

132. Locate and name all the singularities of $\frac{z^6 + 1}{(z-1)^3 (3z+2)^2} \sin\left(\frac{z^2}{z-3}\right)$.

133. By using only properties of infinite series, prove that

(a) $\left\{1 + a + \frac{a^2}{2!} + \frac{a^3}{3!} + \cdots\right\}\left\{1 + b + \frac{b^2}{2!} + \frac{b^3}{3!} + \cdots\right\} = \left\{1 + (a+b) + \frac{(a+b)^2}{2!} + \cdots\right\}$

(b) $\left\{1 - \frac{a^2}{2!} + \frac{a^4}{4!} - \frac{a^6}{6!} + \cdots\right\}^2 + \left\{a - \frac{a^3}{3!} + \frac{a^5}{5!} - \frac{a^7}{7!} + \cdots\right\}^2 = 1$

134. If $f(z) = \sum_{n=0}^{\infty} a_n z^n$ converges for $|z| < R$ and $0 \leqq r < R$, prove that
$$\frac{1}{2\pi} \int_0^{2\pi} |f(re^{i\theta})|^2 \, d\theta = \sum_{n=0}^{\infty} |a_n|^2 r^{2n}$$

135. Use Problem 134 to prove Cauchy's inequality (Page 118), namely
$$|f^{(n)}(0)| \leqq \frac{M \cdot n!}{r^n} \qquad n = 0, 1, 2, \ldots$$

136. If a function has six zeros of order 4, and four poles of orders 3, 4, 7 and 8, but no other singularities in the finite plane, prove that it has a pole of order 2 at $z = \infty$.

137. State whether each of the following functions are entire, meromorphic or neither:
   (a) $z^2 e^{-z}$,   (b) $\cot 2z$,   (c) $(1 - \cos z)/z$,   (d) $\cosh z^2$,   (e) $z \sin(1/z)$,   (f) $z + 1/z$,   (g) $\sin\sqrt{z}/\sqrt{z}$,   (h) $\sqrt{\sin z}$.
   Ans. (a) entire,   (b) meromorphic,   (c) entire,   (d) entire,   (e) neither,   (f) meromorphic,   (g) entire,   (h) neither

138. If $-\pi < \theta < \pi$, prove that
$$\ln(2 \cos \theta/2) = \cos \theta - \frac{1}{2} \cos 2\theta + \frac{1}{3} \cos 3\theta - \frac{1}{4} \cos 4\theta + \cdots$$

139. (a) Expand $1/\ln(1+z)$ in a Laurent series about $z = 0$ and   (b) determine the region of convergence.

Ans. (a) $\dfrac{1}{z} + \dfrac{z}{2} - \dfrac{z}{12} + \dfrac{z^2}{24} + \dfrac{89z^3}{720} + \cdots$ (b) $0 < |z| < 1$

140. If $S(z) = a_0 + a_1 z + a_2 z^2 + \cdots,$ prove that

$$\frac{S(z)}{1-z} = a_0 + (a_0 + a_1)z + (a_0 + a_1 + a_2)z^2 + \cdots$$

giving restrictions if any.

141. Show that the series

$$\frac{1}{1+|z|} - \frac{1}{2+|z|} + \frac{1}{3+|z|} - \frac{1}{4+|z|} + \cdots$$

(a) is not absolutely convergent but   (b) is uniformly convergent for all values of z.

142. Prove that $\displaystyle\sum_{n=1}^{\infty} \frac{z^n}{n}$ converges at all points of $|z| \leqq 1$ except $z = 1$.

143. Prove that the solution of $z = a + \zeta e^z$, which has the value a when $\zeta = 0$, is given by

$$z = a + \sum_{n=1}^{\infty} \frac{n^{n-1} e^{na} \zeta^n}{n!}$$

if $|\zeta| < |e^{-(a+1)}|$.

144. Find the sum of the series $1 + \cos\theta + \dfrac{\cos 2\theta}{2!} + \dfrac{\cos 3\theta}{3!} + \cdots.$ *Ans.* $e^{\cos\theta} \cos(\sin\theta)$

145. Let $F(z)$ be analytic in the finite plane and suppose that $F(z)$ has period 2π, i.e. $F(z + 2\pi) = F(z)$. Prove that

$$F(z) = \sum_{n=-\infty}^{\infty} \alpha_n e^{inz} \qquad \text{where} \qquad \alpha_n = \frac{1}{2\pi} \int_0^{2\pi} F(z)\, e^{-inz}\, dz$$

The series is called the *Fourier series* for $F(z)$.

146. Prove that the series

$$\sin\theta + \tfrac{1}{3}\sin 3\theta + \tfrac{1}{5}\sin 5\theta + \cdots$$

is equal to $\pi/4$ if $0 < \theta < \pi$, and to $-\pi/4$ if $-\pi < \theta < 0$.

147. Prove that $|z| = 1$ is a natural boundary for the series $\displaystyle\sum_{n=0}^{\infty} 2^{-n} z^{3^n}$.

148. If $f(z)$ is analytic and not identically zero in the region $0 < |z - z_0| < R$, and if $\displaystyle\lim_{z \to z_0} f(z) = 0$, prove that there exists a positive integer n such that $f(z) = (z - z_0)^n g(z)$ where $g(z)$ is analytic at z_0 and different from zero.

149. If $f(z)$ is analytic in a deleted neighborhood of z_0 and $\displaystyle\lim_{z \to z_0} |f(z)| = \infty$,   prove that $z = z_0$ is a pole of $f(z)$.

150. Explain why Problem 149 does not hold for $f(x) = e^{1/x^2}$ where x is real.

151. (a) Show that the function $f(z) = e^{1/z}$ can assume any value except zero. (b) Discuss the relationship of the result of (a) to the Casorati-Weierstrass theorem and Picard's theorem.

152. (a) Determine whether the function $g(z) = z^2 - 3z + 2$ can assume any complex value. (b) Is there any relationship of the result in (a) to the theorems of Casorati-Weierstrass and Picard? Explain.

153. Prove the Casorati-Weierstrass theorem stated on Page 145. [*Hint.* Use the fact that if $z = a$ is an essential singularity of $f(z)$, then it is also an essential singularity of $1/\{f(z) - A\}$.]

154. (a) Prove that along any ray through $z = 0$, $|z + e^z| \to \infty$.
(b) Does the result in (a) contradict the Casorati-Weierstrass theorem?

155. (a) Prove that an entire function $f(z)$ can assume any value whatsoever, with perhaps one exception.

(b) Illustrate the result of (a) by considering $f(z) = e^z$ and stating the exception in this case.

(c) What is the relationship of the result to the Casorati-Weierstrass and Picard theorems?

156. Prove that every entire function has a singularity at infinity.   What type of singularity must this be? Justify your answer.

157. Prove that: (a) $\dfrac{\ln(1+z)}{1+z} = z - (1+\tfrac{1}{2})z^2 + (1+\tfrac{1}{2}+\tfrac{1}{3})z^3 - \cdots$, $|z| < 1$

(b) $\{\ln(1+z)\}^2 = z^2 - (1+\tfrac{1}{2})\dfrac{2z^3}{3} + (1+\tfrac{1}{2}+\tfrac{1}{3})\dfrac{2z^4}{4} - \cdots$, $|z| < 1$

158. Find the sum of the following series if $|a| < 1$:

$$(a)\ \sum_{n=1}^{\infty} na^n \sin n\theta, \qquad (b)\ \sum_{n=1}^{\infty} n^2 a^n \sin n\theta$$

159. Show that $e^{\sin z} = 1 + z + \dfrac{z^2}{2} - \dfrac{z^4}{8} - \dfrac{z^5}{15} + \cdots$, $|z| < \infty$.

160. (a) Show that $\displaystyle\sum_{n=1}^{\infty} \dfrac{z^n}{n^2}$ converges for $|z| \leqq 1$.

(b) Show that the function $F(z)$, defined as the collection of all possible analytic continuations of the series in (a), has a singular point at $z = 1$.

(c) Reconcile the results of (a) and (b).

161. Let $\displaystyle\sum_{n=1}^{\infty} a_n z^n$ converge inside a circle of convergence of radius R. There is a theorem which states that the function $F(z)$ defined by the collection of all possible continuations of this series, has at least one singular point on the circle of convergence.   (a) Illustrate the theorem by several examples. (b) Can you prove the theorem?

162. Show that

$$u(r, \theta) = \frac{R^2 - r^2}{2\pi} \int_0^{2\pi} \frac{U(\phi)\, d\phi}{R^2 - 2rR\cos(\theta - \phi) + r^2}$$

$$= \frac{a_0}{2} + \sum_{n=1}^{\infty} \left(\frac{r}{R}\right)^n \{a_n \cos n\theta + b_n \sin n\theta\}$$

where

$$a_n = \frac{1}{\pi}\int_0^{2\pi} U(\phi)\cos n\phi\, d\phi, \qquad b_n = \frac{1}{\pi}\int_0^{2\pi} U(\phi)\sin n\phi\, d\phi$$

163. Let $\dfrac{z}{e^z - 1} = 1 + B_1 z + \dfrac{B_2 z^2}{2!} + \dfrac{B_3 z^3}{3!} + \cdots$.   (a) Show that the numbers B_n, called the *Bernoulli numbers*, satisfy the recursion formula $(B+1)^n = B^n$ where B^k is formally replaced by B_k after expanding.   (b) Using (a) or otherwise, determine $B_1, \ldots, B_6$.

Ans. (b) $B_1 = -\tfrac{1}{2}$, $B_2 = \tfrac{1}{6}$, $B_3 = 0$, $B_4 = -\tfrac{1}{30}$, $B_5 = 0$, $B_6 = \tfrac{1}{42}$

164. (a) Prove that $\dfrac{z}{e^z - 1} = \dfrac{z}{2}\left(\coth\dfrac{z}{2} - 1\right)$.   (b) Use Problem 163 and part (a) to show that $B_{2k+1} = 0$ if $k = 1, 2, 3, \ldots$.

165. Derive the series expansions:

(a) $\coth z = \dfrac{1}{z} + \dfrac{z}{3} - \dfrac{z^3}{45} + \cdots + \dfrac{B_{2n}(2z)^{2n}}{(2n)!\, z} + \cdots$, $|z| < \pi$

(b) $\cot z = \dfrac{1}{z} - \dfrac{z}{3} - \dfrac{z^3}{45} + \cdots (-1)^n \dfrac{B_{2n}(2z)^{2n}}{(2n)!\, z} + \cdots$, $|z| < \pi$

(c) $\tan z = z + \dfrac{z^3}{3} + \dfrac{2z^5}{15} + \cdots (-1)^{n-1} \dfrac{2(2^{2n}-1)B_{2n}(2z)^{2n-1}}{(2n)!}$, $|z| < \pi/2$

(d) $\csc z = \dfrac{1}{z} + \dfrac{z}{6} + \dfrac{7z^3}{360} + \cdots (-1)^{n-1} \dfrac{2(2^{2n-1}-1)B_{2n}z^{2n-1}}{(2n)!} + \cdots$, $|z| < \pi$

[*Hint.* For (a) use Problem 164;   for (b) replace z by iz in (a);   for (c) use $\tan z = \cot z - 2\cot 2z$; for (d) use $\csc z = \cot z + \tan z/2$.]

Chapter 7

The Residue Theorem
Evaluation of Integrals and Series

RESIDUES

Let $f(z)$ be single-valued and analytic inside and on a circle C except at the point $z = a$ chosen as the center of C. Then, as we have seen in Chapter 6, $f(z)$ has a Laurent series about $z = a$ given by

$$f(z) = \sum_{n=-\infty}^{\infty} a_n(z-a)^n$$
$$= a_0 + a_1(z-a) + a_2(z-a)^2 + \cdots + \frac{a_{-1}}{z-a} + \frac{a_{-2}}{(z-a)^2} + \cdots \qquad (1)$$

where
$$a_n = \frac{1}{2\pi i} \oint_C \frac{f(z)}{(z-a)^{n+1}} \, dz \qquad n = 0, \pm 1, \pm 2, \ldots \qquad (2)$$

In the special case $n = -1$, we have from (2)

$$\oint_C f(z) \, dz = 2\pi i \, a_{-1} \qquad (3)$$

Formally we can obtain (3) from (1) by integrating term by term and using the results (Problems 21 and 22, Chapter 4)

$$\oint_C \frac{dz}{(z-a)^p} = \begin{cases} 2\pi i & p = 1 \\ 0 & p = \text{integer} \neq 1 \end{cases} \qquad (4)$$

Because of the fact that (3) involves only the coefficient a_{-1} in (1), we call a_{-1} the *residue* of $f(z)$ at $z = a$.

CALCULATION OF RESIDUES

To obtain the residue of a function $f(z)$ at $z = a$, it may appear from (1) that the Laurent expansion of $f(z)$ about $z = a$ must be obtained. However, in the case where $z = a$ is a pole of order k there is a simple formula for a_{-1} given by

$$a_{-1} = \lim_{z \to a} \frac{1}{(k-1)!} \frac{d^{k-1}}{dz^{k-1}} \{(z-a)^k f(z)\} \qquad (5)$$

If $k = 1$ (simple pole) the result is especially simple and is given by

$$a_{-1} = \lim_{z \to a} (z-a) f(z) \qquad (6)$$

which is a special case of (5) with $k = 1$ if we define $0! = 1$.

Example 1: If $f(z) = \dfrac{z}{(z-1)(z+1)^2}$, then $z = 1$ and $z = -1$ are poles of orders one and two respectively. We have, using (6) and (5) with $k = 2$,

Residue at $z = 1$ is $\quad \lim\limits_{z \to 1} (z-1) \left\{ \dfrac{z}{(z-1)(z+1)^2} \right\} = \dfrac{1}{4}$

Residue at $z = -1$ is $\quad \lim\limits_{z \to -1} \dfrac{1}{1!} \dfrac{d}{dz} \left\{ (z+1)^2 \left(\dfrac{z}{(z-1)(z+1)^2} \right) \right\} = -\dfrac{1}{4}$

If $z = a$ is an essential singularity, the residue can sometimes be found by using known series expansions.

Example 2: If $f(z) = e^{-1/z}$, then $z = 0$ is an essential singularity and from the known expansion for e^u with $u = -1/z$ we find

$$e^{-1/z} = 1 - \frac{1}{z} + \frac{1}{2! \, z^2} - \frac{1}{3! \, z^3} + \cdots$$

from which we see that the residue at $z = 0$ is the coefficient of $1/z$ and equals -1.

THE RESIDUE THEOREM

Let $f(z)$ be single-valued and analytic inside and on a simple closed curve C except at the singularities $a, b, c, \ldots$ inside C which have residues given by $a_{-1}, b_{-1}, c_{-1}, \ldots$ [see Fig. 7-1]. Then the *residue theorem* states that

$$\oint_C f(z)\, dz = 2\pi i (a_{-1} + b_{-1} + c_{-1} + \cdots) \quad (7)$$

i.e. the integral of $f(z)$ around C is $2\pi i$ times the sum of the residues of $f(z)$ at the singularities enclosed by C. Note that (7) is a generalization of (3). Cauchy's theorem and integral formulas are special cases of this theorem (see Problem 75).

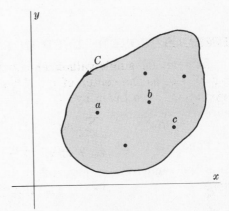

Fig. 7-1

EVALUATION OF DEFINITE INTEGRALS

The evaluation of definite integrals is often achieved by using the residue theorem together with a suitable function $f(z)$ and a suitable closed path or contour C, the choice of which may require great ingenuity. The following types are most common in practice.

1. $\displaystyle\int_{-\infty}^{\infty} F(x)\, dx$, $F(x)$ is a rational function.

   Consider $\displaystyle\oint_C F(z)\, dz$ along a contour C consisting of the line along the x axis from $-R$ to $+R$ and the semicircle Γ above the x axis having this line as diameter [Fig. 7-2]. Then let $R \to \infty$. If $F(x)$ is an even function this can be used to evaluate $\displaystyle\int_0^{\infty} F(x)\, dx$. See Problems 7-10.

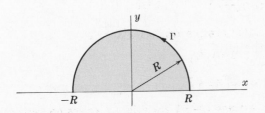

Fig. 7-2

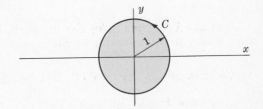

Fig. 7-3

2. $\displaystyle\int_0^{2\pi} G(\sin\theta, \cos\theta)\, d\theta$, $G(\sin\theta, \cos\theta)$ is a rational function of $\sin\theta$ and $\cos\theta$.

   Let $z = e^{i\theta}$. Then $\sin\theta = \dfrac{z - z^{-1}}{2i}$, $\cos\theta = \dfrac{z + z^{-1}}{2}$ and $dz = ie^{i\theta} d\theta$ or $d\theta = dz/iz$. The given integral is equivalent to $\displaystyle\oint_C F(z)\, dz$ where C is the unit circle with center at the origin [Fig. 7-3]. See Problems 11-14.

3. $\int_{-\infty}^{\infty} F(x) \begin{Bmatrix} \cos mx \\ \sin mx \end{Bmatrix} dx$, $F(x)$ is a rational function.

Here we consider $\oint_C F(z)\, e^{imz}\, dz$ where C is the same contour as that in Type 1. See Problems 15-17, and 37.

4. Miscellaneous integrals involving particular contours. See Problems 18-23.

SPECIAL THEOREMS USED IN EVALUATING INTEGRALS

In evaluating integrals such as those of Types 1 and 3 above, it is often necessary to show that $\int_\Gamma F(z)\, dz$ and $\int_\Gamma e^{imz} F(z)\, dz$ approach zero as $R \to \infty$. The following theorems are fundamental.

Theorem 1. If $|F(z)| \leqq \dfrac{M}{R^k}$ for $z = Re^{i\theta}$, where $k > 1$ and M are constants, then if Γ is the semicircle of Fig. 7-2,

$$\lim_{R \to \infty} \int_\Gamma F(z)\, dz = 0$$

See Problem 7.

Theorem 2. If $|F(z)| \leqq \dfrac{M}{R^k}$ for $z = Re^{i\theta}$, where $k > 0$ and M are constants, then if Γ is the semicircle of Fig. 7-2,

$$\lim_{R \to \infty} \int_\Gamma e^{imz} F(z)\, dz = 0$$

See Problem 15.

THE CAUCHY PRINCIPAL VALUE OF INTEGRALS

If $F(x)$ is continuous in $a \leqq x \leqq b$ except at a point x_0 such that $a < x_0 < b$, then if ϵ_1 and ϵ_2 are positive we define

$$\int_a^b F(x)\, dx = \lim_{\substack{\epsilon_1 \to 0 \\ \epsilon_2 \to 0}} \left\{ \int_a^{x_0 - \epsilon_1} F(x)\, dx + \int_{x_0 + \epsilon_2}^b F(x)\, dx \right\}$$

In some cases the above limit does not exist for $\epsilon_1 \neq \epsilon_2$ but does exist if we take $\epsilon_1 = \epsilon_2 = \epsilon$. In such case we call

$$\int_a^b F(x)\, dx = \lim_{\epsilon \to 0} \left\{ \int_a^{x_0 - \epsilon} F(x)\, dx + \int_{x_0 + \epsilon}^b F(x)\, dx \right\}$$

the *Cauchy principal value* of the integral on the left.

Example: $\int_{-1}^1 \dfrac{dx}{x^3} = \lim_{\substack{\epsilon_1 \to 0 \\ \epsilon_2 \to 0}} \left\{ \int_{-1}^{-\epsilon_1} \dfrac{dx}{x^3} + \int_{\epsilon_2}^1 \dfrac{dx}{x^3} \right\} = \lim_{\substack{\epsilon_1 \to 0 \\ \epsilon_2 \to 0}} \left\{ \dfrac{1}{2\epsilon_2^2} - \dfrac{1}{2\epsilon_1^2} \right\}$

does not exist. However, the Cauchy principal value with $\epsilon_1 = \epsilon_2 = \epsilon$ does exist and equals zero.

DIFFERENTIATION UNDER THE INTEGRAL SIGN. LEIBNITZ'S RULE

A useful method for evaluating integrals employs *Leibnitz's rule* for differentiation under the integral sign. This rule states that

$$\frac{d}{d\alpha} \int_a^b F(x, \alpha)\, dx = \int_a^b \frac{\partial F}{\partial \alpha}\, dx$$

The rule is valid if a and b are constants, α is a real parameter such that $\alpha_1 \leqq \alpha \leqq \alpha_2$ where α_1 and α_2 are constants, and $F(x, \alpha)$ is continuous and has a continuous partial derivative with respect to α for $a \leqq x \leqq b$, $\alpha_1 \leqq \alpha \leqq \alpha_2$. It can be extended to cases where the limits a and b are infinite or dependent on α.

SUMMATION OF SERIES

The residue theorem can often be used to sum various types of series. The following results are valid under very mild restrictions on $f(z)$ which are generally satisfied whenever the series converge. See Problems 24-32, and 38.

1. $\displaystyle\sum_{-\infty}^{\infty} f(n)$ $\qquad = -\{\text{sum of residues of } \pi \cot \pi z\, f(z) \text{ at all the poles of } f(z)\}$

2. $\displaystyle\sum_{-\infty}^{\infty} (-1)^n f(n)$ $\qquad = -\{\text{sum of residues of } \pi \csc \pi z\, f(z) \text{ at all the poles of } f(z)\}$

3. $\displaystyle\sum_{-\infty}^{\infty} f\left(\frac{2n+1}{2}\right)$ $\quad = \{\text{sum of residues of } \pi \tan \pi z\, f(z) \text{ at all the poles of } f(z)\}$

4. $\displaystyle\sum_{-\infty}^{\infty} (-1)^n f\left(\frac{2n+1}{2}\right)$ $\quad = \{\text{sum of residues of } \pi \sec \pi z\, f(z) \text{ at all the poles of } f(z)\}$

MITTAG-LEFFLER'S EXPANSION THEOREM

1. Suppose that the only singularities of $f(z)$ in the finite z plane are the simple poles $a_1, a_2, a_3, \ldots$ arranged in order of increasing absolute value.

2. Let the residues of $f(z)$ at $a_1, a_2, a_3, \ldots$ be $b_1, b_2, b_3, \ldots$.

3. Let C_N be circles of radius R_N which do not pass through any poles and on which $|f(z)| < M$, where M is independent of N and $R_N \to \infty$ as $N \to \infty$.

Then *Mittag-Leffler's expansion theorem* states that

$$f(z) \;=\; f(0) + \sum_{n=1}^{\infty} b_n \left\{ \frac{1}{z - a_n} + \frac{1}{a_n} \right\}$$

SOME SPECIAL EXPANSIONS

1. $\csc z = \dfrac{1}{z} - 2z\left(\dfrac{1}{z^2 - \pi^2} - \dfrac{1}{z^2 - 4\pi^2} + \dfrac{1}{z^2 - 9\pi^2} - \cdots\right)$

2. $\sec z = \pi\left(\dfrac{1}{(\pi/2)^2 - z^2} - \dfrac{3}{(3\pi/2)^2 - z^2} + \dfrac{5}{(5\pi/2)^2 - z^2} - \cdots\right)$

3. $\tan z = 2z\left(\dfrac{1}{(\pi/2)^2 - z^2} + \dfrac{1}{(3\pi/2)^2 - z^2} + \dfrac{1}{(5\pi/2)^2 - z^2} + \cdots\right)$

4. $\cot z = \dfrac{1}{z} + 2z\left(\dfrac{1}{z^2 - \pi^2} + \dfrac{1}{z^2 - 4\pi^2} + \dfrac{1}{z^2 - 9\pi^2} + \cdots\right)$

5. $\operatorname{csch} z = \dfrac{1}{z} - 2z\left(\dfrac{1}{z^2 + \pi^2} - \dfrac{1}{z^2 + 4\pi^2} + \dfrac{1}{z^2 + 9\pi^2} - \cdots\right)$

6. $\operatorname{sech} z = \pi\left(\dfrac{1}{(\pi/2)^2 + z^2} - \dfrac{3}{(3\pi/2)^2 + z^2} + \dfrac{5}{(5\pi/2)^2 + z^2} - \cdots\right)$

7. $\tanh z = 2z\left(\dfrac{1}{z^2 + (\pi/2)^2} + \dfrac{1}{z^2 + (3\pi/2)^2} + \dfrac{1}{z^2 + (5\pi/2)^2} + \cdots\right)$

8. $\coth z = \dfrac{1}{z} + 2z\left(\dfrac{1}{z^2 + \pi^2} + \dfrac{1}{z^2 + 4\pi^2} + \dfrac{1}{z^2 + 9\pi^2} + \cdots\right)$

Solved Problems

RESIDUES AND THE RESIDUE THEOREM

1. Let $f(z)$ be analytic inside and on a simple closed curve C except at point a inside C.

   (a) Prove that

   $$f(z) = \sum_{n=-\infty}^{\infty} a_n(z-a)^n \qquad \text{where} \qquad a_n = \frac{1}{2\pi i} \oint_C \frac{f(z)}{(z-a)^{n+1}} \, dz, \quad n = 0, \pm 1, \pm 2, \ldots$$

   i.e. $f(z)$ can be expanded into a converging Laurent series about $z = a$.

   (b) Prove that

   $$\oint_C f(z) \, dz = 2\pi i \, a_{-1}$$

   (a) This follows from Problem 25 of Chapter 6.

   (b) If we let $n = -1$ in the result of (a), we find

   $$a_{-1} = \frac{1}{2\pi i} \oint_C f(z) \, dz, \qquad \text{i.e.} \qquad \oint_C f(z) \, dz = 2\pi i \, a_{-1}$$

   We call a_{-1} the *residue* of $f(z)$ at $z = a$.

2. Prove the *residue theorem*. If $f(z)$ is analytic inside and on a simple closed curve C except at a finite number of points $a, b, c, \ldots$ inside C at which the residues are $a_{-1}, b_{-1}, c_{-1}, \ldots$ respectively, then

   $$\oint_C f(z) \, dz = 2\pi i(a_{-1} + b_{-1} + c_{-1} + \cdots)$$

   i.e. $2\pi i$ times the sum of the residues at all singularities enclosed by C.

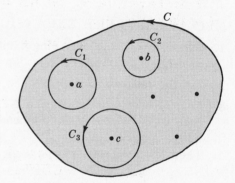

Fig. 7-4

   With centers at $a, b, c, \ldots$ respectively construct circles $C_1, C_2, C_3, \ldots$ which lie entirely inside C as shown in Fig. 7-4. This can be done since $a, b, c, \ldots$ are interior points. By Theorem 5, Page 97, we have

   $$\oint_C f(z) \, dz = \oint_{C_1} f(z) \, dz + \oint_{C_2} f(z) \, dz + \oint_{C_3} f(z) \, dz + \cdots \tag{1}$$

   But by Problem 1,

   $$\oint_{C_1} f(z) \, dz = 2\pi i \, a_{-1}, \qquad \oint_{C_2} f(z) \, dz = 2\pi i \, b_{-1}, \qquad \oint_{C_3} f(z) \, dz = 2\pi i \, c_{-1}, \qquad \ldots \tag{2}$$

   Then from (1) and (2) we have, as required,

   $$\oint_C f(z) \, dz = 2\pi i(a_{-1} + b_{-1} + c_{-1} + \cdots) = 2\pi i \, (\text{sum of residues})$$

   The proof given here establishes the residue theorem for simply-connected regions containing a finite number of singularities of $f(z)$. It can be extended to regions with infinitely many isolated singularities and to multiply-connected regions (see Problems 96 and 97).

3. Let $f(z)$ be analytic inside and on a simple closed curve C except at a pole a of order m inside C. Prove that the residue of $f(z)$ at a is given by

   $$a_{-1} = \lim_{z \to a} \frac{1}{(m-1)!} \frac{d^{m-1}}{dz^{m-1}} \{(z-a)^m f(z)\}$$

   Method 1. If $f(z)$ has a pole a of order m, then the Laurent series of $f(z)$ is

   $$f(z) = \frac{a_{-m}}{(z-a)^m} + \frac{a_{-m+1}}{(z-a)^{m-1}} + \cdots + \frac{a_{-1}}{z-a} + a_0 + a_1(z-a) + a_2(z-a)^2 + \cdots \tag{1}$$

Then multiplying both sides by $(z-a)^m$, we have

$$(z-a)^m f(z) \;=\; a_{-m} + a_{-m+1}(z-a) + \cdots + a_{-1}(z-a)^{m-1} + a_0(z-a)^m + \cdots \qquad (2)$$

This represents the Taylor series about $z = a$ of the analytic function on the left. Differentiating both sides $m-1$ times with respect to z, we have

$$\frac{d^{m-1}}{dz^{m-1}}\{(z-a)^m f(z)\} \;=\; (m-1)!\,a_{-1} + m(m-1)\cdots 2a_0(z-a) + \cdots$$

Thus on letting $z \to a$,

$$\lim_{z\to a}\frac{d^{m-1}}{dz^{m-1}}\{(z-a)^m f(z)\} \;=\; (m-1)!\,a_{-1}$$

from which the required result follows.

Method 2. The required result also follows directly from Taylor's theorem on noting that the coefficient of $(z-a)^{m-1}$ in the expansion (2) is

$$a_{-1} \;=\; \frac{1}{(m-1)!}\frac{d^{m-1}}{dz^{m-1}}\{(z-a)^m f(z)\}\Big|_{z=a}$$

Method 3. See Problem 28, Chapter 5, Page 132.

4. Find the residues of   (a) $f(z) = \dfrac{z^2 - 2z}{(z+1)^2\,(z^2+4)}$ and   (b) $f(z) = e^z\csc^2 z$ at all its poles in the finite plane.

(a) $f(z)$ has a double pole at $z = -1$ and simple poles at $z = \pm 2i$.

   Method 1.
   Residue at $z = -1$ is

$$\lim_{z\to -1}\frac{1}{1!}\frac{d}{dz}\left\{(z+1)^2 \cdot \frac{z^2-2z}{(z+1)^2(z^2+4)}\right\} \;=\; \lim_{z\to -1}\frac{(z^2+4)(2z-2)-(z^2-2z)(2z)}{(z^2+4)^2} \;=\; -\frac{14}{25}$$

   Residue at $z = 2i$ is

$$\lim_{z\to 2i}\left\{(z-2i)\cdot\frac{z^2-2z}{(z+1)^2(z-2i)(z+2i)}\right\} \;=\; \frac{-4-4i}{(2i+1)^2(4i)} \;=\; \frac{7+i}{25}$$

   Residue at $z = -2i$ is

$$\lim_{z\to -2i}\left\{(z+2i)\cdot\frac{z^2-2z}{(z+1)^2(z-2i)(z+2i)}\right\} \;=\; \frac{-4+4i}{(-2i+1)^2(-4i)} \;=\; \frac{7-i}{25}$$

   Method 2.
   Residue at $z = 2i$ is

$$\lim_{z\to 2i}\left\{\frac{(z-2i)(z^2-2z)}{(z+1)^2(z^2+4)}\right\} \;=\; \left\{\lim_{z\to 2i}\frac{z^2-2z}{(z+1)^2}\right\}\left\{\lim_{z\to 2i}\frac{z-2i}{z^2+4}\right\}$$

$$=\; \frac{-4-4i}{(2i+1)^2}\cdot\lim_{z\to 2i}\frac{1}{2z} \;=\; \frac{-4-4i}{(2i+1)^2}\cdot\frac{1}{4i} \;=\; \frac{7+i}{25}$$

   using L'Hospital's rule. In a similar manner, or by replacing i by $-i$ in the result, we can obtain the residue at $z = -2i$.

(b) $f(z) = e^z\csc^2 z = \dfrac{e^z}{\sin^2 z}$ has double poles at $z = 0, \pm\pi, \pm 2\pi, \ldots,$ i.e. $z = m\pi$ where $m = 0, \pm 1, \pm 2, \ldots$.

   Method 1.
   Residue at $z = m\pi$ is

$$\lim_{z\to m\pi}\frac{1}{1!}\frac{d}{dz}\left\{(z-m\pi)^2\frac{e^z}{\sin^2 z}\right\}$$

$$=\; \lim_{z\to m\pi}\frac{e^z[(z-m\pi)^2\sin z + 2(z-m\pi)\sin z - 2(z-m\pi)^2\cos z]}{\sin^3 z}$$

Letting $z - m\pi = u$ or $z = u + m\pi$, this limit can be written

$$\lim_{u \to 0} e^{u + m\pi} \left\{ \frac{u^2 \sin u + 2u \sin u - 2u^2 \cos u}{\sin^3 u} \right\}$$

$$= e^{m\pi} \left\{ \lim_{u \to 0} \frac{u^2 \sin u + 2u \sin u - 2u^2 \cos u}{\sin^3 u} \right\}$$

The limit in braces can be obtained using L'Hospital's rule. However, it is easier to first note

that $\lim\limits_{u \to 0} \dfrac{u^3}{\sin^3 u} = \lim\limits_{u \to 0} \left(\dfrac{u}{\sin u} \right)^3 = 1$ and thus write the limit as

$$e^{m\pi} \lim_{u \to 0} \left(\frac{u^2 \sin u + 2u \sin u - 2u^2 \cos u}{u^3} \cdot \frac{u^3}{\sin^3 u} \right)$$

$$= e^{m\pi} \lim_{u \to 0} \frac{u^2 \sin u + 2u \sin u - 2u^2 \cos u}{u^3} = e^{m\pi}$$

using L'Hospital's rule several times. In evaluating this limit we can instead use the series expansions $\sin u = u - u^3/3! + \cdots$, $\cos u = 1 - u^2/2! + \cdots$.

Method 2 (using Laurent's series).

In this method we expand $f(z) = e^z \csc^2 z$ in a Laurent series about $z = m\pi$ and obtain the coefficient of $1/(z - m\pi)$ as the required residue. To make the calculation easier let $z = u + m\pi$. Then the function to be expanded in a Laurent series about $u = 0$ is $e^{m\pi + u} \csc^2 (m\pi + u) = e^{m\pi} e^u \csc^2 u$. Using the Maclaurin expansions for e^u and $\sin u$, we find using long division

$$e^{m\pi} e^u \csc^2 u = \frac{e^{m\pi} \left(1 + u + \frac{u^2}{2!} + \frac{u^3}{3!} + \cdots \right)}{\left(u - \frac{u^3}{3!} + \frac{u^5}{5!} - \cdots \right)^2} = \frac{e^{m\pi} \left(1 + u + \frac{u^2}{2} + \cdots \right)}{u^2 \left(1 - \frac{u^2}{6} + \frac{u^4}{120} - \cdots \right)^2}$$

$$= \frac{e^{m\pi} \left(1 + u + \frac{u^2}{2!} + \cdots \right)}{u^2 \left(1 - \frac{u^2}{3} + \frac{2u^4}{45} + \cdots \right)} = e^{m\pi} \left(\frac{1}{u^2} + \frac{1}{u} + \frac{5}{6} + \frac{u}{3} + \cdots \right)$$

and so the residue is $e^{m\pi}$.

5. Find the residue of $F(z) = \dfrac{\cot z \coth z}{z^3}$ at $z = 0$.

We have as in Method 2 of Problem 4(b),

$$F(z) = \frac{\cos z \cosh z}{z^3 \sin z \sinh z} = \frac{\left(1 - \frac{z^2}{2!} + \frac{z^4}{4!} - \cdots \right)\left(1 + \frac{z^2}{2!} + \frac{z^4}{4!} + \cdots \right)}{z^3 \left(z - \frac{z^3}{3!} + \frac{z^5}{5!} - \cdots \right)\left(z + \frac{z^3}{3!} + \frac{z^5}{5!} + \cdots \right)}$$

$$= \frac{\left(1 - \frac{z^4}{6} + \cdots \right)}{z^5 \left(1 - \frac{z^4}{90} + \cdots \right)} = \frac{1}{z^5} \left(1 - \frac{7z^4}{45} + \cdots \right)$$

and so the residue (coefficient of $1/z$) is $-7/45$.

Another method. The result can also be obtained by finding

$$\lim_{z \to 0} \frac{1}{4!} \frac{d^4}{dz^4} \left\{ z^5 \frac{\cos z \cosh z}{z^3 \sin z \sinh z} \right\}$$

but this method is much more laborious than that given above.

6. Evaluate $\dfrac{1}{2\pi i} \displaystyle\oint_C \dfrac{e^{zt}}{z^2(z^2 + 2z + 2)} \, dz$ around the circle C with equation $|z| = 3$.

The integrand $\dfrac{e^{zt}}{z^2(z^2 + 2z + 2)}$ has a double pole at $z = 0$ and two simple poles at $z = -1 \pm i$ [roots of $z^2 + 2z + 2 = 0$]. All these poles are inside C.

Residue at $z = 0$ is

$$\lim_{z \to 0} \frac{1}{1!} \frac{d}{dz} \left\{ z^2 \frac{e^{zt}}{z^2(z^2 + 2z + 2)} \right\} = \lim_{z \to 0} \frac{(z^2 + 2z + 2)(te^{zt}) - (e^{zt})(2z + 2)}{(z^2 + 2z + 2)^2} = \frac{t - 1}{2}$$

Residue at $z = -1 + i$ is

$$\lim_{z \to -1+i} \left\{ [z - (-1 + i)] \frac{e^{zt}}{z^2(z^2 + 2z + 2)} \right\} = \lim_{z \to -1+i} \left\{ \frac{e^{zt}}{z^2} \right\} \lim_{z \to -1+i} \left\{ \frac{z + 1 - i}{z^2 + 2z + 2} \right\}$$

$$= \frac{e^{(-1+i)t}}{(-1+i)^2} \cdot \frac{1}{2i} = \frac{e^{(-1+i)t}}{4}$$

Residue at $z = -1 - i$ is

$$\lim_{z \to -1-i} \left\{ [z - (-1 - i)] \frac{e^{zt}}{z^2(z^2 + 2z + 2)} \right\} = \frac{e^{(-1-i)t}}{4}$$

Then by the residue theorem

$$\oint_C \frac{e^{zt}}{z^2(z^2 + 2z + 2)} \, dz = 2\pi i \text{ (sum of residues)}$$

$$= 2\pi i \left\{ \frac{t - 1}{2} + \frac{e^{(-1+i)t}}{4} + \frac{e^{(-1-i)t}}{4} \right\}$$

$$= 2\pi i \left\{ \frac{t - 1}{2} + \frac{1}{2} e^{-t} \cos t \right\}$$

i.e.,
$$\frac{1}{2\pi i} \oint_C \frac{e^{zt}}{z^2(z^2 + 2z + 2)} \, dz = \frac{t - 1}{2} + \frac{1}{2} e^{-t} \cos t$$

DEFINITE INTEGRALS OF THE TYPE $\displaystyle\int_{-\infty}^{\infty} F(x) \, dx$

7. If $|F(z)| \leq M/R^k$ for $z = Re^{i\theta}$ where $k > 1$ and M are constants, prove that $\displaystyle\lim_{R \to \infty} \int_\Gamma F(z) \, dz = 0$ where Γ is the semicircular arc of radius R shown in Fig. 7-5.

By Property 5, Page 93, we have

$$\left| \int_\Gamma F(z) \, dz \right| \leq \frac{M}{R^k} \cdot \pi R = \frac{\pi M}{R^{k-1}}$$

since the length of arc $L = \pi R$.   Then

$$\lim_{R \to \infty} \left| \int_\Gamma F(z) \, dz \right| = 0 \qquad \text{and so} \qquad \lim_{R \to \infty} \int_\Gamma F(z) \, dz = 0$$

8. Show that for $z = Re^{i\theta}$, $|f(z)| \leq \dfrac{M}{R^k}$, $k > 1$ if $f(z) = \dfrac{1}{z^6 + 1}$.

If $z = Re^{i\theta}$, $|f(z)| = \left| \dfrac{1}{R^6 e^{6i\theta} + 1} \right| \leq \dfrac{1}{|R^6 e^{6i\theta}| - 1} = \dfrac{1}{R^6 - 1} \leq \dfrac{2}{R^6}$ if R is large enough (say $R > 2$, for example) so that $M = 2$, $k = 6$.

Note that we have made use of the inequality $|z_1 + z_2| \geq |z_1| - |z_2|$ with $z_1 = R^6 e^{6i\theta}$ and $z_2 = 1$.

9. Evaluate $\displaystyle\int_0^\infty \frac{dx}{x^6 + 1}$.

Consider $\displaystyle\oint_C \frac{dz}{z^6 + 1}$, where C is the closed contour of Fig. 7-5 consisting of the line from $-R$ to R and the semicircle Γ, traversed in the positive (counterclockwise) sense.

Since $z^6 + 1 = 0$ when $z = e^{\pi i/6}$, $e^{3\pi i/6}$, $e^{5\pi i/6}$, $e^{7\pi i/6}$, $e^{9\pi i/6}$, $e^{11\pi i/6}$, these are simple poles of $1/(z^6 + 1)$. Only the poles $e^{\pi i/6}$, $e^{3\pi i/6}$ and $e^{5\pi i/6}$ lie within C. Then using L'Hospital's rule,

$$\text{Residue at } e^{\pi i/6} = \lim_{z \to e^{\pi i/6}} \left\{ (z - e^{\pi i/6}) \frac{1}{z^6 + 1} \right\} = \lim_{z \to e^{\pi i/6}} \frac{1}{6z^5} = \frac{1}{6} e^{-5\pi i/6}$$

$$\text{Residue at } e^{3\pi i/6} = \lim_{z \to e^{3\pi i/6}} \left\{ (z - e^{3\pi i/6}) \frac{1}{z^6 + 1} \right\} = \lim_{z \to e^{3\pi i/6}} \frac{1}{6z^5} = \frac{1}{6} e^{-5\pi i/2}$$

$$\text{Residue at } e^{5\pi i/6} = \lim_{z \to e^{5\pi i/6}} \left\{ (z - e^{5\pi i/6}) \frac{1}{z^6 + 1} \right\} = \lim_{z \to e^{5\pi i/6}} \frac{1}{6z^5} = \frac{1}{6} e^{-25\pi i/6}$$

Thus

$$\oint_C \frac{dz}{z^6 + 1} = 2\pi i \{ \tfrac{1}{6} e^{-5\pi i/6} + \tfrac{1}{6} e^{-5\pi i/2} + \tfrac{1}{6} e^{-25\pi i/6} \} = \frac{2\pi}{3}$$

i.e.,

$$\int_{-R}^{R} \frac{dx}{x^6 + 1} + \int_{\Gamma} \frac{dz}{z^6 + 1} = \frac{2\pi}{3} \tag{1}$$

Taking the limit of both sides of (1) as $R \to \infty$ and using Problems 7 and 8, we have

$$\lim_{R \to \infty} \int_{-R}^{R} \frac{dx}{x^6 + 1} = \int_{-\infty}^{\infty} \frac{dx}{x^6 + 1} = \frac{2\pi}{3} \tag{2}$$

Since $\displaystyle\int_{-\infty}^{\infty} \frac{dx}{x^6 + 1} = 2 \int_{0}^{\infty} \frac{dx}{x^6 + 1}$, the required integral has the value $\pi/3$.

10. Show that $\displaystyle\int_{-\infty}^{\infty} \frac{x^2 \, dx}{(x^2 + 1)^2 (x^2 + 2x + 2)} = \frac{7\pi}{50}$.

The poles of $\dfrac{z^2}{(z^2 + 1)^2 (z^2 + 2z + 2)}$ enclosed by the contour C of Fig. 7-5 are $z = i$ of order 2 and $z = -1 + i$ of order 1.

Residue at $z = i$ is

$$\lim_{z \to i} \frac{d}{dz} \left\{ (z - i)^2 \frac{z^2}{(z + i)^2 (z - i)^2 (z^2 + 2z + 2)} \right\} = \frac{9i - 12}{100}.$$

Residue at $z = -1 + i$ is $\displaystyle\lim_{z \to -1 + i} (z + 1 - i) \frac{z^2}{(z^2 + 1)^2 (z + 1 - i)(z + 1 + i)} = \frac{3 - 4i}{25}.$

Then

$$\oint_C \frac{z^2 \, dz}{(z^2 + 1)^2 (z^2 + 2z + 2)} = 2\pi i \left\{ \frac{9i - 12}{100} + \frac{3 - 4i}{25} \right\} = \frac{7\pi}{50}$$

or

$$\int_{-R}^{R} \frac{x^2 \, dx}{(x^2 + 1)^2 (x^2 + 2x + 2)} + \int_{\Gamma} \frac{z^2 \, dz}{(z^2 + 1)^2 (z^2 + 2z + 2)} = \frac{7\pi}{50}$$

Taking the limit as $R \to \infty$ and noting that the second integral approaches zero by Problem 7, we obtain the required result.

DEFINITE INTEGRALS OF THE TYPE $\displaystyle\int_0^{2\pi} G(\sin\theta, \cos\theta) \, d\theta$

11. Evaluate $\displaystyle\int_0^{2\pi} \frac{d\theta}{3 - 2\cos\theta + \sin\theta}$.

Let $z = e^{i\theta}$. Then $\sin\theta = \dfrac{e^{i\theta} - e^{-i\theta}}{2i} = \dfrac{z - z^{-1}}{2i}$, $\cos\theta = \dfrac{e^{i\theta} + e^{-i\theta}}{2} = \dfrac{z + z^{-1}}{2}$, $dz = iz \, d\theta$ so that

$$\int_0^{2\pi} \frac{d\theta}{3 - 2\cos\theta + \sin\theta} = \oint_C \frac{dz/iz}{3 - 2(z + z^{-1})/2 + (z - z^{-1})/2i} = \oint_C \frac{2 \, dz}{(1 - 2i)z^2 + 6iz - 1 - 2i}$$

where C is the circle of unit radius with center at the origin (Fig. 7-6).

The poles of $\dfrac{2}{(1-2i)z^2 + 6iz - 1 - 2i}$ are the simple poles

$$z = \frac{-6i \pm \sqrt{(6i)^2 - 4(1-2i)(-1-2i)}}{2(1-2i)}$$

$$= \frac{-6i \pm 4i}{2(1-2i)} = 2 - i, \ (2-i)/5$$

Only $(2-i)/5$ lies inside C.

Residue at $(2-i)/5 = \lim_{z \to (2-i)/5} \{z - (2-i)/5\} \left\{ \dfrac{2}{(1-2i)z^2 + 6iz - 1 - 2i} \right\}$

$$= \lim_{z \to (2-i)/5} \frac{2}{2(1-2i)z + 6i} = \frac{1}{2i} \quad \text{by L'Hospital's rule.}$$

Then $\displaystyle\oint_C \frac{2\,dz}{(1-2i)z^2 + 6iz - 1 - 2i} = 2\pi i \left(\frac{1}{2i} \right) = \pi, \quad$ the required value.

12. Show that $\displaystyle\int_0^{2\pi} \frac{d\theta}{a + b\sin\theta} = \frac{2\pi}{\sqrt{a^2 - b^2}}$ if $a > |b|$.

Let $z = e^{i\theta}$. Then $\sin\theta = \dfrac{e^{i\theta} - e^{-i\theta}}{2i} = \dfrac{z - z^{-1}}{2i}, \quad dz = ie^{i\theta}\,d\theta = iz\,d\theta$ so that

$$\int_0^{2\pi} \frac{d\theta}{a + b\sin\theta} = \oint_C \frac{dz/iz}{a + b(z - z^{-1})/2i} = \oint_C \frac{2\,dz}{bz^2 + 2aiz - b}$$

where C is the circle of unit radius with center at the origin, as shown in Fig. 7-6.

The poles of $\dfrac{2}{bz^2 + 2aiz - b}$ are obtained by solving $bz^2 + 2aiz - b = 0$ and are given by

$$z = \frac{-2ai \pm \sqrt{-4a^2 + 4b^2}}{2b} = \frac{-ai \pm \sqrt{a^2 - b^2}\,i}{b}$$

$$= \left\{ \frac{-a + \sqrt{a^2 - b^2}}{b} \right\} i, \ \left\{ \frac{-a - \sqrt{a^2 - b^2}}{b} \right\} i$$

Only $\dfrac{-a + \sqrt{a^2 - b^2}}{b}\, i$ lies inside C, since

$$\left| \frac{-a + \sqrt{a^2 - b^2}}{b}\, i \right| = \left| \frac{\sqrt{a^2 - b^2} - a}{b} \cdot \frac{\sqrt{a^2 - b^2} + a}{\sqrt{a^2 - b^2} + a} \right| = \left| \frac{b}{(\sqrt{a^2 - b^2} + a)} \right| < 1 \quad \text{if } a > |b|$$

Residue at $z_1 = \dfrac{-a + \sqrt{a^2 - b^2}}{b}\, i = \lim_{z \to z_1} (z - z_1) \dfrac{2}{bz^2 + 2aiz - b}$

$$= \lim_{z \to z_1} \frac{2}{2bz + 2ai} = \frac{1}{bz_1 + ai} = \frac{1}{\sqrt{a^2 - b^2}\, i}$$

by L'Hospital's rule.

Then $\displaystyle\oint_C \frac{2\,dz}{bz^2 + 2aiz - b} = 2\pi i \left(\frac{1}{\sqrt{a^2 - b^2}\, i} \right) = \frac{2\pi}{\sqrt{a^2 - b^2}}, \quad$ the required value.

13. Show that $\displaystyle\int_0^{2\pi} \frac{\cos 3\theta}{5 - 4\cos\theta}\,d\theta = \frac{\pi}{12}$.

If $z = e^{i\theta}$, then $\cos\theta = \dfrac{z + z^{-1}}{2}, \quad \cos 3\theta = \dfrac{e^{3i\theta} + e^{-3i\theta}}{2} = \dfrac{z^3 + z^{-3}}{2}, \quad dz = iz\,d\theta$ so that

$$\int_0^{2\pi} \frac{\cos 3\theta}{5 - 4\cos\theta}\,d\theta = \oint_C \frac{(z^3 + z^{-3})/2}{5 - 4(z + z^{-1})/2}\,\frac{dz}{iz} = -\frac{1}{2i} \oint_C \frac{z^6 + 1}{z^3(2z - 1)(z - 2)}\,dz$$

where C is the contour of Fig. 7-6.

The integrand has a pole of order 3 at $z = 0$ and a simple pole $z = \frac{1}{2}$ inside C.

Residue at $z = 0$ is $\lim\limits_{z \to 0} \dfrac{1}{2!} \dfrac{d^2}{dz^2} \left\{ z^3 \cdot \dfrac{z^6 + 1}{z^3(2z - 1)(z - 2)} \right\} = \dfrac{21}{8}$.

Residue at $z = \frac{1}{2}$ is $\lim\limits_{z \to 1/2} \left\{ \left(z - \dfrac{1}{2} \right) \cdot \dfrac{z^6 + 1}{z^3(2z - 1)(z - 2)} \right\} = -\dfrac{65}{24}$.

Then $-\dfrac{1}{2i} \oint_C \dfrac{z^6 + 1}{z^3(2z - 1)(z - 2)} \, dz = -\dfrac{1}{2i}(2\pi i) \left\{ \dfrac{21}{8} - \dfrac{65}{24} \right\} = \dfrac{\pi}{12}$ as required.

14. Show that $\displaystyle\int_0^{2\pi} \dfrac{d\theta}{(5 - 3\sin\theta)^2} = \dfrac{5\pi}{32}$.

Letting $z = e^{i\theta}$, we have $\sin\theta = (z - z^{-1})/2i$, $dz = ie^{i\theta} \, d\theta = iz \, d\theta$ and so

$$\int_0^{2\pi} \frac{d\theta}{(5 - 3\sin\theta)^2} = \oint_C \frac{dz/iz}{\{5 - 3(z - z^{-1})/2i\}^2} = -\frac{4}{i} \oint_C \frac{z \, dz}{(3z^2 - 10iz - 3)^2}$$

where C is the contour of Fig. 7-6.

The integrand has poles of order 2 at $z = \dfrac{10i \pm \sqrt{-100 + 36}}{6} = \dfrac{10i \pm 8i}{6} = 3i, \, i/3$.   Only the pole $i/3$ lies inside C.

Residue at $z = i/3$ $= \lim\limits_{z \to i/3} \dfrac{d}{dz} \left\{ (z - i/3)^2 \cdot \dfrac{z}{(3z^2 - 10iz - 3)^2} \right\}$

$$= \lim_{z \to i/3} \frac{d}{dz} \left\{ (z - i/3)^2 \cdot \frac{z}{(3z - i)^2 \, (z - 3i)^2} \right\} = -\frac{5}{256}.$$

Then $-\dfrac{4}{i} \oint_C \dfrac{z \, dz}{(3z^2 - 10iz - 3)^2} = -\dfrac{4}{i}(2\pi i)\left(\dfrac{-5}{256} \right) = \dfrac{5\pi}{32}$

Another method.

From Problem 12, we have for $a > |b|$,

$$\int_0^{2\pi} \frac{d\theta}{a + b\sin\theta} = \frac{2\pi}{\sqrt{a^2 - b^2}}$$

Then by differentiating both sides with respect to a (considering b as constant) using Leibnitz's rule, we have

$$\frac{d}{da} \int_0^{2\pi} \frac{d\theta}{a + b\sin\theta} = \int_0^{2\pi} \frac{\partial}{\partial a}\left(\frac{1}{a + b\sin\theta} \right) d\theta = -\int_0^{2\pi} \frac{d\theta}{(a + b\sin\theta)^2}$$

$$= \frac{d}{da}\left(\frac{2\pi}{\sqrt{a^2 - b^2}} \right) = \frac{-2\pi a}{(a^2 - b^2)^{3/2}}$$

i.e., $\displaystyle\int_0^{2\pi} \frac{d\theta}{(a + b\sin\theta)^2} = \frac{2\pi a}{(a^2 - b^2)^{3/2}}$

Letting $a = 5$ and $b = -3$, we have

$$\int_0^{2\pi} \frac{d\theta}{(5 - 3\sin\theta)^2} = \frac{2\pi(5)}{(5^2 - 3^2)^{3/2}} = \frac{5\pi}{32}$$

DEFINITE INTEGRALS OF THE TYPE $\displaystyle\int_{-\infty}^{\infty} F(x) \begin{Bmatrix} \cos mx \\ \sin mx \end{Bmatrix} dx$

15. If $|F(z)| \leqq \dfrac{M}{R^k}$ for $z = Re^{i\theta}$ where $k > 0$ and M are constants, prove that

$$\lim_{R \to \infty} \int_\Gamma e^{imz} F(z) \, dz = 0$$

where Γ is the semicircular arc of Fig. 7-5 and m is a positive constant.

If $z = Re^{i\theta}$, $\displaystyle\int_\Gamma e^{imz} F(z) \, dz = \int_0^\pi e^{imRe^{i\theta}} F(Re^{i\theta}) \, iRe^{i\theta} \, d\theta$.   Then

$$\left| \int_0^\pi e^{imRe^{i\theta}} F(Re^{i\theta})\, iRe^{i\theta}\, d\theta \right| \leqq \int_0^\pi |e^{imRe^{i\theta}} F(Re^{i\theta})\, iRe^{i\theta}|\, d\theta$$

$$= \int_0^\pi |e^{imR\cos\theta - mR\sin\theta} F(Re^{i\theta})\, iRe^{i\theta}|\, d\theta$$

$$= \int_0^\pi e^{-mR\sin\theta} |F(Re^{i\theta})|\, R\, d\theta$$

$$\leqq \frac{M}{R^{k-1}} \int_0^\pi e^{-mR\sin\theta}\, d\theta = \frac{2M}{R^{k-1}} \int_0^{\pi/2} e^{-mR\sin\theta}\, d\theta$$

Now $\sin\theta \geqq 2\theta/\pi$ for $0 \leqq \theta \leqq \pi/2$, as can be seen geometrically from Fig. 7-7 or analytically from Prob. 99.

Then the last integral is less than or equal to

$$\frac{2M}{R^{k-1}} \int_0^{\pi/2} e^{-2mR\theta/\pi}\, d\theta = \frac{\pi M}{mR^k}(1 - e^{-mR})$$

As $R \to \infty$ this approaches zero, since m and k are positive, and the required result is proved.

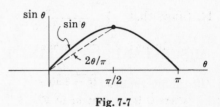

Fig. 7-7

16. Show that $\displaystyle \int_0^\infty \frac{\cos mx}{x^2 + 1}\, dx = \frac{\pi}{2} e^{-m}, \quad m > 0.$

Consider $\displaystyle \oint_C \frac{e^{imz}}{z^2 + 1}\, dz$ where C is the contour of Fig. 7-5. The integrand has simple poles at $z = \pm i$, but only $z = i$ lies inside C.

Residue at $z = i$ is $\displaystyle \lim_{z \to i} \left\{ (z - i)\frac{e^{imz}}{(z-i)(z+i)} \right\} = \frac{e^{-m}}{2i}.$ Then

$$\oint_C \frac{e^{imz}}{z^2 + 1}\, dz = 2\pi i\left(\frac{e^{-m}}{2i} \right) = \pi e^{-m}$$

or

$$\int_{-R}^R \frac{e^{imx}}{x^2 + 1}\, dx + \int_\Gamma \frac{e^{imz}}{z^2 + 1}\, dz = \pi e^{-m}$$

i.e.,

$$\int_{-R}^R \frac{\cos mx}{x^2 + 1}\, dx + i\int_{-R}^R \frac{\sin mx}{x^2 + 1}\, dx + \int_\Gamma \frac{e^{imz}}{z^2 + 1}\, dz = \pi e^{-m}$$

and so

$$2\int_0^R \frac{\cos mx}{x^2 + 1}\, dx + \int_\Gamma \frac{e^{imz}}{z^2 + 1}\, dz = \pi e^{-m}$$

Taking the limit as $R \to \infty$ and using Problem 15 to show that the integral around Γ approaches zero, we obtain the required result.

17. Evaluate $\displaystyle \int_{-\infty}^\infty \frac{x \sin \pi x}{x^2 + 2x + 5}\, dx.$

Consider $\displaystyle \oint_C \frac{z e^{i\pi z}}{z^2 + 2z + 5}\, dz$ where C is the contour of Fig. 7-5. The integrand has simple poles at $z = -1 \pm 2i$, but only $z = -1 + 2i$ lies inside C.

Residue at $z = -1 + 2i$ is $\displaystyle \lim_{z \to -1 + 2i} \left\{ (z + 1 - 2i) \cdot \frac{z e^{i\pi z}}{z^2 + 2z + 5} \right\} = (-1 + 2i)\frac{e^{-i\pi - 2\pi}}{4i}.$ Then

$$\oint_C \frac{z e^{i\pi z}}{z^2 + 2z + 5}\, dz = 2\pi i(-1 + 2i)\left(\frac{e^{-i\pi - 2\pi}}{4i} \right) = \frac{\pi}{2}(1 - 2i)e^{-2\pi}$$

or

$$\int_{-R}^R \frac{x e^{i\pi x}}{x^2 + 2x + 5}\, dx + \int_\Gamma \frac{z e^{i\pi z}}{z^2 + 2z + 5}\, dz = \frac{\pi}{2}(1 - 2i)e^{-2\pi}$$

i.e.,

$$\int_{-R}^R \frac{x \cos \pi x}{x^2 + 2x + 5}\, dx + i\int_{-R}^R \frac{x \sin \pi x}{x^2 + 2x + 5}\, dx + \int_\Gamma \frac{z e^{i\pi z}}{z^2 + 2z + 5}\, dz = \frac{\pi}{2}(1 - 2i)e^{-2\pi}$$

Taking the limit as $R \to \infty$ and using Problem 15 to show that the integral around Γ approaches zero, this becomes

$$\int_{-\infty}^{\infty} \frac{x \cos \pi x}{x^2 + 2x + 5} \, dx \;+\; i \int_{-\infty}^{\infty} \frac{x \sin \pi x}{x^2 + 2x + 5} \, dx \;=\; \frac{\pi}{2} e^{-2\pi} \;-\; i\pi e^{-2\pi}$$

Equating real and imaginary parts,

$$\int_{-\infty}^{\infty} \frac{x \cos \pi x}{x^2 + 2x + 5} \, dx \;=\; \frac{\pi}{2} e^{-2\pi}, \qquad \int_{-\infty}^{\infty} \frac{x \sin \pi x}{x^2 + 2x + 5} \, dx \;=\; -\pi e^{-2\pi}$$

Thus we have obtained the value of another integral in addition to the required one.

MISCELLANEOUS DEFINITE INTEGRALS

18. Show that $\displaystyle\int_0^{\infty} \frac{\sin x}{x} \, dx = \frac{\pi}{2}$.

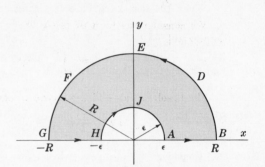

Fig. 7-8

The method of Problem 16 leads us to consider the integral of e^{iz}/z around the contour of Fig. 7-5. However, since $z = 0$ lies on this path of integration and since we cannot integrate through a singularity, we modify that contour by indenting the path at $z = 0$, as shown in Fig. 7-8, which we call contour C' or $ABDEFGHJA$.

Since $z = 0$ is outside C', we have

$$\oint_{C'} \frac{e^{iz}}{z} \, dz \;=\; 0$$

or

$$\int_{-R}^{-\epsilon} \frac{e^{ix}}{x} \, dx \;+\; \int_{HJA} \frac{e^{iz}}{z} \, dz \;+\; \int_{\epsilon}^{R} \frac{e^{ix}}{x} \, dx \;+\; \int_{BDEFG} \frac{e^{iz}}{z} \, dz \;=\; 0$$

Replacing x by $-x$ in the first integral and combining with the third integral, we find

$$\int_{\epsilon}^{R} \frac{e^{ix} - e^{-ix}}{x} \, dx \;+\; \int_{HJA} \frac{e^{iz}}{z} \, dz \;+\; \int_{BDEFG} \frac{e^{iz}}{z} \, dz \;=\; 0$$

or

$$2i \int_{\epsilon}^{R} \frac{\sin x}{x} \, dx \;=\; -\int_{HJA} \frac{e^{iz}}{z} \, dz \;-\; \int_{BDEFG} \frac{e^{iz}}{z} \, dz$$

Let $\epsilon \to 0$ and $R \to \infty$. By Problem 15, the second integral on the right approaches zero. Letting $z = \epsilon e^{i\theta}$ in the first integral on the right, we see that it approaches

$$-\lim_{\epsilon \to 0} \int_{\pi}^{0} \frac{e^{i\epsilon e^{i\theta}}}{\epsilon e^{i\theta}} i\epsilon e^{i\theta} \, d\theta \;=\; -\lim_{\epsilon \to 0} \int_{\pi}^{0} ie^{i\epsilon e^{i\theta}} \, d\theta \;=\; \pi i$$

since the limit can be taken under the integral sign.

Then we have

$$\lim_{\substack{R \to \infty \\ \epsilon \to 0}} 2i \int_{\epsilon}^{R} \frac{\sin x}{x} \, dx \;=\; \pi i \qquad \text{or} \qquad \int_0^{\infty} \frac{\sin x}{x} \, dx \;=\; \frac{\pi}{2}$$

19. Prove that

$$\int_0^{\infty} \sin x^2 \, dx \;=\; \int_0^{\infty} \cos x^2 \, dx \;=\; \frac{1}{2} \sqrt{\frac{\pi}{2}}$$

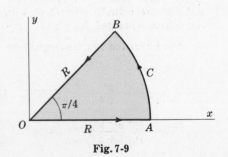

Fig. 7-9

Let C be the contour indicated in Fig. 7-9, where AB is the arc of a circle with center at O and radius R. By Cauchy's theorem,

$$\oint_C e^{iz^2} \, dz \;=\; 0$$

or

$$\int_{OA} e^{iz^2}\, dz \;+\; \int_{AB} e^{iz^2}\, dz \;+\; \int_{BO} e^{iz^2}\, dz \;=\; 0 \tag{1}$$

Now on OA, $z = x$ (from $x = 0$ to $x = R$);   on AB, $z = Re^{i\theta}$ (from $\theta = 0$ to $\theta = \pi/4$);   on BO, $z = re^{\pi i/4}$ (from $r = R$ to $r = 0$).   Hence from (1),

$$\int_0^R e^{ix^2}\, dx \;+\; \int_0^{\pi/4} e^{iR^2 e^{2i\theta}}\, iRe^{i\theta}\, d\theta \;+\; \int_R^0 e^{ir^2 e^{\pi i/2}}\, e^{\pi i/4}\, dr \;=\; 0 \tag{2}$$

i.e.,

$$\int_0^R (\cos x^2 + i\sin x^2)\, dx \;=\; e^{\pi i/4}\int_0^R e^{-r^2}\, dr \;-\; \int_0^{\pi/4} e^{iR^2\cos 2\theta - R^2\sin 2\theta}\, iRe^{i\theta}\, d\theta \tag{3}$$

We consider the limit of (3) as $R \to \infty$. The first integral on the right becomes [see Problem 14, Chapter 10]

$$e^{\pi i/4}\int_0^\infty e^{-r^2}\, dr \;=\; \frac{\sqrt{\pi}}{2}\, e^{\pi i/4} \;=\; \frac{1}{2}\sqrt{\frac{\pi}{2}} + \frac{i}{2}\sqrt{\frac{\pi}{2}} \tag{4}$$

The absolute value of the second integral on the right of (3) is

$$\left| \int_0^{\pi/4} e^{iR^2\cos 2\theta - R^2\sin 2\theta}\, iRe^{i\theta}\, d\theta \right| \;\leqq\; \int_0^{\pi/4} e^{-R^2\sin 2\theta}\, R\, d\theta$$

$$= \frac{R}{2}\int_0^{\pi/2} e^{-R^2\sin\phi}\, d\phi$$

$$\leqq \frac{R}{2}\int_0^{\pi/2} e^{-2R^2\phi/\pi}\, d\phi$$

$$= \frac{\pi}{4R}(1 - e^{-R^2})$$

where we have used the transformation $2\theta = \phi$ and the inequality $\sin\phi \geqq 2\phi/\pi$, $0 \leqq \phi \leqq \pi/2$ (see Problem 15). This shows that as $R \to \infty$ the second integral on the right of (3) approaches zero. Then (3) becomes

$$\int_0^\infty (\cos x^2 + i\sin x^2)\, dx \;=\; \frac{1}{2}\sqrt{\frac{\pi}{2}} + \frac{i}{2}\sqrt{\frac{\pi}{2}}$$

and so equating real and imaginary parts we have, as required,

$$\int_0^\infty \cos x^2\, dx \;=\; \int_0^\infty \sin x^2\, dx \;=\; \frac{1}{2}\sqrt{\frac{\pi}{2}}$$

20. Show that $\displaystyle \int_0^\infty \frac{x^{p-1}}{1+x}\, dx \;=\; \frac{\pi}{\sin p\pi}, \quad 0 < p < 1.$

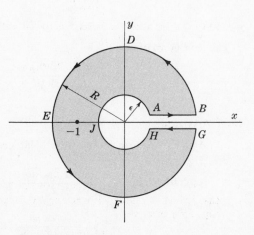

Consider $\displaystyle \oint_C \frac{z^{p-1}}{1+z}\, dz$.   Since $z = 0$ is a branch point, choose C as the contour of Fig. 7-10 where the positive real axis is the branch line and where AB and GH are actually coincident with the x axis but are shown separated for visual purposes.

The integrand has the simple pole $z = -1$ inside C.

Residue at $z = -1 = e^{\pi i}$ is

$$\lim_{z \to -1} (z+1)\frac{z^{p-1}}{1+z} \;=\; (e^{\pi i})^{p-1} \;=\; e^{(p-1)\pi i}$$

Then $\displaystyle \oint_C \frac{z^{p-1}}{1+z}\, dz \;=\; 2\pi i\, e^{(p-1)\pi i}$ or, omitting the integrand,

Fig. 7-10

$$\int_{AB} + \int_{BDEFG} + \int_{GH} + \int_{HJA} = 2\pi i e^{(p-1)\pi i}$$

We thus have

$$\int_{\epsilon}^{R} \frac{x^{p-1}}{1+x} \, dx + \int_{0}^{2\pi} \frac{(Re^{i\theta})^{p-1} \, iRe^{i\theta} \, d\theta}{1+Re^{i\theta}} + \int_{R}^{\epsilon} \frac{(xe^{2\pi i})^{p-1}}{1+xe^{2\pi i}} \, dx$$

$$+ \int_{2\pi}^{0} \frac{(\epsilon e^{i\theta})^{p-1} \, i\epsilon e^{i\theta} \, d\theta}{1+\epsilon e^{i\theta}} = 2\pi i e^{(p-1)\pi i}$$

where we have used $z = xe^{2\pi i}$ for the integral along GH, since the argument of z is increased by 2π in going around the circle $BDEFG$.

Taking the limit as $\epsilon \to 0$ and $R \to \infty$ and noting that the second and fourth integrals approach zero, we find

$$\int_{0}^{\infty} \frac{x^{p-1}}{1+x} \, dx + \int_{\infty}^{0} \frac{e^{2\pi i(p-1)} \, x^{p-1}}{1+x} \, dx = 2\pi e^{(p-1)\pi i}$$

or

$$(1 - e^{2\pi i(p-1)}) \int_{0}^{\infty} \frac{x^{p-1}}{1+x} \, dx = 2\pi i e^{(p-1)\pi i}$$

so that

$$\int_{0}^{\infty} \frac{x^{p-1}}{1+x} \, dx = \frac{2\pi i e^{(p-1)\pi i}}{1 - e^{2\pi i(p-1)}} = \frac{2\pi i}{e^{p\pi i} - e^{-p\pi i}} = \frac{\pi}{\sin p\pi}$$

21. Prove that $\displaystyle \int_{0}^{\infty} \frac{\cosh ax}{\cosh x} \, dx = \frac{\pi}{2\cos(\pi a/2)}$ where $|a| < 1$.

Consider $\displaystyle \oint_{C} \frac{e^{az}}{\cosh z} \, dz$ where C is a rectangle having vertices at $-R, R, R+\pi i, -R+\pi i$ (see Fig. 7-11).

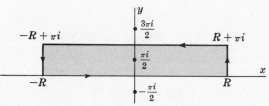

Fig. 7-11

The poles of $e^{az}/\cosh z$ are simple and occur where $\cosh z = 0$, i.e. $z = (n + \frac{1}{2})\pi i$, $n = 0, \pm 1, \pm 2, \ldots$. The only pole enclosed by C is $\pi i/2$.

Residue of $\dfrac{e^{az}}{\cosh z}$ at $z = \pi i/2$ is

$$\lim_{z \to \pi i/2} (z - \pi i/2) \frac{e^{az}}{\cosh z} = \frac{e^{a\pi i/2}}{\sinh(\pi i/2)} = \frac{e^{a\pi i/2}}{i \sin(\pi/2)} = -ie^{a\pi i/2}$$

Then by the residue theorem,

$$\oint_{C} \frac{e^{az}}{\cosh z} \, dz = 2\pi i (-ie^{a\pi i/2}) = 2\pi e^{a\pi i/2}$$

This can be written

$$\int_{-R}^{R} \frac{e^{ax}}{\cosh x} \, dx + \int_{0}^{\pi} \frac{e^{a(R+iy)}}{\cosh(R+iy)} \, i \, dy + \int_{R}^{-R} \frac{e^{a(x+\pi i)}}{\cosh(x+\pi i)} \, dx$$

$$+ \int_{\pi}^{0} \frac{e^{a(-R+iy)}}{\cosh(-R+iy)} \, i \, dy = 2\pi e^{a\pi i/2} \qquad (1)$$

As $R \to \infty$ the second and fourth integrals on the left approach zero. To show this let us consider the second integral. Since

$$|\cosh(R+iy)| = \left| \frac{e^{R+iy} + e^{-R-iy}}{2} \right| \geqq \tfrac{1}{2}\{ |e^{R+iy}| - |e^{-R-iy}| \} = \tfrac{1}{2}(e^R - e^{-R}) \geqq \tfrac{1}{4}e^R$$

we have

$$\left| \int_{0}^{\pi} \frac{e^{a(R+iy)}}{\cosh(R+iy)} \, i \, dy \right| \leqq \int_{0}^{\pi} \frac{e^{aR}}{\frac{1}{4}e^R} \, dy = 4\pi e^{(a-1)R}$$

and the result follows on noting that the right side approaches zero as $R \to \infty$ since $|a| < 1$. In a similar manner we can show that the fourth integral on the left of (1) approaches zero as $R \to \infty$. Hence (1) becomes

$$\lim_{R \to \infty}\left\{ \int_{-R}^{R} \frac{e^{ax}}{\cosh x}\, dx \; + \; e^{a\pi i} \int_{-R}^{R} \frac{e^{ax}}{\cosh x}\, dx \right\} \;=\; 2\pi e^{a\pi i/2}$$

since $\cosh(x + \pi i) = -\cosh x$.   Thus

$$\lim_{R \to \infty} \int_{-R}^{R} \frac{e^{ax}}{\cosh x}\, dx \;=\; \int_{-\infty}^{\infty} \frac{e^{ax}}{\cosh x}\, dx \;=\; \frac{2\pi e^{a\pi i/2}}{1 + e^{a\pi i}} \;=\; \frac{2\pi}{e^{a\pi i/2} + e^{-a\pi i/2}} \;=\; \frac{\pi}{\cos(\pi a/2)}$$

Now

$$\int_{-\infty}^{0} \frac{e^{ax}}{\cosh x}\, dx \;+\; \int_{0}^{\infty} \frac{e^{ax}}{\cosh x}\, dx \;=\; \frac{\pi}{\cos(\pi a/2)}$$

Then replacing x by $-x$ in the first integral, we have

$$\int_{0}^{\infty} \frac{e^{-ax}}{\cosh x}\, dx \;+\; \int_{0}^{\infty} \frac{e^{ax}}{\cosh x}\, dx \;=\; 2 \int_{0}^{\infty} \frac{\cosh ax}{\cosh x}\, dx \;=\; \frac{\pi}{\cos(\pi a/2)}$$

from which the required result follows.

22. Prove that $\displaystyle \int_{0}^{\infty} \frac{\ln(x^2 + 1)}{x^2 + 1}\, dx \;=\; \pi \ln 2.$

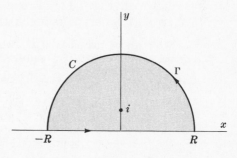

Fig. 7-12

Consider $\displaystyle \oint_C \frac{\ln(z + i)}{z^2 + 1}\, dz$ around the contour C consisting of the real axis from $-R$ to R and the semicircle Γ of radius R (see Fig. 7-12).

The only pole of $\ln(z + i)/(z^2 + 1)$ inside C is the simple pole $z = i$, and the residue is

$$\lim_{z \to i} (z - i) \frac{\ln(z + i)}{(z - i)(z + i)} \;=\; \frac{\ln(2i)}{2i}$$

Hence by the residue theorem,

$$\oint_C \frac{\ln(z + i)}{z^2 + 1}\, dz \;=\; 2\pi i \left\{ \frac{\ln(2i)}{2i} \right\} \;=\; \pi \ln(2i) \;=\; \pi \ln 2 + \tfrac{1}{2}\pi^2 i \tag{1}$$

on writing $\ln(2i) = \ln 2 + \ln i = \ln 2 + \ln e^{\pi i/2} = \ln 2 + \pi i/2$ using principal values of the logarithm. The result can be written

$$\int_{-R}^{R} \frac{\ln(x + i)}{x^2 + 1}\, dx \;+\; \int_{\Gamma} \frac{\ln(z + i)}{z^2 + 1}\, dz \;=\; \pi \ln 2 + \tfrac{1}{2}\pi^2 i$$

or

$$\int_{-R}^{0} \frac{\ln(x + i)}{x^2 + 1}\, dx \;+\; \int_{0}^{R} \frac{\ln(x + i)}{x^2 + 1}\, dx \;+\; \int_{\Gamma} \frac{\ln(z + i)}{z^2 + 1}\, dz \;=\; \pi \ln 2 + \tfrac{1}{2}\pi^2 i$$

Replacing x by $-x$ in the first integral, this can be written

$$\int_{0}^{R} \frac{\ln(i - x)}{x^2 + 1}\, dx \;+\; \int_{0}^{R} \frac{\ln(i + x)}{x^2 + 1}\, dx \;+\; \int_{\Gamma} \frac{\ln(z + i)}{z^2 + 1}\, dz \;=\; \pi \ln 2 + \tfrac{1}{2}\pi^2 i$$

or, since $\ln(i - x) + \ln(i + x) = \ln(i^2 - x^2) = \ln(x^2 + 1) + \pi i$,

$$\int_{0}^{R} \frac{\ln(x^2 + 1)}{x^2 + 1}\, dx \;+\; \int_{0}^{R} \frac{\pi i}{x^2 + 1}\, dx \;+\; \int_{\Gamma} \frac{\ln(z + i)}{z^2 + 1}\, dz \;=\; \pi \ln 2 + \tfrac{1}{2}\pi^2 i \tag{2}$$

As $R \to \infty$ we can show that the integral around Γ approaches zero (see Problem 101). Hence on taking real parts we find, as required,

$$\lim_{R \to \infty} \int_{0}^{R} \frac{\ln(x^2 + 1)}{x^2 + 1}\, dx \;=\; \int_{0}^{\infty} \frac{\ln(x^2 + 1)}{x^2 + 1}\, dx \;=\; \pi \ln 2$$

23. Prove that $\displaystyle\int_0^{\pi/2} \ln \sin x \, dx = \int_0^{\pi/2} \ln \cos x \, dx = -\tfrac{1}{2}\pi \ln 2$

Letting $x = \tan \theta$ in the result of Problem 22, we find

$$\int_0^{\pi/2} \frac{\ln (\tan^2 \theta + 1)}{\tan^2 \theta + 1} \sec^2 \theta \, d\theta = -2 \int_0^{\pi/2} \ln \cos \theta \, d\theta = \pi \ln 2$$

from which

$$\int_0^{\pi/2} \ln \cos \theta \, d\theta = -\tfrac{1}{2}\pi \ln 2 \qquad\qquad (1)$$

which establishes part of the required result. Letting $\theta = \pi/2 - \phi$ in (1), we find

$$\int_0^{\pi/2} \ln \sin \phi \, d\phi = -\tfrac{1}{2}\pi \ln 2$$

SUMMATION OF SERIES

24. Let C_N be a square with vertices at

$$(N+\tfrac{1}{2})(1+i), \quad (N+\tfrac{1}{2})(-1+i),$$
$$(N+\tfrac{1}{2})(-1-i), \quad (N+\tfrac{1}{2})(1-i)$$

as shown in Fig. 7-13. Prove that on C_N, $|\cot \pi z| < A$ where A is a constant.

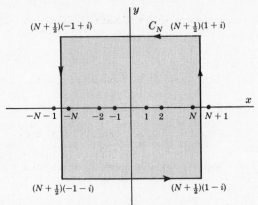

Fig. 7-13

We consider the parts of C_N which lie in the regions $y > \tfrac{1}{2}$, $-\tfrac{1}{2} \leq y \leq \tfrac{1}{2}$ and $y < -\tfrac{1}{2}$.

Case 1: $y > \tfrac{1}{2}$. In this case if $z = x + iy$,

$$|\cot \pi z| = \left| \frac{e^{\pi i z} + e^{-\pi i z}}{e^{\pi i z} - e^{-\pi i z}} \right|$$

$$= \left| \frac{e^{\pi i x - \pi y} + e^{-\pi i x + \pi y}}{e^{\pi i x - \pi y} - e^{-\pi i x + \pi y}} \right|$$

$$\leq \frac{|e^{\pi i x - \pi y}| + |e^{-\pi i x + \pi y}|}{|e^{-\pi i x + \pi y}| - |e^{\pi i x - \pi y}|}$$

$$= \frac{e^{-\pi y} + e^{\pi y}}{e^{\pi y} - e^{-\pi y}} = \frac{1 + e^{-2\pi y}}{1 - e^{-2\pi y}} \leq \frac{1 + e^{-\pi}}{1 - e^{-\pi}} = A_1$$

Case 2: $y < -\tfrac{1}{2}$. Here as in Case 1,

$$|\cot \pi z| \leq \frac{|e^{\pi i x - \pi y}| + |e^{-\pi i x + \pi y}|}{|e^{\pi i x - \pi y}| - |e^{-\pi i x + \pi y}|} = \frac{e^{-\pi y} + e^{\pi y}}{e^{-\pi y} - e^{\pi y}} = \frac{1 + e^{2\pi y}}{1 - e^{2\pi y}} \leq \frac{1 + e^{-\pi}}{1 - e^{-\pi}} = A_1$$

Case 3: $-\tfrac{1}{2} \leq y \leq \tfrac{1}{2}$. Consider $z = N + \tfrac{1}{2} + iy$. Then

$$|\cot \pi z| = |\cot \pi (N + \tfrac{1}{2} + iy)| = |\cot (\pi/2 + \pi iy)| = |\tanh \pi y| \leq \tanh (\pi/2) = A_2$$

If $z = -N - \tfrac{1}{2} + iy$, we have similarly

$$|\cot \pi z| = |\cot \pi (-N - \tfrac{1}{2} + iy)| = |\tanh \pi y| \leq \tanh (\pi/2) = A_2$$

Thus if we choose A as a number greater than the larger of A_1 and A_2, we have $|\cot \pi z| < A$ on C_N where A is independent of N. It is of interest to note that we actually have $|\cot \pi z| \leq A_1 = \coth (\pi/2)$ since $A_2 < A_1$.

25. Let $f(z)$ be such that along the path C_N of Fig. 7-13, $|f(z)| \leq \dfrac{M}{|z|^k}$ where $k > 1$ and M are constants independent of N. Prove that

$$\sum_{-\infty}^{\infty} f(n) = - \{\text{sum of residues of } \pi \cot \pi z \, f(z) \text{ at the poles of } f(z)\}$$

Case 1: $f(z)$ has a finite number of poles.

In this case we can choose N so large that the path C_N of Fig. 7-13 encloses all poles of $f(z)$. The poles of $\cot \pi z$ are simple and occur at $z = 0, \pm 1, \pm 2, \ldots$.

Residue of $\pi \cot \pi z \, f(z)$ at $z = n$, $n = 0, \pm 1, \pm 2, \ldots,$ is

$$\lim_{z \to n} (z - n)\pi \cot \pi z \, f(z) \;\; = \;\; \lim_{z \to n} \pi \left(\frac{z - n}{\sin \pi z} \right) \cos \pi z \, f(z) \;\; = \;\; f(n)$$

using L'Hospital's rule. We have assumed here that $f(z)$ has no poles at $z = n$, since otherwise the given series diverges.

By the residue theorem,

$$\oint_{C_N} \pi \cot \pi z \, f(z) \, dz \;\; = \;\; \sum_{n=-N}^{N} f(n) \; + \; S \tag{1}$$

where S is the sum of the residues of $\pi \cot \pi z \, f(z)$ at the poles of $f(z)$. By Problem 24 and our assumption on $f(z)$, we have

$$\left| \oint_{C_N} \pi \cot \pi z \, f(z) \, dz \right| \;\; \leq \;\; \frac{\pi A M}{N^k} (8N + 4)$$

since the length of path C_N is $8N + 4$. Then taking the limit as $N \to \infty$ we see that

$$\lim_{N \to \infty} \oint_{C_N} \pi \cot \pi z \, f(z) \, dz \;\; = \;\; 0 \tag{2}$$

Thus from (1) we have as required,

$$\sum_{-\infty}^{\infty} f(n) \;\; = \;\; -S \tag{3}$$

Case 2: $f(z)$ has infinitely many poles.

If $f(z)$ has an infinite number of poles, we can obtain the required result by an appropriate limiting procedure. See Problem 103.

26. Prove that $\displaystyle\sum_{n=-\infty}^{\infty} \frac{1}{n^2 + a^2} \;=\; \frac{\pi}{a} \coth \pi a$ **where** $a > 0$.

Let $f(z) = \dfrac{1}{z^2 + a^2}$ which has simple poles at $z = \pm ai$.

Residue of $\dfrac{\pi \cot \pi z}{z^2 + a^2}$ at $z = ai$ is

$$\lim_{z \to ai} (z - ai) \frac{\pi \cot \pi z}{(z - ai)(z + ai)} \;\; = \;\; \frac{\pi \cot \pi ai}{2ai} \;\; = \;\; -\frac{\pi}{2a} \coth \pi a$$

Similarly the residue at $z = -ai$ is $\dfrac{-\pi}{2a} \coth \pi a$, and the sum of the residues is $-\dfrac{\pi}{a} \coth \pi a$. Then by Problem 25,

$$\sum_{n=-\infty}^{\infty} \frac{1}{n^2 + a^2} \;\; = \;\; -(\text{sum of residues}) \;\; = \;\; \frac{\pi}{a} \coth \pi a$$

27. Prove that $\displaystyle\sum_{n=1}^{\infty} \frac{1}{n^2 + a^2} \;=\; \frac{\pi}{2a} \coth \pi a \;-\; \frac{1}{2a^2}$ **where** $a > 0$.

The result of Problem 26 can be written in the form

$$\sum_{n=-\infty}^{-1} \frac{1}{n^2 + a^2} \; + \; \frac{1}{a^2} \; + \; \sum_{n=1}^{\infty} \frac{1}{n^2 + a^2} \;\; = \;\; \frac{\pi}{a} \coth \pi a$$

or

$$2 \sum_{n=1}^{\infty} \frac{1}{n^2 + a^2} \; + \; \frac{1}{a^2} \;\; = \;\; \frac{\pi}{a} \coth \pi a$$

which gives the required result.

28. Prove that $\dfrac{1}{1^2} + \dfrac{1}{2^2} + \dfrac{1}{3^2} + \cdots = \dfrac{\pi^2}{6}$.

We have
$$F(z) \;=\; \frac{\pi \cot \pi z}{z^2} \;=\; \frac{\pi \cos \pi z}{z^2 \sin \pi z} \;=\; \frac{\left(1 - \dfrac{\pi^2 z^2}{2!} + \dfrac{\pi^4 z^4}{4!} - \cdots\right)}{z^3\left(1 - \dfrac{\pi^2 z^2}{3!} + \dfrac{\pi^4 z^4}{5!} - \cdots\right)}$$

$$= \;\frac{1}{z^3}\left(1 - \frac{\pi^2 z^2}{2!} + \cdots\right)\left(1 + \frac{\pi^2 z^2}{3!} + \cdots\right) \;=\; \frac{1}{z^3}\left(1 - \frac{\pi^2 z^2}{3} + \cdots\right)$$

so that the residue at $z = 0$ is $-\pi^2/3$.

Then as in Problems 26 and 27,
$$\oint_{C_N} \frac{\pi \cot \pi z}{z^2}\, dz \;=\; \sum_{n=-N}^{-1} \frac{1}{n^2} + \sum_{n=1}^{N} \frac{1}{n^2} - \frac{\pi^2}{3}$$

$$= \; 2\sum_{n=1}^{N} \frac{1}{n^2} - \frac{\pi^2}{3}$$

Taking the limit as $N \to \infty$ we have, since the left side approaches zero,
$$2\sum_{n=1}^{\infty} \frac{1}{n^2} - \frac{\pi^2}{3} \;=\; 0 \qquad \text{or} \qquad \sum_{n=1}^{\infty} \frac{1}{n^2} \;=\; \frac{\pi^2}{6}$$

Another method. Take the limit as $a \to 0$ in the result of Problem 27. Then using L'Hospital's rule,
$$\lim_{a \to 0} \sum_{n=1}^{\infty} \frac{1}{n^2 + a^2} \;=\; \sum_{n=1}^{\infty} \frac{1}{n^2} \;=\; \lim_{a \to 0} \frac{\pi a \coth \pi a - 1}{2a^2} \;=\; \frac{\pi^2}{6}$$

29. If $f(z)$ satisfies the same conditions given in Problem 25, prove that
$$\sum_{-\infty}^{\infty} (-1)^n f(n) \;=\; - \{\text{sum of residues of } \pi \csc \pi z\, f(z) \text{ at the poles of } f(z)\}$$

We proceed in a manner similar to that in Problem 25. The poles of $\csc \pi z$ are simple and occur at $z = 0, \pm 1, \pm 2, \ldots$.

Residue of $\pi \csc \pi z\, f(z)$ at $z = n$, $n = 0, \pm 1, \pm 2, \ldots$, is
$$\lim_{z \to n} (z - n)\pi \csc \pi z\, f(z) \;=\; \lim_{z \to n} \pi \left(\frac{z - n}{\sin \pi z}\right) f(z) \;=\; (-1)^n f(n)$$

By the residue theorem,
$$\oint_{C_N} \pi \csc \pi z\, f(z)\, dz \;=\; \sum_{n=-N}^{N} (-1)^n f(n) + S \tag{1}$$

where S is the sum of the residues of $\pi \csc \pi z\, f(z)$ at the poles of $f(z)$.

Letting $N \to \infty$, the integral on the left of *(1)* approaches zero (Problem 106) so that, as required, *(1)* becomes
$$\sum_{-\infty}^{\infty} (-1)^n f(n) \;=\; -S \tag{2}$$

30. Prove that $\displaystyle\sum_{n=-\infty}^{\infty} \frac{(-1)^n}{(n + a)^2} = \frac{\pi^2 \cos \pi a}{\sin^2 \pi a}$ where a is real and different from $0, \pm 1, \pm 2, \ldots$.

Let $f(z) = \dfrac{1}{(z + a)^2}$ which has a double pole at $z = -a$.

Residue of $\dfrac{\pi \csc \pi z}{(z + a)^2}$ at $z = -a$ is
$$\lim_{z \to -a} \frac{d}{dz}\left\{(z + a)^2 \cdot \frac{\pi \csc \pi z}{(z + a)^2}\right\} \;=\; -\pi^2 \csc \pi a \cot \pi a$$

Then by Problem 29,
$$\sum_{n=-\infty}^{\infty} \frac{(-1)^n}{(n + a)^2} \;=\; -(\text{sum of residues}) \;=\; \pi^2 \csc \pi a \cot \pi a \;=\; \frac{\pi^2 \cos \pi a}{\sin^2 \pi a}$$

31. Prove that if $a \neq 0, \pm 1, \pm 2, \ldots$, then

$$\frac{a^2+1}{(a^2-1)^2} - \frac{a^2+4}{(a^2-4)^2} + \frac{a^2+9}{(a^2-9)^2} - \cdots = \frac{1}{2a^2} - \frac{\pi^2 \cos \pi a}{2 \sin^2 \pi a}$$

The result of Problem 30 can be written in the form

$$\frac{1}{a^2} - \left\{ \frac{1}{(a+1)^2} + \frac{1}{(a-1)^2} \right\} + \left\{ \frac{1}{(a+2)^2} + \frac{1}{(a-2)^2} \right\} + \cdots = \frac{\pi^2 \cos \pi a}{\sin^2 \pi a}$$

or

$$\frac{1}{a^2} - \frac{2(a^2+1)}{(a^2-1)^2} + \frac{2(a^2+4)}{(a^2-4)^2} - \frac{2(a^2+9)}{(a^2-9)^2} + \cdots = \frac{\pi^2 \cos \pi a}{\sin^2 \pi a}$$

from which the required result follows. Note that the grouping of terms in the infinite series is permissible since the series is absolutely convergent.

32. Prove that $\dfrac{1}{1^3} - \dfrac{1}{3^3} + \dfrac{1}{5^3} - \dfrac{1}{7^3} + \cdots = \dfrac{\pi^3}{32}$.

We have
$$F(z) = \frac{\pi \sec \pi z}{z^3} = \frac{\pi}{z^3 \cos \pi z} = \frac{\pi}{z^3(1 - \pi^2 z^2/2! + \cdots)}$$

$$= \frac{\pi}{z^3}\left(1 + \frac{\pi^2 z^2}{2} + \cdots \right) = \frac{\pi}{z^3} + \frac{\pi^3}{2z} + \cdots$$

so that the residue at $z = 0$ is $\pi^3/2$.

The residue of $F(z)$ at $z = n + \frac{1}{2}$, $n = 0, \pm 1, \pm 2, \ldots$ [which are the simple poles of $\sec \pi z$], is

$$\lim_{z \to n + \frac{1}{2}} \left\{z - (n+\tfrac{1}{2})\right\} \frac{\pi}{z^3 \cos \pi z} = \frac{\pi}{(n+\frac{1}{2})^3} \lim_{z \to n+\frac{1}{2}} \frac{z - (n+\frac{1}{2})}{\cos \pi z} = \frac{-(-1)^n}{(n+\frac{1}{2})^3}$$

If C_N is a square with vertices at $N(1+i)$, $N(1-i)$, $N(-1+i)$, $N(-1-i)$, then

$$\oint_{C_N} \frac{\pi \sec \pi z}{z^3} \, dz = -\sum_{n=-N}^{N} \frac{(-1)^n}{(n+\frac{1}{2})^3} + \frac{\pi^3}{2} = -8\sum_{n=-N}^{N} \frac{(-1)^n}{(2n+1)^3} + \frac{\pi^3}{2}$$

and since the integral on the left approaches zero as $N \to \infty$, we have

$$\sum_{-\infty}^{\infty} \frac{(-1)^n}{(2n+1)^3} = 2\left\{ \frac{1}{1^3} - \frac{1}{3^3} + \frac{1}{5^3} - \cdots \right\} = \frac{\pi^3}{16}$$

from which the required result follows.

MITTAG-LEFFLER'S EXPANSION THEOREM

33. Prove Mittag-Leffler's expansion theorem (see Page 175).

Let $f(z)$ have poles at $z = a_n$, $n = 1, 2, \ldots$, and suppose that $z = \zeta$ is not a pole of $f(z)$. Then the function $\dfrac{f(z)}{z - \zeta}$ has poles at $z = a_n$, $n = 1, 2, 3, \ldots$ and ζ.

Residue of $\dfrac{f(z)}{z-\zeta}$ at $z = a_n$, $n = 1, 2, 3, \ldots$, is $\lim\limits_{z \to a_n} (z - a_n) \dfrac{f(z)}{z-\zeta} = \dfrac{b_n}{a_n - \zeta}$.

Residue of $\dfrac{f(z)}{z-\zeta}$ at $z = \zeta$ is $\lim\limits_{z \to \zeta} (z - \zeta) \dfrac{f(z)}{z-\zeta} = f(\zeta)$.

Then by the residue theorem,

$$\frac{1}{2\pi i} \oint_{C_N} \frac{f(z)}{z-\zeta} \, dz = f(\zeta) + \sum_n \frac{b_n}{a_n - \zeta} \qquad (1)$$

where the last summation is taken over all poles inside circle C_N of radius R_N (Fig. 7-14).

Suppose that $f(z)$ is analytic at $z = 0$. Then putting $\zeta = 0$ in (1), we have

$$\frac{1}{2\pi i} \oint_{C_N} \frac{f(z)}{z} \, dz = f(0) + \sum_n \frac{b_n}{a_n} \qquad (2)$$

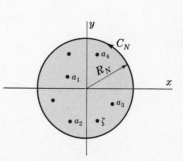

Fig. 7-14

Subtraction of (2) from (1) yields

$$f(\zeta) - f(0) + \sum_n b_n \left(\frac{1}{a_n - \zeta} - \frac{1}{a_n} \right) = \frac{1}{2\pi i} \oint_{C_N} f(z) \left\{ \frac{1}{z - \zeta} - \frac{1}{z} \right\} dz$$

$$= \frac{\zeta}{2\pi i} \oint_{C_N} \frac{f(z)}{z(z - \zeta)} dz \qquad (3)$$

Now since $|z - \zeta| \geqq |z| - |\zeta| = R_N - |\zeta|$ for z on C_N, we have, if $|f(z)| \leqq M$,

$$\left| \oint_{C_N} \frac{f(z)}{z(z - \zeta)} dz \right| \leqq \frac{M \cdot 2\pi R_N}{R_N(R_N - |\zeta|)}$$

As $N \to \infty$ and therefore $R_N \to \infty$, it follows that the integral on the left approaches zero, i.e.,

$$\lim_{N \to \infty} \oint_{C_N} \frac{f(z)}{z(z - \zeta)} dz = 0$$

Hence from (3), letting $N \to \infty$, we have as required

$$f(\zeta) = f(0) + \sum_n b_n \left(\frac{1}{\zeta - a_n} + \frac{1}{a_n} \right)$$

the result on Page 175 being obtained on replacing ζ by z.

34. Prove that $\quad \cot z = \dfrac{1}{z} + \displaystyle\sum_n \left(\dfrac{1}{z - n\pi} + \dfrac{1}{n\pi} \right) \quad$ **where the summation extends over** $n = \pm 1, \pm 2, \ldots$.

Consider the function $\quad f(z) = \cot z - \dfrac{1}{z} = \dfrac{z \cos z - \sin z}{z \sin z}$. Then $f(z)$ has simple poles at $z = n\pi$, $n = \pm 1, \pm 2, \pm 3, \ldots$, and the residue at these poles is

$$\lim_{z \to n\pi} (z - n\pi) \left(\frac{z \cos z - \sin z}{z \sin z} \right) = \lim_{z \to n\pi} \left(\frac{z - n\pi}{\sin z} \right) \lim_{z \to n\pi} \left(\frac{z \cos z - \sin z}{z} \right) = 1$$

At $z = 0$, $f(z)$ has a removable singularity since

$$\lim_{z \to 0} \left(\cot z - \frac{1}{z} \right) = \lim_{z \to 0} \left(\frac{z \cos z - \sin z}{z \sin z} \right) = 0$$

by L'Hospital's rule. Hence we can define $f(0) = 0$.

By Problem 110 it follows that $f(z)$ is bounded on circles C_N having center at the origin and radius $R_N = (N + \frac{1}{2})\pi$. Hence by Problem 33,

$$\cot z - \frac{1}{z} = \sum_n \left(\frac{1}{z - n\pi} + \frac{1}{n\pi} \right)$$

from which the required result follows.

35. Prove that $\quad \cot z = \dfrac{1}{z} + 2z \left\{ \dfrac{1}{z^2 - \pi^2} + \dfrac{1}{z^2 - 4\pi^2} + \cdots \right\}$.

We can write the result of Problem 34 in the form

$$\cot z = \frac{1}{z} + \lim_{N \to \infty} \left\{ \sum_{n=-N}^{-1} \left(\frac{1}{z - n\pi} + \frac{1}{n\pi} \right) + \sum_{n=1}^{N} \left(\frac{1}{z - n\pi} + \frac{1}{n\pi} \right) \right\}$$

$$= \frac{1}{z} + \lim_{N \to \infty} \left\{ \left(\frac{1}{z + \pi} + \frac{1}{z - \pi} \right) + \left(\frac{1}{z + 2\pi} + \frac{1}{z - 2\pi} \right) + \cdots + \left(\frac{1}{z + N\pi} + \frac{1}{z - N\pi} \right) \right\}$$

$$= \frac{1}{z} + \lim_{N \to \infty} \left\{ \frac{2z}{z^2 - \pi^2} + \frac{2z}{z^2 - 4\pi^2} + \cdots + \frac{2z}{z^2 - N^2\pi^2} \right\}$$

$$= \frac{1}{z} + 2z \left\{ \frac{1}{z^2 - \pi^2} + \frac{1}{z^2 - 4\pi^2} + \cdots \right\}$$

MISCELLANEOUS PROBLEMS

36. Evaluate $\dfrac{1}{2\pi i}\displaystyle\int_{a-i\infty}^{a+i\infty}\dfrac{e^{zt}}{\sqrt{z+1}}\,dz$ where a and t are any positive constants.

The integrand has a branch point at $z=-1$. We shall take as branch line that part of the real axis to the left of $z=-1$. Since we cannot cross this branch line, let us consider

$$\oint_C \frac{e^{zt}}{\sqrt{z+1}}\,dz$$

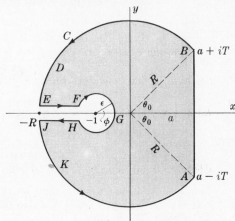

Fig. 7-15

where C is the contour $ABDEFGHJKA$ shown in Fig. 7-15. In this figure EF and HJ actually lie on the real axis but have been shown separated for visual purposes. Also, FGH is a circle of radius ϵ while BDE and JKA represent arcs of a circle of radius R.

Since $e^{zt}/\sqrt{z+1}$ is analytic inside and on C, we have by Cauchy's theorem

$$\oint_C \frac{e^{zt}}{\sqrt{z+1}}\,dz \;=\; 0 \qquad (1)$$

Omitting the integrand, this can be written

$$\int_{AB} + \int_{BDE} + \int_{EF} + \int_{FGH} + \int_{HJ} + \int_{JKA} \;=\; 0 \qquad (2)$$

Now on BDE and JKA, $z=Re^{i\theta}$ where θ goes from θ_0 to π and π to $2\pi-\theta_0$ respectively.

On EF, $z+1=ue^{\pi i}$, $\sqrt{z+1}=\sqrt{u}\,e^{\pi i/2}=i\sqrt{u}$; whereas on HJ, $z+1=ue^{-\pi i}$, $\sqrt{z+1}=\sqrt{u}\,e^{-\pi i/2}=-i\sqrt{u}$. In both cases $z=-u-1$, $dz=-du$, where u varies from $R-1$ to ϵ along EF and ϵ to $R-1$ along HJ.

On FGH, $z+1=\epsilon e^{i\phi}$ where ϕ goes from $-\pi$ to π.

Thus (2) can be written

$$\int_{a-iT}^{a+iT}\frac{e^{zt}}{\sqrt{z+1}}\,dz \;+\; \int_{\theta_0}^{\pi}\frac{e^{Re^{i\theta}t}}{\sqrt{Re^{i\theta}+1}}\,iRe^{i\theta}\,d\theta \;+\; \int_{R-1}^{\epsilon}\frac{e^{-(u+1)t}\,(-du)}{i\sqrt{u}}$$

$$+\; \int_{\pi}^{-\pi}\frac{e^{(\epsilon e^{i\phi}-1)t}}{\sqrt{\epsilon e^{i\phi}+1}}\,i\epsilon e^{i\phi}\,d\phi \;+\; \int_{\epsilon}^{R-1}\frac{e^{-(u+1)t}\,(-du)}{-i\sqrt{u}}$$

$$+\; \int_{\pi}^{2\pi-\theta_0}\frac{e^{Re^{i\theta}t}}{\sqrt{Re^{i\theta}+1}}\,iRe^{i\theta}\,d\theta \;=\; 0 \qquad (3)$$

Let us now take the limit as $R\to\infty$ (and $T=\sqrt{R^2-a^2}\to\infty$) and $\epsilon\to0$. We can show (see Problem 111) that the second, fourth and sixth integrals approach zero. Hence we have

$$\int_{a-i\infty}^{a+i\infty}\frac{e^{zt}}{\sqrt{z+1}}\,dz \;=\; \lim_{\substack{\epsilon\to0\\ R\to\infty}} 2i\int_{\epsilon}^{R-1}\frac{e^{-(u+1)t}}{\sqrt{u}}\,du \;=\; 2i\int_{0}^{\infty}\frac{e^{-(u+1)t}}{\sqrt{u}}\,du$$

or letting $u=v^2$,

$$\frac{1}{2\pi i}\int_{a-i\infty}^{a+i\infty}\frac{e^{zt}}{\sqrt{z+1}}\,dz \;=\; \frac{1}{\pi}\int_{0}^{\infty}\frac{e^{-(u+1)t}}{\sqrt{u}}\,du \;=\; \frac{2e^{-t}}{\pi}\int_{0}^{\infty}e^{-v^2 t}\,dv \;=\; \frac{e^{-t}}{\sqrt{\pi t}}$$

37. Prove that $\displaystyle\int_0^{\infty}\frac{(\ln u)^2}{u^2+1}\,du \;=\; \frac{\pi^3}{8}$.

Let C be the closed curve of Fig. 7-16 below where Γ_1 and Γ_2 are semicircles of radii ϵ and R respectively and center at the origin. Consider

$$\oint_C \frac{(\ln z)^2}{z^2+1}\,dz$$

Since the integrand has a simple pole $z = i$ inside C
and since the residue at this pole is

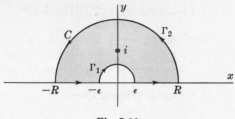

$$\lim_{z \to i} (z - i) \frac{(\ln z)^2}{(z - i)(z + i)} = \frac{(\ln i)^2}{2i}$$
$$= \frac{(\pi i/2)^2}{2i}$$
$$= \frac{-\pi^2}{8i}$$

Fig. 7-16

we have by the residue theorem

$$\oint_C \frac{(\ln z)^2}{z^2 + 1} \, dz = 2\pi i \left(\frac{-\pi^2}{8i} \right) = \frac{-\pi^3}{4} \tag{1}$$

Now

$$\oint_C \frac{(\ln z)^2}{z^2 + 1} \, dz = \int_{-R}^{-\epsilon} \frac{(\ln z)^2}{z^2 + 1} \, dz + \int_{\Gamma_1} \frac{(\ln z)^2}{z^2 + 1} \, dz + \int_{\epsilon}^{R} \frac{(\ln z)^2}{z^2 + 1} \, dz + \int_{\Gamma_2} \frac{(\ln z)^2}{z^2 + 1} \, dz \tag{2}$$

Let $z = -u$ in the first integral on the right so that $\ln z = \ln(-u) = \ln u + \ln(-1) = \ln u + \pi i$
and $dz = -du$. Also let $z = u$ (so that $dz = du$ and $\ln z = \ln u$) in the third integral on the right.
Then using (1), we have

$$\int_{\epsilon}^{R} \frac{(\ln u + \pi i)^2}{u^2 + 1} \, du + \int_{\Gamma_1} \frac{(\ln z)^2}{z^2 + 1} \, dz + \int_{\epsilon}^{R} \frac{(\ln u)^2}{u^2 + 1} \, du + \int_{\Gamma_2} \frac{(\ln z)^2}{z^2 + 1} \, dz = \frac{-\pi^3}{4}$$

Now let $\epsilon \to 0$ and $R \to \infty$. Since the integrals around Γ_1 and Γ_2 approach zero, we have

$$\int_0^{\infty} \frac{(\ln u + \pi i)^2}{u^2 + 1} \, du + \int_0^{\infty} \frac{(\ln u)^2}{u^2 + 1} \, du = \frac{-\pi^3}{4}$$

or

$$2 \int_0^{\infty} \frac{(\ln u)^2}{u^2 + 1} \, du + 2\pi i \int_0^{\infty} \frac{\ln u}{u^2 + 1} \, du - \pi^2 \int_0^{\infty} \frac{du}{u^2 + 1} = \frac{-\pi^3}{4}$$

Using the fact that $\int_0^{\infty} \frac{du}{u^2 + 1} = \tan^{-1} u \Big|_0^{\infty} = \frac{\pi}{2}$,

$$2 \int_0^{\infty} \frac{(\ln u)^2}{u^2 + 1} \, du + 2\pi i \int_0^{\infty} \frac{\ln u}{u^2 + 1} \, du = \frac{\pi^3}{4}$$

Equating real and imaginary parts, we find

$$\int_0^{\infty} \frac{(\ln u)^2}{u^2 + 1} \, du = \frac{\pi^3}{8}, \qquad \int_0^{\infty} \frac{\ln u}{u^2 + 1} \, du = 0$$

the second integral being a by-product of the evaluation.

38. Prove that

$$\frac{\coth \pi}{1^3} + \frac{\coth 2\pi}{2^3} + \frac{\coth 3\pi}{3^3} + \cdots = \frac{7\pi^3}{180}$$

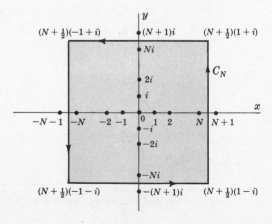

Consider

$$\oint_{C_N} \frac{\pi \cot \pi z \coth \pi z}{z^3} \, dz$$

taken around the square C_N shown in Fig. 7-17.
The poles of the integrand are located at: $z = 0$
(pole of order 5); $z = \pm 1, \pm 2, \ldots$ (simple poles);
$z = \pm i, \pm 2i, \ldots$ (simple poles).

By Problem 5 (replacing z by πz) we see that:

Residue at $z = 0$ is $\dfrac{-7\pi^3}{45}$.

Residue at $z = n$ $(n = \pm 1, \pm 2, \ldots)$ is

Fig. 7-17

$$\lim_{z \to n} \left\{ \frac{(z-n)}{\sin \pi z} \cdot \frac{\pi \cos \pi z \coth \pi z}{z^3} \right\} = \frac{\coth n\pi}{n^3}$$

Residue at $z = ni$ $(n = \pm 1, \pm 2, \ldots)$ is

$$\lim_{z \to ni} \left\{ \frac{(z-ni)}{\sinh \pi z} \cdot \frac{\pi \cot \pi z \cosh \pi z}{z^3} \right\} = \frac{\coth n\pi}{n^3}$$

Hence by the residue theorem,

$$\oint_{C_N} \frac{\pi \cot \pi z \coth \pi z}{z^3} \, dz = \frac{-7\pi^3}{45} + 4 \sum_{n=1}^{N} \frac{\coth n\pi}{n^3}$$

Taking the limit as $N \to \infty$, we find as in Problem 25 that the integral on the left approaches zero and the required result follows.

Supplementary Problems

RESIDUES AND THE RESIDUE THEOREM

39. For each of the following functions determine the poles and the residues at the poles:

(a) $\dfrac{2z+1}{z^2 - z - 2}$,   (b) $\left(\dfrac{z+1}{z-1} \right)^2$,   (c) $\dfrac{\sin z}{z^2}$,   (d) sech z,   (e) cot z.

Ans. (a) $z = -1, 2;\ 1/3,\ 5/3$

       (b) $z = 1;\ 4$ (d) $z = \frac{1}{2}(2k+1)\pi i;\ (-1)^{k+1} i$ where $k = 0, \pm 1, \pm 2, \ldots$

       (c) $z = 0;\ 1$ (e) $z = k\pi i;\ 1$ where $k = 0, \pm 1, \pm 2, \ldots$

40. Prove that $\displaystyle\oint_C \frac{\cosh z}{z^3} \, dz = \pi i$ if C is the square with vertices at $\pm 2 \pm 2i$.

41. Show that the residue of $(\csc z \operatorname{csch} z)/z^3$ at $z = 0$ is $-1/60$.

42. Evaluate $\displaystyle\oint_C \frac{e^z \, dz}{\cosh z}$ around the circle C defined by $|z| = 5$.   Ans. $8\pi i$

43. Find the zeros and poles of $f(z) = \dfrac{z^2 + 4}{z^3 + 2z^2 + 2z}$ and determine the residues at the poles.

Ans. Zeros: $z = \pm 2i$ Res; at $z = 0$ is 2 Res; at $z = -1 + i$ is $-\frac{1}{2}(1 - 3i)$ Res; at $z = -1 - i$ is $-\frac{1}{2}(1 + 3i)$

44. Evaluate $\displaystyle\oint_C e^{-1/z} \sin (1/z) \, dz$ where C is the circle $|z| = 1$.   Ans. $2\pi i$

45. Let C be a square bounded by $x = \pm 2,\ y = \pm 2$.   Evaluate $\displaystyle\oint_C \frac{\sinh 3z}{(z - \pi i/4)^3} \, dz$.   Ans. $-9\pi \sqrt{2}/2$

46. Evaluate $\displaystyle\oint_C \frac{2z^2 + 5}{(z+2)^3 (z^2 + 4)z^2} \, dz$ where C is (a) $|z - 2i| = 6$, (b) the square with vertices at $1 + i,\ 2 + i,\ 2 + 2i,\ 1 + 2i$.

47. Evaluate $\displaystyle\oint_C \frac{2 + 3 \sin \pi z}{z(z-1)^2} \, dz$ where C is a square having vertices at $3 + 3i,\ 3 - 3i,\ -3 + 3i,\ -3 - 3i$.
Ans. $-6\pi i$

48. Evaluate $\dfrac{1}{2\pi i} \displaystyle\oint_C \frac{e^{zt}}{z(z^2 + 1)} \, dz,\ t > 0$ around the square with vertices at $2 + 2i,\ -2 + 2i,\ -2 - 2i,\ 2 - 2i$.   Ans. $1 - \cos t$

DEFINITE INTEGRALS

49. Prove that $\displaystyle\int_0^\infty \frac{dx}{x^4+1} = \frac{\pi}{2\sqrt{2}}$.

50. Evaluate $\displaystyle\int_0^\infty \frac{dx}{(x^2+1)(x^2+4)^2}$.   Ans. $5\pi/288$

51. Evaluate $\displaystyle\int_0^{2\pi} \frac{\sin 3\theta}{5-3\cos\theta}\, d\theta$.   Ans. 0

52. Evaluate $\displaystyle\int_0^{2\pi} \frac{\cos 3\theta}{5+4\cos\theta}\, d\theta$.   53. Prove that $\displaystyle\int_0^{2\pi} \frac{\cos^2 3\theta}{5-4\cos 2\theta}\, d\theta = \frac{3\pi}{8}$.

54. Prove that if $m > 0$, $\displaystyle\int_0^\infty \frac{\cos mx}{(x^2+1)^2}\, dx = \frac{\pi e^{-m}(1+m)}{4}$.

55. (a) Find the residue of $\dfrac{e^{iz}}{(z^2+1)^5}$ at $z=i$.   (b) Evaluate $\displaystyle\int_0^\infty \frac{\cos x}{(x^2+1)^5}\, dx$.

56. If $a^2 > b^2 + c^2$, prove that $\displaystyle\int_0^{2\pi} \frac{d\theta}{a+b\cos\theta+c\sin\theta} = \frac{2\pi}{\sqrt{a^2-b^2-c^2}}$.

57. Prove that $\displaystyle\int_0^{2\pi} \frac{\cos 3\theta}{(5-3\cos\theta)^4}\, d\theta = \frac{135\pi}{16,384}$.

58. Evaluate $\displaystyle\int_0^\infty \frac{dx}{x^4+x^2+1}$.   Ans. $\pi\sqrt{3}/6$

59. Evaluate $\displaystyle\int_{-\infty}^\infty \frac{dx}{(x^2+4x+5)^2}$.   Ans. $\pi/2$

60. Prove that $\displaystyle\int_0^\infty \frac{\sin^2 x}{x^2}\, dx = \frac{\pi}{2}$.

61. Discuss the validity of the following solution to Problem 19.   Let $u = (1+i)x/\sqrt{2}$ in the result $\displaystyle\int_0^\infty e^{-u^2}\, du = \tfrac{1}{2}\sqrt{\pi}$ to obtain $\displaystyle\int_0^\infty e^{-ix^2}\, dx = \tfrac{1}{2}(1-i)\sqrt{\pi/2}$ from which $\displaystyle\int_0^\infty \cos x^2\, dx =$ $\displaystyle\int_0^\infty \sin x^2\, dx = \tfrac{1}{2}\sqrt{\pi/2}$ on equating real and imaginary parts.

62. Show that $\displaystyle\int_0^\infty \frac{\cos 2\pi x}{x^4+x^2+1}\, dx = \frac{-\pi}{2\sqrt{3}} e^{-\pi/\sqrt{3}}$.

SUMMATION OF SERIES

63. Prove that $\displaystyle\sum_{n=1}^\infty \frac{1}{(n^2+1)^2} = \frac{\pi}{4}\coth\pi + \frac{\pi^2}{4}\operatorname{csch}^2\pi - \frac{1}{2}$.

64. Prove that (a) $\displaystyle\sum_{n=1}^\infty \frac{1}{n^4} = \frac{\pi^4}{90}$,   (b) $\displaystyle\sum_{n=1}^\infty \frac{1}{n^6} = \frac{\pi^6}{945}$.

65. Prove that $\displaystyle\sum_{n=1}^\infty \frac{(-1)^{n-1} n \sin n\theta}{n^2+a^2} = \frac{\pi}{2}\frac{\sinh a\theta}{\sinh a\pi}$, $-\pi < \theta < \pi$.

66. Prove that $\dfrac{1}{1^2} - \dfrac{1}{2^2} + \dfrac{1}{3^2} - \dfrac{1}{4^2} + \cdots = \dfrac{\pi^2}{12}$.

67. Prove that $\displaystyle\sum_{n=-\infty}^\infty \frac{1}{n^4+4a^4} = \frac{\pi}{4a^3}\left\{\frac{\sinh 2\pi a + \sin 2\pi a}{\cosh 2\pi a - \cos 2\pi a}\right\}$.

68. Prove that $\displaystyle\sum_{n=-\infty}^\infty \sum_{m=-\infty}^\infty \frac{1}{(m^2+a^2)(n^2+b^2)} = \frac{\pi^2}{ab}\coth\pi a \coth\pi b$.

MITTAG-LEFFLER'S EXPANSION THEOREM

69. Prove that $\csc z = \dfrac{1}{z} - 2z\left(\dfrac{1}{z^2 - \pi^2} - \dfrac{1}{z^2 - 4\pi^2} + \dfrac{1}{z^2 - 9\pi^2} - \cdots\right).$

70. Prove that $\operatorname{sech} z = \pi\left(\dfrac{1}{(\pi/2)^2 + z^2} - \dfrac{3}{(3\pi/2)^2 + z^2} + \dfrac{5}{(5\pi/2)^2 + z^2} - \cdots\right).$

71. (a) Prove that $\tan z = 2z\left(\dfrac{1}{(\pi/2)^2 - z^2} + \dfrac{1}{(3\pi/2)^2 - z^2} + \dfrac{1}{(5\pi/2)^2 - z^2} + \cdots\right).$

   (b) Use the result in (a) to show that $\dfrac{1}{1^2} + \dfrac{1}{3^2} + \dfrac{1}{5^2} + \dfrac{1}{7^2} + \cdots = \dfrac{\pi^2}{8}.$

72. Prove the expansions (a) 2, (b) 4, (c) 5, (d) 7, (e) 8 on Page 175.

73. Prove that $\displaystyle\sum_{k=1}^{\infty}\dfrac{1}{z^2 + 4k^2\pi^2} = \dfrac{1}{2z}\left\{\dfrac{1}{2} - \dfrac{1}{z} + \dfrac{1}{e^z - 1}\right\}.$

74. Prove that $\dfrac{1}{1^4} + \dfrac{1}{3^4} + \dfrac{1}{5^4} + \dfrac{1}{7^4} + \cdots = \dfrac{\pi^4}{96}.$

MISCELLANEOUS PROBLEMS

75. Prove that Cauchy's theorem and integral formulas can be obtained as special cases of the residue theorem.

76. Prove that the sum of the residues of the function $\dfrac{2z^5 - 4z^2 + 5}{3z^6 - 8z + 10}$ at all the poles is 2/3.

77. If n is a positive integer, prove that $\displaystyle\int_0^{2\pi} e^{\cos\theta}\cos(n\theta - \sin\theta)\,d\theta = \dfrac{2\pi}{n!}.$

78. Evaluate $\displaystyle\oint_C z^3 e^{1/z}\,dz$ around the circle C with equation $|z - 1| = 4$.   *Ans.* 1/24

79. Prove that under suitably stated conditions on the function:

   (a) $\displaystyle\int_0^{2\pi} f(e^{i\theta})\,d\theta = 2\pi\,f(0),$ (b) $\displaystyle\int_0^{2\pi} f(e^{i\theta})\cos\theta\,d\theta = -\pi\,f'(0).$

80. Show that   (a) $\displaystyle\int_0^{2\pi}\cos(\cos\theta)\cosh(\sin\theta)\,d\theta = 2\pi$

   (b) $\displaystyle\int_0^{2\pi} e^{\cos\theta}\cos(\sin\theta)\cos\theta\,d\theta = \pi.$

81. Prove that $\displaystyle\int_0^{\infty}\dfrac{\sin ax}{e^{2\pi x} - 1}\,dx = \dfrac{1}{4}\coth\dfrac{a}{2} - \dfrac{1}{2a}.$

   [*Hint.* Integrate $e^{aiz}/(e^{2\pi z} - 1)$ around a rectangle with vertices at $0,\ R,\ R + i,\ i$ and let $R \to \infty$.]

82. Prove that $\displaystyle\int_0^{\infty}\dfrac{\sin ax}{e^x + 1}\,dx = \dfrac{1}{2a} - \dfrac{\pi}{2\sinh\pi a}.$

83. If a, p and t are positive constants, prove that $\displaystyle\int_{a-i\infty}^{a+i\infty}\dfrac{e^{zt}}{z^2 + p^2}\,dz = \dfrac{\sin pt}{p}.$

84. Prove that $\displaystyle\int_0^{\infty}\dfrac{\ln x}{x^2 + a^2}\,dx = \dfrac{\pi\ln a}{2a}.$

85. If $-\pi < a < \pi$, prove that $\displaystyle\int_{-\infty}^{\infty} e^{i\lambda x}\dfrac{\sinh ax}{\sinh\pi x}\,dx = \dfrac{\sin a}{\cos a + \cosh\lambda}.$

86. Prove that $\displaystyle\int_0^\infty \frac{dx}{(4x^2+\pi^2)\cosh x} = \frac{\ln 2}{2\pi}$.

87. Prove that $\displaystyle (a) \int_0^\infty \frac{\ln x}{x^4+1}\,dx = \frac{-\pi^2\sqrt{2}}{16}$, $\displaystyle (b) \int_0^\infty \frac{(\ln x)^2}{x^4+1}\,dx = \frac{3\pi^3\sqrt{2}}{64}$.

   [*Hint.* Consider $\displaystyle\oint_C \frac{(\ln z)^2}{z^4+1}\,dz$ around a semicircle properly indented at $z=0$.]

88. Evaluate $\displaystyle\int_0^\infty \frac{\ln x}{(x^2+1)^2}\,dx$.   *Ans.* $\frac{1}{4}\pi\ln 2$

89. Prove that if $|a|<1$ and $b>0$, $\displaystyle\int_0^\infty \frac{\sinh ax}{\sinh x}\cos bx\,dx = \frac{\pi}{2}\left(\frac{\sin a\pi}{\cos a\pi + \cosh b\pi}\right)$.

90. Prove that if $-1<p<1$, $\displaystyle\int_0^\infty \frac{\cos px}{\cosh x}\,dx = \frac{\pi}{2\cosh(p\pi/2)}$.

91. Prove that $\displaystyle\int_0^\infty \frac{\ln(1+x)}{1+x^2}\,dx = \frac{\pi\ln 2}{2}$.

92. If $\alpha>0$ and $-\pi/2<\beta<\pi/2$, prove that

   (a) $\displaystyle\int_0^\infty e^{-\alpha x^2\cos\beta}\cos(\alpha x^2\sin\beta)\,dx = \frac{1}{2}\sqrt{\pi/\alpha}\,\cos(\beta/2)$.

   (b) $\displaystyle\int_0^\infty e^{-\alpha x^2\cos\beta}\sin(\alpha x^2\sin\beta)\,dx = \frac{1}{2}\sqrt{\pi/\alpha}\,\sin(\beta/2)$.

93. Prove that $\displaystyle\csc^2 z = \sum_{n=-\infty}^\infty \frac{1}{(z-n\pi)^2}$.

94. If α and p are real and such that $0<|p|<1$ and $0<|\alpha|<\pi$, prove that
$$\int_0^\infty \frac{x^{-p}\,dx}{x^2+2x\cos\alpha+1} = \left(\frac{\pi}{\sin p\pi}\right)\left(\frac{\sin p\alpha}{\sin\alpha}\right)$$

95. Prove that $\displaystyle\int_0^1 \frac{dx}{\sqrt[3]{x^2-x^3}} = \frac{2\pi}{\sqrt{3}}$. [*Hint.* Consider the contour of Fig. 7-18.]

96. Prove the residue theorem for multiply-connected regions.

97. Find sufficient conditions under which the residue theorem (Problem 2) is valid if C encloses infinitely many isolated singularities.

98. Let C be a circle with equation $|z|=4$. Determine the value of the integral
$$\oint_C z^2 \csc\frac{1}{z}\,dz$$
if it exists.

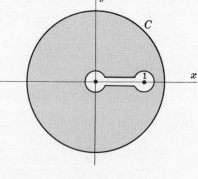

Fig. 7-18

99. Give an analytical proof that $\sin\theta \geq 2\theta/\pi$ for $0\leq\theta\leq\pi/2$.
   [*Hint.* Consider the derivative of $(\sin\theta)/\theta$, showing that it is a decreasing function.]

100. Prove that $\displaystyle\int_0^\infty \frac{x}{\sinh\pi x}\,dx = \frac{1}{4}$.

101. Verify that the integral around Γ in equation (2) of Problem 22 goes to zero as $R\to\infty$.

102. (a) If r is real, prove that $\displaystyle\int_0^\pi \ln(1-2r\cos\theta+r^2)\,d\theta = \begin{cases} 0 & \text{if } |r|\leq 1 \\ \pi\ln r^2 & \text{if } |r|\geq 1 \end{cases}$.

   (b) Use the result in (a) to evaluate $\displaystyle\int_0^{\pi/2} \ln\sin\theta\,d\theta$ (see Problem 23).

103. Complete the proof of Case 2 in Problem 25.

104. If $0 < p < 1$, prove that $\displaystyle\int_0^\infty \frac{x^{-p}}{x-1}\,dx \;=\; \pi \cot p\pi$ in the Cauchy principal value sense.

105. Show that $\displaystyle\sum_{n=-\infty}^{\infty} \frac{1}{n^4 + n^2 + 1} \;=\; \frac{\pi\sqrt{3}}{3}\tanh\left(\frac{\pi\sqrt{3}}{2}\right)$

106. Verify that as $N \to \infty$ the integral on the left of (1) in Problem 29 goes to zero.

107. Prove that $\displaystyle\frac{1}{1^5} - \frac{1}{3^5} + \frac{1}{5^5} - \frac{1}{7^5} + \cdots \;=\; \frac{5\pi^5}{1536}$.

108. Prove the results given on Page 175 for (a) $\displaystyle\sum_{-\infty}^{\infty} f\left(\frac{2n+1}{2}\right)$ and (b) $\displaystyle\sum_{-\infty}^{\infty} (-1)^n f\left(\frac{2n+1}{2}\right)$.

109. If $-\pi \leqq \theta \leqq \pi$, prove that $\displaystyle\sum_{n=1}^{\infty} \frac{(-1)^n \sin n\theta}{n^3} \;=\; \frac{\theta(\pi-\theta)(\pi+\theta)}{12}$.

110. Prove that the function $\cot z - 1/z$ of Problem 34 is bounded on the circles C_N.

111. Show that the second, fourth and sixth integrals in equation (3) of Problem 36 approach zero as $\epsilon \to 0$ and $R \to \infty$.

112. Prove that $\displaystyle\frac{1}{\cosh(\pi/2)} - \frac{1}{3\cosh(3\pi/2)} + \frac{1}{5\cosh(5\pi/2)} - \cdots \;=\; \frac{\pi}{8}$.

113. Prove that $\displaystyle\frac{1}{2\pi i}\int_{a-i\infty}^{a+i\infty} \frac{e^{zt}}{\sqrt{z}}\,dz \;=\; \frac{1}{\sqrt{\pi t}}$ where a and t are any positive constants.

114. Prove that $\displaystyle\sum_{n=1}^{\infty} \frac{\coth n\pi}{n^7} \;=\; \frac{19\pi^7}{56{,}700}$.

115. Prove that $\displaystyle\int_0^\infty \frac{dx}{(x^2+1)\cosh \pi x} \;=\; \frac{4-\pi}{2}$.

116. Prove that $\displaystyle\frac{1}{1^3 \sinh \pi} - \frac{1}{2^3 \sinh 2\pi} + \frac{1}{3^3 \sinh 3\pi} - \cdots \;=\; \frac{\pi^3}{360}$.

117. Prove that if a and t are any positive constants,
$$\frac{1}{2\pi i}\int_{a-i\infty}^{a+i\infty} e^{zt}\cot^{-1} z\,dz \;=\; \frac{\sin t}{t}$$

Chapter 8

Conformal Mapping

TRANSFORMATIONS OR MAPPINGS

The set of equations

$$\begin{aligned} u &= u(x, y) \\ v &= v(x, y) \end{aligned} \Big\}$$

(1)

defines, in general, a *transformation* or *mapping* which establishes a correspondence between points in the uv and xy planes. The equations (1) are called *transformation equations*. If to each point of the uv plane there corresponds one and only one point of the xy plane, and conversely, we speak of a *one to one* transformation or mapping. In such case a set of points in the xy plane [such as a curve or region] is *mapped* into a set of points in the uv plane [curve or region] and conversely. The corresponding sets of points in the two planes are often called *images* of each other.

JACOBIAN OF A TRANSFORMATION

Under the transformation (1) a closed region $\mathcal{R}$ of the xy plane is in general mapped into a closed region $\mathcal{R}'$ of the uv plane. Then if ΔA_{xy} and ΔA_{uv} denote respectively the areas of these regions, we can show that if u and v are continuously differentiable,

$$\lim \frac{\Delta A_{uv}}{\Delta A_{xy}} = \left| \frac{\partial(u, v)}{\partial(x, y)} \right|$$

(2)

where lim denotes the limit as ΔA_{xy} (or ΔA_{uv}) approaches zero and where the determinant

$$\frac{\partial(u, v)}{\partial(x, y)} = \begin{vmatrix} \dfrac{\partial u}{\partial x} & \dfrac{\partial u}{\partial y} \\ \dfrac{\partial v}{\partial x} & \dfrac{\partial v}{\partial y} \end{vmatrix} = \frac{\partial u}{\partial x}\frac{\partial v}{\partial y} - \frac{\partial u}{\partial y}\frac{\partial v}{\partial x}$$

(3)

is called the *Jacobian of the transformation* (1).

If we solve (1) for x and y in terms of u and v, we obtain the transformation $x = x(u, v)$, $y = y(u, v)$, often called the *inverse transformation* corresponding to (1). If x and y are single-valued and continuously differentiable, the Jacobian of this transformation is $\dfrac{\partial(x, y)}{\partial(u, v)}$ and can be shown equal to the reciprocal of $\dfrac{\partial(u, v)}{\partial(x, y)}$ [see Problem 7]. Thus if one Jacobian is different from zero in a region, so also is the other.

Conversely we can show that if u and v are continuously differentiable in a region $\mathcal{R}$ and if the Jacobian $\dfrac{\partial(u, v)}{\partial(x, y)}$ does not vanish in $\mathcal{R}$, then the transformation (1) is one to one.

COMPLEX MAPPING FUNCTIONS

A case of special interest occurs when u and v are real and imaginary parts of an analytic function of a complex variable $z = x + iy$, i.e. $w = u + iv = f(z) = f(x + iy)$.

In such case the Jacobian of the transformation is given by

$$\frac{\partial(u, v)}{\partial(x, y)} \;=\; |f'(z)|^2 \tag{4}$$

(see Problem 5). It follows that the transformation is one to one in regions where $f'(z) \neq 0$. Points where $f'(z) = 0$ are called *critical points*.

CONFORMAL MAPPING

Suppose that under transformation (1) point (x_0, y_0) of the xy plane is mapped into point (u_0, v_0) of the uv plane [Figs. 8-1 and 8-2] while curves C_1 and C_2 [intersecting at (x_0, y_0)] are mapped respectively into curves C_1' and C_2' [intersecting at (u_0, v_0)]. Then if the transformation is such that the angle at (x_0, y_0) between C_1 and C_2 is equal to the angle at (u_0, v_0) between C_1' and C_2' both in magnitude and sense, the transformation or mapping is said to be *conformal* at (x_0, y_0). A mapping which preserves the magnitudes of angles but not necessarily the sense is called *isogonal*.

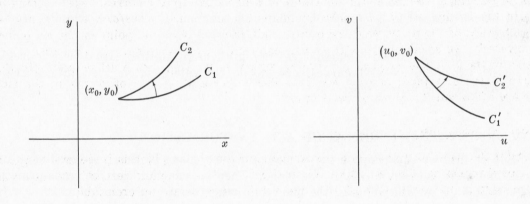

Fig. 8-1   Fig. 8-2

The following theorem is fundamental.

Theorem. If $f(z)$ is analytic and $f'(z) \neq 0$ in a region $\mathcal{R}$, then the mapping $w = f(z)$ is conformal at all points of $\mathcal{R}$.

For conformal mappings or transformations, small figures in the neighborhood of a point z_0 in the z plane map into similar small figures in the w plane and are magnified [or reduced] by an amount given approximately by $|f'(z_0)|^2$, called the *area magnification factor* or simply *magnification factor*. Short distances in the z plane in the neighborhood of z_0 are magnified [or reduced] in the w plane by an amount given approximately by $|f'(z_0)|$, called the *linear magnification factor*. Large figures in the z plane usually map into figures in the w plane which are far from similar.

RIEMANN'S MAPPING THEOREM

Let C [Fig. 8-3] be a simple closed curve in the z plane forming the boundary of a region $\mathcal{R}$. Let C' [Fig. 8-4] be a circle of radius one and center at the origin [the *unit circle*] forming the boundary of region $\mathcal{R}'$ in the w plane. The region $\mathcal{R}'$ is sometimes called the *unit disk*. Then *Riemann's mapping theorem* states that there exists a function $w = f(z)$, analytic in $\mathcal{R}$, which maps each point of $\mathcal{R}$ into a corresponding point of $\mathcal{R}'$ and each point of C into a corresponding point of C', the correspondence being one to one.

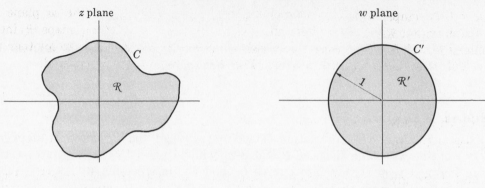

Fig. 8-3   Fig. 8-4

This function $f(z)$ contains three arbitrary real constants which can be determined by making the center of C' correspond to some given point in $\mathcal{R}$, while some point on C' corresponds to a given point on C. It should be noted that while Riemann's mapping theorem demonstrates the *existence* of a mapping function, it does not actually produce this function.

It is possible to extend Riemann's mapping theorem to the case where a region bounded by two simple closed curves, one inside the other, is mapped into a region bounded by two concentric circles.

FIXED OR INVARIANT POINTS OF A TRANSFORMATION

Suppose that we superimpose the w plane on the z plane so that the coordinate axes coincide and there is essentially only one plane. Then we can think of the transformation $w = f(z)$ as taking certain points of the plane into other points. Points for which $z = f(z)$ will however remain fixed, and for this reason we call them the *fixed* or *invariant points* of the transformation.

> **Example:** The fixed or invariant points of the transformation $w = z^2$ are solutions of $z^2 = z$, i.e. $z = 0, 1$.

SOME GENERAL TRANSFORMATIONS

In the following α, β are given complex constants while a, θ_0 are real constants.

1. **Translation.** $w = z + \beta$

   By this transformation, figures in the z plane are *displaced* or *translated* in the direction of vector β.

2. **Rotation.** $w = e^{i\theta_0} z$

   By this transformation, figures in the z plane are rotated through an angle θ_0. If $\theta_0 > 0$ the rotation is counterclockwise, while if $\theta_0 < 0$ the rotation is clockwise.

3. **Stretching.** $w = az$

   By this transformation, figures in the z plane are stretched (or contracted) in the direction z if $a > 1$ (or $0 < a < 1$). We consider contraction as a special case of stretching.

4. **Inversion.** $w = 1/z$

SUCCESSIVE TRANSFORMATIONS

If $w = f_1(\zeta)$ maps region $\mathcal{R}_\zeta$ of the ζ plane into region $\mathcal{R}_w$ of the w plane while $\zeta = f_2(z)$ maps region $\mathcal{R}_z$ of the z plane into region $\mathcal{R}_\zeta$, then $w = f_1[f_2(z)]$ maps $\mathcal{R}_z$ into $\mathcal{R}_w$. The functions f_1 and f_2 define *successive transformations* from one plane to another which are equivalent to a single transformation. These ideas are easily generalized.

THE LINEAR TRANSFORMATION

The transformation

$$ w = \alpha z + \beta \tag{5} $$

where α and β are given complex constants, is called a *linear transformation*. Since we can write (5) in terms of the successive transformations $w = \zeta + \beta$, $\zeta = e^{i\theta_0}\tau$, $\tau = az$ where $\alpha = ae^{i\theta_0}$, we see that a general linear transformation is a combination of the transformations of translation, rotation and stretching.

THE BILINEAR OR FRACTIONAL TRANSFORMATION

The transformation

$$ w = \frac{\alpha z + \beta}{\gamma z + \delta}, \qquad \alpha\delta - \beta\gamma \neq 0 \tag{6} $$

is called a *bilinear* or *fractional transformation*. This transformation can be considered as combinations of the transformations of translation, rotation, stretching and inversion.

The transformation (6) has the property that circles in the z plane are mapped into circles in the w plane, where by circles we include circles of infinite radius which are straight lines. See Problems 14 and 15.

The transformation maps any three distinct points of the z plane into three distinct points of the w plane, one of which may be at infinity.

If z_1, z_2, z_3, z_4 are distinct, then the quantity

$$ \frac{(z_4 - z_1)(z_2 - z_3)}{(z_2 - z_1)(z_4 - z_3)} \tag{7} $$

is called the *cross ratio* of z_1, z_2, z_3, z_4. This ratio is invariant under the bilinear transformation, and this property can be used in obtaining specific bilinear transformations mapping three points into three other points.

MAPPING OF A HALF PLANE ON TO A CIRCLE

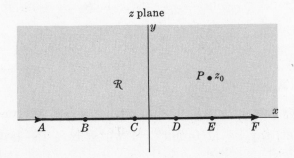

Fig. 8-5

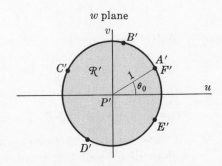

Fig. 8-6

Let z_0 be any point P in the upper half of the z plane denoted by $\mathcal{R}$ in Fig. 8-5 above. Then the transformation

$$w = e^{i\theta_0}\left(\frac{z - z_0}{z - \bar{z}_0}\right) \tag{8}$$

maps this upper half plane in a one to one manner on to the interior $\mathcal{R}'$ of the unit circle $|w| = 1$, and conversely. Each point of the x axis is mapped on to the boundary of the circle. The constant θ_0 can be determined by making one particular point of the x axis correspond to a given point on the circle.

In the above figures we have used the convention that unprimed points such as A, B, C, etc., in the z plane correspond to primed points A', B', C', etc., in the w plane. Also, in the case where points are at infinity we indicate this by an arrow such as at A and F in Fig. 8-5 which correspond respectively to A' and F' (the same point) in Fig. 8-6 above. As point z moves on the boundary of $\mathcal{R}$ [i.e. the real axis] from $-\infty$ (point A) to $+\infty$ (point F), w moves counterclockwise along the unit circle from A' back to A'.

THE SCHWARZ-CHRISTOFFEL TRANSFORMATION

Consider a polygon [Fig. 8-7] in the w plane having vertices at $w_1, w_2, \ldots, w_n$ with corresponding interior angles $\alpha_1, \alpha_2, \ldots, \alpha_n$ respectively. Let the points $w_1, w_2, \ldots, w_n$ map respectively into points $x_1, x_2, \ldots, x_n$ on the real axis of the z plane [Fig. 8-8].

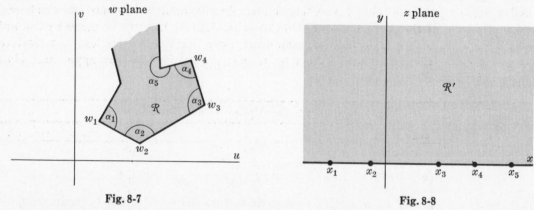

Fig. 8-7   Fig. 8-8

A transformation which maps the interior $\mathcal{R}$ of the polygon of the w plane on to the upper half $\mathcal{R}'$ of the z plane and the boundary of the polygon on to the real axis is given by

$$\frac{dw}{dz} = A\,(z - x_1)^{\alpha_1/\pi - 1}\,(z - x_2)^{\alpha_2/\pi - 1} \cdots (z - x_n)^{\alpha_n/\pi - 1} \tag{9}$$

or

$$w = A \int (z - x_1)^{\alpha_1/\pi - 1}\,(z - x_2)^{\alpha_2/\pi - 1} \cdots (z - x_n)^{\alpha_n/\pi - 1}\,dz \;+\; B \tag{10}$$

where A and B are complex constants.

The following facts should be noted:

1. Any three of the points $x_1, x_2, \ldots, x_n$ can be chosen at will.

2. The constants A and B determine the size, orientation and position of the polygon.

3. It is convenient to choose one point, say x_n, at infinity in which case the last factor of (9) and (10) involving x_n is not present.

4. Infinite open polygons can be considered as limiting cases of closed polygons.

TRANSFORMATIONS OF BOUNDARIES IN PARAMETRIC FORM

Suppose that in the z plane a curve C [Fig. 8-9], which may or may not be closed, has parametric equations given by

$$x = F(t), \quad y = G(t) \qquad (11)$$

where we assume that F and G are continuously differentiable. Then the transformation

$$z = F(w) + iG(w) \qquad (12)$$

maps curve C on to the real axis C' of the w plane [Fig. 8-10].

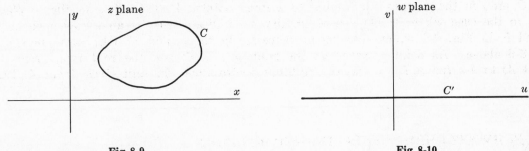

Fig. 8-9   Fig. 8-10

SOME SPECIAL MAPPINGS

For reference purposes we list here some special mappings which are useful in practice. For convenience we have listed separately the mapping functions which map the given region $\mathcal{R}$ of the w or z plane on to the upper half of the z or w plane or the unit circle in the z or w plane, depending on which mapping function is simpler. As we have already seen there exists a transformation [equation (8)] which maps the upper half plane on to the unit circle.

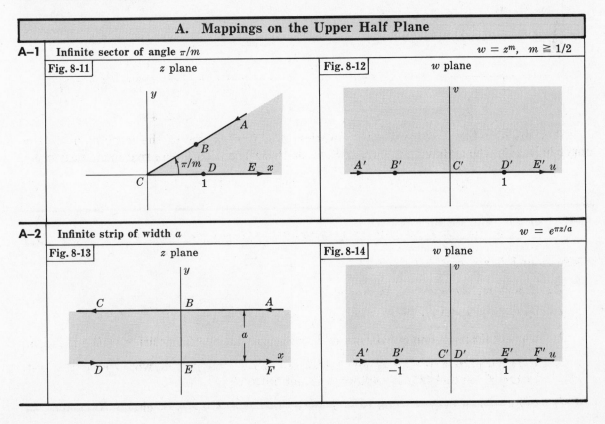

A–3 | Semi-Infinite strip of width a

(a) $w = \sin \dfrac{\pi z}{a}$

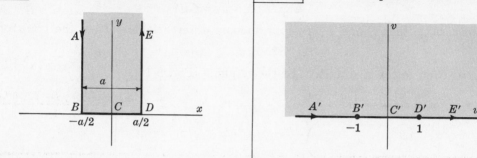

Fig. 8-15 z plane   Fig. 8-16 w plane

(b) $w = \cos \dfrac{\pi z}{a}$

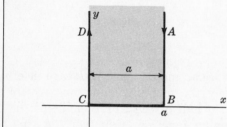
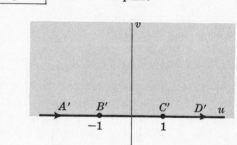

Fig. 8-17 z plane   Fig. 8-18 w plane

(c) $w = \cosh \dfrac{\pi z}{a}$

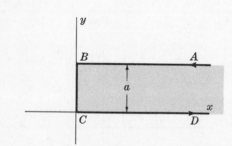
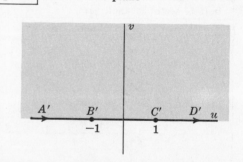

Fig. 8-19 z plane   Fig. 8-20 w plane

A–4 | Half plane with semicircle removed $w = \dfrac{a}{2}\left(z + \dfrac{1}{z}\right)$

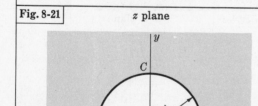
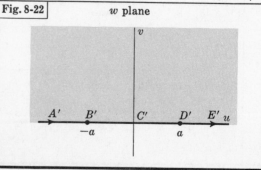

Fig. 8-21 z plane   Fig. 8-22 w plane

A–5   Semicircle $w = \left(\dfrac{1+z}{1-z}\right)^2$

Fig. 8-23 z plane	**Fig. 8-24** w plane
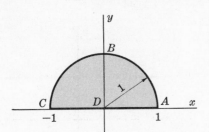	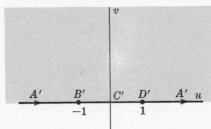

A–6   Sector of a circle $w = \left(\dfrac{1+z^m}{1-z^m}\right)^2, \quad m \geqq \frac{1}{2}$

Fig. 8-25 z plane	**Fig. 8-26** w plane
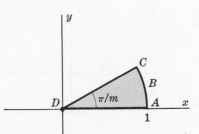	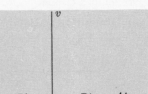

A–7   Lens-shaped region of angle π/m $w = e^{2mi\cot^{-1} p}\left(\dfrac{z+1}{z-1}\right)^m, \quad m \geqq 2$
[ABC and CDA are circular arcs.]

Fig. 8-27 z plane	**Fig. 8-28** w plane
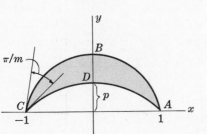	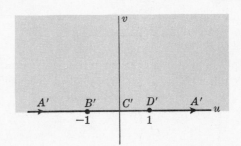

A–8   Half plane with circle removed $w = \coth(\pi/z)$

Fig. 8-29 z plane	**Fig. 8-30** w plane
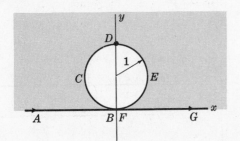	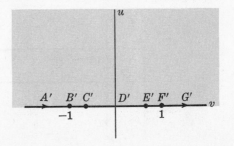

A–9   **Exterior of parabola** $y^2 = 4p(p - x)$ $w = i(\sqrt{z} - \sqrt{p})$

| Fig. 8-31 | z plane | Fig. 8-32 | w plane |

A–10   **Interior of the parabola** $y^2 = 4p(p - x)$ $w = e^{\pi i \sqrt{z/p}}$

| Fig. 8-33 | z plane | Fig. 8-34 | w plane |

A–11   **Plane with two semi-infinite parallel cuts** $w = -\pi i + 2 \ln z - z^2$

| Fig. 8-35 | w plane | Fig. 8-36 | z plane |

A–12   **Channel with right angle bend** $w = \dfrac{2}{\pi}\{\tanh^{-1} p\sqrt{z} - p \tan^{-1} \sqrt{z}\}$

| Fig. 8-37 | w plane | Fig. 8-38 | z plane |

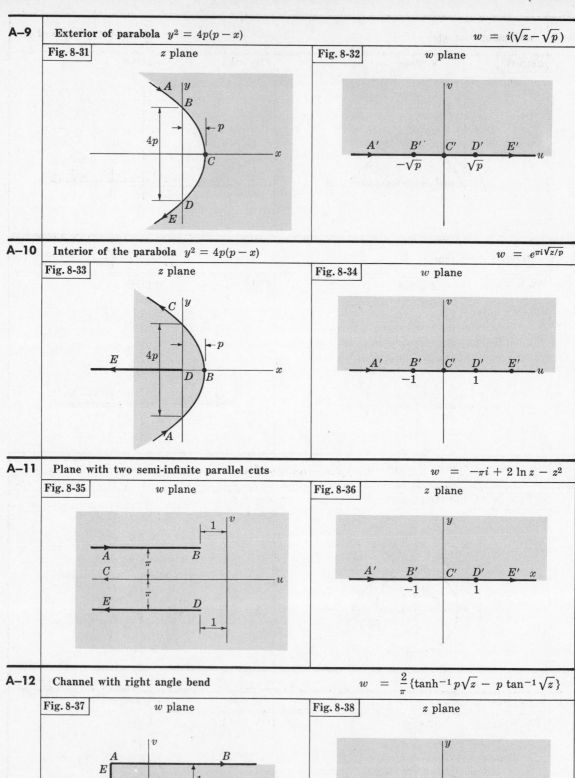

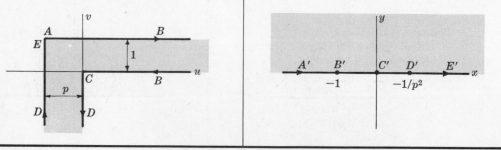

A–13 | Interior of triangle

$$w = \int_0^z t^{\alpha/\pi-1}(1-t)^{\beta/\pi-1}\,dt$$

Fig. 8-39   w plane

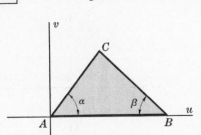

Fig. 8-40   z plane

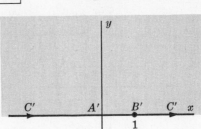

A–14 | Interior of rectangle

$$w = \int_0^z \frac{dt}{\sqrt{(1-t^2)(1-k^2t^2)}}, \quad 0 < k < 1$$

Fig. 8-41   w plane

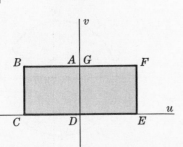

Fig. 8-42   z plane

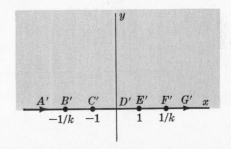

B. Mappings on the Unit Circle

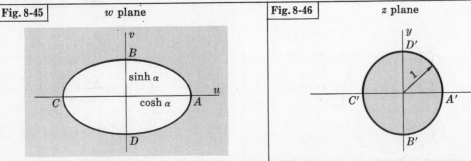

B–1 | Exterior of unit circle

$$w = 1/z$$

Fig. 8-43   w plane

Fig. 8-44   z plane

B–2 | Exterior of ellipse

$$w = \tfrac{1}{2}(ze^{-\alpha} + z^{-1}e^{\alpha})$$

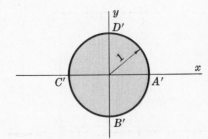

Fig. 8-45   w plane

Fig. 8-46   z plane

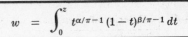
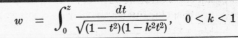

| **B–3** | Exterior of parabola $y^2 = 4p(p-x)$ | $w = 2\sqrt{\dfrac{p}{z}} - 1$ |

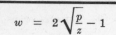

Fig. 8-47 — z plane

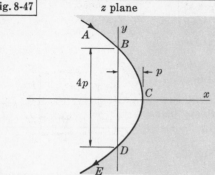

Fig. 8-48 — w plane

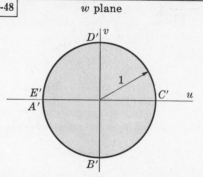

| **B–4** | Interior of parabola $y^2 = 4p(p-x)$ | $w = \tan^2\dfrac{\pi}{4}\sqrt{\dfrac{z}{p}}$ |

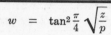

Fig. 8-49 — z plane

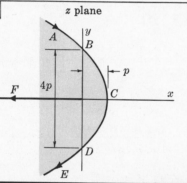

Fig. 8-50 — w plane

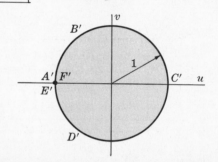

C. Miscellaneous Mappings

| **C–1** | Semi-infinite strip of width a on to quarter plane | $w = \sin\dfrac{\pi z}{2a}$ |

Fig. 8-51 — z plane

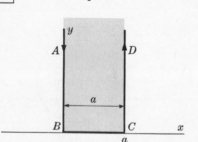

Fig. 8-52 — w plane

| **C–2** | Interior of cardioid on to circle | $w = z^2$ |

Fig. 8-53 — w plane

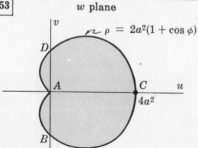

Fig. 8-54 — z plane

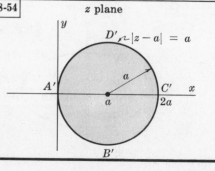

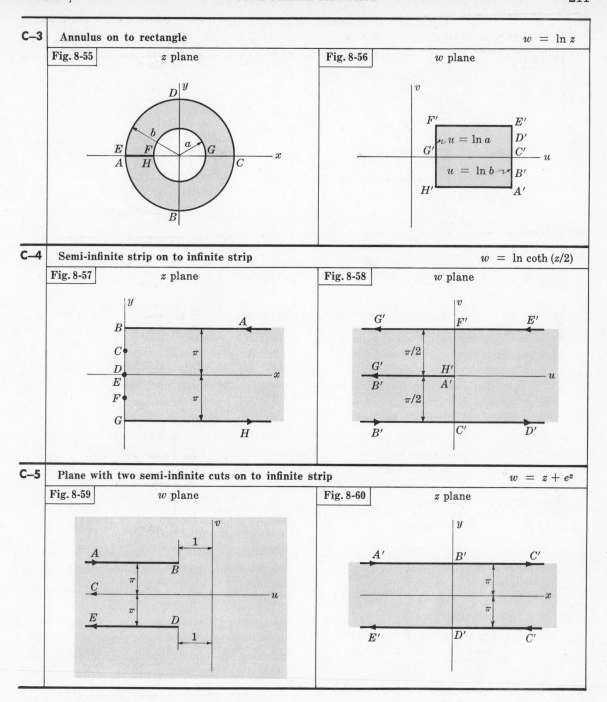

C–3 | Annulus on to rectangle | $w = \ln z$

Fig. 8-55 — z plane

Fig. 8-56 — w plane

$u = \ln a$

$u = \ln b$

C–4 | Semi-infinite strip on to infinite strip | $w = \ln \coth(z/2)$

Fig. 8-57 — z plane

Fig. 8-58 — w plane

C–5 | Plane with two semi-infinite cuts on to infinite strip | $w = z + e^z$

Fig. 8-59 — w plane

Fig. 8-60 — z plane

Solved Problems

TRANSFORMATIONS

1. Let the rectangular region $\mathcal{R}$ [Fig. 8-61 below] in the z plane be bounded by $x=0$, $y=0$, $x=2$, $y=1$. Determine the region $\mathcal{R}'$ of the w plane into which $\mathcal{R}$ is mapped under the transformations:

(a) $w = z + (1-2i)$, (b) $w = \sqrt{2}\,e^{\pi i/4}\,z$, (c) $w = \sqrt{2}\,e^{\pi i/4}\,z + (1-2i)$.

(a) If $w = z + (1-2i)$, then $u + iv = x + iy + 1 - 2i = (x+1) + i(y-2)$ and $u = x+1$, $v = y-2$.

Line $x=0$ is mapped into $u=1$; $y=0$ into $v=-2$; $x=2$ into $u=3$; $y=1$ into $v=-1$ [Fig. 8-62]. Similarly, we can show that each point of $\mathcal{R}$ is mapped into one and only one point of $\mathcal{R}'$ and conversely.

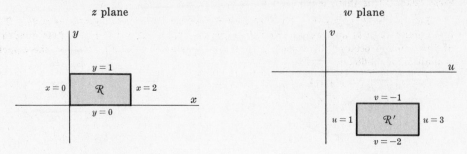

Fig. 8-61   Fig. 8-62

The transformation or mapping accomplishes a *translation* of the rectangle. In general, $w = z + \beta$ accomplishes a translation of any region.

(b) If $w = \sqrt{2}\, e^{\pi i/4}\, z$, then $u + iv = (1+i)(x+iy) = x-y+i(x+y)$ and $u = x-y$, $v = x+y$.

Line $x=0$ is mapped into $u=-y$, $v=y$ or $u=-v$; $y=0$ into $u=x$, $v=x$ or $u=v$; $x=2$ into $u=2-y$, $v=2+y$ or $u+v=4$; $y=1$ into $u=x-1$, $v=x+1$ or $v-u=2$ [Fig. 8-64].

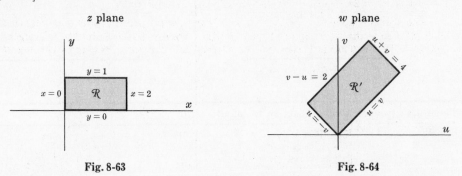

Fig. 8-63   Fig. 8-64

The mapping accomplishes a *rotation* of $\mathcal{R}$ (through angle $\pi/4$ or $45°$) and a *stretching* of lengths (of magnitude $\sqrt{2}$). In general the transformation $w = \alpha z$ accomplishes a rotation and stretching of a region.

(c) If $w = \sqrt{2}\, e^{\pi i/4}\, z + (1-2i)$, then $u + iv = (1+i)(x+iy) + 1 - 2i$ and $u = x-y+1$, $v = x+y-2$.

The lines $x=0$, $y=0$, $x=2$, $y=1$ are mapped respectively into $u+v=-1$, $u-v=3$, $u+v=3$, $u-v=1$ [Fig. 8-66].

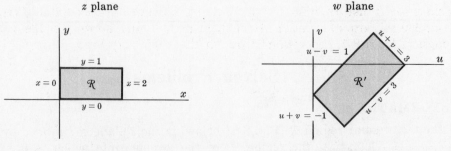

Fig. 8-65   Fig. 8-66

The mapping accomplishes a rotation and stretching as in (b) and a subsequent translation. In general the transformation $w = \alpha z + \beta$ accomplishes a rotation, stretching and translation. This can be considered as two successive mappings $w = \alpha z_1$ (rotation and stretching) and $z_1 = z + \beta/\alpha$ (translation).

2. Determine the region of the w plane into which each of the following is mapped by the transformation $w = z^2$.

(a) **First quadrant of the z plane.**

   Let $z = re^{i\theta}$, $w = \rho e^{i\phi}$. Then if $w = z^2$, $\rho e^{i\phi} = r^2 e^{2i\theta}$ and $\rho = r^2$, $\phi = 2\theta$. Thus points in the z plane at (r, θ) are rotated through angle 2θ. Since all points in the first quadrant [Fig. 8-67] of the z plane occupy the region $0 \leqq \theta \leqq \pi/2$, they map into $0 \leqq \phi \leqq \pi$ or the upper half of the w plane [Fig. 8-68].

Fig. 8-67 Fig. 8-68

(b) **Region bounded by $x = 1$, $y = 1$ and $x + y = 1$.**

   Since $w = z^2$ is equivalent to $u + iv = (x + iy)^2 = x^2 - y^2 + 2ixy$, we see that $u = x^2 - y^2$, $v = 2xy$. Then line $x = 1$ maps into $u = 1 - y^2$, $v = 2y$ or $u = 1 - v^2/4$; line $y = 1$ into $u = x^2 - 1$, $v = 2x$ or $u = v^2/4 - 1$; line $x + y = 1$ or $y = 1 - x$ into $u = x^2 - (1-x)^2 = 2x - 1$, $v = 2x(1-x) = 2x - 2x^2$ or $v = \frac{1}{2}(1 - u^2)$ on eliminating x.

   The regions appear shaded in Figs. 8-69 and 8-70 below where points A, B, C map into A', B', C'. Note that the angles of triangle ABC are equal respectively to the angles of curvilinear triangle $A'B'C'$. This is a consequence of the fact that the mapping is *conformal*.

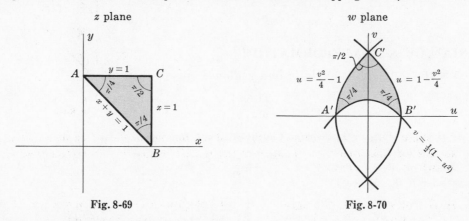

Fig. 8-69 Fig. 8-70

CONFORMAL TRANSFORMATIONS

3. Consider the transformation $w = f(z)$ where $f(z)$ is analytic at z_0 and $f'(z_0) \neq 0$. Prove that under this transformation the tangent at z_0 to any curve C in the z plane passing through z_0 [Fig. 8-71] is rotated through the angle $\arg f'(z_0)$.

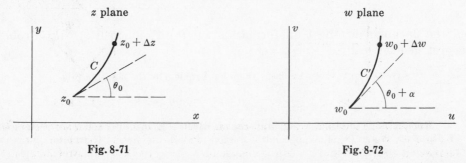

Fig. 8-71 Fig. 8-72

As a point moves from z_0 to $z_0 + \Delta z$ along C, the image point moves along C' in the w plane from w_0 to $w_0 + \Delta w$. If the parameter used to describe the curve is t, then corresponding to the path $z = z(t)$ [or $x = x(t)$, $y = y(t)$] in the z plane, we have the path $w = w(t)$ [or $u = u(t)$, $v = v(t)$] in the w plane.

The derivatives dz/dt and dw/dt represent tangent vectors to corresponding points on C and C'.

Now $\dfrac{dw}{dt} = \dfrac{dw}{dz} \cdot \dfrac{dz}{dt} = f'(z) \dfrac{dz}{dt}$ and, in particular at z_0 and w_0,

$$\left. \frac{dw}{dt} \right|_{w=w_0} = f'(z_0) \left. \frac{dz}{dt} \right|_{z=z_0} \tag{1}$$

provided $f(z)$ is analytic at $z = z_0$. Writing $\left. \dfrac{dw}{dt} \right|_{w=w_0} = \rho_0 e^{i\phi_0}$, $f'(z) = R e^{i\alpha}$, $\left. \dfrac{dz}{dt} \right|_{z=z_0} = r_0 e^{i\theta_0}$, we have from (1)

$$\rho_0 e^{i\phi_0} = R r_0 e^{i(\theta_0 + \alpha)} \tag{2}$$

so that, as required,

$$\phi_0 = \theta_0 + \alpha = \theta_0 + \arg f'(z_0) \tag{3}$$

Note that if $f'(z_0) = 0$, then α is indeterminate. Points where $f'(z) = 0$ are called *critical points*.

4. Prove that the angle between two curves C_1 and C_2 passing through the point z_0 in the z plane [see Figs. 8-1 and 8-2, Page 201] is preserved [in magnitude and sense] under the transformation $w = f(z)$, i.e. the mapping is conformal, if $f(z)$ is analytic at z_0 and $f'(z_0) \neq 0$.

By Problem 3 each curve is rotated through the angle $\arg f'(z_0)$. Hence the angle between the curves must be preserved, both in magnitude and sense, in the mapping.

JACOBIAN OF A TRANSFORMATION

5. If $w = f(z) = u + iv$ is analytic in a region $\mathcal{R}$, prove that

$$\frac{\partial(u, v)}{\partial(x, y)} = |f'(z)|^2$$

If $f(z)$ is analytic in $\mathcal{R}$, then the Cauchy-Riemann equations

$$\frac{\partial u}{\partial x} = \frac{\partial v}{\partial y}, \quad \frac{\partial v}{\partial x} = -\frac{\partial u}{\partial y}$$

are satisfied in $\mathcal{R}$.   Hence

$$\frac{\partial(u, v)}{\partial(x, y)} = \begin{vmatrix} \dfrac{\partial u}{\partial x} & \dfrac{\partial u}{\partial y} \\[2mm] \dfrac{\partial v}{\partial x} & \dfrac{\partial v}{\partial y} \end{vmatrix} = \begin{vmatrix} \dfrac{\partial u}{\partial x} & \dfrac{\partial u}{\partial y} \\[2mm] -\dfrac{\partial u}{\partial y} & \dfrac{\partial u}{\partial x} \end{vmatrix} = \left(\frac{\partial u}{\partial x} \right)^2 + \left(\frac{\partial u}{\partial y} \right)^2$$

$$= \left| \frac{\partial u}{\partial x} + i \frac{\partial u}{\partial y} \right|^2 = |f'(z)|^2$$

using Problem 5, Chapter 3, Page 72.

6. Find the Jacobian of the transformation in (a) Problem 1(c), (b) Problem 2 and interpret geometrically.

(a) If $w = f(z) = \sqrt{2} \, e^{\pi i/4} z + (1 - 2i)$,   then by Problem 5 the Jacobian is

$$\frac{\partial(u, v)}{\partial(x, y)} = |f'(z)|^2 = |\sqrt{2} \, e^{\pi i/4}|^2 = 2$$

Geometrically this shows that any region in the z plane [in particular rectangular region $\mathcal{R}$ of Fig. 8-65, Page 212] is mapped into a region of twice the area. The factor $|f'(z)|^2 = 2$ is called the *magnification factor*.

Another method. The transformation is equivalent to $u = x - y$, $v = x + y$ and so

$$\frac{\partial(u, v)}{\partial(x, y)} = \begin{vmatrix} \dfrac{\partial u}{\partial x} & \dfrac{\partial u}{\partial y} \\[2mm] \dfrac{\partial v}{\partial x} & \dfrac{\partial v}{\partial y} \end{vmatrix} = \begin{vmatrix} 1 & -1 \\ 1 & 1 \end{vmatrix} = 2$$

(b) If $w = f(z) = z^2$, then

$$\frac{\partial(u, v)}{\partial(x, y)} = |f'(z)|^2 = |2z|^2 = |2x + 2iy|^2 = 4(x^2 + y^2)$$

Geometrically, a small region in the z plane having area A and at approximate distance r from the origin would be mapped into a region of the w plane having area $4r^2A$. Thus regions far from the origin would be mapped into regions of greater area than similar regions near the origin.

Note that at the critical point $z = 0$ the Jacobian is zero. At this point the transformation is not conformal.

7. **Prove that** $\dfrac{\partial(u, v)}{\partial(x, y)} \cdot \dfrac{\partial(x, y)}{\partial(u, v)} = 1.$

Corresponding to the transformation *(1)* $u = u(x, y)$, $v = v(x, y)$, with Jacobian $\dfrac{\partial(u, v)}{\partial(x, y)}$, we have the inverse transformation *(2)* $x = x(u, v)$, $y = y(u, v)$, with Jacobian $\dfrac{\partial(x, y)}{\partial(u, v)}$.

From *(1)*, $\qquad du = \dfrac{\partial u}{\partial x} dx + \dfrac{\partial u}{\partial y} dy, \qquad dv = \dfrac{\partial v}{\partial x} dx + \dfrac{\partial v}{\partial y} dy.$

From *(2)*, $\qquad dx = \dfrac{\partial x}{\partial u} du + \dfrac{\partial x}{\partial v} dv, \qquad dy = \dfrac{\partial y}{\partial u} du + \dfrac{\partial y}{\partial v} dv.$

Hence, $\qquad du = \dfrac{\partial u}{\partial x}\left\{ \dfrac{\partial x}{\partial u} du + \dfrac{\partial x}{\partial v} dv \right\} + \dfrac{\partial u}{\partial y}\left\{ \dfrac{\partial y}{\partial u} du + \dfrac{\partial y}{\partial v} dv \right\}$

$$= \left\{ \frac{\partial u}{\partial x}\frac{\partial x}{\partial u} + \frac{\partial u}{\partial y}\frac{\partial y}{\partial u} \right\} du + \left\{ \frac{\partial u}{\partial x}\frac{\partial x}{\partial v} + \frac{\partial u}{\partial y}\frac{\partial y}{\partial v} \right\} dv$$

from which $\qquad \dfrac{\partial u}{\partial x}\dfrac{\partial x}{\partial u} + \dfrac{\partial u}{\partial y}\dfrac{\partial y}{\partial u} = 1, \qquad \dfrac{\partial u}{\partial x}\dfrac{\partial x}{\partial v} + \dfrac{\partial u}{\partial y}\dfrac{\partial y}{\partial v} = 0 \qquad\qquad (3)$

Similarly we find $\qquad \dfrac{\partial v}{\partial x}\dfrac{\partial x}{\partial v} + \dfrac{\partial v}{\partial y}\dfrac{\partial y}{\partial v} = 1, \qquad \dfrac{\partial v}{\partial x}\dfrac{\partial x}{\partial u} + \dfrac{\partial v}{\partial y}\dfrac{\partial y}{\partial u} = 0 \qquad\qquad (4)$

Using *(3)* and *(4)* and the rule for products of determinants (see Problem 94), we have

$$\frac{\partial(u, v)}{\partial(x, y)} \cdot \frac{\partial(x, y)}{\partial(u, v)} = \begin{vmatrix} \dfrac{\partial u}{\partial x} & \dfrac{\partial u}{\partial y} \\[2mm] \dfrac{\partial v}{\partial x} & \dfrac{\partial v}{\partial y} \end{vmatrix} \cdot \begin{vmatrix} \dfrac{\partial x}{\partial u} & \dfrac{\partial x}{\partial v} \\[2mm] \dfrac{\partial y}{\partial u} & \dfrac{\partial y}{\partial v} \end{vmatrix}$$

$$= \begin{vmatrix} \dfrac{\partial u}{\partial x}\dfrac{\partial x}{\partial u} + \dfrac{\partial u}{\partial y}\dfrac{\partial y}{\partial u} & \dfrac{\partial u}{\partial x}\dfrac{\partial x}{\partial v} + \dfrac{\partial u}{\partial y}\dfrac{\partial y}{\partial v} \\[2mm] \dfrac{\partial v}{\partial x}\dfrac{\partial x}{\partial u} + \dfrac{\partial v}{\partial y}\dfrac{\partial y}{\partial u} & \dfrac{\partial v}{\partial x}\dfrac{\partial x}{\partial v} + \dfrac{\partial v}{\partial y}\dfrac{\partial y}{\partial v} \end{vmatrix} = \begin{vmatrix} 1 & 0 \\ 0 & 1 \end{vmatrix} = 1$$

8. **Discuss Problem 7 if u and v are real and imaginary parts of an analytic function $f(z)$.**

In this case $\dfrac{\partial(u, v)}{\partial(x, y)} = |f'(z)|^2$ by Problem 5. If the inverse to $w = f(z)$ is $z = g(w)$ assumed single-valued and analytic, then $\dfrac{\partial(x, y)}{\partial(u, v)} = |g'(w)|^2$. The result of Problem 7 is a consequence of the fact that

$$|f'(z)|^2\,|g'(w)|^2 \;=\; \left|\frac{dw}{dz}\right|^2 \cdot \left|\frac{dz}{dw}\right|^2 \;=\; 1$$

since $dw/dz = 1/(dz/dw)$.

BILINEAR OR FRACTIONAL TRANSFORMATIONS

9. Find a bilinear transformation which maps points z_1, z_2, z_3 of the z plane into points w_1, w_2, w_3 of the w plane respectively.

If w_k corresponds to z_k, $k = 1, 2, 3$, we have

$$w - w_k \;=\; \frac{\alpha z + \beta}{\gamma z + \delta} - \frac{\alpha z_k + \beta}{\gamma z_k + \delta} \;=\; \frac{(\alpha\delta - \beta\gamma)(z - z_k)}{(\gamma z + \delta)(\gamma z_k + \delta)}$$

Then
$$w - w_1 = \frac{(\alpha\delta - \beta\gamma)(z - z_1)}{(\gamma z + \delta)(\gamma z_1 + \delta)}, \qquad w - w_3 = \frac{(\alpha\delta - \beta\gamma)(z - z_3)}{(\gamma z + \delta)(\gamma z_3 + \delta)} \qquad (1)$$

Replacing w by w_2, and z by z_2,

$$w_2 - w_1 = \frac{(\alpha\delta - \beta\gamma)(z_2 - z_1)}{(\gamma z_2 + \delta)(\gamma z_1 + \delta)}, \qquad w_2 - w_3 = \frac{(\alpha\delta - \beta\gamma)(z_2 - z_3)}{(\gamma z_2 + \delta)(\gamma z_3 + \delta)} \qquad (2)$$

By division of (1) and (2), assuming $\alpha\delta - \beta\gamma \neq 0$,

$$\frac{(w - w_1)(w_2 - w_3)}{(w - w_3)(w_2 - w_1)} \;=\; \frac{(z - z_1)(z_2 - z_3)}{(z - z_3)(z_2 - z_1)} \qquad (3)$$

Solving for w in terms of z gives the required transformation. The right hand side of (3) is called the *cross ratio* of z_1, z_2, z_3 and z.

10. Find a bilinear transformation which maps points $z = 0, -i, -1$ into $w = i, 1, 0$ respectively.

Method 1. Since $w = \dfrac{\alpha z + \beta}{\gamma z + \delta}$, we have

$$(1)\quad i = \frac{\alpha(0) + \beta}{\gamma(0) + \delta}, \qquad (2)\quad 1 = \frac{\alpha(-i) + \beta}{\gamma(-i) + \delta}, \qquad (3)\quad 0 = \frac{\alpha(-1) + \beta}{\gamma(-1) + \delta}$$

From (3), $\beta = \alpha$. From (1), $\delta = \beta/i = -i\alpha$. From (2), $\gamma = i\alpha$. Then

$$w \;=\; \frac{\alpha z + \alpha}{i\alpha z - i\alpha} \;=\; \frac{1}{i}\left(\frac{z+1}{z-1}\right) \;=\; -i\left(\frac{z+1}{z-1}\right)$$

Method 2. Use Problem 9. Then

$$\frac{(w-i)(1-0)}{(w-0)(1-i)} \;=\; \frac{(z-0)(-i+1)}{(z+1)(-i-0)}. \qquad \text{Solving,} \quad w = -i\left(\frac{z+1}{z-1}\right).$$

11. If z_0 is in the upper half of the z plane, show that the bilinear transformation $w = e^{i\theta_0}\left(\dfrac{z - z_0}{z - \bar z_0}\right)$ maps the upper half of the z plane into the interior of the unit circle in the w plane, i.e. $|w| \leqq 1$.

We have
$$|w| \;=\; \left| e^{i\theta_0}\left(\frac{z - z_0}{z - \bar z_0}\right) \right| \;=\; \left| \frac{z - z_0}{z - \bar z_0} \right|$$

From Fig. 8-73 if z is in the upper half plane, $|z - z_0| \leqq |z - \bar z_0|$, the equality holding if and only if z is on the x axis. Hence $|w| \leqq 1$, as required.

The transformation can also be derived directly (see Problem 61).

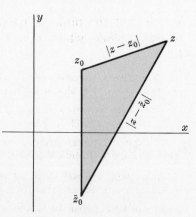

Fig. 8-73

12. Find a bilinear transformation which maps the upper half of the z plane into the unit circle in the w plane in such a way that $z=i$ is mapped into $w=0$ while the point at infinity is mapped into $w=-1$.

We have $w=0$ corresponding to $z=i$, and $w=-1$ corresponding to $z=\infty$. Then from $w = e^{i\theta_0}\left(\dfrac{z-z_0}{z-\bar{z}_0}\right)$ we have $0 = e^{i\theta_0}\left(\dfrac{i-z_0}{i-\bar{z}_0}\right)$ so that $z_0 = i$. Corresponding to $z=\infty$ we have $w = e^{i\theta_0} = -1$. Hence the required transformation is

$$w = (-1)\left(\frac{z-i}{z+i}\right) = \frac{i-z}{i+z}$$

The situation is described graphically in Figures 8-74 and 8-75.

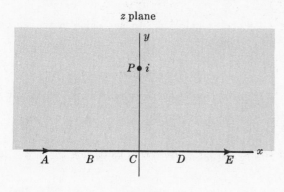

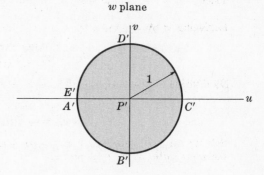

Fig. 8-74   Fig. 8-75

13. Find the fixed or invariant points of the transformation $w = \dfrac{2z-5}{z+4}$.

The fixed points are solutions to $z = \dfrac{2z-5}{z+4}$ or $z^2+2z+5 = 0$, i.e. $z = -1 \pm 2i$.

14. Prove that the bilinear transformation can be considered as a combination of the transformations of translation, rotation, stretching and inversion.

By division, $w = \dfrac{\alpha z + \beta}{\gamma z + \delta} = \dfrac{\alpha}{\gamma} + \dfrac{\beta\gamma - \alpha\delta}{\gamma(\gamma z + \delta)} = \lambda + \dfrac{\mu}{z+\nu}$ where $\lambda = \alpha/\gamma$, $\mu = (\beta\gamma - \alpha\delta)/\gamma^2$ and $\nu = \delta/\gamma$ are constants. The transformation is equivalent to $\zeta = z+\nu$, $\tau = 1/\zeta$ and $w = \lambda + \mu\tau$ which are combinations of the transformations of translation, rotation, stretching and inversion.

15. Prove that the bilinear transformation transforms circles of the z plane into circles of the w plane, where by circles we include circles of infinite radius which are straight lines.

The general equation of a circle in the z plane is by Problem 44, Chapter 1, $Az\bar{z} + Bz + \bar{B}\bar{z} + C = 0$, where $A > 0$, $C > 0$ and B is complex. If $A = 0$ the circle reduces to a straight line.

Under the transformation of inversion, $w = 1/z$ or $z = 1/w$, this equation becomes $Cw\bar{w} + \bar{B}w + B\bar{w} + A = 0$, a circle in the w plane.

Under the transformation of rotation and stretching, $w = az$ or $z = w/a$, this equation becomes $Aw\bar{w} + (B\bar{a})w + (\bar{B}a)\bar{w} + Ca\bar{a} = 0$, also a circle.

Similarly we can show either analytically or geometrically that under the transformation of translation, circles are transformed into circles.

Since by Problem 14 a bilinear transformation can be considered as a combination of translation, rotation, stretching and inversion, the required result follows.

SPECIAL MAPPING FUNCTIONS

16. Verify the entries   (a) A-2, Page 205   (b) A-4, Page 206   (c) B-1, Page 209.

(a) Refer to Figs. 8-13 and 8-14, Page 205.

If $z = x + iy$, then

$$w = u + iv = e^{\pi z/a} = e^{\pi(x+iy)/a} = e^{\pi x/a}(\cos \pi y/a + i \sin \pi y/a)$$

or $u = e^{\pi x/a} \cos \pi y/a, \ v = e^{\pi x/a} \sin \pi y/a$.

The line $y = 0$ [the real axis in the z plane; DEF in Fig. 8-13] maps into $u = e^{\pi x/a}, \ v = 0$ [the positive real axis in the w plane; $D'E'F'$ in Fig. 8-14]. The origin $E \ [z=0]$ maps into $E' \ [w=1]$ while $D \ [x=-\infty, \ y=0]$ and $F \ [x=+\infty, \ y=0]$ map into $D' \ [w=0]$ and $F' \ [w=\infty]$ respectively.

The line $y = a$ [ABC in Fig. 8-13] maps into $u = -e^{\pi x/a}, \ v = 0$ [the negative real axis in the w plane; $A'B'C'$ in Fig. 8-14]. The points $A \ [x=+\infty, \ y=a]$ and $C \ [x=-\infty, \ y=a]$ map into $A' \ [w=-\infty]$ and $C' \ [w=0]$ respectively.

Any point for which $0 < y < a, \ -\infty < x < \infty$ maps uniquely into one point in the uv plane for which $v > 0$.

(b) Refer to Figs. 8-21 and 8-22, Page 206.

If $z = re^{i\theta}$, then

$$w = u + iv = \frac{a}{2}\left(z + \frac{1}{z}\right) = \frac{a}{2}\left(re^{i\theta} + \frac{1}{r}e^{-i\theta}\right) = \frac{a}{2}\left(r + \frac{1}{r}\right)\cos\theta + \frac{ia}{2}\left(r - \frac{1}{r}\right)\sin\theta$$

and $u = \dfrac{a}{2}\left(r + \dfrac{1}{r}\right)\cos\theta, \ v = \dfrac{a}{2}\left(r - \dfrac{1}{r}\right)\sin\theta$.

Semicircle BCD [$r = 1, \ 0 \leqq \theta \leqq \pi$] maps into line segment $B'C'D'$ [$u = a\cos\theta, \ v = 0, \ 0 \leqq \theta \leqq \pi$, i.e. $-a \leqq u \leqq a$].

The line DE [$\theta = 0, \ r > 1$] maps into line $D'E'$ $\left[u = \dfrac{a}{2}\left(r + \dfrac{1}{r}\right), \ v = 0\right]$; line AB [$\theta = \pi, r > 1$] maps into line $A'B'$ $\left[u = -\dfrac{a}{2}\left(r + \dfrac{1}{r}\right), \ v = 0\right]$.

Any point of the z plane for which $r \geqq 1$ and $0 < \theta < \pi$ maps uniquely into one point of the uv plane for which $v \geqq 0$.

(c) Refer to Figs. 8-43 and 8-44, Page 209.

If $z = re^{i\theta}$ and $w = \rho e^{i\phi}$, then $w = 1/z$ becomes $\rho e^{i\phi} = \dfrac{1}{re^{i\theta}} = \dfrac{1}{r}e^{-i\theta}$ from which $\rho = 1/r$, $\phi = -\theta$.

The circle $ABCD$ [$\rho = 1$] in the w plane maps into the circle $A'B'C'D'$ [$r = 1$] of the z plane. Note that if $ABCD$ is described counterclockwise, $A'B'C'D'$ is described clockwise.

Any point exterior to the circle $ABCD$ [$\rho > 1$] is mapped uniquely into a point interior to the circle $A'B'C'D'$ [$r < 1$].

THE SCHWARZ-CHRISTOFFEL TRANSFORMATION

17. Establish the validity of the Schwarz-Christoffel transformation.

We must show that the mapping function obtained from

$$\frac{dw}{dz} = A(z - x_1)^{\alpha_1/\pi - 1}(z - x_2)^{\alpha_2/\pi - 1} \cdots (z - x_n)^{\alpha_n/\pi - 1} \tag{1}$$

maps a given polygon of the w plane [Fig. 8-76 below] into the real axis of the z plane [Fig. 8-77 below].

To show this observe that from (1) we have

$$\arg dw = \arg dz + \arg A + \left(\frac{\alpha_1}{\pi} - 1\right)\arg(z - x_1) + \left(\frac{\alpha_2}{\pi} - 1\right)\arg(z - x_2)$$
$$+ \cdots + \left(\frac{\alpha_n}{\pi} - 1\right)\arg(z - x_n) \tag{2}$$

As z moves along the real axis from the left toward x_1, let us assume that w moves along a side of the polygon toward w_1. When z crosses from the left of x_1 to the right of x_1, $\theta_1 = \arg(z - x_1)$ changes from π to 0 while all other terms in (2) stay constant. Hence $\arg dw$ decreases by $(\alpha_1/\pi - 1)\arg(z - x_1) = (\alpha_1/\pi - 1)\pi = \alpha_1 - \pi$ or, what is the same thing, increases by $\pi - \alpha_1$ [an increase being in the counterclockwise direction].

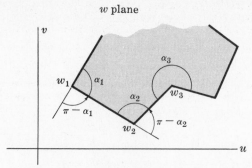

Fig. 8-76

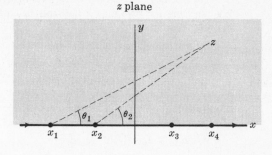

Fig. 8-77

It follows from this that the direction through w_1 turns through the angle $\pi - \alpha_1$, and thus w now moves along the side $w_1 w_2$ of the polygon.

When z moves through x_2, $\theta_1 = \arg(z - x_1)$ and $\theta_2 = \arg(z - x_2)$ change from π to 0 while all other terms stay constant. Hence another turn through angle $\pi - \alpha_2$ in the w plane is made. By continuing the process we see that as z traverses the x axis, w traverses the polygon, and conversely.

We can actually prove that the interior of the polygon (if it is closed) is mapped on to the upper half plane by (1) [see Problem 26].

18. Prove that for closed polygons the sum of the exponents $\dfrac{\alpha_1}{\pi} - 1,\ \dfrac{\alpha_2}{\pi} - 1,\ \ldots,\ \dfrac{\alpha_n}{\pi} - 1$ in the Schwarz-Christoffel transformation (9) or (10), Page 204, is equal to -2.

The sum of the exterior angles of any closed polygon is 2π. Then

$$(\pi - \alpha_1) + (\pi - \alpha_2) + \cdots + (\pi - \alpha_n) = 2\pi$$

and dividing by $-\pi$, we obtain as required,

$$\left(\frac{\alpha_1}{\pi} - 1\right) + \left(\frac{\alpha_2}{\pi} - 1\right) + \cdots + \left(\frac{\alpha_n}{\pi} - 1\right) = -2$$

19. If in the Schwarz-Christoffel transformation (9) or (10), Page 204, one point, say x_n, is chosen at infinity, show that the last factor is not present.

In (9), Page 204, let $A = K/(-x_n)^{\alpha_n/\pi - 1}$ where K is a constant. Then the right side of (9) can be written

$$K\,(z - x_1)^{\alpha_1/\pi - 1}\,(z - x_2)^{\alpha_2/\pi - 1} \cdots (z - x_{n-1})^{\alpha_{n-1}/\pi - 1}\left(\frac{x_n - z}{x_n}\right)^{\alpha_n/\pi - 1}$$

As $x_n \to \infty$, this last factor approaches 1; this is equivalent to removal of the factor.

20. Determine a function which maps each of the indicated regions in the w plane on to the upper half of the z plane.

(a)

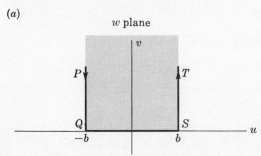

Fig. 8-78

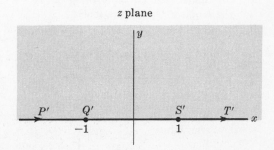

Fig. 8-79

Let points P, Q, S and T [Fig. 8-78 above] map respectively into P', Q', S' and T' [Fig. 8-79 above]. We can consider $PQST$ as a limiting case of a polygon (a triangle) with two vertices at Q and S and the third vertex P or T at infinity.

By the Schwarz-Christoffel transformation, since the angles at Q and S are equal to $\pi/2$, we have

$$\frac{dw}{dz} = A(z+1)^{\frac{\pi/2}{\pi}-1}(z-1)^{\frac{\pi/2}{\pi}-1} = \frac{A}{\sqrt{z^2-1}} = \frac{K}{\sqrt{1-z^2}}$$

Integrating,
$$w = K\int \frac{dz}{\sqrt{1-z^2}} + B = K\sin^{-1}z + B$$

When $z=1$, $w=b$.   Hence (1) $b = K\sin^{-1}(1) + B = K\pi/2 + B.$

When $z=-1$, $w=-b$.   Hence, (2) $-b = K\sin^{-1}(-1) + B = -K\pi/2 + B.$

Solving (1) and (2) simultaneously, we find $B=0$, $K=2b/\pi$.   Then

$$w = \frac{2b}{\pi}\sin^{-1}z \qquad\text{or}\qquad z = \sin\frac{\pi w}{2b}$$

The result is equivalent to entry A-3(a) in the table on Page 206 if we interchange w and z, and let $b = a/2$.

(b)

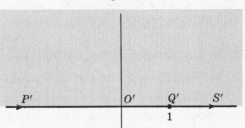

Fig. 8-80   Fig. 8-81

Let points P, O, Q $[w=bi]$ and S map into P', O', Q' $[z=1]$ and S' respectively. Note that P, S, P', S' are at infinity (as indicated by the arrows) while O and O' are the origins $[w=0$ and $z=0]$ of the w and z planes. Since the interior angles at O and Q are $\pi/2$ and $3\pi/2$ respectively, we have by the Schwarz-Christoffel transformation,

$$\frac{dw}{dz} = A(z-0)^{\frac{\pi/2}{\pi}-1}(z-1)^{\frac{3\pi/2}{\pi}-1} = A\sqrt{\frac{z-1}{z}} = K\sqrt{\frac{1-z}{z}}$$

Then
$$w = K\int\sqrt{\frac{1-z}{z}}\,dz$$

To integrate this, let $z=\sin^2\theta$ and obtain

$$w = 2K\int\cos^2\theta\,d\theta = K\int(1+\cos 2\theta)\,d\theta = K(\theta + \tfrac{1}{2}\sin 2\theta) + B$$

$$= K(\theta + \sin\theta\cos\theta) + B = K(\sin^{-1}\sqrt{z} + \sqrt{z(1-z)}) + B$$

When $z=0$, $w=0$ so that $B=0$. When $z=1$, $w=bi$ so that $bi = K\pi/2$ or $K=2bi/\pi$. Then the required transformation is

$$w = \frac{2bi}{\pi}\left(\sin^{-1}\sqrt{z} + \sqrt{z(1-z)}\right)$$

21. Find a transformation which maps a polygon in the w plane on to the unit circle in the ζ plane.

A polygon in the w plane can be mapped on to the x axis of the z plane by the Schwarz-Christoffel transformation

$$w = A\int (z-x_1)^{\alpha_1/\pi-1}(z-x_2)^{\alpha_2/\pi-1}\cdots(z-x_n)^{\alpha_n/\pi-1}\,dz + B \qquad(1)$$

A transformation which maps the upper half of the z plane into the unit circle in the ζ plane is

$$\zeta = \frac{i-z}{i+z} \qquad \text{or} \qquad z = i\left(\frac{1-\zeta}{1+\zeta}\right) \tag{2}$$

on replacing w by ζ and taking $\theta = \pi$, $z_0 = i$ in equation (8), Page 204.

If we let $x_1, x_2, \ldots, x_n$ map into $\zeta_1, \zeta_2, \ldots, \zeta_n$ respectively on the unit circle, then we have for $k = 1, 2, \ldots, n$.

$$z - x_k = i\left(\frac{1-\zeta}{1+\zeta}\right) - i\left(\frac{1-\zeta_k}{1+\zeta_k}\right) = \frac{-2i(\zeta - \zeta_k)}{(1+\zeta)(1+\zeta_k)}$$

Also, $dz = -2i\, d\zeta/(1+\zeta)^2$. Substituting into (1) and simplifying using the fact that the sum of the exponents $\frac{\alpha_1}{\pi} - 1, \frac{\alpha_2}{\pi} - 1, \ldots, \frac{\alpha_n}{\pi} - 1$ is -2, we find the required transformation

$$w = A' \int (\zeta - \zeta_1)^{\alpha_1/\pi - 1}(\zeta - \zeta_2)^{\alpha_2/\pi - 1} \cdots (\zeta - \zeta_n)^{\alpha_n/\pi - 1}\, d\zeta + B$$

where A' is a new arbitrary constant.

TRANSFORMATIONS OF BOUNDARIES IN PARAMETRIC FORM

22. Let C be a curve in the z plane with parametric equations $x = F(t)$, $y = G(t)$. Show that the transformation

$$z = F(w) + iG(w)$$

maps curve C on to the real axis of the w plane.

If $z = x + iy$, $w = u + iv$, the transformation can be written

$$x + iy = F(u + iv) + iG(u + iv)$$

Then $v = 0$ [the real axis of the w plane] corresponds to $x + iy = F(u) + iG(u)$, i.e. $x = F(u)$, $y = G(u)$, which represents the curve C.

23. Find a transformation which maps the ellipse $\dfrac{x^2}{a^2} + \dfrac{y^2}{b^2} = 1$ in the z plane on to the real axis of the w plane.

A set of parametric equations for the ellipse is given by $x = a \cos t$, $y = b \sin t$ where $a > 0$, $b > 0$. Then by Problem 22 the required transformation is $z = a \cos w + ib \sin w$.

MISCELLANEOUS PROBLEMS

24. Find a function which maps the interior of a triangle in the w plane [Fig. 8-82] on to the upper half of the z plane.

Let vertices $P\,[w = 0]$ and $Q\,[w = 1]$ of the triangle map into points $P'\,[z = 0]$ and $Q'\,[z = 1]$ on the z plane while the third vertex R maps into $R'\,[z = \infty]$.

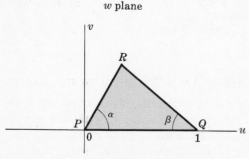

Fig. 8-82

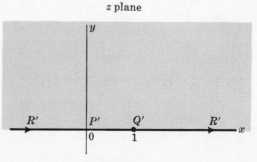

Fig. 8-83

By the Schwarz-Christoffel transformation,

$$\frac{dw}{dz} = A\,z^{\alpha/\pi-1}\,(z-1)^{\beta/\pi-1} = K\,z^{\alpha/\pi-1}\,(1-z)^{\beta/\pi-1}$$

Then by integration,

$$w = K\int_0^z \zeta^{\alpha/\pi-1}\,(1-\zeta)^{\beta/\pi-1}\,d\zeta + B$$

Since $w = 0$ when $z = 0$, we have $B = 0$. Also since $w = 1$ when $z = 1$, we have

$$1 = K\int_0^1 \zeta^{\alpha/\pi-1}\,(1-\zeta)^{\beta/\pi-1}\,d\zeta = \frac{\Gamma(\alpha/\pi)\,\Gamma(\beta/\pi)}{\Gamma\left(\dfrac{\alpha+\beta}{\pi}\right)}$$

using properties of the beta and gamma functions [Chapter 10]. Hence

$$K = \frac{\Gamma\left(\dfrac{\alpha+\beta}{\pi}\right)}{\Gamma(\alpha/\pi)\,\Gamma(\beta/\pi)}$$

and the required transformation is

$$w = \frac{\Gamma\left(\dfrac{\alpha+\beta}{\pi}\right)}{\Gamma(\alpha/\pi)\,\Gamma(\beta/\pi)}\int_0^z \zeta^{\alpha/\pi-1}\,(1-\zeta)^{\beta/\pi-1}\,d\zeta$$

Note that this agrees with entry A-13 on Page 209, since the length of side AB in Fig. 8-39 is

$$\int_0^1 \zeta^{\alpha/\pi-1}\,(1-\zeta)^{\beta/\pi-1}\,d\zeta = \frac{\Gamma(\alpha/\pi)\,\Gamma(\beta/\pi)}{\Gamma\left(\dfrac{\alpha+\beta}{\pi}\right)}$$

25. (a) Find a function which maps the shaded region in the w plane of Fig. 8-84 on to the upper half of the z plane of Fig. 8-85.

   (b) Discuss the case where $b \to 0$.

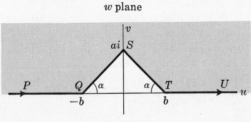

Fig. 8-84

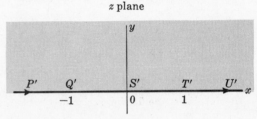

Fig. 8-85

(a) The interior angles at Q and T are each $\pi - \alpha$, while the angle at S is $2\pi - (\pi - 2\alpha) = \pi + 2\alpha$. Then by the Schwarz-Christoffel transformation we have

$$\frac{dw}{dz} = A\,(z+1)^{(\pi-\alpha)/\pi-1}\,z^{(\pi+2\alpha)/\pi-1}\,(z-1)^{(\pi-\alpha)/\pi-1}$$

$$= \frac{A\,z^{2\alpha/\pi}}{(z^2-1)^{\alpha/\pi}} = \frac{K\,z^{2\alpha/\pi}}{(1-z^2)^{\alpha/\pi}}$$

Hence by integration

$$w = K\int_0^z \frac{\zeta^{2\alpha/\pi}}{(1-\zeta^2)^{\alpha/\pi}}\,d\zeta + B$$

When $z = 0$, $w = ai$; then $B = ai$ and

$$w = K\int_0^z \frac{\zeta^{2\alpha/\pi}}{(1-\zeta^2)^{\alpha/\pi}}\,d\zeta + ai \qquad (1)$$

The value of K can be expressed in terms of the gamma function using the fact that $w = b$ when $z = 1$ [Problem 102]. We find

$$K = \frac{(b-ai)\sqrt{\pi}}{\Gamma\left(\dfrac{\alpha}{\pi}+\dfrac{1}{2}\right)\Gamma\left(1-\dfrac{\alpha}{\pi}\right)} \qquad (2)$$

(b) As $b \to 0$, $\alpha \to \pi/2$ and the result in (a) reduces to

$$w = ai - ai \int_0^z \frac{\zeta \, d\zeta}{\sqrt{1 - \zeta^2}} = ai\sqrt{1 - z^2}$$
$$= a\sqrt{z^2 - 1}$$

In this case Fig. 8-84 reduces to Fig. 8-86. The result for this case can be found directly from the Schwarz-Christoffel transformation by considering $PQSTU$ as a polygon with interior angles at Q, S and T equal to $\pi/2, 2\pi$, and $\pi/2$ respectively.

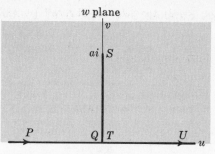

Fig. 8-86

26. **Prove that the Schwarz-Christoffel transformation of Problem 17 maps the interior of the polygon on to the upper half plane.**

It suffices to prove that the transformation maps the interior on to the unit circle, since we already know [Problem 11] that the unit circle can be mapped on to the upper half plane.

Suppose that the function mapping polygon P in the w plane on to the unit circle C in the z plane is given by $w = f(z)$ where $f(z)$ is analytic inside C.

We must now show that to each point a inside P there corresponds one and only one point, say z_0, such that $f(z_0) = a$.

Now by Cauchy's integral formula, since a is inside P,

$$\frac{1}{2\pi i} \oint_P \frac{dw}{w - a} = 1$$

Then since $w - a = f(z) - a$,

$$\frac{1}{2\pi i} \oint_C \frac{f'(z)}{f(z) - a} \, dz = 1$$

But $f(z) - a$ is analytic inside C. Hence from Problem 17, Chapter 5, we have shown that there is only one zero (say z_0) of $f(z) - a$ inside C, i.e. $f(z_0) = a$, as required.

27. **Let C be a circle in the z plane having its center on the real axis, and suppose further that it passes through $z = 1$ and has $z = -1$ as an interior point. Determine the image of C in the w plane under the transformation $w = f(z) = \frac{1}{2}(z + 1/z)$.**

We have $dw/dz = \frac{1}{2}(1 - 1/z^2)$. Since $dw/dz = 0$ at $z = 1$, it follows that $z = 1$ is a critical point. From the Taylor series of $f(z) = \frac{1}{2}(z + 1/z)$ about $z = 1$, we have

$$w - 1 = \frac{1}{2}[(z - 1)^2 - (z - 1)^3 + (z - 1)^4 - \cdots]$$

By Problem 100 we see that angles with vertices at $z = 1$ are doubled under the transformation. In particular, since the angle at $z = 1$ exterior to C is π, the angle at $w = 1$ exterior to the image C' is 2π. Hence C' has a sharp tail at $w = 1$ (see Fig. 8-88). Other points of C' can be found directly.

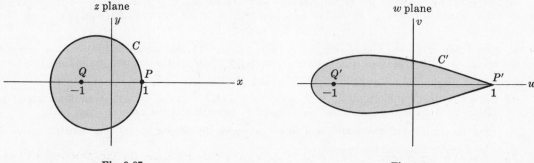

Fig. 8-87   Fig. 8-88

It is of interest to note that in this case C encloses the circle $|z| = 1$ which under the transformation is mapped into the slit from $w = -1$ to $w = 1$. Thus as C approaches $|z| = 1$, C' approaches the straight line joining $w = -1$ to $w = 1$.

28. Suppose the circle C of Problem 27 is moved so that its center is in the upper half plane but that it still passes through $z = 1$ and encloses $z = -1$. Determine the image of C under the transformation $w = \frac{1}{2}(z + 1/z)$.

As in Problem 27, since $z = 1$ is a critical point, we will obtain the sharp tail at $w = 1$ [Fig. 8-90]. If C does not entirely enclose the circle $|z| = 1$ [as shown in Fig. 8-89], the image C' will not entirely enclose the image of $|z| = 1$ [which is the slit from $w = -1$ to $w = 1$]. Instead, C' will only enclose that portion of the slit which corresponds to the part of $|z| = 1$ inside C. The appearance of C' is therefore as shown in Fig. 8-90. By changing C appropriately, other shapes similar to C' can be obtained.

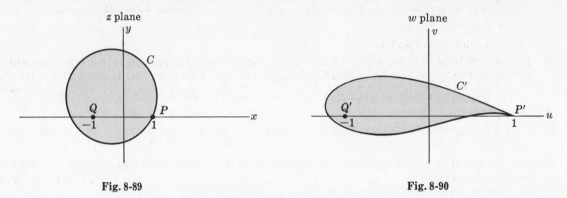

Fig. 8-89   Fig. 8-90

The fact that C' resembles the cross-section of the wing of an airplane, sometimes called an *airfoil*, is important in aerodynamic theory (see Chapter 10) and was first used by *Joukowski*. For this reason shapes such as C' are called *Joukowski airfoils* or *profiles* and $w = \frac{1}{2}(z + 1/z)$ is called a *Joukowski transformation*.

Supplementary Problems

TRANSFORMATIONS

29. Given triangle T in the z plane with vertices at i, $1 - i$, $1 + i$. Determine the triangle T' into which T is mapped under the transformations (a) $w = 3z + 4 - 2i$, (b) $w = iz + 2 - i$, (c) $w = 5e^{\pi i/3}z - 2 + 4i$. What is the relationship between T and T' in each case?

30. Sketch the region of the w plane into which the interior of triangle T of Problem 29 is mapped under the transformations (a) $w = z^2$, (b) $w = iz^2 + (2 - i)z$, (c) $w = z + 1/z$.

31. (a) Show that by means of the transformation $w = 1/z$ the circle C given by $|z - 3| = 5$ is mapped into the circle $|w + 3/16| = 5/16$. (b) Into what region is the interior of C mapped?

32. (a) Prove that under the transformation $w = (z - i)/(iz - 1)$ the region $\text{Im}\{z\} \geqq 0$ is mapped into the region $|w| \leqq 1$. (b) Into what region is $\text{Im}\{z\} \leqq 0$ mapped under the transformation?

33. (a) Show that the transformation $w = \frac{1}{2}(ze^{-\alpha} + z^{-1}e^{\alpha})$ where α is real, maps the interior of the circle $|z| = 1$ on to the exterior of an ellipse [see entry B-2 in the table on Page 209].

(b) Find the lengths of the major and minor axes of the ellipse in (a) and construct the ellipse.

Ans. (b) $2\cosh\alpha$ and $2\sinh\alpha$ respectively.

34. Determine the equation of the curve in the w plane into which the straight line $x + y = 1$ is mapped under the transformations (a) $w = z^2$, (b) $w = 1/z$.

Ans. (a) $u^2 + 2v = 1$, (b) $u^2 + 2uv + 2v^2 = u + v$

35. Show that $w = \left(\dfrac{1+z}{1-z}\right)^{2/3}$ maps the unit circle on to a wedge-shaped region and illustrate graphically.

36. (a) Show that the transformation $w = 2z - 3i\bar{z} + 5 - 4i$ is equivalent to $u = 2x + 3y + 5$, $v = 2y - 3x - 4$.

    (b) Determine the triangle in the uv plane into which triangle T of Problem 29 is mapped under the transformation in (a). Are the triangles similar?

37. Express the transformations (a) $u = 4x^2 - 8y$, $v = 8x - 4y^2$ and (b) $u = x^3 - 3xy^2$, $v = 3x^2y - y^3$ in the form $w = F(z, \bar{z})$.   Ans. (a) $w = (1+i)(z^2 + \bar{z}^2) + (2 - 2i)z\bar{z} + 8iz$, (b) $w = z^3$

CONFORMAL TRANSFORMATIONS

38. The straight lines $y = 2x$, $x + y = 6$ in the xy plane are mapped on to the w plane by means of the transformation $w = z^2$. (a) Show graphically the images of the straight lines in the w plane. (b) Show analytically that the angle of intersection of the straight lines is the same as the angle of intersection of their images and explain why this is so.

39. Work Problem 38 if the transformation is (a) $w = \dfrac{1}{z}$, (b) $w = \dfrac{z-1}{z+1}$.

40. The interior of the square $\mathcal{S}$ with vertices at 1, 2, $1+i$, $2+i$ is mapped into a region $\mathcal{S}'$ by means of the transformations (a) $w = 2z + 5 - 3i$, (b) $w = z^2$, (c) $w = \sin \pi z$. In each case sketch the regions and verify directly that the interior angles of $\mathcal{S}'$ are right angles.

41. (a) Sketch the images of the circle $(x-3)^2 + y^2 = 2$ and the line $2x + 3y = 7$ under the transformation $w = 1/z$. (b) Determine whether the images of the circle and line of (a) intersect at the same angles as the circle and line. Explain.

42. Work Problem 41 for the case of the circle $(x-3)^2 + y^2 = 5$ and the line $2x + 3y = 14$.

43. (a) Work Problem 38 if the transformation is $w = 3z - 2i\bar{z}$.

    (b) Is your answer to part (b) the same? Explain.

44. Prove that a necessary and sufficient condition for the transformation $w = F(z, \bar{z})$ to be conformal in a region $\mathcal{R}$ is that $\partial F/\partial \bar{z} = 0$ and $\partial F/\partial z \neq 0$ in $\mathcal{R}$ and explain the significance of this.

JACOBIANS

45. (a) For each part of Problem 29, determine the ratio of the areas T and T'. (b) Compare your findings in part (a) with the magnification factor $|dw/dz|^2$ and explain the significance.

46. Find the Jacobian of the transformations (a) $w = 2z^2 - iz + 3 - i$, (b) $u = x^2 - xy + y^2$, $v = x^2 + xy + y^2$. Ans. (a) $|4z - i|^2$, (b) $4(x^2 + y^2)$

47. Prove that a polygon in the z plane is mapped into a similar polygon in the w plane by means of the transformation $w = F(z)$ if and only if $F'(z)$ is a constant different from zero.

48. The analytic function $F(z)$ maps the interior $\mathcal{R}$ of a circle C defined by $|z| = 1$ into a region $\mathcal{R}'$ bounded by a simple closed curve C'. Prove that (a) the length of C' is $\displaystyle\oint_C |F'(z)|\, |dz|$, (b) the area of $\mathcal{R}'$ is $\displaystyle\iint_{\mathcal{R}} |F'(z)|^2\, dx\, dy$.

49. Prove the result (2) on Page 200.

50. Find the ratio of areas of the triangles in Problem 36(b) and compare with the magnification factor as obtained from the Jacobian.

51. Let $u = u(x, y)$, $v = v(x, y)$ and $x = x(\xi, \eta)$, $y = y(\xi, \eta)$.

(a) Prove that $\dfrac{\partial(u, v)}{\partial(\xi, \eta)} = \dfrac{\partial(u, v)}{\partial(x, y)} \cdot \dfrac{\partial(x, y)}{\partial(\xi, \eta)}$.

(b) Interpret the result of (a) geometrically.

(c) Generalize the result in (a).

52. Show that if $w = u + iv = F(z)$, $z = x + iy = G(\zeta)$ and $\zeta = \xi + i\eta$, the result in Problem 51(a) is equivalent to the relation

$$\left|\frac{dw}{d\zeta}\right| = \left|\frac{dw}{dz}\right|\left|\frac{dz}{d\zeta}\right|$$

BILINEAR OR FRACTIONAL TRANSFORMATIONS

53. Find a bilinear transformation which maps the points $i, -i, 1$ of the z plane into $0, 1, \infty$ of the w plane respectively.   *Ans.* $w = (1 - i)(z - i)/2(z - 1)$

54. (a) Find a bilinear transformation which maps the vertices $1 + i$, $-i$, $2 - i$ of a triangle T of the z plane into the points $0, 1, i$ of the w plane.

(b) Sketch the region into which the interior of triangle T is mapped under the transformation obtained in (a).

Ans. (a) $w = (2z - 2 - 2i)/\{(i - 1)z - 3 - 5i\}$

55. Prove that the result of (a) two successive bilinear transformations, (b) any number of successive bilinear transformations is also a bilinear transformation.

56. If $a \neq b$ are the two fixed points of the bilinear transformation, show that it can be written in the form

$$\frac{w - a}{w - b} = K\left(\frac{z - a}{z - b}\right)$$

where K is a constant.

57. If $a = b$ in Problem 56, show that the transformation can be written in the form

$$\frac{1}{w - a} = \frac{1}{z - a} + k$$

where k is a constant.

58. Prove that the most general bilinear transformation which maps $|z| = 1$ on to $|w| = 1$ is

$$w = e^{i\theta}\left(\frac{z - p}{\bar{p}z - 1}\right)$$

where p is a constant,

59. Show that the transformation of Problem 58 maps $|z| < 1$ on to   (a) $|w| < 1$ if $|p| < 1$ and   (b) $|w| > 1$ if $|p| > 1$.

60. Discuss Problem 58 if $|p| = 1$.

61. Work Problem 11 directly.

62. (a) If z_1, z_2, z_3, z_4 are any four different points of a circle, prove that the cross ratio is real.

(b) Is the converse of part (a) true?   *Ans.* (b) Yes

THE SCHWARZ-CHRISTOFFEL TRANSFORMATION

63. Use the Schwarz-Christoffel transformation to determine a function which maps each of the indicated regions in the w plane on to the upper half of the z plane.

(a)

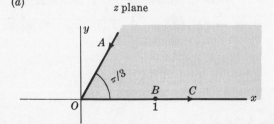

Fig. 8-91

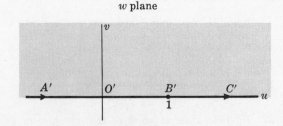

Fig. 8-92

(b)   z plane   w plane

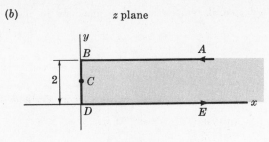

Fig. 8-93

Fig. 8-94

(c)   z plane   w plane

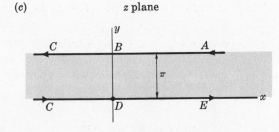

Fig. 8-95

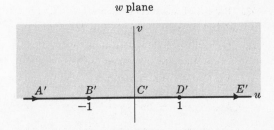

Fig. 8-96

(d)   z plane   w plane

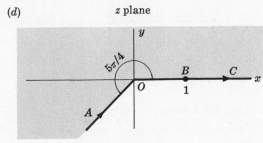

Fig. 8-97

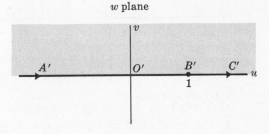

Fig. 8-98

Ans. (a) $w = z^3$, (b) $w = \cosh(\pi z/2)$, (c) $w = e^z$, (d) $w = z^{4/5}$

64. Verify entry A-14 in the table on Page 209 by using the Schwarz-Christoffel transformation.

65. Find a function which maps the infinite shaded region of Fig. 8-99 on to the upper half of the z plane [Fig. 8-100] so that P, Q, R map into P', Q', R' respectively [where P, R, P', R' are at infinity as indicated by the arrows].   *Ans.* $z = (w + \pi - \pi i)^2$

w plane   z plane

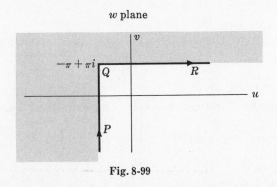

Fig. 8-99

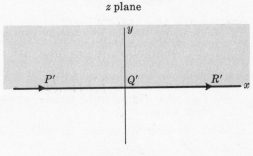

Fig. 8-100

66. Verify entry A-12 in the table on Page 208 by using the Schwarz-Christoffel transformation.

67. Find a function which maps each of the indicated shaded regions in the w plane on to the upper half of the z plane.

(a)

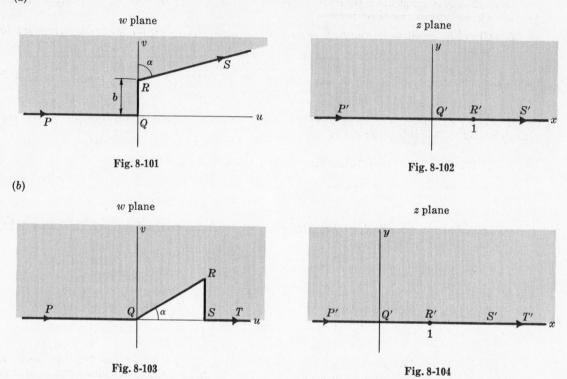

Fig. 8-101   Fig. 8-102

(b)

Fig. 8-103   Fig. 8-104

68. (a) Verify entry A-11 of the table on Page 208 by using the Schwarz-Christoffel transformation.

   (b) Use the result of (a) together with entry A-2 of the table on Page 205 to arrive at the entry C-5 in the table on Page 211.

TRANSFORMATIONS OF BOUNDARIES IN PARAMETRIC FORM

69. (a) Find a transformation which maps the parabola $y^2 = 4p(p-x)$ into a straight line.

   (b) Discuss the relationship of your answer to entry A-9 in the table on Page 208.

   Ans. (a) One possibility is $z = p - pw^2 + 2piw = p(1+iw)^2$ obtained by using the parametric equations $x = p(1-t^2)$, $y = 2pt$.

70. Find a transformation which maps the hyperbola $x = a \cosh t$, $y = a \sinh t$ into a straight line.
   Ans. $z = a(\cosh w + i \sinh w)$

71. Find a transformation which maps the cycloid $x = a(t - \sin t)$, $y = a(1 - \cos t)$ into a straight line.
   Ans. $z = a(w + i - ie^{-iw})$

72. (a) Find a transformation which maps the hypocycloid $x^{2/3} + y^{2/3} = a^{2/3}$ into a straight line.

   (b) Into what region is the interior of the hypocycloid mapped under the transformation? Justify your answer.

   [Hint. Parametric equations for the hypocycloid are $x = a \cos^3 t$, $y = a \sin^3 t$, $0 \leqq t < 2\pi$.]
   Ans. (a) $z = a(\cos^3 w + i \sin^3 w)$

73. Two sets of parametric equations for the parabola $y = x^2$ are (a) $x = t$, $y = t^2$ and (b) $x = \pm e^t$, $y = e^{2t}$. Use these parametric equations to arrive at two possible transformations mapping the parabola into a straight line and determine whether there is any advantage in using one rather than the other.

MISCELLANEOUS PROBLEMS

74. (a) Show that the transformation $w = 1/z$ maps the circle $|z - a| = a$, where $a > 0$, into a straight line. Illustrate graphically showing the region into which the interior of the circle is mapped, as well as various points of the circle.

   (b) Show how the result in (a) can be used to derive the transformation for the upper half plane into the unit circle.

75. Prove that the function $w = (z^2/a^2) - 1$ maps one loop of the lemniscate $r^2 = 2a^2 \cos 2\theta$ on to the unit circle.

76. Prove that the function $w = z^2$ maps the circle $|z - a| = a, \ a > 0$, on to the cardioid $\rho = 2a^2(1 + \cos \phi)$ [see entry C-2 in the table on Page 210].

77. Show that the Joukowsky transformation $w = z + k^2/z$ can be written as

$$\frac{w - 2k}{w + 2k} \ = \ \left(\frac{z - k}{z + k} \right)^2$$

78. (a) Let $w = F(z)$ be a bilinear transformation. Show that the most general linear transformation for which $F\{F(z)\} = z$ is given by

$$\frac{w - p}{w - q} \ = \ k \frac{z - p}{z - q}$$

   where $k^2 = 1$.

   (b) What is the result in (a) if $F\{F[F(z)]\} = z$?

   (c) Generalize the results in (a) and (b).

   Ans. (b) Same as (a) with $k^3 = 1$.

79. (a) Determine a transformation which rotates the ellipse $x^2 + xy + y^2 = 5$ so that the major and minor axes are parallel to the coordinate axes. (b) What are the lengths of the major and minor axes?

80. Find a bilinear transformation which maps the circle $|z - 1| = 2$ on to the line $x + y = 1$.

81. Verify the transformations (a) A-6, (b) A-7, (c) A-8, in the table on Page 207.

82. Consider the stereographic projection of the complex plane on to a unit sphere tangent to it [see Page 6]. Let an XYZ rectangular coordinate system be constructed so that the Z axis coincides with NS while the X and Y axes coincide with the x and y axes of Fig. 1-6, Page 6. Prove that the point (X, Y, Z) of the sphere corresponding to (x, y) on the plane is such that

$$X \ = \ \frac{x}{x^2 + y^2 + 1}, \qquad Y \ = \ \frac{y}{x^2 + y^2 + 1}, \qquad Z \ = \ \frac{x^2 + y^2}{x^2 + y^2 + 1}$$

83. Prove that a mapping by means of stereographic projection is conformal.

84. (a) Prove that by means of a stereographic projection, arc lengths of the sphere are magnified in the ratio $(x^2 + y^2 + 1) : 1$.

   (b) Discuss what happens to regions in the vicinity of the north pole. What effect does this produce on navigational charts?

85. Let $u = u(x, y), \ v = v(x, y)$ be a transformation of points of the xy plane on to points of the uv plane.

   (a) Show that in order that the transformation preserve angles, it is necessary and sufficient that

$$\left(\frac{\partial u}{\partial x} \right)^2 + \left(\frac{\partial v}{\partial x} \right)^2 \ = \ \left(\frac{\partial u}{\partial y} \right)^2 + \left(\frac{\partial v}{\partial y} \right)^2, \qquad \frac{\partial u}{\partial x} \frac{\partial u}{\partial y} + \frac{\partial v}{\partial x} \frac{\partial v}{\partial y} \ = \ 0$$

   (b) Deduce from (a) that we must have either

$$\text{(i)} \quad \frac{\partial u}{\partial x} = \frac{\partial v}{\partial y}, \ \frac{\partial u}{\partial y} = -\frac{\partial v}{\partial x} \qquad \text{or} \qquad \text{(ii)} \quad \frac{\partial u}{\partial x} = -\frac{\partial v}{\partial y}, \ \frac{\partial u}{\partial y} = \frac{\partial v}{\partial x}$$

   Thus conclude that $u + iv$ must be an analytic function of $x + iy$.

86. Find the area of the ellipse $ax^2 + bxy + cy^2 = 1$ where $a > 0$, $c > 0$ and $b^2 < 4ac$.

Ans. $2\pi/\sqrt{4ac - b^2}$

87. A transformation $w = f(z)$ of points in a plane is called *involutory* if $z = f(w)$. In this case a single repetition of the transformation restores each point to its original position. Find conditions on $\alpha, \beta, \gamma, \delta$ in order that the bilinear transformation $w = (\alpha z + \beta)/(\gamma z + \delta)$ be involutory.   Ans. $\delta = -\alpha$

88. Show that the transformations (a) $w = (z+1)/(z-1)$, (b) $w = \ln \coth(z/2)$ are involutory.

89. Find a bilinear transformation which maps $|z| \leqq 1$ on to $|w - 1| \leqq 1$ so that the points $1, -i$ correspond to $2, 0$ respectively.

90. Discuss the significance of the vanishing of the Jacobian for a bilinear transformation.

91. Prove that the bilinear transformation $w = (\alpha z + \beta)/(\gamma z + \delta)$ has one fixed point if and only if $(\delta + \alpha)^2 = 4(\alpha\delta - \beta\gamma) \neq 0$.

92. (a) Show that the transformation $w = (\alpha z + \overline{\gamma})/(\gamma z + \overline{\alpha})$ where $|\alpha|^2 - |\gamma|^2 = 1$ transforms the unit circle and its interior into itself.

(b) Show that if $|\gamma|^2 - |\alpha|^2 = 1$ the interior is mapped into the exterior.

93. Suppose under the transformation $w = F(z, \bar{z})$ any intersecting curves C_1 and C_2 in the z plane map respectively into corresponding intersecting curves C_1' and C_2' in the w plane. Prove that if the transformation is conformal then (a) $F(z, \bar{z})$ is a function of z alone, say $f(z)$, and (b) $f(z)$ is analytic.

94. (a) Prove the multiplication rule for determinants [see Problem 7]:

$$\begin{vmatrix} a_1 & b_1 \\ c_1 & d_1 \end{vmatrix} \begin{vmatrix} a_2 & b_2 \\ c_2 & d_2 \end{vmatrix} = \begin{vmatrix} a_1 a_2 + b_1 c_2 & a_1 b_2 + b_1 d_2 \\ c_1 a_2 + c_1 c_2 & c_1 b_2 + d_1 d_2 \end{vmatrix}$$

(b) Show how to generalize the result in (a) to third order and higher order determinants.

95. Find a function which maps on to each other the shaded regions of Figures 8-105 and 8-106, where QS has length b.

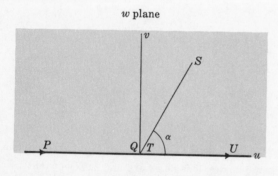

w plane

Fig. 8-105

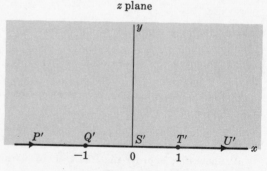

z plane

Fig. 8-106

96. (a) Show that the function $w = \displaystyle\int_0^z \frac{dt}{(1 - t^6)^{1/3}}$ maps a regular hexagon into the unit circle.

(b) What is the length of a side of the hexagon in (a)?

Ans. (b) $\frac{1}{6}\sqrt[3]{2}\,\Gamma(\frac{1}{3})$

97. Show that the transformation $w = (Az^2 + Bz + C)/(Dz^2 + Ez + F)$ can be considered as a combination of two bilinear transformations separated by a transformation of the type $\tau = \zeta^2$.

98. Find a function which maps a regular polygon of n sides into the unit circle.

99. Verify the entries: (a) A-9, Page 208; (b) A-10, Page 208; (c) B-3, Page 210; (d) B-4, Page 210; (e) C-3, Page 211; (f) C-4, Page 211.

100. Suppose the mapping function $w = f(z)$ has the Taylor series expansion

$$w \;=\; f(z) \;=\; f(a) + f'(a)\,(z-a) + \cdots + \frac{f^{(n)}(a)}{n!}(z-a)^n + \cdots$$

Show that if $f^{(k)}(a) = 0$ for $k = 0, 1, \ldots, n-1$ while $f^{(n)}(a) \neq 0$, then angles in the z plane with vertices at $z = a$ are multiplied by n in the w plane.

101. Determine a function which maps the infinite strip $-\pi/4 \leqq x \leqq \pi/4$ on to the interior of the unit circle $|w| \leqq 1$ so that $z = 0$ corresponds to $w = 0$. 　　*Ans.* $w = \tan z$

102. Verify the value of K obtained in equation (2) of Problem 25.

103. Find a function which maps the upper half plane on to the interior of a triangle with vertices at $w = 0, 1, i$ corresponding to $z = 0, 1, \infty$ respectively.

　　Ans. $w \;=\; \dfrac{\Gamma(3/4)}{\sqrt{\pi}\,\Gamma(1/4)} \displaystyle\int_0^z t^{-1/2}\,(1-t)^{-3/4}\,dt$

Chapter 9

Physical Applications of Conformal Mapping

BOUNDARY VALUE PROBLEMS

Many problems of science and engineering when formulated mathematically lead to *partial differential equations* and associated conditions called *boundary conditions*. The problem of determining solutions to a partial differential equation which satisfy the boundary conditions is called a *boundary-value problem*.

It is of fundamental importance, from a mathematical as well as physical viewpoint, that one should not only be able to find such solutions (i.e. that solutions *exist*) but that for any given problem there should be only one solution (i.e. the solution is *unique*).

HARMONIC AND CONJUGATE FUNCTIONS

A function satisfying *Laplace's equation*

$$\nabla^2 \Phi = \frac{\partial^2 \Phi}{\partial x^2} + \frac{\partial^2 \Phi}{\partial y^2} = 0 \tag{1}$$

in a region $\mathcal{R}$ is called *harmonic* in $\mathcal{R}$. As we have already seen, if $f(z) = u(x,y) + iv(x,y)$ is analytic in $\mathcal{R}$, then u and v are harmonic in $\mathcal{R}$.

> **Example:** If $f(z) = 4z^2 - 3iz = 4(x+iy)^2 - 3i(x+iy) = 4x^2 - 4y^2 + 3y + i(8xy - 3x)$, then $u = 4x^2 - 4y^2 + 3y$, $v = 8xy - 3x$. Since u and v satisfy Laplace's equation, they are harmonic.

The functions u and v are called *conjugate functions*; and given one, the other can be determined within an arbitrary additive constant [see Chapter 3].

DIRICHLET AND NEUMANN PROBLEMS

Let $\mathcal{R}$ [Fig. 9-1] be a simply-connected region bounded by a simple closed curve C. Two types of boundary-value problems are of great importance.

1. **Dirichlet's problem** seeks the determination of a function Φ which satisfies Laplace's equation *(1)* [i.e. is harmonic] in $\mathcal{R}$ and takes prescribed values on the boundary C.

2. **Neumann's problem** seeks the determination of a function Φ which satisfies Laplace's equation *(1)* in $\mathcal{R}$ and whose normal derivative $\partial\Phi/\partial n$ takes prescribed values on the boundary C.

The region $\mathcal{R}$ may be unbounded. For example $\mathcal{R}$ can be the upper half plane with the x axis as the boundary C.

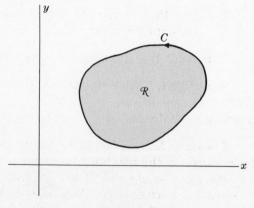

Fig. 9-1

232

It can be shown that solutions to both the Dirichlet and Neumann problems exist and are unique [the Neumann problem within an arbitrary additive constant] under very mild restrictions on the boundary conditions [see Problems 29 and 80].

It is of interest that a Neumann problem can be stated in terms of an appropriately stated Dirichlet problem (see Problem 79). Hence if we can solve the Dirichlet problem we can (at least theoretically) solve a corresponding Neumann problem.

THE DIRICHLET PROBLEM FOR THE UNIT CIRCLE. POISSON'S FORMULA

Let C be the unit circle $|z| = 1$ and $\mathcal{R}$ be its interior. A function which satisfies Laplace's equation [i.e. is harmonic] at each point (r, θ) in $\mathcal{R}$ and takes on the prescribed value $F(\theta)$ on C [i.e. $\Phi(1, \theta) = F(\theta)$], is given by

$$\Phi(r, \theta) \;=\; \frac{1}{2\pi} \int_0^{2\pi} \frac{(1 - r^2)\, F(\phi)\, d\phi}{1 - 2r \cos(\theta - \phi) + r^2} \tag{2}$$

This is called *Poisson's formula for a circle* [see Chapter 5, Page 119].

THE DIRICHLET PROBLEM FOR THE HALF PLANE

A function which is harmonic in the half plane $y > 0$ [Im $\{z\} > 0$] and which takes on the prescribed value $G(x)$ on the x axis [i.e. $\Phi(x, 0) = G(x)$, $-\infty < x < \infty$], is given by

$$\Phi(x, y) \;=\; \frac{1}{\pi} \int_{-\infty}^{\infty} \frac{y\, G(\eta)\, d\eta}{y^2 + (x - \eta)^2} \tag{3}$$

This is sometimes called *Poisson's formula for the half plane* [see Chapter 5, Page 120].

SOLUTIONS TO DIRICHLET AND NEUMANN PROBLEMS
BY CONFORMAL MAPPING

The Dirichlet and Neumann problems can be solved for any simply-connected region $\mathcal{R}$ which can be mapped conformally by an analytic function on to the interior of a unit circle or half plane. [By Riemann's mapping theorem this can always be accomplished, at least in theory.] The basic ideas involved are as follows.

(a) Use the mapping function to transform the boundary-value problem for the region $\mathcal{R}$ into a corresponding one for the unit circle or half plane.

(b) Solve the problem for the unit circle or half plane.

(c) Use the solution in (b) to solve the given problem by employing the inverse mapping function.

Important theorems used in this connection are as follows.

Theorem 1. Let $w = f(z)$ be analytic in a region $\mathcal{R}$ of the z plane. Then there exists a unique inverse $z = g(w)$ in $\mathcal{R}$, provided $f'(z) \neq 0$ in $\mathcal{R}$ [thus insuring that the mapping is conformal at each point of $\mathcal{R}$].

Theorem 2. Let $\Phi(x, y)$ be harmonic in $\mathcal{R}$ and suppose that $\mathcal{R}$ is mapped into $\mathcal{R}'$ of the w plane by the mapping function $w = f(z)$ where $f(z)$ is analytic and $f'(z) \neq 0$ so that $x = x(u, v)$, $y = y(u, v)$. Then $\Phi(x, y) = \Phi[x(u, v), y(u, v)] \equiv \Psi(u, v)$ is harmonic in $\mathcal{R}'$. In words, a harmonic function is transformed into a harmonic function under a transformation $w = f(z)$ which is analytic [see Problem 4].

Theorem 3. If $\Phi = a$ [a constant] on the boundary or part of the boundary C of a region in the z plane, then $\Psi = a$ on its image C' in the w plane. Similarly if the normal derivative of Φ is zero, i.e. $\partial\Phi/\partial n = 0$ on C, then the normal derivative of Ψ is zero on C'.

Applications to Fluid Flow

BASIC ASSUMPTIONS

The solution of many important problems in fluid flow, also referred to as *fluid dynamics, hydrodynamics* or *aerodynamics,* is often achieved by complex variable methods under the following assumptions.

1. **The fluid flow is two dimensional,** i.e. the basic flow pattern and characteristics of the fluid motion in any plane are essentially the same as in any parallel plane. This permits us to confine our attention to just a single plane which we take to be the z plane. Figures constructed in this plane are interpreted as cross-sections of corresponding infinite cylinders perpendicular to the plane. For example, in Fig. 9-7, Page 237, the circle represents an infinite cylindrical obstacle around which the fluid flows. Naturally, an infinite cylinder is nothing more than a *mathematical model* of a physical cylinder which is so long that end effects can be reasonably neglected.

2. **The flow is stationary or steady,** i.e. the velocity of the fluid at any point depends only on the position (x, y) and not on time.

3. **The velocity components are derivable from a potential,** i.e. if V_x and V_y denote the components of velocity of the fluid at (x, y) in the positive x and y directions respectively, there exists a function Φ, called the *velocity potential,* such that

$$V_x = \frac{\partial\Phi}{\partial x}, \qquad V_y = \frac{\partial\Phi}{\partial y} \tag{4}$$

   An equivalent assumption is that if C is any simple closed curve in the z plane and V_t is the tangential component of velocity on C, then

$$\oint_C V_t\, ds = \oint_C V_x\, dx + V_y\, dy = 0 \tag{5}$$

   See Problem 48.

   Either of the integrals in (5) is called the *circulation* of the fluid along C. When the circulation is zero the flow is called *irrotational* or *circulation free.*

4. **The fluid is incompressible,** i.e. the density, or mass per unit volume of the fluid, is constant. If V_n is the normal component of velocity on C this leads to the conclusion (see Problem 48) that

$$\oint_C V_n\, ds = \oint_C V_x\, dy - V_y\, dx = 0 \tag{6}$$

   or

$$\frac{\partial V_x}{\partial x} + \frac{\partial V_y}{\partial y} = 0 \tag{7}$$

   which expresses the condition that the quantity of fluid contained inside C is a constant, i.e. the quantity entering C is equal to the quantity leaving C. For this reason equation (6), or the equivalent (7), is called the *equation of continuity.*

5. **The fluid is non-viscous,** i.e. has no viscosity or internal friction. A moving viscous fluid tends to adhere to the surface of an obstacle placed in its path. If there is no viscosity, the pressure forces on the surface are perpendicular to the surface. A fluid which is non-viscous and incompressible is often called an *ideal fluid.* It must of course be realized that such a fluid is only a mathematical model of a real fluid in which such effects can be safely assumed negligible.

THE COMPLEX POTENTIAL

From (4) and (7) it is seen that the velocity potential Φ is harmonic, i.e. satisfies Laplace's equation

$$\frac{\partial^2 \Phi}{\partial x^2} + \frac{\partial^2 \Phi}{\partial y^2} = 0 \tag{8}$$

It follows that there must exist a conjugate harmonic function, say $\Psi(x, y)$, such that

$$\Omega(z) = \Phi(x, y) + i\Psi(x, y) \tag{9}$$

is analytic. By differentiation we have, using (4),

$$\frac{d\Omega}{dz} = \Omega'(z) = \frac{\partial \Phi}{\partial x} + i\frac{\partial \Psi}{\partial x} = \frac{\partial \Phi}{\partial x} - i\frac{\partial \Phi}{\partial y} = V_x - iV_y \tag{10}$$

Thus the velocity [sometimes called the *complex velocity*] is given by

$$\mathcal{V} = V_x + iV_y = \overline{d\Omega/dz} = \overline{\Omega'(z)} \tag{11}$$

and has magnitude

$$V = |\mathcal{V}| = \sqrt{V_x^2 + V_y^2} = |\overline{\Omega'(z)}| = |\Omega'(z)| \tag{12}$$

Points at which the velocity is zero, i.e. $\Omega'(z) = 0$, are called *stagnation points*.

The function $\Omega(z)$, of fundamental importance in characterizing a flow, is called the *complex potential*.

EQUIPOTENTIAL LINES AND STREAMLINES

The one parameter families of curves

$$\Phi(x, y) = \alpha, \qquad \Psi(x, y) = \beta \tag{13}$$

where α and β are constants, are orthogonal families called respectively the *equipotential lines* and *streamlines* of the flow [although the more appropriate terms *equipotential curves* and *stream curves* are sometimes used]. In steady motion, streamlines represent the actual paths of fluid particles in the flow pattern.

The function Ψ is called the *stream function* while, as already seen, the function Φ is called the *velocity potential function* or briefly the *velocity potential*.

SOURCES AND SINKS

In the above development of theory we assumed that there were no points in the z plane [i.e. lines in the fluid] at which fluid appears or disappears. Such points are called *sources* and *sinks* respectively [also called *line sources* and *line sinks*]. At such points, which are singular points, the equation of continuity (7), and hence (8), fail to hold. In particular the circulation integral in (5) may not be zero around closed curves C which include such points.

No difficulty arises in using the above theory, however, provided we introduce the proper singularities into the complex potential $\Omega(z)$ and note that equations such as (7) and (8) then hold in any region which excludes these singular points.

SOME SPECIAL FLOWS

Theoretically, any complex potential $\Omega(z)$ can be associated with, or interpreted as, a particular two-dimensional fluid flow. The following are some simple cases arising in practice. [Note that a constant can be added to all complex potentials without affecting the flow pattern.]

1. **Uniform Flow.** The complex potential corresponding to the flow of a fluid at constant speed V_0 in a direction making an angle δ with the positive x direction is (Fig. 9-2 below)

$$\Omega(z) = V_0 e^{-i\delta} z \qquad (14)$$

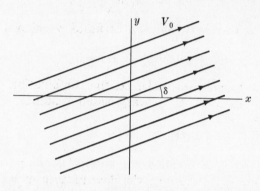

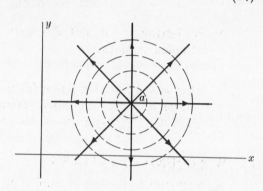

Fig. 9-2   Fig. 9-3

2. **Source at $z = a$.** If fluid is emerging at constant rate from a line source at $z = a$ (Fig. 9-3 above), the complex potential is

$$\Omega(z) = k \ln(z - a) \qquad (15)$$

where $k > 0$ is called the *strength* of the source. The streamlines are shown heavy while the equipotential lines are dashed.

3. **Sink at $z = a$.** In this case the fluid is disappearing at $z = a$ (Fig. 9-4 below) and the complex potential is obtained from that of the source by replacing k by $-k$, giving

$$\Omega(z) = -k \ln(z - a) \qquad (16)$$

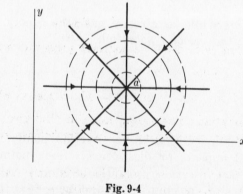

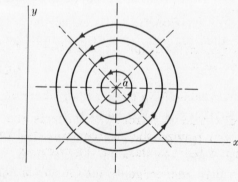

Fig. 9-4   Fig. 9-5

4. **Flow with Circulation.** The flow corresponding to the complex potential

$$\Omega(z) = -ik \ln(z - a) \qquad (17)$$

is as indicated in Fig. 9-5 above. The magnitude of the velocity of fluid at any point is in this case inversely proportional to the distance from a.

The point $z = a$ is called a *vortex* and k is called its *strength*. The circulation [see equation (5)] about any simple closed curve C enclosing $z = a$ is equal in magnitude to $2\pi k$. Note that by changing k to $-k$ in (17) the complex potential corresponding to a "clockwise" vortex is obtained.

5. **Superposition of Flows.** By addition of complex potentials, more complicated flow patterns can be described. An important example is obtained by considering the flow due to a source at $z = -a$ and a sink of equal strength at $z = a$. Then the complex potential is

$$\Omega(z) = k \ln(z + a) - k \ln(z - a) = k \ln\left(\frac{z+a}{z-a}\right) \tag{18}$$

By letting $a \to 0$ and $k \to \infty$ in such a way that $2ka = \mu$ is finite we obtain the complex potential

$$\Omega(z) = \frac{\mu}{z} \tag{19}$$

This is the complex potential due to a *doublet* or *dipole*, i.e. the combination of a source and sink of equal strengths separated by a very small distance. The quantity μ is called the *dipole moment*.

FLOW AROUND OBSTACLES

An important problem in fluid flow is that of determining the flow pattern of a fluid initially moving with uniform velocity V_0 in which an obstacle has been placed.

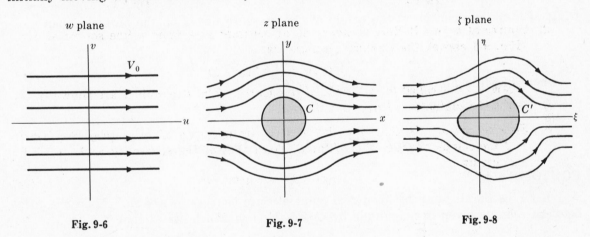

w plane　　　　　z plane　　　　　ζ plane

Fig. 9-6　　　　　**Fig. 9-7**　　　　　**Fig. 9-8**

A general principle involved in this type of problem is to design a complex potential having the form

$$\Omega(z) = V_0 z + G(z) \tag{20}$$

(if the flow is in the z plane) where $G(z)$ is such that $\lim_{|z| \to \infty} G'(z) = 0$, which means physically that far from the obstacle the velocity has constant magnitude (in this case V_0). Furthermore, the complex potential must be chosen so that one of the streamlines represents the boundary of the obstacle.

A knowledge of conformal mapping functions is often useful in obtaining complex potentials. For example, the complex potential corresponding to the uniform flow in the w plane of Fig. 9-6 is given by $V_0 w$. By use of the mapping function $w = z + a^2/z$ [see entry A-4, Page 206] the upper half w plane of Fig. 9-6 is transformed into the upper half z plane exterior to circle C, and the complex potential for the flow of Fig. 9-7 is given by

$$\Omega(z) = V_0\left(z + \frac{a^2}{z}\right) \tag{21}$$

Similarly if $z = F(\zeta)$ maps C and its exterior on to C' and its exterior [see Fig. 9-8], then the complex potential for the flow of Fig. 9-8 is obtained by replacing z by $F(\zeta)$ in (21). The complex potential can also be obtained on going directly from the w to the ζ plane by means of a suitable mapping function.

Using the above and introducing other physical phenomena such as circulation, we can describe the flow pattern about variously shaped airfoils and thus describe the motion of an airplane in flight.

BERNOULLI'S THEOREM

If P denotes the pressure in a fluid and V is the speed of the fluid, then *Bernoulli's theorem* states that

$$P + \tfrac{1}{2}\sigma V^2 = K \tag{22}$$

where σ is the fluid density and K is a constant along any streamline.

THEOREMS OF BLASIUS

1. Let X and Y be the net forces, in the positive x and y directions respectively, due to fluid pressure on the surface of an obstacle bounded by a simple closed curve C. Then if Ω is the complex potential for the flow,

$$X - iY = \tfrac{1}{2}i\sigma \oint_C \left(\frac{d\Omega}{dz}\right)^2 dz \tag{23}$$

2. If M is the moment about the origin of the pressure forces on the obstacle, then

$$M = \mathrm{Re}\left\{-\tfrac{1}{2}\sigma \oint_C z\left(\frac{d\Omega}{dz}\right)^2 dz\right\} \tag{24}$$

where "Re" denotes as usual "real part of".

Applications to Electrostatics

COULOMB'S LAW

Let r be the distance between two point electric charges q_1 and q_2. Then the force between them is given in magnitude by *Coulomb's law* which states that

$$F = \frac{q_1 q_2}{\kappa r^2} \tag{25}$$

and is one of repulsion or attraction according as the charges are like (both positive or both negative) or unlike (one positive and the other negative). The constant κ in (25), which is called the *dielectric constant*, depends on the medium; in a vacuum $\kappa = 1$, in other cases $\kappa > 1$. In the following we assume $\kappa = 1$ unless otherwise specified.

ELECTRIC FIELD INTENSITY. ELECTROSTATIC POTENTIAL

Suppose we are given a charge distribution which may be continuous, discrete, or a combination. This charge distribution sets up an electric field. If a unit positive charge (small enough so as not to affect the field appreciably) is placed at any point A not already occupied by charge, the force acting on this charge is called the *electric field intensity* at A and is denoted by $\mathcal{E}$. This force is derivable from a potential Φ which is sometimes called the *electrostatic potential*. In symbols,

$$\mathcal{E} = -\mathrm{grad}\,\Phi = -\nabla\Phi \tag{26}$$

If the charge distribution is two dimensional, which is our main concern here, then

$$\mathcal{E} \;=\; E_x + iE_y \;=\; -\frac{\partial \Phi}{\partial x} - i\frac{\partial \Phi}{\partial y} \quad \text{where} \quad E_x = -\frac{\partial \Phi}{\partial x}, \; E_y = -\frac{\partial \Phi}{\partial y} \tag{27}$$

In such case if E_t denotes the component of the electric field intensity tangential to any simple closed curve C in the z plane,

$$\oint_C E_t \, ds \;=\; \oint_C E_x \, dx + E_y \, dy \;=\; 0 \tag{28}$$

GAUSS' THEOREM

Let us confine ourselves to charge distributions which can be considered two dimensional. If C is any simple closed curve in the z plane having a net charge q in its interior (actually an infinite cylinder enclosing a net charge q) and E_n is the normal component of the electric field intensity, then *Gauss' theorem* states that

$$\oint_C E_n \, ds \;=\; 4\pi q \tag{29}$$

If C does not enclose any net charge, this reduces to

$$\oint_C E_n \, ds \;=\; \oint_C E_x \, dy - E_y \, dx \;=\; 0 \tag{30}$$

It follows that in any region not occupied by charge,

$$\frac{\partial E_x}{\partial x} + \frac{\partial E_y}{\partial y} \;=\; 0 \tag{31}$$

From (27) and (31), we have

$$\frac{\partial^2 \Phi}{\partial x^2} + \frac{\partial^2 \Phi}{\partial y^2} \;=\; 0 \tag{32}$$

i.e. Φ is harmonic at all points not occupied by charge.

THE COMPLEX ELECTROSTATIC POTENTIAL

From the above it is evident that there must exist a harmonic function Ψ conjugate to Φ such that
$$\Omega(z) \;=\; \Phi(x,y) + i\Psi(x,y) \tag{33}$$
is analytic in any region not occupied by charge. We call $\Omega(z)$ the *complex electrostatic potential* or, briefly, *complex potential*. In terms of this, (27) becomes

$$\mathcal{E} \;=\; -\frac{\partial \Phi}{\partial x} - i\frac{\partial \Phi}{\partial y} \;=\; -\frac{\partial \Phi}{\partial x} + i\frac{\partial \Psi}{\partial y} \;=\; -\overline{\frac{d\Omega}{dz}} \;=\; -\overline{\Omega'(z)} \tag{34}$$

and the magnitude of $\mathcal{E}$ is given by $E = |\mathcal{E}| = |-\overline{\Omega'(z)}| = |\Omega'(z)|$.

The curves (cylindrical surfaces in three dimensions)

$$\Phi(x,y) = \alpha, \quad \Psi(x,y) = \beta \tag{35}$$

are called *equipotential lines* and *flux lines* respectively.

LINE CHARGES

The analogy of the above with fluid flow is quite apparent. The electric field in electrostatic problems corresponds to the velocity field in fluid flow problems, the only difference being a change of sign in the corresponding complex potentials.

The idea of sources and sinks of fluid flow have corresponding analogs for electrostatics. Thus the complex (electrostatic) potential due to a line charge q per unit length at z_0 (in a vacuum) is given by

$$\Omega(z) \;=\; -\,2q \ln(z - z_0) \tag{36}$$

and represents a source or sink according as $q < 0$ or $q > 0$. Similarly we talk about doublets or dipoles, etc. If the medium is not a vacuum, we replace q in (36) by q/κ.

CONDUCTORS

If a solid is perfectly conducting, i.e. is a *perfect conductor*, all charge is located on its surface. Thus if we consider the surface represented by the simple closed curve C in the z plane, the charges are in equilibrium on C and hence C is an equipotential line.

An important problem is the calculation of potential due to a set of charged cylinders. This can be accomplished by use of conformal mapping.

CAPACITANCE

Two conductors having charges of equal magnitude q but of opposite sign, have a difference of potential, say V. The quantity C defined by

$$q \;=\; CV \tag{37}$$

depends only on the geometry of the conductors and is called the *capacitance*. The conductors themselves form what is called a *condenser* or *capacitor*.

Applications to Heat Flow

HEAT FLUX

Consider a solid having a temperature distribution which may be varying. We are often interested in the quantity of heat conducted per unit area per unit time across a surface located in the solid. This quantity, sometimes called the *heat flux* across the surface, is given by

$$\mathcal{Q} \;=\; -\,K \operatorname{grad} \Phi \tag{38}$$

where Φ is the temperature and K, assumed to be a constant, is called the *thermal conductivity* and depends on the material of which the solid is made.

THE COMPLEX TEMPERATURE

If we restrict ourselves to problems of two dimensional type, then

$$\mathcal{Q} \;=\; -\,K\!\left(\frac{\partial \Phi}{\partial x} + i\frac{\partial \Phi}{\partial y}\right) \;=\; Q_x + iQ_y \qquad \text{where} \quad Q_x = -K\frac{\partial \Phi}{\partial x}, \; Q_y = -K\frac{\partial \Phi}{\partial y} \tag{39}$$

Let C be any simple closed curve in the z plane (representing the cross section of a cylinder). If Q_t and Q_n are the tangential and normal components of the heat flux and if *steady state* conditions prevail so that there is no net accumulation of heat inside C, then we have

$$\oint_C Q_n\, ds \;=\; \oint_C Q_x\, dy - Q_y\, dx \;=\; 0, \qquad \oint_C Q_t\, ds \;=\; \oint_C Q_x\, dx + Q_y\, dy \;=\; 0 \tag{40}$$

assuming no sources or sinks inside C. The first equation of (40) yields

$$\frac{\partial Q_x}{\partial x} + \frac{\partial Q_y}{\partial y} = 0 \qquad (41)$$

which becomes on using (39),

$$\frac{\partial^2 \Phi}{\partial x^2} + \frac{\partial^2 \Phi}{\partial y^2} = 0$$

i.e. Φ is harmonic. Introducing the harmonic conjugate function Ψ, we see that

$$\Omega(z) = \Phi(x, y) + i\,\Psi(x, y) \qquad (42)$$

is analytic. The families of curves

$$\Phi(x, y) = \alpha, \qquad \Psi(x, y) = \beta \qquad (43)$$

are called *isothermal lines* and *flux lines* respectively, while $\Omega(z)$ is called the *complex temperature*.

The analogies with fluid flow and electrostatics are evident and procedures used in these fields can be similarly employed in solving various temperature problems.

Solved Problems

HARMONIC FUNCTIONS

1. Show that the functions (a) $x^2 - y^2 + 2y$ and (b) $\sin x \cosh y$ are harmonic in any finite region $\mathcal{R}$ of the z plane.

   (a) If $\Phi = x^2 - y^2 + 2y$, we have $\frac{\partial^2 \Phi}{\partial x^2} = 2$, $\frac{\partial^2 \Phi}{\partial y^2} = -2$. Then $\frac{\partial^2 \Phi}{\partial x^2} + \frac{\partial^2 \Phi}{\partial y^2} = 0$ and Φ is harmonic in $\mathcal{R}$.

   (b) If $\Phi = \sin x \cosh y$, we have $\frac{\partial^2 \Phi}{\partial x^2} = -\sin x \cosh y$, $\frac{\partial^2 \Phi}{\partial y^2} = \sin x \cosh y$. Then $\frac{\partial^2 \Phi}{\partial x^2} + \frac{\partial^2 \Phi}{\partial y^2} = 0$ and Φ is harmonic in $\mathcal{R}$.

2. Show that the functions of Problem 1 are harmonic in the w plane under the transformation $z = w^3$.

   If $z = w^3$, then $x + iy = (u + iv)^3 = u^3 - 3uv^2 + i(3u^2v - v^3)$ and $x = u^3 - 3uv^2$, $y = 3u^2v - v^3$.

   (a) $\Phi = x^2 - y^2 + 2y = (u^3 - 3uv^2)^2 - (3u^2v - v^3)^2 + 2(3u^2v - v^3)$
   $= u^6 - 15u^4v^2 + 15u^2v^4 - v^6 + 6u^2v - 2v^3$

   Then $\frac{\partial^2 \Phi}{\partial u^2} = 30u^4 - 180u^2v^2 + 30v^4 + 12v$, $\qquad \frac{\partial^2 \Phi}{\partial v^2} = -30u^4 + 180u^2v^2 - 30v^4 - 12v$

   and $\frac{\partial^2 \Phi}{\partial u^2} + \frac{\partial^2 \Phi}{\partial v^2} = 0$ as required.

   (b) We must show that $\Phi = \sin(u^3 - 3uv^2) \cosh(3u^2v - v^3)$ satisfies $\frac{\partial^2 \Phi}{\partial u^2} + \frac{\partial^2 \Phi}{\partial v^2} = 0$. This can readily be established by straightforward but tedious differentiation.

   This problem illustrates a general result proved in Problem 4.

3. Prove that $\dfrac{\partial^2 \Phi}{\partial x^2} + \dfrac{\partial^2 \Phi}{\partial y^2} = |f'(z)|^2 \left(\dfrac{\partial^2 \Phi}{\partial u^2} + \dfrac{\partial^2 \Phi}{\partial v^2} \right)$ where $w = f(z)$ is analytic and $f'(z) \neq 0$.

The function $\Phi(x, y)$ is transformed into a function $\Phi[x(u, v), y(u, v)]$ by the transformation. By differentiation we have

$$\frac{\partial \Phi}{\partial x} = \frac{\partial \Phi}{\partial u}\frac{\partial u}{\partial x} + \frac{\partial \Phi}{\partial v}\frac{\partial v}{\partial x}, \qquad \frac{\partial \Phi}{\partial y} = \frac{\partial \Phi}{\partial u}\frac{\partial u}{\partial y} + \frac{\partial \Phi}{\partial v}\frac{\partial v}{\partial y}$$

$$\frac{\partial^2 \Phi}{\partial x^2} = \frac{\partial \Phi}{\partial u}\frac{\partial^2 u}{\partial x^2} + \frac{\partial u}{\partial x}\frac{\partial}{\partial x}\left(\frac{\partial \Phi}{\partial u}\right) + \frac{\partial \Phi}{\partial v}\frac{\partial^2 v}{\partial x^2} + \frac{\partial v}{\partial x}\frac{\partial}{\partial x}\left(\frac{\partial \Phi}{\partial v}\right)$$

$$= \frac{\partial \Phi}{\partial u}\frac{\partial^2 u}{\partial x^2} + \frac{\partial u}{\partial x}\left[\frac{\partial}{\partial u}\left(\frac{\partial \Phi}{\partial u}\right)\frac{\partial u}{\partial x} + \frac{\partial}{\partial v}\left(\frac{\partial \Phi}{\partial u}\right)\frac{\partial v}{\partial x}\right]$$

$$+ \frac{\partial \Phi}{\partial v}\frac{\partial^2 v}{\partial x^2} + \frac{\partial v}{\partial x}\left[\frac{\partial}{\partial u}\left(\frac{\partial \Phi}{\partial v}\right)\frac{\partial u}{\partial x} + \frac{\partial}{\partial v}\left(\frac{\partial \Phi}{\partial v}\right)\frac{\partial v}{\partial x}\right]$$

$$= \frac{\partial \Phi}{\partial u}\frac{\partial^2 u}{\partial x^2} + \frac{\partial u}{\partial x}\left[\frac{\partial^2 \Phi}{\partial u^2}\frac{\partial u}{\partial x} + \frac{\partial^2 \Phi}{\partial v \partial u}\frac{\partial v}{\partial x}\right] + \frac{\partial \Phi}{\partial v}\frac{\partial^2 v}{\partial x^2} + \frac{\partial v}{\partial x}\left[\frac{\partial^2 \Phi}{\partial u \partial v}\frac{\partial u}{\partial x} + \frac{\partial^2 \Phi}{\partial v^2}\frac{\partial v}{\partial x}\right]$$

Similarly,

$$\frac{\partial^2 \Phi}{\partial y^2} = \frac{\partial \Phi}{\partial u}\frac{\partial^2 u}{\partial y^2} + \frac{\partial u}{\partial y}\left[\frac{\partial^2 \Phi}{\partial u^2}\frac{\partial u}{\partial y} + \frac{\partial^2 \Phi}{\partial v \partial u}\frac{\partial v}{\partial y}\right] + \frac{\partial \Phi}{\partial v}\frac{\partial^2 v}{\partial y^2} + \frac{\partial v}{\partial y}\left[\frac{\partial^2 \Phi}{\partial u \partial v}\frac{\partial u}{\partial y} + \frac{\partial^2 \Phi}{\partial v^2}\frac{\partial v}{\partial y}\right]$$

Adding,

$$\frac{\partial^2 \Phi}{\partial x^2} + \frac{\partial^2 \Phi}{\partial y^2} = \frac{\partial \Phi}{\partial u}\left(\frac{\partial^2 u}{\partial x^2} + \frac{\partial^2 u}{\partial y^2}\right) + \frac{\partial \Phi}{\partial v}\left(\frac{\partial^2 v}{\partial x^2} + \frac{\partial^2 v}{\partial y^2}\right) + \frac{\partial^2 \Phi}{\partial u^2}\left[\left(\frac{\partial u}{\partial x}\right)^2 + \left(\frac{\partial u}{\partial y}\right)^2\right]$$

$$+ 2\frac{\partial^2 \Phi}{\partial u \partial v}\left[\frac{\partial u}{\partial x}\frac{\partial v}{\partial x} + \frac{\partial u}{\partial y}\frac{\partial v}{\partial y}\right] + \frac{\partial^2 \Phi}{\partial v^2}\left[\left(\frac{\partial v}{\partial x}\right)^2 + \left(\frac{\partial v}{\partial y}\right)^2\right]$$

$$(1)$$

Since u and v are harmonic, $\dfrac{\partial^2 u}{\partial x^2} + \dfrac{\partial^2 u}{\partial y^2} = 0$, $\dfrac{\partial^2 v}{\partial x^2} + \dfrac{\partial^2 v}{\partial y^2} = 0$. Also, by the Cauchy-Riemann equations, $\dfrac{\partial u}{\partial x} = \dfrac{\partial v}{\partial y}$, $\dfrac{\partial v}{\partial x} = -\dfrac{\partial u}{\partial y}$. Then

$$\left(\frac{\partial u}{\partial x}\right)^2 + \left(\frac{\partial u}{\partial y}\right)^2 = \left(\frac{\partial v}{\partial x}\right)^2 + \left(\frac{\partial v}{\partial y}\right)^2 = \left(\frac{\partial u}{\partial x}\right)^2 + \left(\frac{\partial v}{\partial x}\right)^2 = \left|\frac{\partial u}{\partial x} + i\frac{\partial v}{\partial x}\right|^2 = |f'(z)|^2$$

$$\frac{\partial u}{\partial x}\frac{\partial v}{\partial x} + \frac{\partial u}{\partial y}\frac{\partial v}{\partial y} = 0$$

Hence (1) becomes $\qquad \dfrac{\partial^2 \Phi}{\partial x^2} + \dfrac{\partial^2 \Phi}{\partial y^2} = |f'(z)|^2 \left(\dfrac{\partial^2 \Phi}{\partial u^2} + \dfrac{\partial^2 \Phi}{\partial v^2} \right)$

4. Prove that a harmonic function $\Phi(x, y)$ remains harmonic under the transformation $w = f(z)$ where $f(z)$ is analytic and $f'(z) \neq 0$.

This follows at once from Problem 3, since if $\dfrac{\partial^2 \Phi}{\partial x^2} + \dfrac{\partial^2 \Phi}{\partial y^2} = 0$ and $f'(z) \neq 0$, then $\dfrac{\partial^2 \Phi}{\partial u^2} + \dfrac{\partial^2 \Phi}{\partial v^2} = 0$.

5. If a is real, show that the real and imaginary parts of $w = \ln(z - a)$ are harmonic functions in any region $\mathcal{R}$ not containing $z = a$.

Method 1.

If $\mathcal{R}$ does not contain a, then $w = \ln(z - a)$ is analytic in $\mathcal{R}$. Hence the real and imaginary parts are harmonic in $\mathcal{R}$.

Method 2.

Let $z - a = re^{i\theta}$. Then if principal values are used for θ, $w = u + iv = \ln(z - a) = \ln r + i\theta$ so that $u = \ln r$, $v = \theta$.

In the polar coordinates (r, θ), Laplace's equation is $\dfrac{\partial^2 \Phi}{\partial r^2} + \dfrac{1}{r} \dfrac{\partial \Phi}{\partial r} + \dfrac{1}{r^2} \dfrac{\partial^2 \Phi}{\partial \theta^2} = 0$ and by direct substitution we find that $u = \ln r$ and $v = \theta$ are solutions if $\mathcal{R}$ does not contain $r = 0$, i.e. $z = a$.

Method 3.

If $z - a = re^{i\theta}$, then $x - a = r\cos\theta$, $y = r\sin\theta$ and $r = \sqrt{(x - a)^2 + y^2}$, $\theta = \tan^{-1}\{y/(x - a)\}$. Then $w = u + iv = \frac{1}{2}\ln\{(x - a)^2 + y^2\} + i\tan^{-1}\{y/(x - a)\}$ and $u = \frac{1}{2}\ln\{(x - a)^2 + y^2\}$, $v = \tan^{-1}\{y/(x - a)\}$. Substituting these into Laplace's equation $\dfrac{\partial^2 \Phi}{\partial x^2} + \dfrac{\partial^2 \Phi}{\partial y^2} = 0$, we find after straightforward differentiation that u and v are solutions if $z \neq a$.

DIRICHLET AND NEUMANN PROBLEMS

6. Find a function harmonic in the upper half of the z plane, $\text{Im}\{z\} > 0$, which takes the prescribed values on the x axis given by $G(x) = \begin{cases} 1 & x > 0 \\ 0 & x < 0 \end{cases}$.

We must solve for $\Phi(x, y)$ the boundary-value problem

$$\frac{\partial^2 \Phi}{\partial x^2} + \frac{\partial^2 \Phi}{\partial y^2} = 0, \quad y > 0; \qquad \lim_{y \to 0+} \Phi(x, y) = G(x) = \begin{cases} 1 & x > 0 \\ 0 & x < 0 \end{cases}$$

This is a Dirichlet problem for the upper half plane [see Fig. 9-9].

The function $A\theta + B$, where A and B are real constants, is harmonic since it is the imaginary part of $A \ln z + B$.

To determine A and B note that the boundary conditions are $\Phi = 1$ for $x > 0$, i.e. $\theta = 0$ and $\Phi = 0$ for $x < 0$, i.e. $\theta = \pi$. Thus

    (1) $1 = A(0) + B$,   (2) $0 = A(\pi) + B$

from which $A = -1/\pi$, $B = 1$.

Then the required solution is

$$\Phi = A\theta + B = 1 - \frac{\theta}{\pi} = 1 - \frac{1}{\pi}\tan^{-1}\left(\frac{y}{x}\right)$$

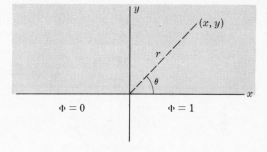

Fig. 9-9

Another method, using Poisson's formula for the half plane.

$$\Phi(x, y) = \frac{1}{\pi}\int_{-\infty}^{\infty} \frac{y\, G(\eta)\, d\eta}{y^2 + (x - \eta)^2} = \frac{1}{\pi}\int_{-\infty}^{0} \frac{y[0]\, d\eta}{y^2 + (x - \eta)^2} + \frac{1}{\pi}\int_{0}^{\infty} \frac{y[1]\, d\eta}{y^2 + (x - \eta)^2}$$

$$= \frac{1}{\pi}\tan^{-1}\left(\frac{\eta - x}{y}\right)\Big|_{0}^{\infty} = \frac{1}{2} + \frac{1}{\pi}\tan^{-1}\left(\frac{x}{y}\right) = 1 - \frac{1}{\pi}\tan^{-1}\left(\frac{y}{x}\right)$$

7. Solve the boundary-value problem

$$\frac{\partial^2 \Phi}{\partial x^2} + \frac{\partial^2 \Phi}{\partial y^2} = 0, \quad y > 0;$$

$$\lim_{y \to 0+} \Phi(x, y) = G(x) = \begin{cases} T_0 & x < -1 \\ T_1 & -1 < x < 1 \\ T_2 & x > 1 \end{cases}$$

where T_0, T_1, T_2 are constants.

This is a Dirichlet problem for the upper half plane [see Fig. 9-10].

The function $A\theta_1 + B\theta_2 + C$ where A, B and C are real constants, is harmonic since it is the imaginary part of $A \ln(z + 1) + B \ln(z - 1) + C$.

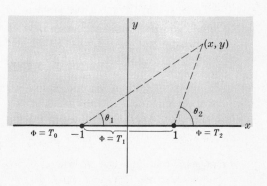

Fig. 9-10

To determine A, B, C note that the boundary conditions are: $\Phi = T_2$ for $x > 1$, i.e. $\theta_1 = \theta_2 = 0$; $\Phi = T_1$ for $-1 < x < 1$, i.e. $\theta_1 = 0$, $\theta_2 = \pi$; $\Phi = T_0$ for $x < -1$, i.e. $\theta_1 = \pi$, $\theta_2 = \pi$. Thus

(1) $\quad T_2 = A(0) + B(0) + C$ $\qquad$ (2) $\quad T_1 = A(0) + B(\pi) + C$ $\qquad$ (3) $\quad T_0 = A(\pi) + B(\pi) + C$

from which $\quad C = T_2$, $\;\; B = (T_1 - T_2)/\pi$, $\;\; A = (T_0 - T_1)/\pi$.

Then the required solution is

$$\Phi = A\theta_1 + B\theta_2 + C = \frac{T_0 - T_1}{\pi} \tan^{-1}\left(\frac{y}{x+1}\right) + \frac{T_1 - T_2}{\pi} \tan^{-1}\left(\frac{y}{x-1}\right) + T_2$$

Another method, using Poisson's formula for the half plane.

$$\begin{aligned}
\Phi(x, y) &= \frac{1}{\pi} \int_{-\infty}^{\infty} \frac{y\, G(\eta)\, d\eta}{y^2 + (x - \eta)^2} \\
&= \frac{1}{\pi} \int_{-\infty}^{-1} \frac{y\, T_0\, d\eta}{y^2 + (x - \eta)^2} + \frac{1}{\pi} \int_{-1}^{1} \frac{y\, T_1\, d\eta}{y^2 + (x - \eta)^2} + \frac{1}{\pi} \int_{1}^{\infty} \frac{y\, T_2\, d\eta}{y^2 + (x - \eta)^2} \\
&= \frac{T_0}{\pi} \tan^{-1}\left(\frac{\eta - x}{y}\right)\Big|_{-\infty}^{-1} + \frac{T_1}{\pi} \tan^{-1}\left(\frac{\eta - x}{y}\right)\Big|_{-1}^{1} + \frac{T_2}{\pi} \tan^{-1}\left(\frac{\eta - x}{y}\right)\Big|_{1}^{\infty} \\
&= \frac{T_0 - T_1}{\pi} \tan^{-1}\left(\frac{y}{x+1}\right) + \frac{T_1 - T_2}{\pi} \tan^{-1}\left(\frac{y}{x-1}\right) + T_2
\end{aligned}$$

8. **Find a function harmonic inside the unit circle** $|z| = 1$ **and taking the prescribed values given by** $\;F(\theta) = \begin{cases} 1 & 0 < \theta < \pi \\ 0 & \pi < \theta < 2\pi \end{cases}\;$ **on its circumference.**

This is a Dirichlet problem for the unit circle [Fig. 9-11] in which we seek a function satisfying Laplace's equation inside $|z| = 1$ and taking the values 0 on arc ABC and 1 on arc CDE.

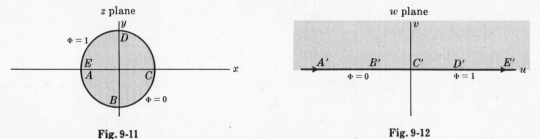

Fig. 9-11 $\qquad\qquad\qquad\qquad\qquad\qquad\qquad\qquad$ Fig. 9-12

Method 1, using conformal mapping.

We map the interior of the circle $|z| = 1$ on to the upper half of the w plane [Fig. 9-12] by using the mapping function $\;z = \dfrac{i - w}{i + w}\;$ or $\;w = i\left(\dfrac{1 - z}{1 + z}\right)\;$ [see Problem 12, Chapter 8, Page 217, and interchange z and w].

Under this transformation, arcs ABC and CDE are mapped on to the negative and positive real axis $A'B'C'$ and $C'D'E'$ respectively of the w plane. Then by Problem 81, the boundary conditions $\Phi = 0$ on arc ABC and $\Phi = 1$ on arc CDE become respectively $\Phi = 0$ on $A'B'C'$ and $\Phi = 1$ on $C'D'E'$.

Thus we have reduced the problem to finding a function Φ harmonic in the upper half w plane and taking the values 0 for $u < 0$ and 1 for $u > 0$. But this problem has already been solved in Problem 6 and the solution (replacing x by u and y by v) is given by

$$\Phi = 1 - \frac{1}{\pi} \tan^{-1}\left(\frac{v}{u}\right) \tag{1}$$

Now from $\;w = i\left(\dfrac{1 - z}{1 + z}\right)$, we find $\;u = \dfrac{2y}{(1 + x)^2 + y^2}$, $\;v = \dfrac{1 - (x^2 + y^2)}{(1 + x)^2 + y^2}$. Then substituting these in (1), we find the required solution

$$\Phi = 1 - \frac{1}{\pi} \tan^{-1}\left(\frac{2y}{1 - [x^2 + y^2]}\right) \tag{2}$$

or in polar coordinates (r, θ), where $x = r\cos\theta$, $y = r\sin\theta$,

$$\Phi = 1 - \frac{1}{\pi} \tan^{-1}\left(\frac{2r\sin\theta}{1 - r^2}\right) \tag{3}$$

Method 2, using Poisson's formula.

$$\Phi(r,\theta) = \frac{1}{2\pi}\int_0^{2\pi}\frac{F(\phi)\,d\phi}{1-2r\cos(\theta-\phi)+r^2}$$

$$= \frac{1}{2\pi}\int_0^{\pi}\frac{d\phi}{1-2r\cos(\theta-\phi)+r^2} = 1-\frac{1}{\pi}\tan^{-1}\left(\frac{2r\sin\theta}{1-r^2}\right)$$

by direct integration [see Problem 69(*b*), Chapter 5, Page 136].

APPLICATIONS TO FLUID FLOW

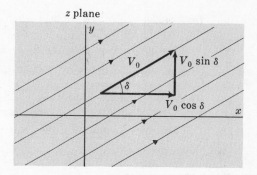

z plane

Fig. 9-13

9. (*a*) Find the complex potential for a fluid moving with constant speed V_0 in a direction making an angle δ with the positive x axis [see Fig. 9-13].

   (*b*) Determine the velocity potential and stream function.

   (*c*) Determine the equations for the streamlines and equipotential lines.

(*a*) The x and y components of velocity are

$$V_x = V_0\cos\delta, \qquad V_y = V_0\sin\delta$$

The complex velocity is

$$\mathcal{V} = V_x + iV_y = V_0\cos\delta + iV_0\sin\delta = V_0\,e^{i\delta}$$

The complex potential $\Omega(z)$ is given by

$$\frac{d\Omega}{dz} = \overline{\mathcal{V}} = V_0\,e^{-i\delta}$$

Then integrating, $\qquad\qquad \Omega(z) = V_0\,e^{-i\delta}\,z$

omitting the constant of integration.

(*b*) The velocity potential Φ and stream function Ψ are the real and imaginary parts of the complex potential. Thus

$$\Omega(z) = \Phi + i\Psi = V_0\,e^{-i\delta}z = V_0(x\cos\delta + y\sin\delta) + iV_0(y\cos\delta - x\sin\delta)$$

and $\qquad\qquad \Phi = V_0(x\cos\delta + y\sin\delta), \qquad \Psi = V_0(y\cos\delta - x\sin\delta)$

Another method.

$$(1)\ \ \frac{\partial\Phi}{\partial x} = V_x = V_0\cos\delta \qquad\qquad (2)\ \ \frac{\partial\Phi}{\partial y} = V_y = V_0\sin\delta$$

Solving for Φ in (1), $\Phi = (V_0\cos\delta)x + G(y)$. Substituting in (2), $G'(y) = V_0\sin\delta$ and $G(y) = (V_0\sin\delta)y$, omitting the constant of integration. Then

$$\Phi = (V_0\cos\delta)x + (V_0\sin\delta)y$$

From the Cauchy-Riemann equations,

$$(3)\ \ \frac{\partial\Psi}{\partial y} = \frac{\partial\Phi}{\partial x} = V_x = V_0\cos\delta \qquad\qquad (4)\ \ \frac{\partial\Psi}{\partial x} = -\frac{\partial\Phi}{\partial y} = -V_y = -V_0\sin\delta$$

Solving for Ψ in (3), $\Psi = (V_0\cos\delta)y + H(x)$. Substituting in (4), $H'(x) = -V_0\sin\delta$ and $H(x) = -(V_0\sin\delta)x$, omitting the constant of integration. Then

$$\Psi = (V_0\cos\delta)y - (V_0\sin\delta)x$$

(*c*) The streamlines are given by $\Psi = V_0(y\cos\delta - x\sin\delta) = \beta$ for different values of β. Physically, under steady-state conditions, a streamline represents the path actually taken by a fluid particle, in this case a straight line path.

The equipotential lines are given by $\Phi = V_0(x\cos\delta + y\sin\delta) = \alpha$ for different values of α. Geometrically they are lines perpendicular to the streamlines; all points on an equipotential line are at equal potential.

10. The complex potential of a fluid flow is given by $\Omega(z) = V_0\left(z + \dfrac{a^2}{z}\right)$ where V_0 and a are positive constants. (a) Obtain equations for the streamlines and equipotential lines, represent them graphically and interpret physically. (b) Show that we can interpret the flow as that around a circular obstacle of radius a. (c) Find the velocity at any point and determine its value far from the obstacle. (d) Find the stagnation points.

(a) Let $z = re^{i\theta}$. Then

$$\Omega(z) = \Phi + i\Psi = V_0\left(re^{i\theta} + \frac{a^2}{r}e^{-i\theta}\right) = V_0\left(r + \frac{a^2}{r}\right)\cos\theta + iV_0\left(r - \frac{a^2}{r}\right)\sin\theta$$

from which $\qquad \Phi = V_0\left(r + \dfrac{a^2}{r}\right)\cos\theta, \qquad \Psi = V_0\left(r - \dfrac{a^2}{r}\right)\sin\theta$

The streamlines are given by $\Psi = \text{constant} = \beta$, i.e.,

$$V_0\left(r - \frac{a^2}{r}\right)\sin\theta = \beta$$

These are indicated by the heavy curves of Fig. 9-14 and show the actual paths taken by fluid particles. Note that $\Psi = 0$ corresponds to $r = a$ and $\theta = 0$ or π.

The equipotential lines are given by $\Phi = \text{constant} = \alpha$, i.e.,

$$V_0\left(r + \frac{a^2}{r}\right)\cos\theta = \alpha$$

These are indicated by the dashed curves of Fig. 9-14 and are orthogonal to the family of streamlines.

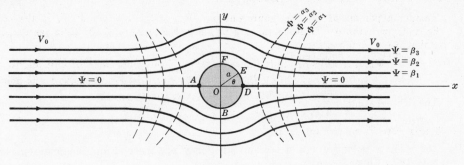

Fig. 9-14

(b) The circle $r = a$ represents a streamline; and since there cannot be any flow across a streamline, it can be considered as a circular obstacle of radius a placed in the path of the fluid.

(c) We have

$$\Omega'(z) = V_0\left(1 - \frac{a^2}{z^2}\right) = V_0\left(1 - \frac{a^2}{r^2}e^{-2i\theta}\right) = V_0\left(1 - \frac{a^2}{r^2}\cos 2\theta\right) + i\frac{V_0 a^2}{r^2}\sin 2\theta$$

Then the complex velocity is

$$\mathcal{V} = \overline{\Omega'(z)} = V_0\left(1 - \frac{a^2}{r^2}\cos 2\theta\right) - i\frac{V_0 a^2}{r^2}\sin 2\theta \tag{1}$$

and its magnitude is

$$V = |\mathcal{V}| = \sqrt{\left\{V_0\left(1 - \frac{a^2}{r^2}\cos 2\theta\right)\right\}^2 + \left\{\frac{V_0 a^2}{r^2}\sin 2\theta\right\}^2}$$

$$= V_0\sqrt{1 - \frac{2a^2\cos 2\theta}{r^2} + \frac{a^4}{r^4}} \tag{2}$$

Far from the obstacle, we see from (1) that $\mathcal{V} = V_0$ approximately, i.e. the fluid is traveling in the direction of the positive x axis with constant speed V_0.

(d) The stagnation points, i.e. points at which the velocity is zero, are given by

$$\Omega'(z) = 0, \quad \text{i.e.} \quad V_0\left(1 - \frac{a^2}{z^2}\right) = 0 \quad \text{or} \quad z = a \text{ and } z = -a$$

The stagnation points are therefore at A and D in Fig. 9-14.

11. Show that under the transformation $w = z + \dfrac{a^2}{z}$ the fluid flow in the z plane considered in Problem 10 is mapped into a uniform flow with constant velocity V_0 in the w plane.

The complex potential for the flow in the w plane is given by

$$V_0\left(z + \frac{a^2}{z}\right) = V_0 w$$

which represents uniform flow with constant velocity V_0 in the w plane [compare entry A-4 in the table on Page 206].

In general, the transformation $w = \Omega(z)$ maps the fluid flow in the z plane with complex potential $\Omega(z)$ into a uniform flow in the w plane. This is very useful in determining complex potentials of complicated fluid patterns through a knowledge of mapping functions.

12. Fluid emanates at a constant rate from an infinite line source perpendicular to the z plane at $z = 0$ [Fig. 9-15]. (a) Show that the speed of the fluid at a distance r from the source is $V = k/r$ where k is a constant. (b) Show that the complex potential is $\Omega(z) = k \ln z$. (c) What modification should be made in (b) if the line source is at $z = a$? (d) What modification is made in (b) if the source is replaced by a sink in which fluid is disappearing at a constant rate?

(a) Consider a portion of the line source of unit length [Fig. 9-16]. If V_r is the radial velocity of the fluid at distance r from the source and σ is the density of the fluid (assumed incompressible so that σ is constant), then:

Mass of fluid per unit time emanating from line source of unit length
=   Mass of fluid crossing surface of cylinder of radius r and height 1
=   (Surface area)(Radial velocity)(Fluid density)
= $(2\pi r \cdot 1)(V_r)(\sigma)$ = $2\pi r V_r \sigma$

If this is to be a constant κ, then

$$V_r = \frac{\kappa}{2\pi\sigma r} = \frac{k}{r}$$

where $k = \kappa/2\pi\sigma$ is called the *strength* of the source.

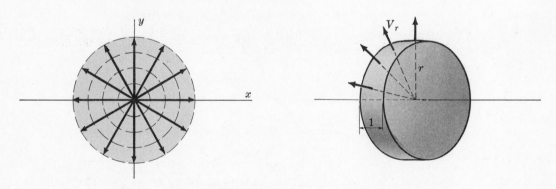

Fig. 9-15   Fig. 9-16

(b) Since $V_r = \dfrac{\partial\Phi}{\partial r} = \dfrac{k}{r}$, we have on integrating and omitting the constant of integration, $\Phi = k \ln r$. But this is the real part of $\Omega(z) = k \ln z$ which is therefore the required complex potential.

(c) If the line source is at $z = a$ instead of $z = 0$, replace z by $z - a$ to obtain the complex potential $\Omega(z) = k \ln (z - a)$.

(d) If the source is replaced by a sink, the complex potential is $\Omega(z) = -k \ln z$, the minus sign arising from the fact that the velocity is directed toward $z = 0$.

Similarly, $\Omega(z) = -k \ln (z - a)$ is the complex potential for a sink at $z = a$.

13. (a) Find the complex potential due to a source at $z = -a$ and a sink at $z = a$ of equal strengths k. (b) Determine the equipotential lines and streamlines and represent graphically. (c) Find the speed of the fluid at any point.

(a) Complex potential due to source at $z = -a$ of strength k is $k \ln (z + a)$.

Complex potential due to sink at $z = a$ of strength k is $-k \ln (z - a)$.

Then by superposition:

Complex potential due to source at $z = -a$ and sink at $z = a$ of strengths k is

$$\Omega(z) = k \ln (z + a) - k \ln (z - a) = k \ln \left(\frac{z + a}{z - a}\right)$$

(b) Let $z + a = r_1 e^{i\theta_1}$, $z - a = r_2 e^{i\theta_2}$. Then

$$\Omega(z) = \Phi + i\Psi = k \ln \left(\frac{r_1 e^{i\theta_1}}{r_2 e^{i\theta_2}}\right) = k \ln \left(\frac{r_1}{r_2}\right) + ik(\theta_1 - \theta_2)$$

so that $\Phi = k \ln (r_1/r_2)$, $\Psi = k(\theta_1 - \theta_2)$. The equipotential lines and streamlines are thus given by

$$\Phi = k \ln (r_1/r_2) = \alpha, \qquad \Psi = k(\theta_1 - \theta_2) = \beta$$

Using $r_1 = \sqrt{(x + a)^2 + y^2}$, $r_2 = \sqrt{(x - a)^2 + y^2}$, $\theta_1 = \tan^{-1}\left(\frac{y}{x + a}\right)$, $\theta_2 = \tan^{-1}\left(\frac{y}{x - a}\right)$, the equipotential lines are given by

$$\frac{\sqrt{(x + a)^2 + y^2}}{\sqrt{(x - a)^2 + y^2}} = e^{\alpha/k}$$

This can be written in the form

$$[x - a \coth (\alpha/k)]^2 + y^2 = a^2 \operatorname{csch}^2 (\alpha/k)$$

which for different values of α are circles having centers at $a \coth (\alpha/k)$ and radii equal to $a \left|\operatorname{csch} (\alpha/k)\right|$.

These circles are shown by the dashed curves of Fig. 9-17.

The streamlines are given by

$$\tan^{-1}\left(\frac{y}{x + a}\right) - \tan^{-1}\left(\frac{y}{x - a}\right) = \beta/k$$

or taking the tangent of both sides and simplifying,

$$x^2 + [y + a \cot (\beta/k)]^2 = a^2 \csc^2 (\beta/k)$$

which for different values of β are circles having centers at $-a \cot (\beta/k)$ and radii $a \left|\csc (\beta/k)\right|$. These circles, which pass through $(-a, 0)$ and $(a, 0)$, are shown heavy in Fig. 9-17.

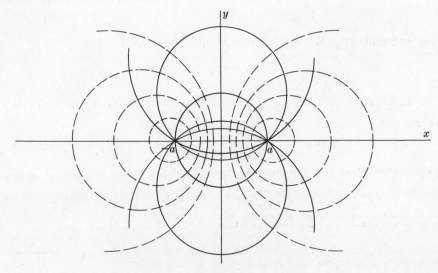

Fig. 9-17

(c) Speed $= |\Omega'(z)| = \left| \dfrac{k}{z+a} - \dfrac{k}{z-a} \right| = \dfrac{2ka}{|z^2 - a^2|}$

$$= \dfrac{2ka}{|a^2 - r^2 e^{2i\theta}|} = \dfrac{2ka}{\sqrt{a^4 - 2a^2 r^2 \cos 2\theta + r^4}}$$

14. Discuss the motion of a fluid having complex potential $\Omega(z) = ik \ln z$ where $k > 0$.

If $z = re^{i\theta}$, then $\Omega(z) = \Phi + i\Psi = ik(\ln r + i\theta) = ik \ln r - k\theta$ or $\Phi = -k\theta$, $\Psi = k \ln r$.

The streamlines are given by

$\Psi = \text{constant}$ or $r = \text{constant}$

which are circles having common center at $z = 0$ [shown heavy in Fig. 9-18].

The equipotential lines, given by $\theta = \text{constant}$, are shown dashed in Fig. 9-18.

Since $\Omega'(z) = \dfrac{ik}{z} = \dfrac{ik}{r} e^{-i\theta} = \dfrac{k \sin \theta}{r} + \dfrac{ik \cos \theta}{r}$,

the complex velocity is given by

$$\mathcal{V} = \overline{\Omega'(z)} = \dfrac{k \sin \theta}{r} - \dfrac{ik \cos \theta}{r}$$

and shows that the direction of fluid flow is clockwise as indicated in the figure. The speed is given by $V = |\mathcal{V}| = k/r$.

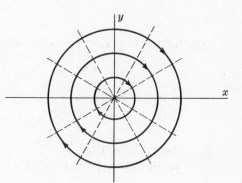

Fig. 9-18

Thus the complex potential describes the flow of a fluid which is rotating around $z = 0$. The flow is sometimes referred to as a *vortex flow* and $z = 0$ is called a *vortex*.

15. Show that the circulation about the vortex in Problem 14 is given by $\gamma = 2\pi k$.

If curve C encloses $z = 0$, the circulation integral is given by

$$\gamma = \oint_C V_t \, ds = \oint_C V_x \, dx + V_y \, dy = \oint_C -\dfrac{\partial \Phi}{\partial x} dx - \dfrac{\partial \Phi}{\partial y} dy = \oint_C -d\Phi$$

$$= \int_0^{2\pi} k \, d\theta = 2\pi k$$

In terms of the circulation the complex potential can be written $\Omega(z) = \dfrac{i\gamma}{2\pi} \ln z$.

16. Discuss the motion of a fluid having complex potential

$$\Omega(z) = V_0 \left(z + \dfrac{a^2}{z} \right) + \dfrac{i\gamma}{2\pi} \ln z$$

This complex potential has the effect of superimposing a circulation on the flow of Problem 10.

If $z = re^{i\theta}$,

$$\Omega(z) = \Phi + i\Psi = V_0 \left(r + \dfrac{a^2}{r} \right) \cos \theta - \dfrac{\gamma\theta}{2\pi} + i \left\{ V_0 \left(r - \dfrac{a^2}{r} \right) \sin \theta + \dfrac{\gamma}{2\pi} \ln r \right\}$$

Then the equipotential lines and streamlines are given by

$$V_0 \left(r + \dfrac{a^2}{r} \right) \cos \theta - \dfrac{\gamma\theta}{2\pi} = \alpha, \qquad V_0 \left(r - \dfrac{a^2}{r} \right) \sin \theta + \dfrac{\gamma}{2\pi} \ln r = \beta$$

There are in general two stagnation points occurring where $\Omega'(z) = 0$, i.e.,

$$V_0 \left(1 - \dfrac{a^2}{z^2} \right) + \dfrac{i\gamma}{2\pi z} = 0 \qquad \text{or} \qquad z = \dfrac{-i\gamma}{4\pi V_0} \pm \sqrt{a^2 - \dfrac{\gamma^2}{16\pi^2 V_0^2}}$$

In case $\gamma = 4\pi a V_0$, there is only one stagnation point.

Since $r = a$ is a streamline corresponding to $\beta = \frac{\gamma}{2\pi} \ln a$, the flow can be considered as one about a circular obstacle as in Problem 10. Far from this obstacle the fluid has velocity V_0 since $\lim\limits_{|z| \to \infty} \Omega'(z) = V_0$.

The flow pattern changes, depending on the magnitude of γ. In Figures 9-19 and 9-20 we have shown two of the many possible ones. Fig. 9-19 corresponds to $\gamma < 4\pi a V_0$; the stagnation points are situated at A and B. Fig. 9-20 corresponds to $\gamma > 4\pi a V_0$ and there is only one stagnation point in the fluid at C.

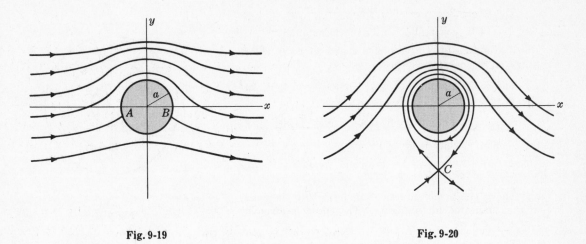

Fig. 9-19   Fig. 9-20

THEOREMS OF BLASIUS

17. Let $\Omega(z)$ be the complex potential describing the flow about a cylindrical obstacle of unit length whose boundary in the z plane is a simple closed curve C. Prove that the net fluid force on the obstacle is given by

$$\bar{F} = X - iY = \tfrac{1}{2} i\sigma \oint_C \left(\frac{d\Omega}{dz}\right)^2 dz$$

where X and Y are the components of force in the positive x and y directions respectively and σ is the fluid density.

The force acting on the element of area ds in Fig. 9-21 is normal to ds and given in magnitude by $P\,ds$ where P is the pressure. On resolving this force into components parallel to the x and y axes, we see that it is given by

$$\begin{aligned}
dF &= dX + i\,dY \\
&= -P\,ds \sin\theta + iP\,ds \cos\theta \\
&= iP\,ds\,(\cos\theta + i\sin\theta) \\
&= iP\,ds\,e^{i\theta} \\
&= iP\,dz
\end{aligned}$$

using the fact that

$$\begin{aligned}
dz &= dx + i\,dy \\
&= ds \cos\theta + i\,ds \sin\theta \\
&= ds\,e^{i\theta}
\end{aligned}$$

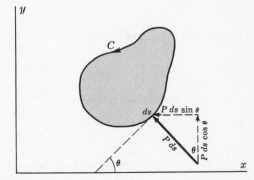

Fig. 9-21

Since C represents a streamline, we have by Bernoulli's theorem, $P + \tfrac{1}{2}\sigma V^2 = K$ or $P = K - \tfrac{1}{2}\sigma V^2$, where V is the fluid speed on the streamline. Also by Problem 49 we have, $\dfrac{d\Omega}{dz} = V e^{-i\theta}$.

Then, integrating over C, we find

$$F = X + iY = \oint_C iP\,dz = i\oint_C (K - \tfrac{1}{2}\sigma V^2)\,dz$$

$$= -\tfrac{1}{2}i\sigma \oint_C V^2\,dz = -\tfrac{1}{2}i\sigma \oint_C V^2 e^{i\theta}\,ds$$

$$= -\tfrac{1}{2}i\sigma \oint_C (V^2\, e^{2i\theta})(e^{-i\theta}\,ds)$$

or

$$\bar{F} = X - iY = \tfrac{1}{2}i\sigma \oint_C (V^2\, e^{-2i\theta})(e^{i\theta}\,ds)$$

$$= \tfrac{1}{2}i\sigma \oint_C \left(\frac{d\Omega}{dz}\right)^2 dz$$

18. Let M denote the total moment about the origin of the pressure forces on the obstacle in Problem 17. Prove that

$$M = \mathrm{Re}\left\{-\tfrac{1}{2}\sigma \oint_C z\left(\frac{d\Omega}{dz}\right)^2 dz\right\}$$

We consider counterclockwise moments as positive. The moment about the origin of the force acting on element ds of Fig. 9-21 is

$$dM = (P\,ds\,\sin\theta)y + (P\,ds\,\cos\theta)x = P(y\,dy + x\,dx)$$

since $ds\,\sin\theta = dy$ and $ds\,\cos\theta = dx$. Then on using Bernoulli's equation, the total moment is

$$M = \oint_C P(y\,dy + x\,dx) = \oint_C (K - \tfrac{1}{2}\sigma V^2)(y\,dy + x\,dx)$$

$$= K\oint_C (y\,dy + x\,dx) - \tfrac{1}{2}\sigma \oint_C V^2\,(y\,dy + x\,dx)$$

$$= 0 \qquad\qquad - \tfrac{1}{2}\sigma \oint_C V^2\,(x\cos\theta + y\sin\theta)\,ds$$

where we have used the fact that $\oint_C (y\,dy + x\,dx) = 0$ since $y\,dy + x\,dx$ is an exact differential. Hence

$$M = -\tfrac{1}{2}\sigma \oint_C V^2\,(x\cos\theta + y\sin\theta)\,ds$$

$$= \mathrm{Re}\left\{-\tfrac{1}{2}\sigma \oint_C V^2\,(x + iy)(\cos\theta - i\sin\theta)\,ds\right\}$$

$$= \mathrm{Re}\left\{-\tfrac{1}{2}\sigma \oint_C V^2 z e^{-i\theta}\,ds\right\} = \mathrm{Re}\left\{-\tfrac{1}{2}\sigma \oint_C z(V^2\, e^{-2i\theta})(e^{i\theta}\,ds)\right\}$$

$$= \mathrm{Re}\left\{-\tfrac{1}{2}\sigma \oint_C z\left(\frac{d\Omega}{dz}\right)^2 dz\right\}$$

Sometimes we write this result in the form $M + iN = -\tfrac{1}{2}\sigma \oint_C z\left(\frac{d\Omega}{dz}\right)^2 dz$ where N has no simple physical significance.

19. Find the net force acting on the cylindrical obstacle of Problem 16.

The complex potential for the flow in Problem 16 is

$$\Omega = V_0\left(z + \frac{a^2}{z}\right) + \frac{i\gamma}{2\pi}\ln z$$

where V_0 is the speed of the fluid at distances far from the obstacle and γ is the circulation. By Problem 17 the net force acting on the cylindrical obstacle is given by F, where

$$\bar{F} \;=\; X - iY \;=\; \tfrac{1}{2}i\sigma \oint_C \left(\frac{d\Omega}{dz}\right)^2 dz \;=\; \tfrac{1}{2}i\sigma \oint_C \left\{ V_0\left(1 - \frac{a^2}{z^2}\right) + \frac{i\gamma}{2\pi z} \right\}^2 dz$$

$$=\; \tfrac{1}{2}i\sigma \oint_C \left\{ V_0^2\left(1 - \frac{a^2}{z^2}\right)^2 + \frac{2iV_0\gamma}{2\pi z}\left(1 - \frac{a^2}{z^2}\right) - \frac{\gamma^2}{4\pi^2 z^2} \right\} dz \;=\; -\sigma V_0\gamma$$

Then $X = 0$, $Y = \sigma V_0\gamma$ and it follows that there is a net force in the positive y direction of magnitude $\sigma V_0\gamma$. In the case where the cylinder is horizontal and the flow takes place in a vertical plane this force is called the *lift* on the cylinder.

APPLICATIONS TO ELECTROSTATICS

20. (a) Find the complex potential due to a line of charge q per unit length perpendicular to the z plane at $z = 0$.

(b) What modification should be made in (a) if the line is at $z = a$?

(c) Discuss the similarity with the complex potential for a line source or sink in fluid flow.

(a) The electric field due to a line charge q per unit length is radial and the normal component of the electric vector is constant and equal to E_r while the tangential component is zero (see Fig. 9-22). If C is any cylinder of radius r with axis at $z = 0$, then by Gauss' theorem,

$$\oint_C E_n \, ds \;=\; E_r \oint_C ds \;=\; E_r \cdot 2\pi r \;=\; 4\pi q$$

and
$$E_r \;=\; \frac{2q}{r}$$

Since $E_r = -\dfrac{\partial \Phi}{\partial r}$ we have $\Phi = -2q \ln r$, omitting the constant of integration. This is the real part of $\Omega(z) = -2q \ln z$ which is the required complex potential.

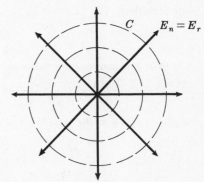

Fig. 9-22

(b) If the line of charge is at $z = a$, the complex potential is $\Omega(z) = -2q \ln(z - a)$.

(c) The complex potential has the same form as that for a line source of fluid if $k = -2q$ [see Problem 12]. If q is a positive charge, this corresponds to a line sink.

21. (a) Find the potential at any point of the region shown in Fig. 9-23 if the potentials on the x axis are given by V_0 for $x > 0$ and $-V_0$ for $x < 0$.

(b) Determine the equipotential and flux lines.

(a) We must find a function, harmonic in the plane, which takes on the values V_0 for $x > 0$, i.e. $\theta = 0$, and $-V_0$ for $x < 0$, i.e. $\theta = \pi$. As in Problem 6, if A and B are real constants $A\theta + B$ is harmonic. Then $A(0) + B = V_0$, $A(\pi) + B = -V_0$ from which $A = -2V_0/\pi$, $B = V_0$ so that the required potential is

$$V_0\left(1 - \frac{2}{\pi}\theta\right) \;=\; V_0\left(1 - \frac{2}{\pi}\tan^{-1}\frac{y}{x}\right)$$

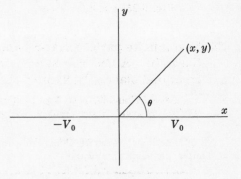

Fig. 9-23

in the upper half plane $y > 0$. The potential in the lower half plane is obtained by symmetry.

(b) The equipotential lines are given by $V_0\left(1 - \dfrac{2}{\pi}\tan^{-1}\dfrac{y}{x}\right) = \alpha$, i.e. $y = mx$ where m is a constant. These are straight lines passing through the origin.

The flux lines are the orthogonal trajectories of the lines $y = mx$ and are given by $x^2 + y^2 = \beta$. They are circles with center at the origin.

Another method. A function conjugate to $V_0\left(1 - \dfrac{2}{\pi}\tan^{-1}\dfrac{y}{x}\right)$ is $-\dfrac{2V_0}{\pi}\ln r$. Then the flux lines are given by $r = \sqrt{x^2 + y^2} = $ constant, which are circles with center at the origin.

22. (a) Find the potential due to a line charge q per unit length at $z = z_0$ and a line charge $-q$ per unit length at $z = \bar{z}_0$.

   (b) Show that the potential due to an infinite plane [*ABC* in Fig. 9-25] kept at zero potential (ground potential) and a line charge q per unit length parallel to this plane can be found from the result in (a).

   (a) The complex potential due to the two line charges [Fig. 9-24] is

$$\Omega(z) \;=\; -2q \ln (z - z_0) \;+\; 2q \ln (z - \bar{z}_0) \;=\; 2q \ln \left(\frac{z - \bar{z}_0}{z - z_0}\right)$$

Then the required potential is the real part of this, i.e.,

$$\Phi \;=\; 2q \; \mathrm{Re}\left\{\ln\left(\frac{z - \bar{z}_0}{z - z_0}\right)\right\} \tag{1}$$

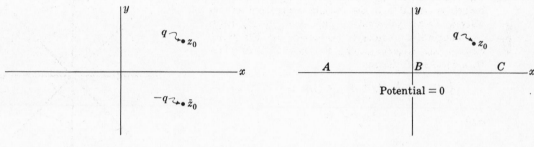

Fig. 9-24   Fig. 9-25

   (b) To prove this we must show that the potential (1) reduces to $\Phi = 0$ on the x axis, i.e. *ABC* in Fig. 9-25 is at potential zero. This follows at once from the fact that on the x axis, $z = x$ so that

$$\Omega \;=\; 2q \ln \left(\frac{x - \bar{z}_0}{x - z_0}\right) \qquad \text{and} \qquad \bar{\Omega} \;=\; 2q \ln \left(\frac{x - z_0}{x - \bar{z}_0}\right) \;=\; -\Omega$$

i.e. $\Phi = \mathrm{Re}\{\Omega\} = 0$ on the x axis.

   Thus we can replace the charge $-q$ at $\bar{z}_0$ [Fig. 9-24] by a plane *ABC* at potential zero [Fig. 9-25] and conversely.

23. Two infinite parallel planes, separated by a distance a, are grounded (i.e. are at potential zero). A line charge q per unit length is located between the planes at a distance b from one plane. Determine the potential at any point between the planes.

   Let *ABC* and *DEF* in Fig. 9-26 represent the two planes perpendicular to the z plane, and suppose the line charge passes through the imaginary axis at the point $z = bi$.

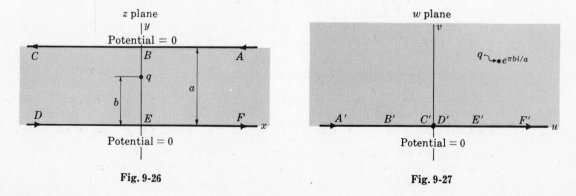

Fig. 9-26   Fig. 9-27

From entry A-2 in the table on Page 205 we see that the transformation $w = e^{\pi z/a}$ maps the shaded region of Fig. 9-26 on to the upper half w plane of Fig. 9-27. The line charge q at $z = bi$ in Fig. 9-26 is mapped into the line charge q at $w = e^{\pi bi/a}$. The boundary $ABCDEF$ of Fig. 9-26 (at potential zero) is mapped into the x axis $A'B'C'D'E'F'$ (at potential zero) where C' and D' are coincident at $w = 0$.

By Problem 22 the potential at any point of the shaded region in Fig. 9-27 above is

$$\Phi = 2q \, \mathrm{Re} \left\{ \frac{w - e^{-\pi bi/a}}{w - e^{\pi bi/a}} \right\}$$

Then the potential at any point of the shaded region in Fig. 9-26 is

$$\Phi = 2q \, \mathrm{Re} \left\{ \frac{e^{\pi z/a} - e^{-\pi bi/a}}{e^{\pi z/a} - e^{\pi bi/a}} \right\}$$

APPLICATIONS TO HEAT FLOW

24. A semi-infinite slab (shaded in Fig. 9-28) has its boundaries maintained at the indicated temperatures where T is constant. Find the steady-state temperature.

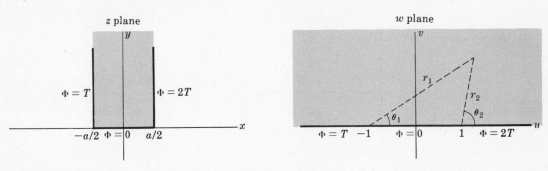

Fig. 9-28   Fig. 9-29

The shaded region of the z plane is mapped into the upper half of the w plane [Fig. 9-29] by the mapping function $w = \sin(\pi z/a)$ which is equivalent to $u = \sin(\pi x/a)\cosh(\pi y/a)$, $v = \cos(\pi x/a)\sinh(\pi y/a)$ [see entry A-3(a) in the table on Page 206].

We must now solve the equivalent problem in the w plane. We use the method of Problem 7 to find that the solution in the w plane is

$$\Phi = \frac{T}{\pi} \tan^{-1}\left(\frac{v}{u+1}\right) - \frac{2T}{\pi}\tan^{-1}\left(\frac{v}{u-1}\right) + 2T$$

and the required solution to the problem in the z plane is therefore

$$\Phi = \frac{T}{\pi} \tan^{-1}\left\{\frac{\cos(\pi x/a)\sinh(\pi y/a)}{\sin(\pi x/a)\cosh(\pi y/a) + 1}\right\} - \frac{2T}{\pi}\tan^{-1}\left\{\frac{\cos(\pi x/a)\sinh(\pi y/a)}{\sin(\pi x/a)\cosh(\pi y/a) - 1}\right\} + 2T$$

25. Find the steady-state temperature at any point of the region shown shaded in Fig. 9-30 if the temperatures are maintained as indicated.

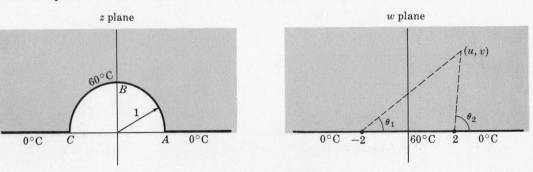

Fig. 9-30   Fig. 9-31

The shaded region of the z plane is mapped on to the upper half of the w plane by means of the mapping function $w = z + \dfrac{1}{z}$ [entry A-4 in the table on Page 206] which is equivalent to

$$u + iv = x + iy + \frac{1}{x + iy} = x + \frac{x}{x^2 + y^2} + i\left(y - \frac{y}{x^2 + y^2}\right), \quad \text{i.e.} \quad u = x + \frac{x}{x^2 + y^2}, \quad v = y - \frac{y}{x^2 + y^2}$$

The solution to the problem in the w plane is, using the method of Problem 7,

$$\frac{60}{\pi} \tan^{-1}\left(\frac{v}{u - 2}\right) \quad - \quad \frac{60}{\pi} \tan^{-1}\left(\frac{v}{u + 2}\right)$$

Then substituting the values of u and v, the solution to the required problem in the z plane is

$$\frac{60}{\pi} \tan^{-1}\left\{\frac{y(x^2 + y^2 - 1)}{(x^2 + y^2 + 1)x - 2(x^2 + y^2)}\right\} \quad - \quad \frac{60}{\pi} \tan^{-1}\left\{\frac{y(x^2 + y^2 - 1)}{(x^2 + y^2 + 1)x + 2(x^2 + y^2)}\right\}$$

or, in polar coordinates,

$$\frac{60}{\pi} \tan^{-1}\left\{\frac{(r^2 - 1)\sin\theta}{(r^2 + 1)\cos\theta - 2r}\right\} \quad - \quad \frac{60}{\pi} \tan^{-1}\left\{\frac{(r^2 - 1)\sin\theta}{(r^2 + 1)\cos\theta + 2r}\right\}$$

MISCELLANEOUS PROBLEMS

26. A region is bounded by two infinitely long concentric cylindrical conductors of radii r_1 and r_2 $(r_2 > r_1)$ which are charged to potentials Φ_1 and Φ_2 respectively [see Fig. 9-32]. Find the (a) potential and (b) electric field vector everywhere in the region.

(a) Consider the function $\Omega = A \ln z + B$ where A and B are real constants. If $z = re^{i\theta}$, then

$$\Omega = \Phi + i\Psi = A \ln r + Ai\theta + B$$

or

$$\Phi = A \ln r + B, \quad \Psi = A\theta$$

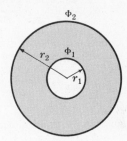

Now Φ satisfies Laplace's equation, i.e. is harmonic, everywhere in the region $r_1 < r < r_2$ and reduces to $\Phi = \Phi_1$ and $\Phi = \Phi_2$ on $r = r_1$ and $r = r_2$ provided A and B are chosen so that

$$\Phi_1 = A \ln r_1 + B, \qquad \Phi_2 = A \ln r_2 + B$$

i.e., $\qquad A = \dfrac{\Phi_2 - \Phi_1}{\ln (r_2/r_1)}, \qquad B = \dfrac{\Phi_1 \ln r_2 - \Phi_2 \ln r_1}{\ln (r_2/r_1)}$

Fig. 9-32

Then the required potential is

$$\Phi = \frac{(\Phi_2 - \Phi_1)}{\ln (r_2/r_1)} \ln r + \frac{\Phi_1 \ln r_2 - \Phi_2 \ln r_1}{\ln (r_2/r_1)}$$

(b) $\quad$ Electric field vector $= \mathcal{E} = -\operatorname{grad} \Phi = -\dfrac{\partial \Phi}{\partial r}$

$$= \frac{\Phi_1 - \Phi_2}{\ln (r_2/r_1)} \cdot \frac{1}{r}$$

Note that the lines of force, or flux lines, are orthogonal to the equipotential lines, and some of these are indicated by the dashed lines of Fig. 9-33.

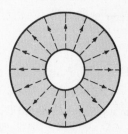

Fig. 9-33

27. Find the capacitance of the condenser formed by the two cylindrical conductors in Problem 26.

If Γ is any simple closed curve containing the inner cylinder and q is the charge on this cylinder, then by Gauss' theorem and the results of Problem 26 we have

$$\oint_{\Gamma} E_n \, ds = \int_{\theta=0}^{2\pi} \left\{\frac{\Phi_1 - \Phi_2}{\ln (r_2/r_1)} \cdot \frac{1}{r}\right\} r \, d\theta = \frac{2\pi(\Phi_1 - \Phi_2)}{\ln (r_2/r_1)} = 4\pi q$$

Then $q = \dfrac{\Phi_1 - \Phi_2}{2 \ln (r_2/r_1)}$ and so

$$\text{Capacitance } C \; = \; \frac{\text{charge}}{\text{difference in potential}} \; = \; \frac{q}{\Phi_1 - \Phi_2} \; = \; \frac{1}{2 \ln (r_2/r_1)}$$

which depends only on the geometry of the condensers, as it should.

The above result holds if there is a vacuum between the conductors. If there is a medium of dielectric constant κ between the conductors, we must replace q by q/κ and in this case the capacitance is $1/[2\kappa \ln (r_2/r_1)]$.

28. Two circular cylindrical conductors of equal radius R and centers at distance D from each other [Fig. 9-34] are charged to potentials V_0 and $-V_0$ respectively. (a) Determine the charge per unit length needed to accomplish this. (b) Find an expression for the capacitance.

(a) We use the results of Problem 13, since we can replace any of the equipotential curves (surfaces) by circular conductors at the specified potentials. Placing $\alpha = -V_0$ and $\alpha = V_0$ and noting that $k = 2q$, we find that the centers of the circles are at

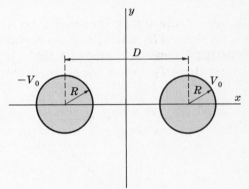

$$x = - a \coth (V_0/2q) \quad \text{and} \quad x = a \coth (V_0/2q)$$

so that   (1) $D = 2a \coth (V_0/2q)$

The radius R of the circles is

(2) $R = a \operatorname{csch} (V_0/2q)$

Division of (1) by (2) yields $2 \cosh (V_0/2q) = D/R$
so that the required charge is

$$q \; = \; \frac{V_0}{2 \cosh^{-1} (D/2R)}$$

Fig. 9-34

(b) $$\text{Capacitance } C \; = \; \frac{\text{charge}}{\text{difference in potential}} \; = \; \frac{q}{2V_0} \; = \; \frac{1}{4 \cosh^{-1} (D/2R)}$$

The result holds for a vacuum. If there is a medium of dielectric constant κ, we must divide the result by κ.

Note that the capacitance depends as usual only on the geometry. The result is fundamental in the theory of transmission line cables.

29. Prove the uniqueness of the solution to Dirichlet's problem.

Dirichlet's problem is the problem of determining a function Φ which satisfies $\dfrac{\partial^2 \Phi}{\partial x^2} + \dfrac{\partial^2 \Phi}{\partial y^2} = 0$ in a simply-connected region $\mathcal{R}$ and which takes on a prescribed value $\Phi = f(x,y)$ on the boundary C of $\mathcal{R}$. To prove the uniqueness, we must show that if such a solution exists it is the only one. To do this suppose that there are two different solutions, say Φ_1 and Φ_2. Then

$$\frac{\partial^2 \Phi_1}{\partial x^2} + \frac{\partial^2 \Phi_1}{\partial y^2} = 0 \quad \text{in } \mathcal{R} \qquad \text{and} \qquad \Phi_1 = f(x,y) \quad \text{on } C \tag{1}$$

$$\frac{\partial^2 \Phi_2}{\partial x^2} + \frac{\partial^2 \Phi_2}{\partial y^2} = 0 \quad \text{in } \mathcal{R} \qquad \text{and} \qquad \Phi_2 = f(x,y) \quad \text{on } C \tag{2}$$

Subtracting and letting $G = \Phi_1 - \Phi_2$, we have

$$\frac{\partial^2 G}{\partial x^2} + \frac{\partial^2 G}{\partial y^2} = 0 \quad \text{in } \mathcal{R} \qquad \text{and} \qquad G = 0 \quad \text{on } C \tag{3}$$

To show that $\Phi_1 = \Phi_2$ identically, we must show that $G = 0$ identically in $\mathcal{R}$.

Let $F = G$ in Problem 31, Chapter 4, Page 112 to obtain

$$\oint_C G\left(\frac{\partial G}{\partial x}\,dx - \frac{\partial G}{\partial y}\,dy\right) = -\iint_{\mathcal{R}}\left[G\left(\frac{\partial^2 G}{\partial x^2} + \frac{\partial^2 G}{\partial y^2}\right) + \left(\frac{\partial G}{\partial x}\right)^2 + \left(\frac{\partial G}{\partial y}\right)^2\right]dx\,dy \qquad (4)$$

Suppose that G is not identically equal to a constant in $\mathcal{R}$. From the fact that $G = 0$ on C, and $\frac{\partial^2 G}{\partial x^2} + \frac{\partial^2 G}{\partial y^2} = 0$ identically in $\mathcal{R}$, (4) becomes

$$\iint_{\mathcal{R}}\left[\left(\frac{\partial G}{\partial x}\right)^2 + \left(\frac{\partial G}{\partial y}\right)^2\right]dx\,dy = 0$$

But this contradicts the assumption that G is not identically equal to a constant in $\mathcal{R}$, since in such case

$$\iint_{\mathcal{R}}\left[\left(\frac{\partial G}{\partial x}\right)^2 + \left(\frac{\partial G}{\partial y}\right)^2\right]dx\,dy > 0$$

It follows that G must be constant in $\mathcal{R}$, and by continuity we must have $G = 0$. Thus $\Phi_1 = \Phi_2$ and there is only one solution.

30. An infinite wedge shaped region $ABDE$ of angle $\pi/4$ [shaded in Fig. 9-35] has one of its sides (AB) maintained at constant temperature T_1. The other side BDE has part BD [of unit length] insulated while the remaining part DE is maintained at constant temperature T_2. Find the temperature everywhere in the region.

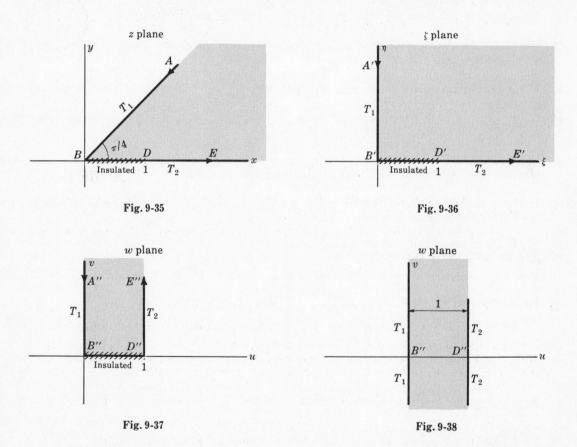

Fig. 9-35

Fig. 9-36

Fig. 9-37

Fig. 9-38

By the transformation $\zeta = z^2$, the shaded region of the z plane [Fig. 9-35] is mapped into the region shaded in Fig. 9-36 with the indicated boundary conditions [see entry A-1 in the table on Page 205].

By the transformation $\zeta = \sin(\pi w/2)$, the shaded region of the ζ plane [Fig. 9-36] is mapped into the region shaded in Fig. 9-37 with the indicated boundary conditions [see entry C-1 in the table on Page 210].

Now the temperature problem represented by Fig. 9-37 with $B''D''$ insulated is equivalent to the temperature problem represented by Fig. 9-38 since, by symmetry, no heat transfer can take place across $B''D''$. But this is the problem of determining the temperature between two parallel planes kept at constant temperatures T_1 and T_2 respectively. In this case the temperature variation is linear and so must be given by $T_1 + (T_2 - T_1)u$.

From $\zeta = z^2$ and $\zeta = \sin(\pi w/2)$ we have on eliminating ζ, $w = \dfrac{2}{\pi}\sin^{-1}z^2$ or $u = \dfrac{2}{\pi}\operatorname{Re}\{\sin^{-1}z^2\}$. Then the required temperature is

$$T_1 + \frac{2(T_2 - T_1)}{\pi}\operatorname{Re}\{\sin^{-1}z^2\}$$

In polar coordinates (r, θ) this can be written as [see Problem 95],

$$T_1 + \frac{2(T_2 - T_1)}{\pi}\sin^{-1}\{\tfrac{1}{2}\sqrt{r^4 + 2r^2\cos 2\theta + 1} - \tfrac{1}{2}\sqrt{r^4 - 2r^2\cos 2\theta + 1}\}$$

Supplementary Problems

HARMONIC FUNCTIONS

31. Show that the functions (a) $2xy + y^3 - 3x^2y$, (b) $e^{-x}\sin y$ are harmonic.

32. Show that the functions of Problem 31 remain harmonic under the transformations (a) $z = w^2$, (b) $z = \sin w$.

33. If $\Phi(x, y)$ is harmonic, prove that $\Phi(x + a, y + b)$, where a and b are any constants, is also harmonic.

34. If $\Phi_1, \Phi_2, \ldots, \Phi_n$ are harmonic in a region $\mathcal{R}$ and $c_1, c_2, \ldots, c_n$ are any constants, prove that $c_1\Phi_1 + c_2\Phi_2 + \cdots + c_n\Phi_n$ is harmonic in $\mathcal{R}$.

35. Prove that all the harmonic functions which depend only on the distance r from a fixed point must have the form $A\ln r + B$ where A and B are any constants.

36. If $F(z)$ is analytic and different from zero in a region $\mathcal{R}$, prove that the real and imaginary parts of $\ln F(z)$ are harmonic in $\mathcal{R}$.

DIRICHLET AND NEUMANN PROBLEMS

37. Find a function harmonic in the upper half z plane $\operatorname{Im}\{z\} > 0$ which takes the prescribed values on the x axis given by $G(x) = \begin{cases} 1 & x > 0 \\ -1 & x < 0 \end{cases}$. $Ans.$ $1 - (2/\pi)\tan^{-1}(y/x)$

38. Work Problem 37 if $G(x) = \begin{cases} 1 & x < -1 \\ 0 & -1 < x < 1 \\ -1 & x > 1 \end{cases}$.

    $Ans.$ $1 - \dfrac{1}{\pi}\tan^{-1}\left(\dfrac{y}{x-1}\right) - \dfrac{1}{\pi}\tan^{-1}\left(\dfrac{y}{x+1}\right)$

39. Find a function harmonic inside the circle $|z| = 1$ and taking the values $F(\theta) = \begin{cases} T & 0 < \theta < \pi \\ -T & \pi < \theta < 2\pi \end{cases}$

    on its circumference. $Ans.$ $T\left\{1 - \dfrac{2}{\pi}\tan^{-1}\left(\dfrac{2r\sin\theta}{1 - r^2}\right)\right\}$

40. Work Problem 39 if $F(\theta) = \begin{cases} T & 0 < \theta < \pi/2 \\ 0 & \pi/2 < \theta < 3\pi/2 \\ -T & 3\pi/2 < \theta < 2\pi \end{cases}$.

41. Work Problem 39 if $F(\theta) = \begin{cases} \sin \theta & 0 < \theta < \pi \\ 0 & \pi < \theta < 2\pi \end{cases}$.

42. Find a function harmonic inside the circle $|z| = 2$ and taking the values $F(\theta) = \begin{cases} 10 & 0 < \theta < \pi \\ 0 & \pi < \theta < 2\pi \end{cases}$.

    Ans. $10 \left\{ 1 - \dfrac{1}{\pi} \tan^{-1} \left(\dfrac{4r \sin \theta}{4 - r^2} \right) \right\}$

43. Show by direct substitution that the answers obtained in (a) Problem 6, (b) Problem 7, (c) Problem 8 are actually solutions to the corresponding boundary-value problems.

44. Find a function $\Phi(x, y)$ harmonic in the first quadrant $x > 0$, $y > 0$ which takes on the values

    $V(x, 0) = -1$, $V(0, y) = 2$.   Ans. $\dfrac{3}{\pi} \tan^{-1} \left(\dfrac{2xy}{x^2 - y^2} \right) - 1$

45. Find a function $\Phi(x, y)$ which is harmonic in the first quadrant $x > 0$, $y > 0$ and which satisfies the boundary conditions $\Phi(x, 0) = e^{-x}$, $\partial \Phi / \partial x \big|_{x=0} = 0$.

APPLICATIONS TO FLUID FLOW

46. Sketch the streamlines and equipotential lines for fluid motion in which the complex potential is given by (a) $z^2 + 2z$, (b) z^4, (c) e^{-z}, (d) $\cos z$.

47. Discuss the fluid flow corresponding to the complex potential $\Omega(z) = V_0(z + 1/z^2)$.

48. Verify the statements made before equations (5) and (6) on Page 234.

49. Derive the relation $d\Omega/dz = Ve^{-i\theta}$, where V and θ are defined as in Problem 17.

50. Referring to Problem 10, (a) show that the speed of the fluid at any point E [Fig. 9-14] is given by $2V_0 |\sin \theta|$ and (b) determine at what points on the cylinder the speed is greatest.

51. (a) If P is the pressure at point E of the obstacle in Fig. 9-14 of Problem 10 and P_∞ is the pressure far from the obstacle, show that
    $$P - P_\infty = \tfrac{1}{2}\sigma V_0^2 (1 - 4 \sin^2 \theta)$$

    (b) Show that a vacuum is created at points B and F if the speed of the fluid is equal to or greater than $V_0 = \sqrt{2P_\infty/3\sigma}$. This is often called *cavitation*.

52. Derive equation (19), Page 237, by a limiting procedure applied to equation (18).

53. Discuss the fluid flow due to three sources of equal strength k located at $z = -a, 0, a$.

54. Discuss the fluid flow due to two sources at $z = \pm a$ and a sink at $z = 0$ if the strengths all have equal magnitude.

55. Prove that under the transformation $w = F(z)$ where $F(z)$ is analytic, a source (or sink) in the z plane at $z = z_0$ is mapped into a source (or sink) of equal strength in the w plane at $w = w_0 = F(z_0)$.

56. Show that the total moment on the cylindrical obstacle of Problem 10 is zero and explain physically.

57. If $\Psi(x, y)$ is the stream function, prove that the mass rate of flow of fluid across an arc C joining points (x_1, y_1) and (x_2, y_2) is $\sigma\{\Psi(x_2, y_2) - \Psi(x_1, y_1)\}$.

58. (a) Show that the complex potential due to a source of strength $k > 0$ in a fluid moving with speed V_0 is $\Omega = V_0 z + k \ln z$ and (b) discuss the motion.

59. A source and sink of equal strengths m are located at $z = \pm 1$ between the parallel lines $y = \pm 1$. Show that the complex potential for the fluid motion is

$$\Omega = m \ln \left\{ \frac{e^{\pi(z+1)} - 1}{e^{\pi(z-1)} - 1} \right\}$$

60. Given a source of fluid at $z = z_0$ and a wall $x = 0$. Prove that the resulting flow is equivalent to removing the wall and introducing another source of equal strength at $z = -z_0$.

61. Fluid flows between the two branches of the hyperbola $ax^2 - by^2 = 1$, $a > 0$, $b > 0$. Prove that the complex potential for the flow is given by $K \cosh^{-1} \alpha z$ where K is a positive constant and $\alpha = \sqrt{ab/(a+b)}$.

APPLICATIONS TO ELECTROSTATICS

62. Two semi-infinite plane conductors, as indicated in Fig. 9-39 below, are charged to constant potentials Φ_1 and Φ_2 respectively. Find the (a) potential Φ and (b) electric field $\mathcal{E}$ everywhere in the shaded region between them.   Ans. (a) $\Phi = \Phi_2 + \left(\dfrac{\Phi_1 - \Phi_2}{\alpha} \right) \theta$ (b) $\mathcal{E} = (\Phi_2 - \Phi_1)/\alpha r$

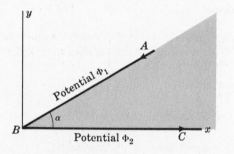

Fig. 9-39

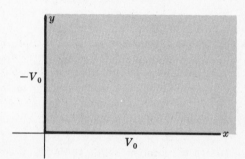

Fig. 9-40

63. Find the (a) potential and (b) electric field everywhere in the shaded region of Fig. 9-40 above if the potentials on the positive x and y axes are constant and equal to V_0 and $-V_0$ respectively.

Ans. $V_0 \left\{ 1 - \dfrac{2}{\pi} \tan^{-1} \left(\dfrac{2xy}{x^2 - y^2} \right) \right\}$

64. An infinite region has in it 3 wires located at $z = -1, 0, 1$ and maintained at constant potentials $-V_0, 2V_0, -V_0$ respectively. Find the (a) potential and (b) electric field everywhere.
Ans. (a) $V_0 \ln \{ z(z^2 - 1) \}$

65. Prove that the capacity of a capacitor is invariant under a conformal transformation.

66. The semi-infinite plane conductors AB and BC which intersect at angle α are grounded [Fig. 9-41]. A line charge q per unit length is located at point z_1 in the shaded region at equal distances a from AB and BC.

Find the potential.   Ans. $\text{Im} \left\{ -2qi \ln \left(\dfrac{z^{\pi/\alpha} - z_1^{\pi/\alpha}}{z^{\pi/\alpha} - \bar{z}_1^{\pi/\alpha}} \right) \right\}$

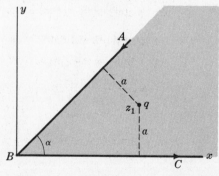

Fig. 9-41

67. Work Problem 66 if q is at a distance a from AB and b from BC.

68. Work Problem 23 if there are two line charges, q per unit length and $-q$ per unit length, located at $z = bi$ and $z = ci$ respectively, where $0 < b < a$, $0 < c < a$ and $b \neq c$.

69. An infinitely long circular cylinder has half of its surface charged to constant potential V_0 while the other half is grounded, the two halves being insulated from each other. Find the potential everywhere.

APPLICATIONS TO HEAT FLOW

70. (a) Find the steady-state temperature at any point of the region shown shaded in Fig. 9-42 below and (b) determine the isothermal and flux lines.   Ans. (a) $60 - (120/\pi) \tan^{-1}(y/x)$

71. Find the steady-state temperature at the point $(2,1)$ of the region shown shaded in Fig. 9-43 below.

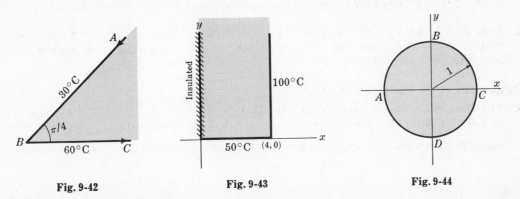

Fig. 9-42   Fig. 9-43   Fig. 9-44

72. The convex portions ABC and ADC of a unit cylinder [Fig. 9-44 above] are maintained at temperatures $40°$ C and $80°$ C respectively. (a) Find the steady-state temperature at any point inside. (b) Determine the isothermal and flux lines.

73. Find the steady-state temperature at the point $(5,2)$ in the shaded region of Fig. 9-45 below if the temperatures are maintained as shown.   Ans. $45.9°$ C

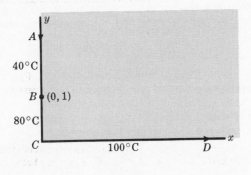

Fig. 9-45

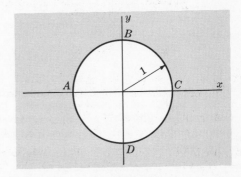

Fig. 9-46

74. An infinite conducting plate has in it a circular hole $ABCD$ of unit radius [Fig. 9-46 above]. Temperatures of $20°$ C and $80°$ C are applied to arcs ABC and ADC and maintained indefinitely. Find the steady-state temperature at any point of the plate.

MISCELLANEOUS PROBLEMS

75. If $\Phi(x, y)$ is harmonic, prove that $\Phi(x/r^2, y/r^2)$ where $r = \sqrt{x^2 + y^2}$ is also harmonic.

76. Prove that if U and V are continuously differentiable, then

$$(a) \quad \frac{\partial U}{\partial n} = \frac{\partial U}{\partial x}\frac{dx}{ds} + \frac{\partial U}{\partial y}\frac{dy}{ds} \qquad (b) \quad \frac{\partial V}{\partial s} = -\frac{\partial V}{\partial x}\frac{dy}{ds} + \frac{\partial V}{\partial y}\frac{dx}{ds}$$

where n and s denote the outward drawn normal and arc length parameter respectively to a simple closed curve C.

77. If U and V are conjugate harmonic functions, prove that $(a)\ \dfrac{\partial U}{\partial n} = \dfrac{\partial V}{\partial s},\ \ (b)\ \dfrac{\partial U}{\partial s} = -\dfrac{\partial V}{\partial n}.$

78. Prove that the function $\dfrac{1 - r^2}{1 - 2r\cos\theta + r^2}$ is harmonic in every region which does not include the point $r = 1,\ \theta = 0$.

79. Let it be required to solve the Neumann problem, i.e. to find a function V harmonic in a region $\mathcal{R}$ such that on the boundary C of $\mathcal{R}$, $\partial V/\partial n = G(s)$ where s is the arc length parameter. Let $H(s) = \int_a^s G(s)\,ds$ where a is any point of C, and suppose that $\oint_C G(s)\,ds = 0$. Show that to find V we must find the conjugate harmonic function U which satisfies the condition $U = -H(s)$ on C. This is an equivalent Dirichlet problem. [*Hint.* Use Problem 77.]

80. Prove that, apart from an arbitrary additive constant, the solution to the Neumann problem is unique.

81. Prove Theorem 3, Page 234.

82. How must Theorem 3, Page 234, be modified if the boundary condition $\Phi = a$ on C is replaced by $\Phi = f(x, y)$ on C?

83. How must Theorem 3, Page 234, be modified if the boundary condition $\partial\Phi/\partial n = 0$ on C is replaced by $\partial\Phi/\partial n = g(x, y)$ on C?

84. If a fluid motion is due to some distribution of sources, sinks and doublets and if C is some curve such that no flow takes place across it, then the distribution of sources, sinks and doublets to one side of C is called the *image* of the distribution of sources, sinks and doublets on the other side of C. Prove that the image of a source inside a circle C is a source of equal strength at the inverse point together with a sink of equal strength at the center of C. [Point P is called the *inverse* of point Q with respect to a circle C with center at O if OPQ is a straight line and $OP \cdot OQ = a^2$ where a is the radius of C.]

85. A source of strength $k > 0$ is located at point z_0 in a fluid which is contained in the first quadrant where the x and y axes are considered as rigid barriers. Prove that the speed of the fluid at any point is given by

$$k\,|\,(z - z_0)^{-1} + (z - \bar{z}_0)^{-1} + (z + z_0)^{-1} + (z + \bar{z}_0)^{-1}\,|$$

86. Two infinitely long cylindrical conductors having cross sections which are confocal ellipses with foci at $(-c, 0)$ and $(c, 0)$ [see Fig. 9-47] are charged to constant potentials Φ_1 and Φ_2 respectively. Show that the capacitance per unit length is equal to

$$\frac{2\pi}{\cosh^{-1}(R_2/c) - \cosh^{-1}(R_1/c)}$$

[*Hint.* Use the transformation $z = c \cosh w$.]

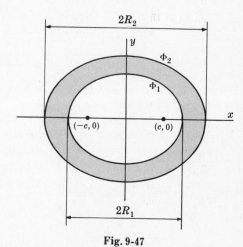

Fig. 9-47

87. In Problem 86 suppose that Φ_1 and Φ_2 represent constant temperatures applied to the elliptic cylinders. Find the steady-state temperature at any point in the conducting region between the cylinders.

88. A circular cylinder obstacle of radius a rests at the bottom of a channel of fluid which at distances far from the obstacle flows with velocity V_0 [see Fig. 9-48].

(a) Prove that the complex potential is given by

$$\Omega(z) = \pi a V_0 \coth(\pi a/z)$$

(b) Show that the speed at the top of the cylinder is $\frac{1}{4}\pi^2 V_0$ and compare with that for a circular obstacle in the middle of a fluid.

(c) Show that the difference in pressure between top and bottom points of the cylinder is $\sigma\pi^4 V_0^2/32$.

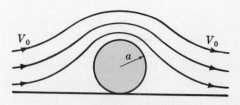

Fig. 9-48

89. (a) Show that the complex potential for fluid flow past the elliptic cylinder of Fig. 9-49 is given by

$$\Omega(z) = V_0 \left\{ \zeta + \frac{(a+b)^2}{4\zeta} \right\}$$

where $\zeta = \frac{1}{2}(z + \sqrt{z^2 - c^2})$ and $c^2 = a^2 - b^2$.

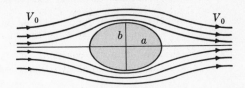

(b) Prove that the fluid speed at the top and bottom of the cylinder is $V_0(1 + b/a)$. Discuss the case $a = b$. [Hint. Express the complex potential in terms of elliptic coordinates (ξ, η) where $z = x + iy = c \cosh(\xi + i\eta) = c \cosh \zeta$.]

Fig. 9-49

90. Show that if the flow in Problem 89 is in a direction making an angle δ with the positive x axis, the complex potential is given by the result in (a) with $\zeta = \frac{1}{2}(z + \sqrt{z^2 - c^2})e^{i\delta}$.

91. In the *theory of elasticity*, the equation

$$\nabla^4 \Phi = \nabla^2(\nabla^2 \Phi) = \frac{\partial^4 \Phi}{\partial x^4} + 2\frac{\partial^4 \Phi}{\partial x^2 \partial y^2} + \frac{\partial^4 \Phi}{\partial y^4} = 0$$

called the *biharmonic equation*, is of fundamental importance. Solutions to this equation are called *biharmonic*. Prove that if $F(z)$ and $G(z)$ are analytic in a region $\mathcal{R}$, then the real part of $\bar{z}F(z) + G(z)$ is biharmonic in $\mathcal{R}$.

92. Show that biharmonic functions (see Problem 91) do not, in general, remain biharmonic under a conformal transformation.

93. (a) Show that $\Omega(z) = K \ln \sinh(\pi z/a)$, $k > 0$, $a > 0$ represents the complex potential due to a row of fluid sources at $z = 0, \pm ai, \pm 2ai, \ldots$.

(b) Show that, apart from additive constants, the potential and stream functions are given by

$$\Phi = K \ln \{\cosh(2\pi x/a) - \cos(2\pi y/a)\}, \qquad \Psi = K \tan^{-1}\left\{ \frac{\tan(\pi y/a)}{\tanh(\pi x/a)} \right\}$$

(c) Graph some of the streamlines for the flow.

94. Prove that the complex potential of Problem 93 is the same as that due to a source located halfway between the parallel lines $y = \pm 3a/2$.

95. Verify the statement made at the end of Problem 30 [compare Problem 137, Chapter 2, Page 62].

96. A condenser is formed from an elliptic cylinder, with major and minor axes of lengths $2a$ and $2b$ respectively, together with a flat plate AB of length $2h$ [see Fig. 9-50 below]. Show that the capacitance is equal to $\dfrac{2\pi}{\cosh^{-1}(a/h)}$.

97. A fluid flows with uniform velocity V_0 through a semi-infinite channel of width D and emerges through the opening AB [Fig. 9-51 below]. (a) Find the complex potential for the flow. (b) Determine the streamlines and equipotential lines and obtain graphs of some of these.
[Hint. Use entry C-5 in the table on Page 211.]

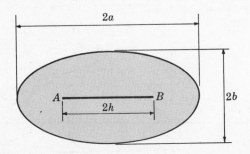

Fig. 9-50

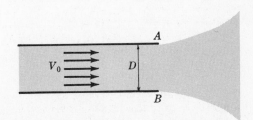

Fig. 9-51

98. Give a potential theory interpretation to Problem 30.

99. (*a*) Show that in a vacuum the capacitance of the parallel cylindrical conductors in Fig. 9-52 is

$$\frac{1}{2\cosh^{-1}\left(\dfrac{D^2 - R_1^2 - R_2^2}{2R_1R_2}\right)}$$

(*b*) Examine the case $R_1 = R_2 = R$ and compare with Problem 28.

100. Show that in a vacuum the capacitance of the two parallel cylindrical conductors in Fig. 9-53 is

$$\frac{1}{2\cosh^{-1}\left(\dfrac{R_1^2 + R_2^2 - D^2}{2R_1R_2}\right)}$$

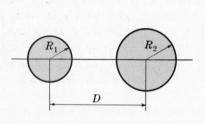

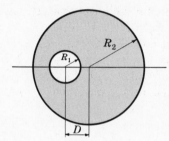

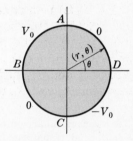

Fig. 9-52   Fig. 9-53   Fig. 9-54

101. Find the potential at any point of the unit cylinder of Fig. 9-54 if AB, BC, CD and DA are kept at potentials $V_0, 0, -V_0$ and 0 respectively.

Ans. $\dfrac{V_0}{\pi}\left(\tan^{-1}\dfrac{2r\sin\theta}{1-r^2} + \tan^{-1}\dfrac{2r\cos\theta}{1-r^2}\right)$

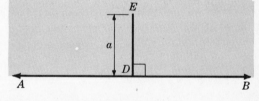

102. The shaded region of Fig. 9-55 represents an infinite conducting half plane in which lines AD, DE and DB are maintained at temperatures $0, T$ and $2T$ respectively, where T is a constant. (*a*) Find the temperature everywhere. (*b*) Give an interpretation involving potential theory.

Fig. 9-55

103. Work the preceding problem if (*a*) DE is insulated, (*b*) AB is insulated.

104. In Fig. 9-55 suppose that DE represents an obstacle perpendicular to the base of an infinite channel in which a fluid is flowing from left to right so that far from the obstacle the speed of the fluid is V_0. Find (*a*) the speed and (*b*) the pressure at any point of the fluid.

105. Find the steady-state temperature at the point $(3, 2)$ in the shaded region of Fig. 9-56.

106. An infinite wedge shaped region $ABCD$ of angle $\pi/4$ [shaded in Fig. 9-57] has one of its sides (CD) maintained at $50°$ C; the other side ABC has the part AB at temperature $25°$ C while part BC, of unit length, is insulated. Find the steady-state temperature at any point.

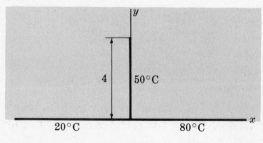

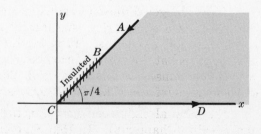

Fig. 9-56   Fig. 9-57

Chapter 10

Special Topics

ANALYTIC CONTINUATION

Let $F_1(z)$ be a function of z which is analytic in a region $\mathcal{R}_1$ [Fig. 10-1]. Suppose that we can find a function $F_2(z)$ which is analytic in a region $\mathcal{R}_2$ and which is such that $F_1(z) = F_2(z)$ in the region common to $\mathcal{R}_1$ and $\mathcal{R}_2$. Then we say that $F_2(z)$ is an *analytic continuation* of $F_1(z)$. This means that there is a function $F(z)$ analytic in the combined regions $\mathcal{R}_1$ and $\mathcal{R}_2$ such that $F(z) = F_1(z)$ in $\mathcal{R}_1$ and $F(z) = F_2(z)$ in $\mathcal{R}_2$. Actually it suffices for $\mathcal{R}_1$ and $\mathcal{R}_2$ to have only a small arc in common, such as LMN in Fig. 10-2.

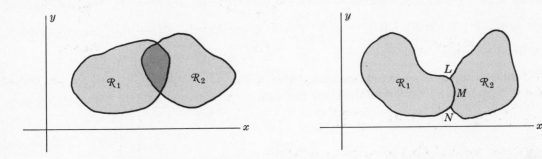

Fig. 10-1	Fig. 10-2

By analytic continuation to regions $\mathcal{R}_3$, $\mathcal{R}_4$, etc., we can extend the original region of definition to other parts of the complex plane. The functions $F_1(z)$, $F_2(z)$, $F_3(z)$, ..., defined in $\mathcal{R}_1$, $\mathcal{R}_2$, $\mathcal{R}_3$, ... respectively, are sometimes called *function elements* or briefly *elements*. It is sometimes impossible to extend a function analytically beyond the boundary of a region. We then call the boundary a *natural boundary*.

If a function $F_1(z)$ defined in $\mathcal{R}_1$ is continued analytically to region $\mathcal{R}_n$ along two different paths [Fig. 10-3], then the two analytic continuations will be identical if there is no singularity between the paths. This is the *uniqueness theorem for analytic continuation*.

If we do get different results, we can show that there is a singularity (specifically a *branch point*) between the paths. It is in this manner that we arrive at the various branches of multiple-valued functions. In this connection the concept of Riemann surfaces [Chapter 2] proves valuable.

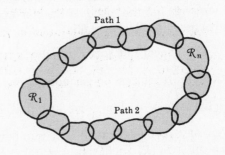

Fig. 10-3

We have already seen how functions represented by power series may be continued analytically (Chapter 6). In this chapter we consider how functions with other representations (such as integrals) may be continued analytically.

SCHWARZ'S REFLECTION PRINCIPLE

Suppose that $F_1(z)$ is analytic in the region $\mathcal{R}_1$ [Fig. 10-4] and that $F_1(z)$ assumes real values on the part LMN of the real axis.

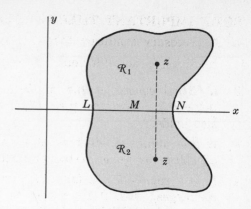

Then *Schwarz's reflection principle* states that the analytic continuation of $F_1(z)$ into region $\mathcal{R}_2$ (considered as a mirror image or reflection of $\mathcal{R}_1$ with LMN as the mirror) is given by

$$F_2(z) = \overline{F_1(\bar{z})} \qquad (1)$$

The result can be extended to cases where LMN is a curve instead of a straight line segment.

Fig. 10-4

INFINITE PRODUCTS

Let $P_n = (1+w_1)(1+w_2)\cdots(1+w_n)$ be denoted by $\prod_{k=1}^{n}(1+w_k)$ where we suppose that for all k, $w_k \neq -1$. If there exists a value $P \neq 0$ such that $\lim_{n\to\infty} P_n = P$, we say that the *infinite product* $(1+w_1)(1+w_2)\cdots \equiv \prod_{k=1}^{\infty}(1+w_k)$, or briefly $\Pi(1+w_k)$, *converges* to P; otherwise it *diverges*. The quantities w_k may be constants or functions of z.

If only a finite number of the quantities $w_k = -1$ while the rest of the infinite product omitting these factors converges, the infinite product is said to *converge to zero*.

ABSOLUTE, CONDITIONAL AND UNIFORM CONVERGENCE OF INFINITE PRODUCTS

If the infinite product $\Pi(1+|w_k|)$ converges, we say that $\Pi(1+w_k)$ is *absolutely convergent*.

If $\Pi(1+w_k)$ converges but $\Pi(1+|w_k|)$ diverges, we say that $\Pi(1+w_k)$ is *conditionally convergent*.

An important theorem, analogous to one for infinite series, states that an absolutely convergent infinite product is convergent, i.e. if $\Pi(1+|w_k|)$ converges then $\Pi(1+w_k)$ converges (see Problem 65).

The concept of *uniform convergence* of infinite products is easily defined by analogy with infinite series or sequences in general. Thus if $\prod_{k=1}^{n}\{1+w_k(z)\} = P_n(z)$ and $\prod_{k=1}^{\infty}\{1+w_k(z)\} = P(z)$, we say that $P_n(z)$ *converges uniformly* to $P(z)$ in a region $\mathcal{R}$ if, given any $\epsilon > 0$, we can find a number N, depending only on ϵ and not on the particular value of z in $\mathcal{R}$, such that $|P_n(z) - P(z)| < \epsilon$ for all $n > N$.

As in the case of infinite series, certain things can be done with absolutely or uniformly convergent infinite products that cannot necessarily be done for infinite products in general. Thus, for example, we can rearrange factors in an absolutely convergent infinite product without changing the value.

SOME IMPORTANT THEOREMS ON INFINITE PRODUCTS

1. A necessary condition that $\Pi(1 + w_k)$ converge is that $\lim\limits_{n \to \infty} w_n = 0$. However, the condition is not sufficient, i.e. even if $\lim\limits_{n \to \infty} w_n = 0$ the infinite product may diverge.

2. If $\Sigma\,|w_k|$ converges [i.e. if Σw_k converges absolutely], then $\Pi(1 + |w_k|)$, and thus $\Pi(1 + w_k)$, converges [i.e. $\Pi(1 + w_k)$ converges absolutely]. The converse theorem also holds.

3. If an infinite product is absolutely convergent, its factors can be altered without affecting the value of the product.

4. If in a region $\mathcal{R}$, $|w_k(z)| < M_k$, $k = 1, 2, 3, \ldots$, where M_k are constants such that ΣM_k converges, then $\Pi\{1 + w_k(z)\}$ is uniformly (and absolutely) convergent. This is the analog of the Weierstrass M test for series.

5. If $w_k(z)$, $k = 1, 2, 3, \ldots$, are analytic in a region $\mathcal{R}$ and $\Sigma w_k(z)$ is uniformly convergent in $\mathcal{R}$, then $\Pi\{1 + w_k(z)\}$ converges to an analytic function in $\mathcal{R}$.

WEIERSTRASS' THEOREM FOR INFINITE PRODUCTS

Let $f(z)$ be analytic for all z [i.e. $f(z)$ is an *entire function*] and suppose that it has simple zeros at $a_1, a_2, a_3, \ldots$ where $0 < |a_1| < |a_2| < |a_3| < \cdots$ and $\lim\limits_{n \to \infty} |a_n| = \infty$. Then $f(z)$ can be expressed as an infinite product of the form

$$f(z) \;=\; f(0)\, e^{f'(0)z/f(0)} \prod_{k=1}^{\infty} \left\{ \left(1 - \frac{z}{a_k} \right) e^{z/a_k} \right\} \tag{2}$$

A generalization of this states that if $f(z)$ has zeros at $a_k \neq 0$, $k = 1, 2, 3, \ldots$, of respective multiplicities or orders μ_k, and if for some integer N, $\sum\limits_{k=1}^{\infty} 1/a_k^N$ is absolutely convergent, then

$$f(z) \;=\; f(0)\, e^{G(z)} \prod_{k=1}^{\infty} \left\{ \left(1 - \frac{z}{a_k} \right) e^{\frac{z}{a_k} + \frac{1}{2}\frac{z^2}{a_k^2} + \cdots + \frac{1}{N-1}\frac{z^{N-1}}{a_k^{N-1}}} \right\}^{\mu_k} \tag{3}$$

where $G(z)$ is an entire function. The result is also true if some of the a_k's are poles, in which case their multiplicities are negative.

The results (2) and (3) are sometimes called *Weierstrass' factor theorems*.

SOME SPECIAL INFINITE PRODUCTS

1. $\sin z \;=\; z\left\{ 1 - \dfrac{z^2}{\pi^2} \right\}\left\{ 1 - \dfrac{z^2}{(2\pi)^2} \right\} \cdots \;=\; z \displaystyle\prod_{k=1}^{\infty} \left(1 - \dfrac{z^2}{k^2\pi^2} \right)$

2. $\cos z \;=\; \left\{ 1 - \dfrac{z^2}{(\pi/2)^2} \right\}\left\{ 1 - \dfrac{z^2}{(3\pi/2)^2} \right\} \cdots \;=\; \displaystyle\prod_{k=1}^{\infty} \left(1 - \dfrac{4z^2}{(2k-1)^2\pi^2} \right)$

3. $\sinh z \;=\; z\left\{ 1 + \dfrac{z^2}{\pi^2} \right\}\left\{ 1 + \dfrac{z^2}{(2\pi)^2} \right\} \cdots \;=\; \displaystyle\prod_{k=1}^{\infty} \left(1 + \dfrac{z^2}{k^2\pi^2} \right)$

4. $\cosh z \;=\; \left\{ 1 + \dfrac{z^2}{(\pi/2)^2} \right\}\left\{ 1 + \dfrac{z^2}{(3\pi/2)^2} \right\} \cdots \;=\; \displaystyle\prod_{k=1}^{\infty} \left(1 + \dfrac{4z^2}{(2k-1)^2\pi^2} \right)$

THE GAMMA FUNCTION

For $\mathrm{Re}\,\{z\} > 0$, we define the gamma function by

$$\Gamma(z) \;=\; \int_0^{\infty} t^{z-1} e^{-t}\, dt \tag{4}$$

Then (see Problem 11) we have the *recursion formula*

$$\Gamma(z+1) = z\,\Gamma(z) \qquad \text{where } \Gamma(1) = 1 \tag{5}$$

If z is a positive integer n, we see from (5) that

$$\Gamma(n+1) = n(n-1)\cdots(1) = n! \tag{6}$$

so that the gamma function is a generalization of the factorial. For this reason the gamma function is also called the *factorial function* and is written as $z!$ rather than $\Gamma(z+1)$, in which case we define $0! = 1$.

From (5) we also see that if z is real and positive, then $\Gamma(z)$ can be determined by knowing the values of $\Gamma(z)$ for $0 < z < 1$. If $z = \frac{1}{2}$, we have [Problem 14]

$$\Gamma(\tfrac{1}{2}) = \sqrt{\pi} \tag{7}$$

For $\operatorname{Re}\{z\} \leqq 0$, the definition (4) breaks down since the integral diverges. By analytic continuation, however, we can define $\Gamma(z)$ in the left hand plane. Essentially this amounts to use of (5) [see Problem 15]. At $z = 0, -1, -2, \ldots$, $\Gamma(z)$ has simple poles [see Problem 16].

PROPERTIES OF THE GAMMA FUNCTION

The following list shows some important properties of the gamma function. The first two can be taken as definitions from which all other properties can be deduced.

1.
$$\Gamma(z+1) = \lim_{k \to \infty} \frac{1 \cdot 2 \cdot 3 \cdots k}{(z+1)(z+2)\cdots(z+k)}\,k^z = \lim_{k \to \infty} \Pi(z, k)$$

where $\Pi(z, k)$ is sometimes called *Gauss' Π function*.

2.
$$\frac{1}{\Gamma(z)} = z e^{\gamma z} \prod_{k=1}^{\infty} \left\{1 + \frac{z}{k}\right\} e^{-z/k}$$

where $\gamma = \lim_{p \to \infty}\left\{1 + \frac{1}{2} + \frac{1}{3} + \cdots + \frac{1}{p} - \ln p\right\} = .5772157\ldots$ is called *Euler's constant*.

3.
$$\Gamma(z)\,\Gamma(1-z) = \frac{\pi}{\sin \pi z}$$

In particular if $z = \frac{1}{2}$, $\Gamma(\frac{1}{2}) = \sqrt{\pi}$.

4.
$$2^{2z-1}\,\Gamma(z)\,\Gamma(z+\tfrac{1}{2}) = \sqrt{\pi}\,\Gamma(2z)$$

This is sometimes called the *duplication formula* for the gamma function.

5. If $m = 1, 2, 3, \ldots$,
$$\Gamma(z)\,\Gamma\!\left(z+\frac{1}{m}\right)\Gamma\!\left(z+\frac{2}{m}\right)\cdots\Gamma\!\left(z+\frac{m-1}{m}\right) = m^{1/2-mz}\,(2\pi)^{(m-1)/2}\,\Gamma(mz)$$

Property 4 is a special case of this with $m = 2$.

6.
$$\frac{\Gamma'(z)}{\Gamma(z)} = -\gamma + \left(\frac{1}{1} - \frac{1}{z}\right) + \left(\frac{1}{2} - \frac{1}{z+1}\right) + \cdots + \left(\frac{1}{n} - \frac{1}{z+n-1}\right) + \cdots$$

7.
$$\Gamma'(1) = \int_0^\infty e^{-t} \ln t\, dt = -\gamma$$

8.
$$\Gamma(z) = \frac{1}{e^{2\pi i z} - 1} \oint_C t^{z-1} e^{-t}\, dt$$

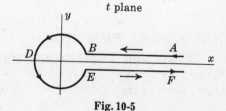

where C is the contour in Fig. 10-5. This is an analytic continuation to the left hand half plane of the gamma function defined in (4).

Fig. 10-5

9. Another contour integral using contour C [Fig. 10-5] is given by

$$\Gamma(z) \;=\; \frac{i}{2 \sin \pi z} \oint_C (-t)^{z-1} e^{-t}\, dt \;=\; -\frac{1}{2\pi i} \oint_C (-t)^{-z} e^{-t}\, dt$$

THE BETA FUNCTION

For $\text{Re}\{m\} > 0$, $\text{Re}\{n\} > 0$, we define the *beta function* by

$$B(m, n) \;=\; \int_0^1 t^{m-1}(1-t)^{n-1}\, dt \tag{8}$$

As seen in Problem 18, this is related to the gamma function according to

$$B(m, n) \;=\; \frac{\Gamma(m)\,\Gamma(n)}{\Gamma(m+n)} \tag{9}$$

Various integrals can be expressed in terms of the beta function and thus in terms of the gamma function. Two interesting results are

$$\int_0^{\pi/2} \sin^{2m-1}\theta \, \cos^{2n-1}\theta \; d\theta \;=\; \tfrac{1}{2} B(m, n) \;=\; \frac{\Gamma(m)\,\Gamma(n)}{2\,\Gamma(m+n)} \tag{10}$$

$$\int_0^\infty \frac{t^{p-1}}{1+t}\, dt \;=\; B(p, 1-p) \;=\; \Gamma(p)\,\Gamma(1-p) \;=\; \frac{\pi}{\sin p\pi} \tag{11}$$

the first holding for $\text{Re}\{m\} > 0$ and $\text{Re}\{n\} > 0$, and the second holding for $0 < \text{Re}\{p\} < 1$.

For $\text{Re}\{m\} \leqq 0$ and $\text{Re}\{n\} \leqq 0$, the definition (8) can be extended by use of analytic continuation.

DIFFERENTIAL EQUATIONS

Suppose we are given the *linear differential equation*

$$Y'' \;+\; p(z)\, Y' \;+\; q(z)\, Y \;=\; 0 \tag{12}$$

If $p(z)$ and $q(z)$ are analytic at a point a, then a is called an *ordinary point* of the differential equation. Points at which $p(z)$ or $q(z)$ or both are not analytic are called *singular points* of the differential equation.

> **Example 1:** For $Y'' + zY' + (z^2 - 4)Y = 0$, every point is an ordinary point.

> **Example 2:** For $(1 - z^2)Y'' - 2zY' + 6Y = 0$ or $Y'' - \dfrac{2z}{1-z^2}Y' + \dfrac{6}{1-z^2}Y = 0$, $z = \pm 1$ are singular points; all other points are ordinary points.

If $z = a$ is a singular point but $(z - a)\, p(z)$ and $(z - a)^2\, q(z)$ are analytic at $z = a$, then $z = a$ is called a *regular singular point*. If $z = a$ is neither an ordinary point or a regular singular point, it is called an *irregular singular point*.

> **Example 3:** In Example 2, $z = 1$ is a regular singular point since $(z - 1)\left(-\dfrac{2z}{1-z^2}\right) = \dfrac{2z}{z+1}$ and $(z - 1)^2\left(\dfrac{6}{1-z^2}\right) = \dfrac{6 - 6z}{z+1}$ are analytic at $z = 1$. Similarly, $z = -1$ is a regular singular point.

> **Example 4:** $z^3 Y'' + (1 - z)Y' - 2Y = 0$ has $z = 0$ as a singular point. Also, $z\left(\dfrac{1-z}{z^3}\right) = \dfrac{1-z}{z^2}$ and $z^2\left(-\dfrac{2}{z^3}\right) = -\dfrac{2}{z}$ are not analytic at $z = 0$, so that $z = 0$ is an irregular singular point.

If $Y_1(z)$ and $Y_2(z)$ are two solutions of (12) which are not constant multiples of each other, we call the solutions *linearly independent*. In such case, if A and B are any constants the general solution of (12) is

$$Y \;=\; AY_1 + BY_2 \tag{13}$$

The following theorems are fundamental.

Theorem 1. If $z = a$ is an ordinary point of (12), then there exist two linearly independent solutions of (12) having the form

$$\sum_{k=0}^{\infty} a_k(z-a)^k \tag{14}$$

where the constants a_k are determined by substitution in (12). In doing this it may be necessary to expand $p(z)$ and $q(z)$ in powers of $(z-a)$. In practice it is desirable to replace $(z-a)$ by a new variable.

The solutions (14) converge in a circle with center at a which extends up to the nearest singularity of the differential equation.

> **Example 5:** The equation $(1-z^2)Y'' - 2zY' + 6Y = 0$ [see Example 2] has a solution of the form $\sum a_k z^k$ which converges inside the circle $|z| = 1$.

Theorem 2. If $z = a$ is a regular singular point, then there exists at least one solution having the form

$$(z-a)^c \sum_{k=0}^{\infty} a_k(z-a)^k \tag{15}$$

where c is a constant. By substituting into (12) and equating the lowest power of $(z-a)$ to zero, a quadratic equation for c (called the *indicial equation*) is obtained. If we call the solutions of this quadratic equation c_1 and c_2, the following situations arise.

1. $c_1 - c_2 \neq$ *an integer*. In this case there are two linearly independent solutions having the form (15).

2. $c_1 = c_2$. Here one solution has the form (15) while the other linearly independent solution has the form

$$\ln (z-a) \sum_{k=0}^{\infty} b_k(z-a)^{k+c} \tag{16}$$

3. $c_1 - c_2 =$ *an integer* $\neq 0$. In this case there is either one solution of the form (15) or two linearly independent solutions having this form. If only one solution of the form (15) can be found, the other linearly independent solution has the form (16).

All solutions obtained converge in a circle with center at a which extends up to the nearest singularity of the differential equation.

SOLUTION OF DIFFERENTIAL EQUATIONS BY CONTOUR INTEGRALS

It is often desirable to seek a solution of a linear differential equation in the form

$$Y(z) = \oint_C K(z, t)\, G(t)\, dt \tag{17}$$

where $K(z, t)$ is called the *kernel*. One useful possibility occurs if $K(z, t) = e^{zt}$, in which case

$$Y(z) = \oint_C e^{zt}\, G(t)\, dt \tag{18}$$

Such solutions may occur where the coefficients in the differential equation are rational functions (see Problems 25 and 26).

BESSEL FUNCTIONS

Bessel's differential equation of order n is given by

$$z^2 Y'' + zY' + (z^2 - n^2)Y = 0 \tag{19}$$

A solution of this equation if $n \geqq 0$ is

$$J_n(z) = \frac{z^n}{2^n\,\Gamma(n+1)}\left\{1 - \frac{z^2}{2(2n+2)} + \frac{z^4}{2\cdot 4(2n+2)(2n+4)} - \cdots\right\} \qquad (20)$$

and is called *Bessel's function of the first kind of order n.*

If n is not an integer, the general solution of (*18*) is

$$Y = A\,J_n(z) + B\,J_{-n}(z) \qquad (21)$$

where A and B are arbitrary constants. However, if n is an integer then $J_{-n}(z) = (-1)^n J_n(z)$ and (*20*) fails to yield the general solution. The general solution in this case can be found as in Problems 182 and 183.

Bessel functions have many interesting and important properties, among them being the following.

1.
$$e^{z(t-1/t)/2} = \sum_{n=-\infty}^{\infty} J_n(z)\,t^n$$

  The left side is often called the *generating function* for the Bessel functions of the first kind for integer values of n.

2.
$$z J_{n-1}(z) - 2n J_n(z) + z J_{n+1}(z) = 0$$

  This is called the *recursion formula for Bessel functions* [see Problem 27].

3.
$$\frac{d}{dz}\{z^n J_n(z)\} = z^n J_{n-1}(z), \qquad \frac{d}{dz}\{z^{-n} J_n(z)\} = -z^{-n} J_{n+1}(z)$$

4.
$$J_n(z) = \frac{1}{\pi}\int_0^\pi \cos(n\phi - z\sin\phi)\,d\phi, \qquad n = \text{integer}$$

5.
$$J_n(z) = \frac{1}{\pi}\int_0^\pi \cos(n\phi - z\sin\phi)\,d\phi - \frac{\sin n\pi}{\pi}\int_0^\infty e^{-n\phi - z\sinh\phi}\,d\phi$$

6.
$$\int_0^z t J_n(at) J_n(bt)\,dt = \frac{z\{a J_n(bz) J_n'(az) - b J_n(az) J_n'(bz)\}}{b^2 - a^2}, \qquad a \neq b$$

7.
$$\int_0^z t J_n(at) J_n(bt)\,dt = \frac{az J_n(bz) J_{n-1}(az) - bz J_n(az) J_{n-1}(bz)}{b^2 - a^2}, \qquad a \neq b$$

8.
$$\int_0^z t\{J_n(at)\}^2\,dt = \frac{z^2}{2}[\{J_n(az)\}^2 - J_{n-1}(az) J_{n+1}(az)]$$

9.
$$J_n(z) = \frac{1}{2\pi i}\oint_C t^{-n-1}\,e^{\frac{1}{2}z(t-1/t)}\,dt, \qquad n = 0, \pm 1, \pm 2, \ldots$$

  where C is any simple closed curve enclosing $t = 0$.

10.
$$J_n(z) = \frac{z^n}{1\cdot 3\cdot 5\cdots(2n-1)\pi}\int_{-1}^1 e^{izt}\,(1-t^2)^{n-1/2}\,dt$$

$$= \frac{z^n}{1\cdot 3\cdot 5\cdots(2n-1)\pi}\int_0^\pi \cos(z\cos\phi)\,\sin^{2n}\phi\,d\phi$$

A second solution to Bessel's differential equation if n is a positive integer, is called *Bessel's function of the second kind of order n* or *Neumann's function* and is given by

$$Y_n(z) = J_n(z)\ln z - \frac{1}{2}\sum_{k=0}^{n-1}\frac{(n-k-1)!}{k!}\left(\frac{z}{2}\right)^{2k-n}$$

$$- \frac{1}{2}\sum_{k=0}^{\infty}\frac{(-1)^k}{(k!)(n+k)!}\left(\frac{z}{2}\right)^{2k+n}\{G(k) + G(n+k)\} \qquad (22)$$

where $G(k) = 1 + \dfrac{1}{2} + \dfrac{1}{3} + \cdots + \dfrac{1}{k}$ and $G(0) = 0$.

If $n = 0$, we have

$$Y_0(z) = J_0(z) \ln z + \frac{z^2}{2^2} - \frac{z^4}{2^2 4^2}(1 + \tfrac{1}{2}) + \frac{z^6}{2^2 4^2 6^2}(1 + \tfrac{1}{2} + \tfrac{1}{3}) - \cdots \qquad (23)$$

In terms of these the general solution of (19) if n is a positive integer can be written

$$Y = A J_n(z) + B Y_n(z) \qquad (24)$$

LEGENDRE FUNCTIONS

Legendre's differential equation of order n is given by

$$(1 - z^2)Y'' - 2zY' + n(n+1)Y = 0 \qquad (25)$$

The general solution of this equation is

$$Y = A\left\{ 1 - \frac{n(n+1)}{2!}z^2 + \frac{n(n-2)(n+1)(n+3)}{4!}z^4 - \cdots \right\}$$
$$+ B\left\{ z - \frac{(n-1)(n+2)}{3!}z^3 + \frac{(n-1)(n-3)(n+2)(n+4)}{5!}z^5 - \cdots \right\} \qquad (26)$$

If n is not an integer, these series solutions converge for $|z| < 1$. If n is zero or a positive integer, polynomial solutions of degree n are obtained. We call these polynomial solutions *Legendre polynomials* and denote them by $P_n(z)$, $n = 0, 1, 2, 3, \ldots$. By choosing these so that $P_n(1) = 1$, we find that they can be expressed by *Rodrigues' formula*

$$P_n(z) = \frac{1}{2^n n!} \frac{d^n}{dz^n}(z^2 - 1)^n \qquad (27)$$

from which $P_0(z) = 1$, $P_1(z) = z$, $P_2(z) = \tfrac{1}{2}(3z^2 - 1)$, $P_3(z) = \tfrac{1}{2}(5z^3 - 3z)$, etc.

The following are some properties of Legendre polynomials.

1.
$$\frac{1}{\sqrt{1 - 2zt + t^2}} = \sum_{n=0}^{\infty} P_n(z) t^n$$

This is called the *generating function* for Legendre polynomials.

2.
$$P_n(z) = \frac{(2n)!}{2^n (n!)^2}\left\{ z^n - \frac{n(n-1)}{2(2n-1)}z^{n-2} + \frac{n(n-1)(n-2)(n-3)}{2 \cdot 4(2n-1)(2n-3)}z^{n-4} - \cdots \right\}$$

3.
$$P_n(z) = \frac{1}{2\pi i} \oint_C \frac{(t^2 - 1)^n}{2^n (t - z)^{n+1}} dt$$

where C is any simple closed curve enclosing the pole $t = z$.

4.
$$\int_{-1}^{1} P_m(z) P_n(z) \, dz = \begin{cases} 0 & \text{if } m \neq n \\ \dfrac{2}{2n+1} & \text{if } m = n \end{cases}$$

[See Problems 30 and 31.]

5.
$$P_n(z) = \frac{1}{\pi} \int_0^{\pi} [z + \sqrt{z^2 - 1} \cos \phi]^n \, d\phi$$

[See Problem 34, Chapter 6.]

6.
$$(n+1) P_{n+1}(z) - (2n+1)z P_n(z) + n P_{n-1}(z) = 0$$

This is called the *recursion formula for Legendre polynomials* [see Prob. 32].

7.
$$(2n+1) P_n(z) = P'_{n+1}(z) - P'_{n-1}(z)$$

If n is a positive integer or zero, the general solution of Legendre's equation can be written as

$$Y = AP_n(z) + BQ_n(z) \tag{28}$$

where $Q_n(z)$ is an infinite series convergent for $|z| < 1$ obtained from (26). If n is not a positive integer, there are two infinite series solutions obtained from (26) which are convergent for $|z| < 1$. These solutions to Legendre's equation are called *Legendre functions*. They have properties analogous to those of the Legendre polynomials.

THE HYPERGEOMETRIC FUNCTION

The function defined by

$$F(a,b;c;z) = 1 + \frac{a \cdot b}{1 \cdot c} z + \frac{a(a+1)b(b+1)}{1 \cdot 2 \cdot c(c+1)} z^2 + \cdots \tag{29}$$

is called the *hypergeometric function* and is a solution to *Gauss' differential equation* or the *hypergeometric equation*

$$z(1-z)Y'' + \{c - (a+b+1)z\}Y' - abY = 0 \tag{30}$$

The series (29) is absolutely convergent for $|z| < 1$ and divergent for $|z| > 1$. For $|z| = 1$ it converges absolutely if $\text{Re}\{c - a - b\} > 0$.

If $|z| < 1$ and $\text{Re}\{c\} > \text{Re}\{b\} > 0$, we have

$$F(a,b;c;z) = \frac{\Gamma(c)}{\Gamma(b)\,\Gamma(c-b)} \int_0^1 t^{b-1}(1-t)^{c-b-1}(1-tz)^{-a} dt \tag{31}$$

For $|z| > 1$ the function can be defined by analytic continuation.

THE ZETA FUNCTION

The *zeta function*, studied extensively by Riemann in connection with the theory of numbers, is defined for $\text{Re}\{z\} > 1$ by

$$\zeta(z) = \frac{1}{1^z} + \frac{1}{2^z} + \frac{1}{3^z} + \cdots = \sum_{k=1}^{\infty} \frac{1}{k^z} \tag{32}$$

It can be extended by analytic continuation to other values of z. This extended definition of $\zeta(z)$ has the interesting property that

$$\zeta(1-z) = 2^{1-z} \pi^{-z} \Gamma(z) \cos(\pi z/2) \zeta(z) \tag{33}$$

Other interesting properties are as follows.

1. $$\zeta(z) = \frac{1}{\Gamma(z)} \int_0^\infty \frac{t^{z-1}}{e^t + 1} dt \qquad \text{Re}\{z\} > 0$$

2. The only singularity of $\zeta(z)$ is a simple pole at $z = 1$ having residue 1.

3. If B_k, $k = 1, 2, 3, \ldots$, is the coefficient of z^{2k} in the expansion

$$\tfrac{1}{2}z \cot(\tfrac{1}{2}z) = 1 - \sum_{k=1}^{\infty} \frac{B_k z^{2k}}{(2k)!}$$

then $$\zeta(2k) = \frac{2^{2k-1} \pi^{2k} B_k}{(2k)!} \qquad k = 1, 2, 3, \ldots$$

We have, for example, $B_1 = 1/6$, $B_2 = 1/30$, $\ldots$, from which $\zeta(2) = \pi^2/6$, $\zeta(4) = \pi^4/90$, $\ldots$. The numbers B_k are called *Bernoulli numbers*. For another definition of the Bernoulli numbers see Problem 163, Page 171.

4.
$$\frac{1}{\zeta(z)} = \left(1 - \frac{1}{2^z}\right)\left(1 - \frac{1}{3^z}\right)\left(1 - \frac{1}{5^z}\right)\left(1 - \frac{1}{7^z}\right)\cdots = \prod_p \left(1 - \frac{1}{p^z}\right)$$

where the product is taken over all positive primes p.

Riemann conjectured that all zeros of $\zeta(z)$ are situated on the line $\operatorname{Re}\{z\} = \frac{1}{2}$, but as yet this has neither been proved nor disproved. It has, however, been shown by Hardy that there are infinitely many zeros which do lie on this line.

ASYMPTOTIC SERIES

A series
$$a_0 + \frac{a_1}{z} + \frac{a_2}{z^2} + \cdots = \sum_{n=0}^{\infty} \frac{a_n}{z^n} \tag{34}$$

is called an *asymptotic series* for a function $F(z)$ if for any specified positive integer M,

$$\lim_{z \to \infty} z^M \left\{ F(z) - \sum_{n=0}^{M} \frac{a_n}{z^n} \right\} = 0 \tag{35}$$

In such case we write

$$F(z) \sim \sum_{n=0}^{\infty} \frac{a_n}{z^n} \tag{36}$$

Asymptotic series, and formulas involving them, are very useful in evaluation of functions for large values of the variable, which might otherwise be difficult. In practice, an asymptotic series may diverge. However, by taking the sum of successive terms of the series, stopping just before the terms begin to increase, we may obtain a good approximation for $F(z)$.

Various operations with asymptotic series are permissible. For example, asymptotic series may be added, multiplied or integrated term by term to yield another asymptotic series. However, differentiation is not always possible. For a given range of values of z an asymptotic series, if it exists, is unique.

THE METHOD OF STEEPEST DESCENTS

Let $I(z)$ be expressible in the form

$$I(z) = \int_C e^{z F(t)} dt \tag{37}$$

where C is some path in the t plane. Since $F(t)$ is complex, we can consider z to be real.

The method of steepest descents is a method for finding an asymptotic formula for (37) valid for large z. Where applicable, it consists of the following steps.

1. Determine the points at which $F'(t) = 0$. Such points are called *saddle points*, and for this reason the method is also called the *saddle point method*.

    We shall assume that there is only one saddle point, say t_0. The method can be extended if there is more than one.

2. Assuming $F(t)$ analytic in a neighborhood of t_0, obtain the Taylor series expansion

$$F(t) = F(t_0) + \frac{F''(t_0)(t - t_0)^2}{2!} + \cdots = F(t_0) - u^2 \tag{38}$$

Now deform contour C so that it passes through the saddle point t_0, and is such that $\operatorname{Re}\{F(t)\}$ is largest at t_0 while $\operatorname{Im}\{F(t)\}$ can be considered equal to the constant $\operatorname{Im}\{F(t_0)\}$ in the neighborhood of t_0. With these assumptions, the variable u defined by (38) is real and we obtain to a high degree of approximation

$$I(z) = e^{z F(t_0)} \int_{-\infty}^{\infty} e^{-zu^2} \left(\frac{dt}{du}\right) du \tag{39}$$

where from (38), we can find constants $b_0, b_1, \ldots$ such that

$$\frac{dt}{du} = b_0 + b_1 u + b_2 u^2 + \cdots \tag{40}$$

3. Substitute (40) into (39) and perform the integrations to obtain the required asymptotic expansion

$$I(z) \sim \sqrt{\frac{\pi}{z}} \, e^{z\,F(t_0)} \left\{ b_0 + \frac{1}{2} \frac{b_2}{z} + \frac{1 \cdot 3}{2 \cdot 2} \frac{b_4}{z^2} + \frac{1 \cdot 3 \cdot 5}{2 \cdot 2 \cdot 2} \frac{b_6}{z^3} + \cdots \right\} \tag{41}$$

For many practical purposes the first term provides enough accuracy and we find

$$I(z) \sim \sqrt{\frac{-2\pi}{z\,F''(t_0)}} \, e^{z\,F(t_0)} \tag{42}$$

Methods similar to the above are also known as *Laplace's method* and the *method of stationary phase*.

SPECIAL ASYMPTOTIC EXPANSIONS

1. The Gamma Function

$$\Gamma(z+1) \sim \sqrt{2\pi z} \, z^z \, e^{-z} \left\{ 1 + \frac{1}{12z} + \frac{1}{288z^2} - \frac{139}{51{,}840z^3} + \cdots \right\} \tag{43}$$

This is sometimes called *Stirling's asymptotic formula for the gamma function*. It holds for large values of $|z|$ such that $-\pi < \arg z < \pi$.

If n is real and large, we have

$$\Gamma(n+1) = \sqrt{2\pi n} \, n^n \, e^{-n} \, e^{\theta/12n} \qquad \text{where } 0 < \theta < 1 \tag{44}$$

In particular, if n is a large positive integer we have

$$n! \sim \sqrt{2\pi n} \, n^n \, e^{-n} \tag{45}$$

called *Stirling's asymptotic formula for* $n!$.

2. Bessel Functions

$$J_n(z) \sim \sqrt{\frac{2}{\pi z}} \left\{ P(z) \cos\left(z - \tfrac{1}{2}n\pi - \tfrac{1}{4}\pi\right) + Q(z) \sin\left(z - \tfrac{1}{2}n\pi - \tfrac{1}{4}\pi\right) \right\} \tag{46}$$

where

$$\begin{aligned}
P(z) &= 1 + \sum_{k=1}^{\infty} \frac{(-1)^k [4n^2 - 1^2][4n^2 - 3^2] \cdots [4n^2 - (4k-1)^2]}{(2k)! \, 2^{6k} \, z^{2k}} \\[2mm]
Q(z) &= \sum_{k=1}^{\infty} \frac{(-1)^k [4n^2 - 1^2][4n^2 - 3^2] \cdots [4n^2 - (4k-3)^2]}{(2k-1)! \, 2^{6k-3} \, z^{2k-1}}
\end{aligned} \tag{47}$$

This holds for large values of $|z|$ such that $-\pi < \arg z < \pi$.

3. The Error Function

$$\operatorname{erf}(z) = \frac{2}{\sqrt{\pi}} \int_0^z e^{-t^2} \, dt \sim 1 + \frac{ze^{-z^2}}{\pi} \sum_{k=1}^{\infty} (-1)^k \frac{\Gamma(k - \tfrac{1}{2})}{z^{2k}} \tag{48}$$

This result holds for large values of $|z|$ such that $-\pi/2 < \arg z < \pi/2$. For $\pi/2 < \arg z < 3\pi/2$ the result holds if we replace z by $-z$ on the right.

4. The Exponential Integral

$$\operatorname{Ei}(z) = \int_z^{\infty} \frac{e^{-t}}{t} \, dt \sim e^{-z} \sum_{k=0}^{\infty} \frac{(-1)^k k!}{z^{k+1}} \tag{49}$$

This result holds for large values of $|z|$ such that $-\pi < \arg z < \pi$.

ELLIPTIC FUNCTIONS

The integral

$$z = \int_0^w \frac{dt}{\sqrt{(1-t^2)(1-k^2t^2)}} \qquad |k| < 1 \qquad (50)$$

is called an *elliptic integral of the first kind*. The integral exists if w is real and such that $|w| < 1$. By analytic continuation we can extend it to other values of w. If $t = \sin \theta$ and $w = \sin \phi$, the integral (50) assumes an equivalent form

$$z = \int_0^\phi \frac{d\theta}{\sqrt{1 - k^2 \sin^2 \theta}} \qquad (51)$$

where we often write $\phi = \operatorname{am} z$.

If $k = 0$, (50) becomes $z = \sin^{-1} w$ or, equivalently, $w = \sin z$. By analogy, we denote the integral in (50) when $k \neq 0$ by $\operatorname{sn}^{-1}(w; k)$ or briefly $\operatorname{sn}^{-1} w$ when k does not change during a given discussion. Thus

$$z = \operatorname{sn}^{-1} w = \int_0^w \frac{dt}{\sqrt{(1-t^2)(1-k^2t^2)}} \qquad (52)$$

This leads to the function $w = \operatorname{sn} z$ which is called an *elliptic function* or sometimes a *Jacobian elliptic function.*

By analogy with the trigonometric functions, it is convenient to define other elliptic functions

$$\operatorname{cn} z = \sqrt{1 - \operatorname{sn}^2 z}, \qquad \operatorname{dn} z = \sqrt{1 - k^2 \operatorname{sn}^2 z} \qquad (53)$$

Another function which is sometimes used is $\operatorname{tn} z = (\operatorname{sn} z)/(\operatorname{cn} z)$.

The following list shows various properties of these functions.

1. $\operatorname{sn}(0) = 0$, $\operatorname{cn}(0) = 1$, $\operatorname{dn}(0) = 1$, $\operatorname{sn}(-z) = -\operatorname{sn} z$, $\operatorname{cn}(-z) = \operatorname{cn} z$, $\operatorname{dn}(-z) = \operatorname{dn} z$

2. $\dfrac{d}{dz} \operatorname{sn} z = \operatorname{cn} z \operatorname{dn} z$, $\quad \dfrac{d}{dz} \operatorname{cn} z = -\operatorname{sn} z \operatorname{dn} z$, $\quad \dfrac{d}{dz} \operatorname{dn} z = -k^2 \operatorname{sn} z \operatorname{cn} z$

3. $\operatorname{sn} z = \sin(\operatorname{am} z)$, $\quad \operatorname{cn} z = \cos(\operatorname{am} z)$

4.
$$\operatorname{sn}(z_1 + z_2) = \frac{\operatorname{sn} z_1 \operatorname{cn} z_2 \operatorname{dn} z_2 + \operatorname{cn} z_1 \operatorname{dn} z_1 \operatorname{sn} z_2}{1 - k^2 \operatorname{sn}^2 z_1 \operatorname{sn}^2 z_2} \qquad (54)$$

$$\operatorname{cn}(z_1 + z_2) = \frac{\operatorname{cn} z_1 \operatorname{cn} z_2 - \operatorname{sn} z_1 \operatorname{sn} z_2 \operatorname{dn} z_1 \operatorname{dn} z_2}{1 - k^2 \operatorname{sn}^2 z_1 \operatorname{sn}^2 z_2} \qquad (55)$$

$$\operatorname{dn}(z_1 + z_2) = \frac{\operatorname{dn} z_1 \operatorname{dn} z_2 - k^2 \operatorname{sn} z_1 \operatorname{sn} z_2 \operatorname{cn} z_1 \operatorname{cn} z_2}{1 - k^2 \operatorname{sn}^2 z_1 \operatorname{sn}^2 z_2} \qquad (56)$$

These are called *addition formulas* for the elliptic functions.

5. The elliptic functions have two periods, and for this reason they are often called *doubly-periodic functions.* Let us write

$$K = \int_0^1 \frac{dt}{\sqrt{(1-t^2)(1-k^2t^2)}} = \int_0^{\pi/2} \frac{d\theta}{\sqrt{1 - k^2 \sin^2 \theta}} \qquad (57)$$

$$K' = \int_0^1 \frac{dt}{\sqrt{(1-t^2)(1-k'^2t^2)}} = \int_0^{\pi/2} \frac{d\theta}{\sqrt{1 - k'^2 \sin^2 \theta}} \qquad (58)$$

where k and k', called the *modulus* and *complementary modulus* respectively, are such that $k' = \sqrt{1-k^2}$. Then the periods of $\operatorname{sn} z$ are $4K$ and $2iK'$, the periods of $\operatorname{cn} z$ are $4K$ and $2K + 2iK'$, and the periods of $\operatorname{dn} z$ are $2K$ and $4iK'$. It follows that there exists a periodic set of parallelograms [often called *period parallelograms*] in the complex plane in which the values of an elliptic function repeat. The smallest of these is often referred to as a *unit cell* or briefly a *cell.*

The above ideas can be extended to other elliptic functions.   Thus there exist *elliptic integrals of the second* and *third kinds* defined respectively by

$$z \;=\; \int_0^w \sqrt{\frac{1-k^2t^2}{1-t^2}}\,dt \;=\; \int_0^\phi \sqrt{1-k^2\sin^2\theta}\,d\theta \tag{59}$$

$$z \;=\; \int_0^w \frac{dt}{(1+nt^2)\sqrt{(1-t^2)(1-k^2t^2)}} \;=\; \int_0^\phi \frac{d\theta}{(1+n\sin^2\theta)\sqrt{1-k^2\sin^2\theta}} \tag{60}$$

Solved Problems

ANALYTIC CONTINUATION

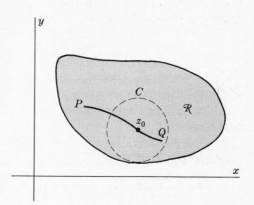

1. Let $F(z)$ be analytic in a region $\mathcal{R}$ and suppose that $F(z) = 0$ at all points on an arc PQ inside $\mathcal{R}$ [Fig. 10-6].   Prove that $F(z) = 0$ throughout $\mathcal{R}$.

    Choose any point, say z_0, on arc PQ.   Then in some circle of convergence C with center at z_0 [this circle extending at least to the boundary of $\mathcal{R}$ where a singularity may exist], $F(z)$ has a Taylor series expansion

    $$F(z) \;=\; F(z_0) + F'(z_0)(z-z_0) + \tfrac{1}{2}F''(z_0)(z-z_0)^2 + \cdots$$

    But by hypothesis $F(z_0) = F'(z_0) = F''(z_0) = \cdots = 0$.   Hence $F(z) = 0$ inside C.

    By choosing another arc inside C, we can continue the process.   In this manner we can show that $F(z) = 0$ throughout $\mathcal{R}$.

Fig. 10-6

2. Given that the identity $\sin^2 z + \cos^2 z = 1$ holds for real values of z, prove that it also holds for all complex values of z.

    Let $F(z) = \sin^2 z + \cos^2 z - 1$ and let $\mathcal{R}$ be a region of the z plane containing a portion of the x axis [Fig. 10-7].

    Since $\sin z$ and $\cos z$ are analytic in $\mathcal{R}$, it follows that $F(z)$ is analytic in $\mathcal{R}$.   Also $F(z) = 0$ on the x axis.   Hence by Problem 1, $F(z) = 0$ identically in $\mathcal{R}$, which shows that $\sin^2 z + \cos^2 z = 1$ for all z in $\mathcal{R}$.   Since $\mathcal{R}$ is arbitrary, we obtain the required result.

    This method is useful in proving for complex values many of the results true for real values.

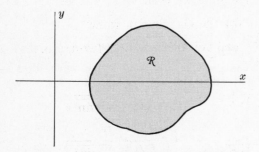

Fig. 10-7

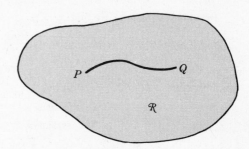

Fig. 10-8

3. Let $F_1(z)$ and $F_2(z)$ be analytic in a region $\mathcal{R}$ [Fig. 10-8] and suppose that on an arc PQ in $\mathcal{R}$, $F_1(z) = F_2(z)$.   Prove that $F_1(z) = F_2(z)$ in $\mathcal{R}$.

    This follows from Problem 1 by choosing $F(z) = F_1(z) - F_2(z)$.

4. Let $F_1(z)$ be analytic in region $\mathcal{R}_1$ [Fig. 10-9] and on the boundary $JKLM$. Suppose that we can find a function $F_2(z)$ analytic in region $\mathcal{R}_2$ and on the boundary $JKLM$ such that $F_1(z) = F_2(z)$ on $JKLM$. Prove that the function

$$F(z) \;=\; \begin{cases} F_1(z) \text{ for } z \text{ in } \mathcal{R}_1 \\ F_2(z) \text{ for } z \text{ in } \mathcal{R}_2 \end{cases}$$

is analytic in the region $\mathcal{R}$ which is composed of $\mathcal{R}_1$ and $\mathcal{R}_2$ [sometimes written $\mathcal{R} = \mathcal{R}_1 + \mathcal{R}_2$].

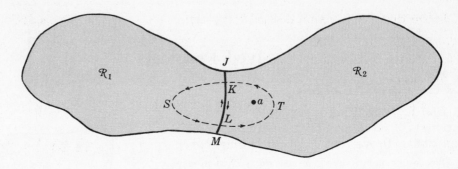

Fig. 10-9

Method 1.

This follows from Problem 3, since there can be only one function $F_2(z)$ in $\mathcal{R}_2$ satisfying the required properties.

Method 2, using Cauchy's integral formulas.

Construct the simple closed curve $SLTKS$ (dashed in Fig. 10-9) and let a be any point inside. From Cauchy's integral formula, we have (since $F_2(z)$ is analytic inside and on $LTKL$ and since $F_2(z) = F(z)$ on LTK)

$$F_2(a) \;=\; \frac{1}{2\pi i} \oint_{LTKL} \frac{F_2(z)}{z-a}\, dz \;=\; \frac{1}{2\pi i} \int_{LTK} \frac{F(z)}{z-a}\, dz \;+\; \frac{1}{2\pi i} \int_{KL} \frac{F(z)}{z-a}\, dz$$

Also we have by Cauchy's theorem (since $F_1(z)/(z-a)$ is analytic inside and on $KSLK$ and since $F_1(z) = F(z)$ on KSL)

$$0 \;=\; \frac{1}{2\pi i} \oint_{KSLK} \frac{F_1(z)}{z-a}\, dz \;=\; \frac{1}{2\pi i} \int_{KSL} \frac{F(z)}{z-a}\, dz \;+\; \frac{1}{2\pi i} \int_{LK} \frac{F(z)}{z-a}\, dz$$

Adding, using the fact that $F(z) = F_1(z) = F_2(z)$ on LK so that the integrals along KL and LK cancel, we have since $F(a) = F_2(a)$

$$F(a) \;=\; \frac{1}{2\pi i} \oint_{LTKSL} \frac{F(z)}{z-a}\, dz$$

In a similar manner we find

$$F^{(n)}(a) \;=\; \frac{n!}{2\pi i} \oint_{LTKSL} \frac{F(z)}{(z-a)^{n+1}}\, dz$$

so that $F(z)$ is analytic at a. But since we can choose a to be any point in the region $\mathcal{R}$ by suitably modifying the dashed contour of Fig. 10-9, it follows that $F(z)$ is analytic in $\mathcal{R}$.

Method 3, using Morera's theorem.

Referring to Fig. 10-9, we have

$$\oint_{KSLTK} F(z)\, dz \;=\; \int_{KSL} F(z)\, dz \;+\; \int_{LK} F(z)\, dz \;+\; \int_{KL} F(z)\, dz \;+\; \int_{LTK} F(z)\, dz$$

$$=\; \oint_{KSLK} F_1(z)\, dz \;+\; \oint_{KLTK} F_2(z)\, dz \;=\; 0$$

by Cauchy's theorem. Thus the integral around any simple closed path in $\mathcal{R}$ is zero, and so by Morera's theorem $F(z)$ must be analytic.

The function $F_2(z)$ is called an *analytic continuation* of $F_1(z)$.

5. (a) Prove that the function defined by $F_1(z) = z - z^2 + z^3 - z^4 + \cdots$ is analytic in the region $|z| < 1$. (b) Find a function which represents all possible analytic continuations of $F_1(z)$.

   (a) By the ratio test, the series converges for $|z| < 1$. Then the series represents an analytic function in this region.

   (b) For $|z| < 1$, the sum of the series is $F_2(z) = z/(1+z)$. But this function is analytic at all points except $z = -1$. Since $F_2(z) = F_1(z)$ inside $|z| = 1$, it is the required function.

6. (a) Prove that the function defined by $F_1(z) = \int_0^\infty t^3 e^{-zt} dt$ is analytic at all points z for which $\mathrm{Re}\{z\} > 0$. (b) Find a function which is the analytic continuation of $F_1(z)$ into the left hand plane $\mathrm{Re}\{z\} < 0$.

   (a) On integrating by parts, we have

$$\int_0^\infty t^3 e^{-zt} dt = \lim_{M \to \infty} \int_0^M t^3 e^{-zt} dt$$

$$= \lim_{M \to \infty} \left\{ (t^3)\left(\frac{e^{-zt}}{-z}\right) - (3t^2)\left(\frac{e^{-zt}}{z^2}\right) + (6t)\left(\frac{e^{-zt}}{-z^3}\right) - (6)\left(\frac{e^{-zt}}{z^4}\right) \right\}\Bigg|_0^M$$

$$= \lim_{M \to \infty} \left\{ \frac{6}{z^4} - \frac{M^3 e^{-Mz}}{z} - \frac{3M^2 e^{-Mz}}{z^2} - \frac{6M e^{-Mz}}{z^3} - \frac{6 e^{-Mz}}{z^4} \right\}$$

$$= \frac{6}{z^4} \quad \text{if} \quad \mathrm{Re}\{z\} > 0$$

   (b) For $\mathrm{Re}\{z\} > 0$, the integral has the value $F_2(z) = 6/z^4$. But this function is analytic at all points except $z = 0$. Since $F_2(z) = F_1(z)$ for $\mathrm{Re}\{z\} > 0$, we see that $F_2(z) = 6/z^4$ must be the required analytic continuation.

SCHWARZ'S REFLECTION PRINCIPLE

7. Prove Schwarz's reflection principle (see Page 266).

   Refer to Fig. 10-4, Page 266. On the real axis $[y = 0]$ we have $F_1(z) = F_1(x) = \overline{F_1(x)} = \overline{F_1(\bar{z})}$. Then by Problem 3 we have only to prove that $\overline{F_1(\bar{z})} = F_2(z)$ is analytic in $\mathcal{R}_2$.

   Let $F_1(z) = U_1(x, y) + i V_1(x, y)$. Since this is analytic in $\mathcal{R}_1$ [i.e. $y > 0$], we have by the Cauchy-Riemann equations,

$$\frac{\partial U_1}{\partial x} = \frac{\partial V_1}{\partial y}, \qquad \frac{\partial V_1}{\partial x} = -\frac{\partial U_1}{\partial y} \tag{1}$$

where these partial derivatives are continuous.

   Now $F_1(\bar{z}) = F_1(x - iy) = U_1(x, -y) + i V_1(x, -y)$, and so $\overline{F_1(\bar{z})} = U_1(x, -y) - i V_1(x, -y)$. If this is to be analytic in $\mathcal{R}_2$ we must have, for $y > 0$,

$$\frac{\partial U_1}{\partial x} = \frac{\partial(-V_1)}{\partial(-y)}, \qquad \frac{\partial(-V_1)}{\partial x} = -\frac{\partial U_1}{\partial(-y)} \tag{2}$$

But these are equivalent to (1), since $\dfrac{\partial(-V_1)}{\partial(-y)} = \dfrac{\partial V_1}{\partial y}, \; \dfrac{\partial(-V_1)}{\partial x} = -\dfrac{\partial V_1}{\partial x}$ and $\dfrac{\partial U_1}{\partial(-y)} = -\dfrac{\partial U_1}{\partial y}$. Hence the required result follows.

INFINITE PRODUCTS

8. Prove that a necessary and sufficient condition for $\displaystyle\prod_{k=1}^\infty (1 + |w_k|)$ to converge is that $\Sigma\,|w_k|$ converges.

   Sufficiency. If $x > 0$, then $1 + x \leqq e^x$ so that

$$P_n = \prod_{k=1}^n (1 + |w_k|) = (1 + |w_1|)(1 + |w_2|)\cdots(1 + |w_n|) \leqq e^{|w_1|} e^{|w_2|} \cdots e^{|w_n|} = e^{|w_1| + |w_2| + \cdots + |w_n|}$$

If $\sum\limits_{k=1}^{\infty} |w_k|$ converges, it follows that P_n is a bounded monotonic increasing sequence and so has

a limit, i.e. $\prod\limits_{k=1}^{\infty} (1 + |w_k|)$, converges.

Necessity. If $S_n = \sum\limits_{k=1}^{n} |w_k|$, we have

$$P_n = (1 + |w_1|)(1 + |w_2|) \cdots (1 + |w_n|) \geqq 1 + |w_1| + |w_2| + \cdots + |w_n| = 1 + S_n \geqq 1$$

If $\lim\limits_{n \to \infty} P_n$ exists, i.e. the infinite product converges, it follows that S_n is a bounded monotonic increasing sequence and so has a limit, i.e. $\sum\limits_{k=1}^{\infty} |w_k|$ converges.

9. Prove that $\prod\limits_{k=1}^{\infty} \left(1 - \dfrac{z^2}{k^2}\right)$ converges.

Let $w_k = -\dfrac{z^2}{k^2}$. Then $|w_k| = \dfrac{|z|^2}{k^2}$ and $\Sigma\, |w_k| = |z|^2\, \Sigma\, \dfrac{1}{k^2}$ converges. Hence by Problem 8, the infinite product is absolutely convergent and thus convergent.

10. Prove that $\quad \sin z = z \left(1 - \dfrac{z^2}{\pi^2}\right)\left(1 - \dfrac{z^2}{4\pi^2}\right)\left(1 - \dfrac{z^2}{9\pi^2}\right) \cdots = z \prod\limits_{k=1}^{\infty} \left(1 - \dfrac{z^2}{k^2 \pi^2}\right).$

From Problem 35, Chapter 7, Page 192, we have

$$\int_0^z \left(\cot t - \frac{1}{t}\right) dt = \ln\left(\frac{\sin t}{t}\right)\Big|_0^z = \ln\left(\frac{\sin z}{z}\right)$$

$$= \int_0^z \left(\frac{2t}{t^2 - \pi^2} + \frac{2t}{t^2 - 4\pi^2} + \cdots\right) dt$$

$$= \sum_{k=1}^{\infty} \ln\left(1 - \frac{z^2}{k^2 \pi^2}\right) = \ln \prod_{k=1}^{\infty} \left(1 - \frac{z^2}{k^2 \pi^2}\right)$$

Then $\quad \sin z = z \prod\limits_{k=1}^{\infty} \left(1 - \dfrac{z^2}{k^2 \pi^2}\right).$

THE GAMMA FUNCTION

11. Prove that $\quad \Gamma(z+1) = z\,\Gamma(z)\quad$ using definition (4), Page 267.

Integrating by parts, we have if $\text{Re}\,\{z\} > 0$,

$$\Gamma(z+1) = \int_0^{\infty} t^z\, e^{-t}\, dt = \lim_{M \to \infty} \int_0^{M} t^z\, e^{-t}\, dt$$

$$= \lim_{M \to \infty} \left\{ (t^z)(-e^{-t})\Big|_0^{M} - \int_0^{M} (zt^{z-1})(-e^{-t})\, dt \right\}$$

$$= z \int_0^{\infty} t^{z-1}\, e^{-t}\, dt = z\,\Gamma(z)$$

12. Prove that $\quad \Gamma(m) = 2 \int_0^{\infty} x^{2m-1}\, e^{-x^2}\, dx, \quad m > 0.$

If $t = x^2$, we have

$$\Gamma(m) = \int_0^{\infty} t^{m-1}\, e^{-t}\, dt = \int_0^{\infty} (x^2)^{m-1}\, e^{-x^2}\, 2x\, dx = 2 \int_0^{\infty} x^{2m-1}\, e^{-x^2}\, dx$$

The result also holds if $\text{Re}\,\{m\} > 0$.

13. Prove that $\quad \Gamma(z)\,\Gamma(1-z) = \dfrac{\pi}{\sin \pi z}.$

We first prove it for real values of z such that $0 < z < 1$. By analytic continuation we can then extend it to other values of z.

From Problem 12, we have for $0 < m < 1$,

$$\Gamma(m)\,\Gamma(1-m) \;=\; \left\{ 2\int_0^\infty x^{2m-1}\,e^{-x^2}\,dx \right\}\left\{ 2\int_0^\infty y^{1-2m}\,e^{-y^2}\,dy \right\}$$

$$=\; 4\int_0^\infty \int_0^\infty x^{2m-1}\,y^{1-2m}\,e^{-(x^2+y^2)}\,dx\,dy$$

In terms of polar coordinates (r,θ) with $\quad x = r\cos\theta,\; y = r\sin\theta\quad$ this becomes

$$4\int_{\theta=0}^{\pi/2} \int_{r=0}^\infty (\tan^{1-2m}\theta)(re^{-r^2})\,dr\,d\theta \;=\; 2\int_0^{\pi/2} \tan^{1-2m}\theta\,d\theta \;=\; \frac{\pi}{\sin m\pi}$$

using Problem 20, Page 185, with $x = \tan^2\theta$ and $p = 1 - m$.

14. Prove that $\quad \Gamma(\tfrac{1}{2}) \;=\; 2\int_0^\infty e^{-u^2}\,du \;=\; \sqrt{\pi}.$

From Problem 12, letting $m = \tfrac{1}{2}$, we have

$$\Gamma(\tfrac{1}{2}) \;=\; 2\int_0^\infty e^{-x^2}\,dx$$

From Problem 13, letting $z = \tfrac{1}{2}$, we have

$$\{\Gamma(\tfrac{1}{2})\}^2 \;=\; \pi \qquad \text{or} \qquad \Gamma(\tfrac{1}{2}) \;=\; \sqrt{\pi}$$

since $\Gamma(\tfrac{1}{2}) > 0$. Thus the required result follows.

Another method. As in Problem 13,

$$\{\Gamma(\tfrac{1}{2})\}^2 \;=\; \left\{ 2\int_0^\infty e^{-x^2}\,dx \right\}\left\{ 2\int_0^\infty e^{-y^2}\,dy \right\}$$

$$=\; 4\int_0^\infty \int_0^\infty e^{-(x^2+y^2)}\,dx\,dy \;=\; 4\int_{\theta=0}^{\pi/2} \int_{r=0}^\infty e^{-r^2}r\,dr\,d\theta \;=\; \pi$$

from which $\Gamma(\tfrac{1}{2}) = \sqrt{\pi}$.

15. By use of analytic continuation, show that $\quad \Gamma(-\tfrac{1}{2}) = -2\sqrt{\pi}.$

If $\mathrm{Re}\,\{z\} > 0$, $\Gamma(z)$ is defined by (4), Page 267, but this definition cannot be used for $\mathrm{Re}\,\{z\} \le 0$. However, we can use the recursion formula $\Gamma(z+1) = z\,\Gamma(z)$, which holds for $\mathrm{Re}\,\{z\} > 0$, to extend the definition for $\mathrm{Re}\,\{z\} \le 0$, i.e. it provides an analytic continuation into the left hand plane.

Substituting $z = -\tfrac{1}{2}$ in $\Gamma(z+1) = z\,\Gamma(z)$, we find $\Gamma(\tfrac{1}{2}) = -\tfrac{1}{2}\Gamma(-\tfrac{1}{2})$ or $\Gamma(-\tfrac{1}{2}) = -2\sqrt{\pi}$ using Problem 14.

16. (*a*) Prove that $\quad \Gamma(z) \;=\; \dfrac{\Gamma(z+n+1)}{z(z+1)(z+2)\cdots(z+n)}.$

(*b*) Use (*a*) to show that $\Gamma(z)$ is an analytic function except for simple poles in the left hand plane at $z = 0, -1, -2, -3, \ldots$.

(*a*) We have $\Gamma(z+1) = z\,\Gamma(z),\; \Gamma(z+2) = (z+1)\,\Gamma(z+1) = (z+1)z\,\Gamma(z),\; \Gamma(z+3) = (z+2)\,\Gamma(z+2) = (z+2)(z+1)z\,\Gamma(z)$ and, in general, $\Gamma(z+n+1) = (z+n)(z+n-1)\cdots(z+2)(z+1)z\,\Gamma(z)$ from which the required result follows.

(*b*) We know that $\Gamma(z)$ is analytic for $\mathrm{Re}\,\{z\} > 0$, from definition (4), Page 267. Also, it is clear from the result in (*a*) that $\Gamma(z)$ is defined and analytic for $\mathrm{Re}\,\{z\} \ge -n$ except for the simple poles at $z = 0, -1, -2, \ldots, -n$. Since this is the case for any positive integer n, the required result follows.

17. Use Weierstrass' factor theorem for infinite products [equation (2), Page 267] to obtain the infinite product for the gamma function [Property 2, Page 268].

Let $f(z) = 1/\Gamma(z+1)$. Then $f(z)$ is analytic everywhere and has simple zeros at $z = -1, -2, -3, \cdots$. By Weierstrass' factor theorem, we find

$$\frac{1}{\Gamma(z+1)} = e^{f'(0)z} \prod_{k=1}^{\infty} \left(1 + \frac{z}{k}\right) e^{-z/k}$$

To determine $f'(0)$, let $z = 1$. Then since $\Gamma(2) = 1$, we have

$$1 = e^{f'(0)} \prod_{k=1}^{\infty} \left(1 + \frac{1}{k}\right) e^{-1/k}$$

$$= e^{f'(0)} \lim_{M \to \infty} \prod_{k=1}^{M} \left(1 + \frac{1}{k}\right) e^{-1/k}$$

Taking logarithms, we see that

$$f'(0) = \lim_{M \to \infty} \left\{ \frac{1}{1} + \frac{1}{2} + \frac{1}{3} + \cdots + \frac{1}{M} - \ln\left[\left(1 + \frac{1}{1}\right)\left(1 + \frac{1}{2}\right) \cdots \left(1 + \frac{1}{M}\right)\right]\right\}$$

$$= \lim_{M \to \infty} \left\{ 1 + \frac{1}{2} + \frac{1}{3} + \cdots + \frac{1}{M} - \ln M \right\} = \gamma$$

where γ is Euler's constant. Then the required result follows on noting that $\Gamma(z+1) = z\,\Gamma(z)$.

THE BETA FUNCTION

18. Prove that $B(m, n) = B(n, m)$.

Letting $t = 1 - u$,

$$B(m, n) = \int_0^1 t^{m-1}(1-t)^{n-1}\, dt = \int_0^1 (1-u)^{m-1} u^{n-1}\, du = B(n, m)$$

19. Prove that $B(m, n) = 2\int_0^{\pi/2} \sin^{2m-1}\theta \cos^{2n-1}\theta\, d\theta = 2\int_0^{\pi/2} \cos^{2m-1}\theta \sin^{2n-1}\theta\, d\theta$.

Let $t = \sin^2\theta$. Then

$$B(m, n) = \int_0^1 t^{m-1}(1-t)^{n-1}\, dt = \int_0^{\pi/2} (\sin^2\theta)^{m-1}(\cos^2\theta)^{n-1}\, 2\sin\theta \cos\theta\, d\theta$$

$$= 2\int_0^{\pi/2} \sin^{2m-1}\theta \cos^{2n-1}\theta\, d\theta = 2\int_0^{\pi/2} \cos^{2m-1}\theta \sin^{2n-1}\theta\, d\theta$$

by Problem 18.

20. Prove that $B(m, n) = \int_0^1 t^{m-1}(1-t)^{n-1}\, dt = \dfrac{\Gamma(m)\,\Gamma(n)}{\Gamma(m+n)}$.

From Problem 12, we have on transforming to polar coordinates,

$$\Gamma(m)\,\Gamma(n) = \left\{ 2\int_0^{\infty} x^{2m-1} e^{-x^2}\, dx \right\}\left\{ 2\int_0^{\infty} y^{2n-1} e^{-y^2}\, dy \right\}$$

$$= 4\int_0^{\infty}\int_0^{\infty} x^{2m-1} y^{2n-1} e^{-(x^2+y^2)}\, dx\, dy$$

$$= 4\int_{\theta=0}^{\pi/2}\int_{r=0}^{\infty} (\cos^{2m-1}\theta \sin^{2n-1}\theta)(r^{2m+2n-1} e^{-r^2})\, dr\, d\theta$$

$$= \left\{ 2\int_0^{\pi/2} \cos^{2m-1}\theta \sin^{2n-1}\theta\, d\theta \right\}\left\{ \int_0^{\infty} r^{2(m+n)-1} e^{-r^2}\, dr \right\}$$

$$= B(m, n)\,\Gamma(m+n)$$

where we have used Problem 19 and Problem 12 with r replacing t and $m + n$ replacing m. From this the required result follows.

21. Evaluate (a) $\displaystyle\int_0^2 \sqrt{x(2-x)}\,dx$, (b) $\displaystyle\int_0^{\pi/2} \sqrt{\tan\theta}\,d\theta$.

(a) Letting $x = 2t$, the integral becomes

$$\int_0^1 \sqrt{4t(1-t)}\,2\,dt \;=\; 4\int_0^1 t^{1/2}(1-t)^{1/2}\,dt \;=\; 4\,B(3/2,3/2)$$

$$=\; 4\,\frac{\Gamma(3/2)\,\Gamma(3/2)}{\Gamma(3)} \;=\; \frac{4(\tfrac{1}{2}\sqrt{\pi})(\tfrac{1}{2}\sqrt{\pi})}{2} \;=\; \frac{\pi}{2}$$

(b)
$$\int_0^{\pi/2} \sqrt{\tan\theta}\,d\theta \;=\; \int_0^{\pi/2} \sin^{1/2}\theta \cos^{-1/2}\theta\,d\theta \;=\; \tfrac{1}{2}B(\tfrac{3}{4},\tfrac{1}{4})$$

$$=\; \tfrac{1}{2}\,\Gamma(\tfrac{3}{4})\,\Gamma(\tfrac{1}{4}) \;=\; \frac{1}{2}\,\frac{\pi}{\sin(\pi/4)} \;=\; \frac{\pi\sqrt{2}}{2}$$

using Problems 13, 19 and 20.

22. Show that $\displaystyle\int_0^4 y^{3/2}(16-y^2)^{1/2}\,dy \;=\; \frac{64}{21}\sqrt{\frac{2}{\pi}}\,\{\Gamma(\tfrac{1}{4})\}^2$.

Let $y^2 = 16t$, i.e. $y = 4t^{1/2}$, $dy = 2t^{-1/2}\,dt$. Then the integral becomes

$$\int_0^1 \{8t^{3/4}\}\{4(1-t)^{1/2}\}\{2t^{-1/2}\,dt\} \;=\; 64\int_0^1 t^{1/4}(1-t)^{1/2}\,dt$$

$$=\; 64\,B(\tfrac{5}{4},\tfrac{3}{2}) \;=\; \frac{64\,\Gamma(\tfrac{5}{4})\,\Gamma(\tfrac{3}{2})}{\Gamma(\tfrac{11}{4})} \;=\; \frac{64(\tfrac{1}{4})\,\Gamma(\tfrac{1}{4})\,(\tfrac{1}{2})\,\Gamma(\tfrac{1}{2})}{\tfrac{7}{4}\cdot\tfrac{3}{4}\,\Gamma(\tfrac{3}{4})}$$

$$=\; \frac{128\sqrt{\pi}}{21}\,\frac{\Gamma(\tfrac{1}{4})}{\Gamma(\tfrac{3}{4})} \;=\; \frac{128\sqrt{\pi}}{21}\,\frac{\{\Gamma(\tfrac{1}{4})\}^2}{\Gamma(\tfrac{1}{4})\,\Gamma(\tfrac{3}{4})} \;=\; \frac{64}{21}\sqrt{\frac{2}{\pi}}\,\{\Gamma(\tfrac{1}{4})\}^2$$

using the fact that $\Gamma(\tfrac{1}{4})\,\Gamma(\tfrac{3}{4}) = \pi/[\sin(\pi/4)] = \pi\sqrt{2}$ [Problem 13].

DIFFERENTIAL EQUATIONS

23. Determine the singular points of each of the following differential equations and specify whether they are regular or irregular.

(a) $z^2Y'' + zY' + (z^2 - n^2)Y = 0$ or $Y'' + \dfrac{1}{z}Y' + \left(\dfrac{z^2-n^2}{z^2}\right)Y = 0$.

$z = 0$ is a singular point. Since $z(1/z) = 1$ and $z^2\left(\dfrac{z^2-n^2}{z^2}\right) = z^2-n^2$ are analytic at $z = 0$, it is a regular singular point.

(b) $(z-1)^4Y'' + 2(z-1)^3Y' + Y = 0$ or $Y'' + \dfrac{2}{z-1}Y' + \dfrac{1}{(z-1)^4}Y = 0$.

At the singular point $z = 1$, $(z-1)\left(\dfrac{2}{z-1}\right) = 2$ is analytic but $(z-1)^2 \cdot \dfrac{1}{(z-1)^4} = \dfrac{1}{(z-1)^2}$ is not analytic. Then $z = 1$ is an irregular singular point.

(c) $z^2(1-z)Y'' + Y' - Y = 0$ or $Y'' + \dfrac{1}{z^2(1-z)}Y' - \dfrac{1}{z^2(1-z)}Y = 0$.

At the singular point $z = 0$, $z\left\{\dfrac{1}{z^2(1-z)}\right\} = \dfrac{1}{z(1-z)}$ and $z^2\left\{\dfrac{-1}{z^2(1-z)}\right\} = \dfrac{-1}{1-z}$ are not both analytic. Hence $z = 0$ is an irregular singular point.

At the singular point $z = 1$, $(z-1)\cdot\left\{\dfrac{1}{z^2(1-z)}\right\} = \dfrac{-1}{z^2}$ and $(z-1)^2\left\{\dfrac{-1}{z^2(1-z)}\right\} = \dfrac{z-1}{z^2}$ are both analytic. Hence $z = 1$ is a regular singular point.

24. Find the general solution of Bessel's differential equation

$$z^2 Y'' + z Y' + (z^2 - n^2) Y \; = \; 0 \qquad \text{where} \quad n \ne 0, \pm 1, \pm 2, \dots$$

The point $z = 0$ is a regular singular point. Hence there is a series solution of the form $Y = \sum\limits_{k=-\infty}^{\infty} a_k z^{k+c}$ where $a_k = 0$ for $k = -1, -2, -3, \dots$. By differentiation, omitting the summation limits, we have

$$Y' \; = \; \Sigma (k+c) a_k z^{k+c-1}, \qquad Y'' \; = \; \Sigma (k+c)(k+c-1) a_k z^{k+c-2}$$

Then

$$z^2 Y'' \; = \; \Sigma (k+c)(k+c-1)\, a_k\, z^{k+c}$$

$$z Y' \; = \; \Sigma (k+c)\, a_k\, z^{k+c}$$

$$(z^2 - n^2) Y \; = \; \Sigma a_k z^{k+c+2} \; - \; \Sigma n^2 a_k z^{k+c}$$

$$= \; \Sigma a_{k-2} z^{k+c} \; - \; \Sigma n^2 a_k z^{k+c}$$

Adding, $z^2 Y'' + z Y' + (z^2 - n^2) Y \; = \; \Sigma \{[(k+c)^2 - n^2]\, a_k \, + \, a_{k-2}\} z^{k+c} \; = \; 0$

from which we obtain

$$[(k+c)^2 - n^2]\, a_k \, + \, a_{k-2} \; = \; 0 \tag{1}$$

If $k = 0$, $(c^2 - n^2) a_0 = 0$; and if $a_0 \ne 0$, we obtain the *indicial equation* $c^2 - n^2 = 0$ with roots $c = \pm n$.

Case 1: $c = n$.

From (1), $[(k+n)^2 - n^2]\, a_k \, + \, a_{k-2} \; = \; 0$ or $k(2n+k)\, a_k \, + \, a_{k-2} \; = \; 0$.

If $k = 1$, $a_1 = 0$. If $k = 2$, $a_2 = -\dfrac{a_0}{2(2n+2)}$. If $k = 3$, $a_3 = 0$. If $k = 4$, $a_4 = -\dfrac{a_2}{4(2n+4)} = \dfrac{a_0}{2 \cdot 4 (2n+2)(2n+4)}$, etc. Then

$$Y \; = \; \Sigma a_k z^{k+c} \; = \; a_0 z^n \left\{ 1 \, - \, \frac{z^2}{2(2n+2)} \, + \, \frac{z^4}{2 \cdot 4(2n+2)(2n+4)} \, - \, \cdots \right\} \tag{2}$$

Case 2: $c = -n$.

The result obtained is

$$Y \; = \; a_0 z^{-n} \left\{ 1 \, - \, \frac{z^2}{2(2-2n)} \, + \, \frac{z^4}{2 \cdot 4(2n+2)(2n+4)} \, - \, \cdots \right\} \tag{3}$$

which can be obtained formally from Case 1 on replacing n by $-n$.

The general solution if $n \ne 0, \pm 1, \pm 2, \dots$ is given by

$$Y \; = \; A z^n \left\{ 1 \, - \, \frac{z^2}{2(2n+2)} \, + \, \frac{z^4}{2 \cdot 4(2n+2)(2n+4)} \, - \, \cdots \right\}$$

$$+ \; B z^{-n} \left\{ 1 \, - \, \frac{z^2}{2(2-2n)} \, + \, \frac{z^4}{2 \cdot 4(2-2n)(4-2n)} \, - \, \cdots \right\} \tag{4}$$

If $n = 0, \pm 1, \pm 2, \dots$ only one solution is obtained. To find the general solution in this case we must proceed as in Problems 175 and 176.

Since the singularity nearest to $z = 0$ is at infinity, the solutions should converge for all z. This is easily shown by the ratio test.

SOLUTION OF DIFFERENTIAL EQUATIONS BY CONTOUR INTEGRALS

25. (*a*) Obtain a solution of the equation $\; z Y'' + (2n+1) Y' + z Y = 0 \;$ having the form $Y = \displaystyle\oint_C e^{zt} G(t)\, dt$. (*b*) By letting $Y = z^r U$ and choosing the constant r appropriately, obtain a contour integral solution of $\; z^2 U'' + z U' + (z^2 - n^2) U = 0$.

(*a*) If $\; Y = \displaystyle\oint_C e^{zt} G(t)\, dt$, we find $\; Y' = \displaystyle\oint_C t e^{zt} G(t)\, dt$, $\; Y'' = \displaystyle\oint_C t^2 e^{zt} G(t)\, dt$.

Then integrating by parts, assuming that C is chosen so that the functional values at the initial and final points P are equal [and the integrated part is zero], we have

$$zY = \oint_C ze^{zt}G(t)\,dt = e^{zt}G(t)\Big|_P^P - \oint_C e^{zt}G'(t)\,dt = -\oint_C e^{zt}G'(t)\,dt$$

$$(2n+1)Y' = \oint_C (2n+1)t\,e^{zt}G(t)\,dt$$

$$zY'' = \oint_C zt^2e^{zt}G(t)\,dt = \oint_C (ze^{zt})\{t^2G(t)\}\,dt$$

$$= e^{zt}\{t^2G(t)\}\Big|_P^P - \oint_C e^{zt}\{t^2G(t)\}'\,dt$$

$$= -\oint_C e^{zt}\{t^2G(t)\}'\,dt$$

Thus

$$zY'' + (2n+1)Y' + zY = 0 = \oint_C e^{zt}[-G'(t) + (2n+1)t\,G(t) - \{t^2G(t)\}']\,dt$$

This is satisfied if we choose $G(t)$ so that the integrand is zero, i.e.,

$$-G'(t) + (2n+1)t\,G(t) - \{t^2G(t)\}' = 0 \quad\text{or}\quad G'(t) = \frac{(2n-1)t}{t^2+1}G(t)$$

Solving, $G(t) = A\,(t^2+1)^{n-\frac{1}{2}}$ where A is any constant. Hence a solution is

$$Y = A\oint_C e^{zt}(t^2+1)^{n-1/2}\,dt$$

(b) If $Y = z^rU$, then $Y' = z^rU' + rz^{r-1}U$ and $Y'' = z^rU'' + 2rz^{r-1}U' + r(r-1)z^{r-2}U$. Hence

$$zY'' + (2n+1)Y' + zY = z^{r+1}U'' + 2rz^rU' + r(r-1)z^{r-1}U$$
$$+ (2n+1)z^rU' + (2n+1)rz^{r-1}U + z^{r+1}U$$

$$= z^{r+1}U'' + [2rz^r + (2n+1)z^r]U'$$
$$+ [r(r-1)z^{r-1} + (2n+1)rz^{r-1} + z^{r+1}]U$$

The given differential equation is thus equivalent to

$$z^2U'' + (2r+2n+1)zU' + [z^2+r^2+2nr]U = 0$$

Letting $r = -n$, this becomes $z^2U'' + zU' + (z^2-n^2)U = 0$.

Hence a contour integral solution is

$$U = z^nY = Az^n\oint_C e^{zt}(t^2+1)^{n-1/2}\,dt$$

26. Obtain the general solution of $Y'' - 3Y' + 2Y = 0$ by the method of contour integrals.

Let $Y = \oint_C e^{zt}G(t)\,dt$, $Y' = \oint_C te^{zt}G(t)\,dt$, $Y'' = \oint_C t^2e^{zt}G(t)\,dt$. Then

$$Y'' - 3Y' + 2Y = \oint_C e^{zt}(t^2-3t+2)\,G(t)\,dt = 0$$

is satisfied if we choose $G(t) = 1/(t^2-3t+2)$. Hence

$$Y = \oint_C \frac{e^{zt}}{t^2-3t+2}\,dt$$

If we choose C so that the simple pole $t = 1$ lies inside C while $t = 2$ lies outside C, the integral has the value $2\pi ie^z$. If $t = 2$ lies inside C while $t = 1$ lies outside C, the integral has the value $2\pi ie^{2z}$.

The general solution is given by $Y = Ae^z + Be^{2z}$.

BESSEL FUNCTIONS

27. Prove that $z J_{n-1}(z) - 2n J_n(z) + z J_{n+1}(z) = 0.$

Differentiating with respect to t both sides of the identity

$$e^{\frac{1}{2}z(t-1/t)} = \sum_{n=-\infty}^{\infty} J_n(z) \, t^n$$

yields

$$e^{\frac{1}{2}z(t-1/t)} \left\{ \frac{z}{2}\left(1+\frac{1}{t^2}\right)\right\} = \sum_{n=-\infty}^{\infty} \frac{z}{2}\left(1+\frac{1}{t^2}\right) J_n(z) \, t^n = \sum_{n=-\infty}^{\infty} n J_n(z) \, t^{n-1}$$

i.e.,

$$\sum_{n=-\infty}^{\infty} z J_n(z) \, t^n + \sum_{n=-\infty}^{\infty} z J_n(z) \, t^{n-2} = \sum_{n=-\infty}^{\infty} 2n J_n(z) \, t^{n-1}$$

Equating coefficients of t^n on both sides, we have

$$z J_n(z) + z J_{n+2}(z) = 2(n+1) J_{n+1}(z)$$

and the required result follows on replacing n by $n-1$.

Since we have used the generating function, the above result is established only for integral values of n. The result also holds for non-integral values of n [see Problem 114].

28. Prove $J_n(z) = \dfrac{1}{2\pi i} \oint_C t^{-n-1} \, e^{\frac{1}{2}z(t-1/t)} \, dt,$ where C is a simple closed curve enclosing $t = 0$.

We have

$$e^{\frac{1}{2}z(t-1/t)} = \sum_{m=-\infty}^{\infty} J_m(z) \, t^m$$

so that

$$t^{-n-1} \, e^{\frac{1}{2}z(t-1/t)} = \sum_{m=-\infty}^{\infty} t^{m-n-1} J_m(z)$$

and

$$\oint_C t^{-n-1} \, e^{\frac{1}{2}z(t-1/t)} \, dt = \sum_{m=-\infty}^{\infty} J_m(z) \oint_C t^{m-n-1} \, dt \qquad (1)$$

Now by Problems 21 and 22, Chapter 4, Page 108, we have

$$\oint_C t^{m-n-1} \, dt = \begin{cases} 2\pi i & \text{if } m = n \\ 0 & \text{if } m \neq n \end{cases} \qquad (2)$$

Thus the series on the right of (1) reduces to $2\pi i J_n(z)$, from which the required result follows.

29. Prove that if $a \neq b$,

$$\int_0^z t J_n(at) J_n(bt) \, dt = \frac{z\{a J_n(bz) J_n'(az) - b J_n(az) J_n'(bz)\}}{b^2 - a^2}$$

$Y_1 = J_n(at)$ and $Y_2 = J_n(bt)$ satisfy the respective differential equations

$$(1) \quad t^2 Y_1'' + t Y_1' + (a^2 t^2 - n^2) Y_1 = 0$$

$$(2) \quad t^2 Y_2'' + t Y_2' + (b^2 t^2 - n^2) Y_2 = 0$$

Multiplying (1) by Y_2, (2) by Y_1 and subtracting, we find

$$t^2(Y_2 Y_1'' - Y_1 Y_2'') + t(Y_2 Y_1' - Y_1 Y_2') = (b^2 - a^2) t^2 Y_1 Y_2$$

This can be written

$$t \frac{d}{dt}(Y_2 Y_1' - Y_1 Y_2') + (Y_2 Y_1' - Y_1 Y_2') = (b^2 - a^2) t Y_1 Y_2$$

or

$$\frac{d}{dt}\{t(Y_2 Y_1' - Y_1 Y_2')\} = (b^2 - a^2) t Y_1 Y_2$$

Integrating with respect to t from 0 to z yields

$$(b^2 - a^2) \int_0^z t Y_1 Y_2 \, dt = t(Y_2 Y_1' - Y_1 Y_2') \Big|_0^z$$

or since $a \neq b$

$$\int_0^z t J_n(at) J_n(bt) \, dt = \frac{z\{a J_n(bz) J_n'(az) - b J_n(az) J_n'(bz)\}}{b^2 - a^2}$$

LEGENDRE FUNCTIONS

30. Prove that $\displaystyle\int_{-1}^{1} P_m(z)\, P_n(z)\, dz \;=\; 0$ **if** $m \neq n$.

We have

$$(1) \quad (1-z^2)\, P_m'' \;-\; 2z\, P_m' \;+\; m(m+1)\, P_m \;=\; 0$$

$$(2) \quad (1-z^2)\, P_n'' \;-\; 2z\, P_n' \;+\; n(n+1)\, P_n \;=\; 0$$

Multiplying (1) by P_n, (2) by P_m, and subtracting, we obtain

$$(1-z^2)\{P_n P_m'' - P_m P_n''\} \;-\; 2z\{P_n P_m' - P_m P_n'\} \;=\; \{n(n+1) - m(m+1)\}\, P_m P_n$$

which can be written

$$(1-z^2)\frac{d}{dz}\{P_n P_m' - P_m P_n'\} \;-\; 2z\{P_n P_m' - P_m P_n'\} \;=\; \{n(n+1) - m(m+1)\}\, P_m P_n$$

or

$$\frac{d}{dz}\{(1-z^2)(P_n P_m' - P_m P_n')\} \;=\; \{n(n+1) - m(m+1)\}\, P_m P_n$$

Integrating from -1 to 1, we have

$$\{n(n+1) - m(m+1)\} \int_{-1}^{1} P_m(z)\, P_n(z)\, dz \;=\; (1-z^2)(P_n P_m' - P_m P_n')\Big|_{-1}^{1} \;=\; 0$$

from which the required result follows, since $m \neq n$.

The result is often called the *orthogonality principle* for Legendre polynomials and we say that the Legendre polynomials form an *orthogonal set*.

31. Prove that $\displaystyle\int_{-1}^{1} P_m(z)\, P_n(z)\, dz \;=\; \frac{2}{2n+1}$ **if** $m = n$.

Squaring both sides of the identity,

$$\frac{1}{\sqrt{1-2zt+t^2}} \;=\; \sum_{n=0}^{\infty} P_n(z)\, t^n$$

we obtain

$$\frac{1}{1-2zt+t^2} \;=\; \sum_{m=0}^{\infty} \sum_{n=0}^{\infty} P_m(z)\, P_n(z)\, t^{m+n}$$

Integrating from -1 to 1 and using Problem 30, we find

$$\int_{-1}^{1} \frac{dz}{1-2zt+t^2} \;=\; \sum_{m=0}^{\infty} \sum_{n=0}^{\infty} \left\{ \int_{-1}^{1} P_m(z)\, P_n(z)\, dz \right\} t^{m+n}$$

$$\;=\; \sum_{n=0}^{\infty} \left\{ \int_{-1}^{1} \{P_n(z)\}^2\, dz \right\} t^{2n} \tag{1}$$

But the left side is equal to

$$-\frac{1}{2t} \ln(1-2zt+t^2)\Big|_{-1}^{1} \;=\; \frac{1}{t} \ln\left(\frac{1+t}{1-t}\right) \;=\; \sum_{n=0}^{\infty} \left\{\frac{2}{2n+1}\right\} t^{2n} \tag{2}$$

using Problem 23(c), Chapter 6, Page 155. Equating coefficients of t^{2n} in the series (1) and (2) yields the required result.

32. Prove that $(n+1)\, P_{n+1}(z) \;-\; (2n+1)z\, P_n(z) \;+\; n\, P_{n-1}(z) \;=\; 0$.

Differentiating with respect to t both sides of the identity

$$\frac{1}{\sqrt{1-2zt+t^2}} \;=\; \sum_{n=0}^{\infty} P_n(z)\, t^n$$

we have

$$\frac{z-t}{(1-2zt+t^2)^{3/2}} \;=\; \sum_{n=0}^{\infty} n\, P_n(z)\, t^{n-1}$$

Then multiplying by $1 - 2zt + t^2$, we have

$$(z - t) \sum_{n=0}^{\infty} P_n(z) \, t^n \; = \; (1 - 2zt + t^2) \sum_{n=0}^{\infty} n \, P_n(z) \, t^{n-1}$$

or $\qquad \sum_{n=0}^{\infty} z \, P_n(z) \, t^n \; - \; \sum_{n=0}^{\infty} P_n(z) \, t^{n+1} \; = \; \sum_{n=0}^{\infty} n \, P_n(z) \, t^{n-1} \; - \; \sum_{n=0}^{\infty} 2nz \, P_n(z) \, t^n$

$$+ \; \sum_{n=0}^{\infty} n \, P_n(z) \, t^{n+1}$$

Equating coefficients of t^n on each side, we obtain

$$z \, P_n(z) \; - \; P_{n-1}(z) \; = \; (n+1) \, P_{n+1}(z) \; - \; 2nz \, P_n(z) \; + \; (n-1) \, P_{n-1}(z)$$

which yields the required result on simplifying.

THE HYPERGEOMETRIC FUNCTION

33. Show that $\quad F(1/2, 1/2; 3/2; z^2) = \dfrac{\sin^{-1} z}{z}$.

Since $\quad F(a, b; c; z) \; = \; 1 + \dfrac{a \cdot b}{1 \cdot c} z + \dfrac{a(a+1) \, b(b+1)}{1 \cdot 2 \cdot c(c+1)} z^2 + \cdots \quad$ we have

$$F(1/2, 1/2; 3/2; z^2) \; = \; 1 + \dfrac{(1/2)(1/2)}{1 \cdot (3/2)} z^2 + \dfrac{(1/2)(3/2)(1/2)(3/2)}{1 \cdot 2 \cdot (3/2)(5/2)} z^4$$

$$+ \; \dfrac{(1/2)(3/2)(5/2)(1/2)(3/2)(5/2)}{1 \cdot 2 \cdot 3 \cdot (3/2)(5/2)(7/2)} z^6 + \cdots$$

$$= \; 1 + \dfrac{1}{2} \dfrac{z^2}{3} + \dfrac{1 \cdot 3}{2 \cdot 4} \dfrac{z^4}{5} + \dfrac{1 \cdot 3 \cdot 5}{2 \cdot 4 \cdot 6} \dfrac{z^6}{7} + \cdots \; = \; \dfrac{\sin^{-1} z}{z}$$

using Problem 89, Chapter 6, Page 166.

THE ZETA FUNCTION

34. Prove that the zeta function $\quad \zeta(z) = \displaystyle\sum_{k=1}^{\infty} \dfrac{1}{k^z} \quad$ is analytic in the region of the z plane for which $\quad \text{Re}\,\{z\} \geqq 1 + \delta \quad$ where δ is any fixed positive number.

Each term $1/k^z$ of the series is an analytic function. Also, if $x = \text{Re}\,\{z\} \geqq 1 + \delta$ then,

$$\left| \dfrac{1}{k^z} \right| \; = \; \left| \dfrac{1}{e^{z \ln k}} \right| \; = \; \dfrac{1}{e^{x \ln k}} \; = \; \dfrac{1}{k^x} \; \leqq \; \dfrac{1}{k^{1+\delta}}$$

Since $\Sigma \, 1/k^{1+\delta}$ converges, we see by the Weierstrass M test that $\displaystyle\sum_{k=1}^{\infty} \dfrac{1}{k^z}$ converges uniformly for $\text{Re}\,\{z\} \geqq 1 + \delta$. Hence by Theorem 21, Page 142, $\zeta(z)$ is analytic in this region.

ASYMPTOTIC EXPANSIONS AND THE METHOD OF STEEPEST DESCENTS

35. (a) If $p > 0$, prove that

$$F(z) \; = \; \int_z^{\infty} \dfrac{e^{-t}}{t^p} \, dt \; = \; e^{-z} \left\{ \dfrac{1}{z^p} - \dfrac{p}{z^{p+1}} + \dfrac{p(p+1)}{z^{p+2}} - \cdots (-1)^n \dfrac{p(p+1) \cdots (p+n-1)}{z^{p+n}} \right\}$$

$$+ \; (-1)^{n+1} p(p+1) \cdots (p+n) \int_z^{\infty} \dfrac{e^{-t}}{t^{p+n+1}} \, dt$$

(b) Use (a) to prove that

$$F(z) \; = \; \int_z^{\infty} \dfrac{e^{-t}}{t^p} \, dt \; \sim \; e^{-z} \left\{ \dfrac{1}{z^p} - \dfrac{p}{z^{p+1}} + \dfrac{p(p+1)}{z^{p+2}} - \cdots \right\} \; = \; S(z)$$

i.e. the series on the right is an asymptotic expansion of the function on the left.

(a) Integrating by parts, we have

$$I_p = \int_z^\infty \frac{e^{-t}}{t^p}\,dt = \lim_{M\to\infty}\int_z^M e^{-t}\,t^{-p}\,dt$$

$$= \lim_{M\to\infty}\left\{(-e^{-t})(t^{-p})\Big|_z^M - \int_z^M (-e^{-t})(-pt^{-p-1})\,dt\right\}$$

$$= \lim_{M\to\infty}\left\{\frac{e^{-z}}{z^p} - \frac{e^{-M}}{M^p} - p\int_z^M \frac{e^{-t}}{t^{p+1}}\,dt\right\}$$

$$= \frac{e^{-z}}{z^p} - p\int_z^\infty \frac{e^{-t}}{t^{p+1}}\,dt = \frac{e^{-z}}{z^p} - p\,I_{p+1}$$

Similarly, $I_{p+1} = \dfrac{e^{-z}}{z^{p+1}} - (p+1)I_{p+2}$ so that

$$I_p = \frac{e^{-z}}{z^p} - p\left\{\frac{e^{-z}}{z^{p+1}} - (p+1)I_{p+2}\right\} = \frac{e^{-z}}{z^p} - \frac{pe^{-z}}{z^{p+1}} + p(p+1)I_{p+2}$$

By continuing in this manner, the result follows.

(b) Let $\quad S_n(z) = e^{-z}\left\{\dfrac{1}{z^p} - \dfrac{p}{z^{p+1}} + \dfrac{p(p+1)}{z^{p+2}} - \cdots (-1)^n \dfrac{p(p+1)\cdots(p+n-1)}{z^{p+n}}\right\}.\quad$ Then

$$R_n(z) = F(z) - S_n(z) = (-1)^{n+1}p(p+1)\cdots(p+n)\int_z^\infty \frac{e^{-t}}{t^{p+n+1}}\,dt$$

Now for real $z>0$,

$$|R_n(z)| = p(p+1)\cdots(p+n)\int_z^\infty \frac{e^{-t}}{t^{p+n+1}}\,dt \;\leqq\; p(p+1)\cdots(p+n)\int_z^\infty \frac{e^{-t}}{z^{p+n+1}}\,dt$$

$$\leqq \frac{p(p+1)\cdots(p+n)}{z^{p+n+1}}$$

since

$$\int_z^\infty e^{-t}\,dt \;\leqq\; \int_0^\infty e^{-t}\,dt = 1$$

Thus

$$\lim_{z\to\infty}|z^n R_n(z)| \;\leqq\; \lim_{z\to\infty}\frac{p(p+1)\cdots(p+n)}{z^p} = 0$$

and it follows that $\lim\limits_{z\to\infty} z^n R_n(z) = 0$. Hence the required result is proved for real $z>0$. The result can also be extended to complex values of z.

Note that since $\left|\dfrac{u_{n+1}}{u_n}\right| = \left|\dfrac{p(p+1)\cdots(p+n)/z^{p+n+1}}{p(p+1)\cdots(p+n-1)/z^{p+n}}\right| = \dfrac{p+n}{|z|},\;$ where u_n is the nth term of the series, we have for all fixed z, $\lim\limits_{n\to\infty}\left|\dfrac{u_{n+1}}{u_n}\right| = \infty\;$ and the series diverges for all z by the ratio test.

36. Show that $\quad \Gamma(z+1) \sim \sqrt{2\pi z}\,z^z\,e^{-z}\left\{1 + \dfrac{1}{12z} + \dfrac{1}{288z^2} - \dfrac{139}{51{,}840z^3} + \cdots\right\}.$

We have $\Gamma(z+1) = \displaystyle\int_0^\infty \tau^z e^{-\tau}\,d\tau.$ By letting $\tau = zt$, this becomes

$$\Gamma(z+1) = z^{z+1}\int_0^\infty t^z e^{-zt}\,dt = z^{z+1}\int_0^\infty e^{z(\ln t - t)}\,dt \tag{1}$$

which has the form (37), Page 274, where $F(t) = \ln t - t$.

$F'(t) = 0$ when $t = 1$. Letting $t = 1 + w$ we find, using Problem 23, Page 154, or otherwise, the Taylor series

$$F(t) = \ln t - t = \ln(1+w) - (1+w) = \left(w - \frac{w^2}{2} + \frac{w^3}{3} - \frac{w^4}{4} + \cdots\right) - 1 - w$$

$$= -1 - \frac{w^2}{2} + \frac{w^3}{3} - \frac{w^4}{4} + \cdots = -1 - \frac{(t-1)^2}{2} + \frac{(t-1)^3}{3} - \frac{(t-1)^4}{4} + \cdots$$

Hence from (1), $\quad \Gamma(z+1) = z^{z+1}e^{-z}\displaystyle\int_0^\infty e^{-z(t-1)^2/2}\,e^{z(t-1)^3/3 - z(t-1)^4/4 + \cdots}\,dt$

$$= z^{z+1}e^{-z}\int_{-1}^\infty e^{-zw^2/2}\,e^{zw^3/3 - zw^4/4 + \cdots}\,dw \tag{2}$$

Letting $w = \sqrt{2/z}\, v$, this becomes

$$\Gamma(z+1) \;=\; \sqrt{2}\, z^{z+1/2}\, e^{-z} \int_{-\sqrt{z/2}}^{\infty} e^{-v^2}\, e^{(2/3)\sqrt{2}\, z^{-1/2} v^3 - z^{-1} v^4 + \cdots}\, dv \tag{3}$$

For large values of z the lower limit can be replaced by $-\infty$, and on expanding the exponential we have

$$\Gamma(z+1) \;\sim\; \sqrt{2}\, z^{z+1/2}\, e^{-z} \int_{-\infty}^{\infty} e^{-v^2}\, \{1 + (\tfrac{2}{3}\sqrt{2}\, z^{-1/2}\, v^3 - z^{-1}\, v^4) + \cdots\}\, dv \tag{4}$$

or

$$\Gamma(z+1) \;\sim\; \sqrt{2\pi z}\, z^z\, e^{-z} \left\{ 1 + \frac{1}{12z} + \frac{1}{288z^2} - \frac{139}{51{,}840z^3} + \cdots \right\} \tag{5}$$

Although we have proceeded above in a formal manner, the analysis can be justified rigorously.

Another method.

If $F(t) \;=\; -1 - \dfrac{(t-1)^2}{2} + \dfrac{(t-1)^3}{3} - \dfrac{(t-1)^4}{4} + \cdots \;=\; -1 - u^2$,   then

$$u^2 \;=\; \frac{(t-1)^2}{2} - \frac{(t-1)^3}{3} + \cdots$$

and by reversion of series or by using the fact that $F(t) = \ln t - t$,   we find

$$\frac{dt}{du} \;=\; b_0 + b_1 u + b_2 u^2 + \cdots \;=\; \sqrt{2} + \frac{\sqrt{2}}{6}u^2 + \frac{\sqrt{2}}{216}u^4 + \cdots$$

Then from (41), Page 275, we find

$$\Gamma(z+1) \;\sim\; \sqrt{\frac{\pi}{z}}\, z^{z+1}\, e^{z(\ln 1 - 1)} \left\{ \sqrt{2} + \frac{1}{2}\left(\frac{\sqrt{2}}{6}\right)\frac{1}{z} + \frac{1\cdot 3}{2\cdot 2}\left(\frac{\sqrt{2}}{216}\right)\frac{1}{z^2} + \cdots \right\}$$

or

$$\Gamma(z+1) \;\sim\; \sqrt{2\pi z}\, z^z\, e^{-z} \left\{ 1 + \frac{1}{12z} + \frac{1}{288z^2} + \cdots \right\}$$

Note that since $F''(1) = -1$, we find on using (42), Page 275,

$$\Gamma(z+1) \;\sim\; \sqrt{2\pi z}\, z^z\, e^{-z}$$

which is the first term. For many purposes this first term provides sufficient accuracy.

ELLIPTIC FUNCTIONS

37. Prove   (a) $\dfrac{d}{dz}\operatorname{sn} z = \operatorname{cn} z\, \operatorname{dn} z$,   (b) $\dfrac{d}{dz}\operatorname{cn} z = -\operatorname{sn} z\, \operatorname{dn} z$.

By definition, if $z = \displaystyle\int_0^w \frac{dt}{\sqrt{(1-t^2)(1-k^2t^2)}}$,   then $w = \operatorname{sn} z$.   Hence

(a) $\dfrac{d}{dz}(\operatorname{sn} z) \;=\; \dfrac{dw}{dz} \;=\; 1/(dz/dw) \;=\; \sqrt{(1-w^2)(1-k^2w^2)} \;=\; \operatorname{cn} z\, \operatorname{dn} z$

(b) $\dfrac{d}{dz}(\operatorname{cn} z) \;=\; \dfrac{d}{dz}(1 - \operatorname{sn}^2 z)^{1/2} \;=\; \tfrac{1}{2}(1 - \operatorname{sn}^2 z)^{-1/2}\dfrac{d}{dz}(-\operatorname{sn}^2 z)$

           $= \;\tfrac{1}{2}(1 - \operatorname{sn}^2 z)^{-1/2}(-2\operatorname{sn} z)(\operatorname{cn} z\, \operatorname{dn} z) \;=\; -\operatorname{sn} z\, \operatorname{dn} z$

38. Prove   (a) $\operatorname{sn}(-z) = -\operatorname{sn} z$,   (b) $\operatorname{cn}(-z) = \operatorname{cn} z$,   (c) $\operatorname{dn}(-z) = \operatorname{dn} z$.

(a)   If $z = \displaystyle\int_0^w \frac{dt}{\sqrt{(1-t^2)(1-k^2t^2)}}$,   then $w = \operatorname{sn} z$.   Let $t = -\tau$; then

$$z = -\int_0^{-w} \frac{d\tau}{\sqrt{(1-\tau^2)(1-k^2\tau^2)}} \quad \text{or} \quad -z = \int_0^{-w} \frac{d\tau}{\sqrt{(1-\tau^2)(1-k^2\tau^2)}},$$

i.e. $\operatorname{sn}(-z) = -w = -\operatorname{sn} z$

(b) $\operatorname{cn}(-z) \;=\; \sqrt{1 - \operatorname{sn}^2(-z)} \;=\; \sqrt{1 - \operatorname{sn}^2 z} \;=\; \operatorname{cn} z$

(c) $\operatorname{dn}(-z) \;=\; \sqrt{1 - k^2 \operatorname{sn}^2(-z)} \;=\; \sqrt{1 - k^2 \operatorname{sn}^2 z} \;=\; \operatorname{dn} z$

39. Prove that   (a) $\operatorname{sn}(z+2K) = -\operatorname{sn} z$,   (b) $\operatorname{cn}(z+2K) = -\operatorname{cn} z$.

We have $z = \displaystyle\int_0^\phi \frac{d\theta}{\sqrt{1-k^2\sin^2\theta}}$ so that $\phi = \operatorname{am} z$ and $\sin\phi = \operatorname{sn} z$, $\cos\phi = \operatorname{cn} z$.   Now

$$\int_0^{\phi+\pi} \frac{d\theta}{\sqrt{1-k^2\sin^2\theta}} \;=\; \int_0^{\pi} \frac{d\theta}{\sqrt{1-k^2\sin^2\theta}} \;+\; \int_\pi^{\phi+\pi} \frac{d\theta}{\sqrt{1-k^2\sin^2\theta}}$$

$$=\; 2\int_0^{\pi/2} \frac{d\theta}{\sqrt{1-k^2\sin^2\theta}} \;+\; \int_0^{\phi} \frac{d\psi}{\sqrt{1-k^2\sin^2\psi}}$$

using the transformation $\theta = \pi + \psi$. Hence $\phi + \pi = \operatorname{am}(z+2K)$.

Thus we have

(a) $\operatorname{sn}(z+2K) = \sin\{\operatorname{am}(z+2K)\} = \sin(\phi+\pi) = -\sin\phi = -\operatorname{sn} z$

(b) $\operatorname{cn}(z+2K) = \cos\{\operatorname{am}(z+2K)\} = \cos(\phi+\pi) = -\cos\phi = -\operatorname{cn} z$

40. Prove that   (a) $\operatorname{sn}(z+4K) = \operatorname{sn} z$,   (b) $\operatorname{cn}(z+4K) = \operatorname{cn} z$,   (c) $\operatorname{dn}(z+2K) = \operatorname{dn} z$.

From Problem 39,

(a) $\operatorname{sn}(z+4K) = -\operatorname{sn}(z+2K) = \operatorname{sn} z$

(b) $\operatorname{cn}(z+4K) = -\operatorname{cn}(z+2K) = \operatorname{cn} z$

(c) $\operatorname{dn}(z+2K) = \sqrt{1-k^2\operatorname{sn}^2(z+2K)} = \sqrt{1-k^2\operatorname{sn}^2 z} = \operatorname{dn} z$

Another method. The integrand $\dfrac{1}{\sqrt{(1-t^2)(1-k^2t^2)}}$ has branch points at $t = \pm 1$ and $t = \pm 1/k$ in the t plane [Fig. 10-10]. Consider the integral from 0 to w along two paths C_1 and C_2. We can deform C_2 into the path $ABDEFGHJA + C_1$, where BDE and GHJ are circles of radius ϵ while JAB and EFG, drawn separately for visual purposes, are actually coincident with the x axis.

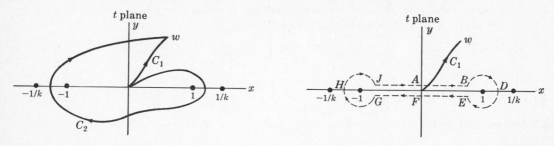

Fig. 10-10   Fig. 10-11

We then have

$$\int_{C_2\,0}^{w} \frac{dt}{\sqrt{(1-t^2)(1-k^2t^2)}} \;=\; \int_0^{1-\epsilon} \frac{dx}{\sqrt{(1-x^2)(1-k^2x^2)}} \;+\; \int_{BDE} \frac{dt}{\sqrt{(1-t^2)(1-k^2t^2)}}$$

$$+\; \int_{1-\epsilon}^{0} \frac{dx}{-\sqrt{(1-x^2)(1-k^2x^2)}} \;+\; \int_0^{-1+\epsilon} \frac{dx}{-\sqrt{(1-x^2)(1-k^2x^2)}}$$

$$+\; \int_{GHJ} \frac{dt}{-\sqrt{(1-t^2)(1-k^2t^2)}} \;+\; \int_{-1+\epsilon}^{0} \frac{dx}{\sqrt{(1-x^2)(1-k^2x^2)}}$$

$$+\; \int_{C_1\,0}^{w} \frac{dt}{\sqrt{(1-t^2)(1-k^2t^2)}}$$

$$=\; 4\int_0^{1-\epsilon} \frac{dx}{\sqrt{(1-x^2)(1-k^2x^2)}} \;+\; \int_{C_1\,0}^{w} \frac{dt}{\sqrt{(1-t^2)(1-k^2t^2)}}$$

$$+\; \int_{BDE} \frac{dt}{\sqrt{(1-t^2)(1-k^2t^2)}} \;+\; \int_{GHJ} \frac{dt}{-\sqrt{(1-t^2)(1-k^2t^2)}}$$

where we have used the fact that in encircling a branch point the sign of the radical is changed.

On BDE and GHJ we have $t = 1 - \epsilon e^{i\theta}$ and $t = -1 + \epsilon e^{i\theta}$ respectively. Then the corresponding integrals equal

$$\int_0^{2\pi} \frac{-i\epsilon e^{i\theta}\, d\theta}{\sqrt{(2 - \epsilon e^{i\theta})(\epsilon e^{i\theta})\{1 - k^2(1 - \epsilon e^{i\theta})^2\}}} \;=\; -i\sqrt{\epsilon} \int_0^{2\pi} \frac{e^{i\theta/2}\, d\theta}{\sqrt{(2 - \epsilon e^{i\theta})\{1 - k^2(1 - \epsilon e^{i\theta})^2\}}}$$

$$\int_0^{2\pi} \frac{i\epsilon e^{i\theta}\, d\theta}{\sqrt{(\epsilon e^{i\theta})(2 - \epsilon e^{i\theta})\{1 - k^2(-1 + \epsilon e^{i\theta})^2\}}} \;=\; i\sqrt{\epsilon} \int_0^{2\pi} \frac{e^{i\theta/2}\, d\theta}{\sqrt{(2 - \epsilon e^{i\theta})\{1 - k^2(-1 + \epsilon e^{i\theta})^2\}}}$$

As $\epsilon \to 0$, these integrals approach zero and we obtain

$$\int_{C_2}^w {}_0 \frac{dt}{\sqrt{(1 - t^2)(1 - k^2 t^2)}} \;=\; 4 \int_0^1 \frac{dx}{\sqrt{(1 - x^2)(1 - k^2 x^2)}} + \int_{C_1}^w {}_0 \frac{dt}{\sqrt{(1 - t^2)(1 - k^2 t^2)}}$$

Now if we write $\quad z = \displaystyle\int_{C_1}^w {}_0 \frac{dt}{\sqrt{(1 - t^2)(1 - k^2 t^2)}} ,\qquad$ i.e. $\quad w = \operatorname{sn} z$

then $\qquad z + 4K = \displaystyle\int_{C_2}^w {}_0 \frac{dt}{\sqrt{(1 - t^2)(1 - k^2 t^2)}} ,\qquad$ i.e. $\quad w = \operatorname{sn}(z + 4K)$

and since the value of w is the same in both cases, $\operatorname{sn}(z + 4K) = \operatorname{sn} z$.

Similarly we can establish the other results.

41. Prove that $\quad (a)\ \operatorname{sn}(K + iK') = 1/k, \quad (b)\ \operatorname{cn}(K + iK') = -ik'/k, \quad (c)\ \operatorname{dn}(K + iK') = 0.$

(a) We have $\quad K' = \displaystyle\int_0^1 \frac{dt}{\sqrt{(1 - t^2)(1 - k'^2 t^2)}} ,\quad$ where $\quad k' = \sqrt{1 - k^2}$.

Let $u = 1/\sqrt{1 - k'^2 t^2}$. When $t = 0$, $u = 1$; when $t = 1$, $u = 1/k$. Thus as t varies from 0 to 1, u varies from 1 to $1/k$. By Problem 43, Page 56, with $p = 1/k$, it follows that $\sqrt{1 - t^2} = -ik'u/\sqrt{1 - k'^2 u^2}$. Thus we have by substitution

$$K' = -i \int_1^{1/k} \frac{du}{\sqrt{(1 - u^2)(1 - k^2 u^2)}}$$

from which

$$K + iK' = \int_0^1 \frac{du}{\sqrt{(1 - u^2)(1 - k^2 u^2)}} + \int_1^{1/k} \frac{du}{\sqrt{(1 - u^2)(1 - k^2 u^2)}} = \int_0^{1/k} \frac{du}{\sqrt{(1 - u^2)(1 - k^2 u^2)}}$$

i.e. $\quad \operatorname{sn}(K + iK') = 1/k$.

(b) From Part (a),

$$\operatorname{cn}(K + iK') = \sqrt{1 - \operatorname{sn}^2(K + iK')} = \sqrt{1 - 1/k^2} = -i\sqrt{1 - k^2}/k = -ik'/k$$

(c) $\operatorname{dn}(K + iK') = \sqrt{1 - k^2 \operatorname{sn}^2(K + iK')} = 0 \quad$ by Part (a).

42. Prove that $\quad (a)\ \operatorname{sn}(2K + 2iK') = 0, \quad (b)\ \operatorname{cn}(2K + 2iK') = 1, \quad (c)\ \operatorname{dn}(2K + 2iK') = -1.$

From the addition formulas with $z_1 = z_2 = K + iK'$, we have

$(a)\quad \operatorname{sn}(2K + 2iK') = \dfrac{2\operatorname{sn}(K + iK')\operatorname{cn}(K + iK')\operatorname{dn}(K + iK')}{1 - k^2 \operatorname{sn}^4(K + iK')} = 0$

$(b)\quad \operatorname{cn}(2K + 2iK') = \dfrac{\operatorname{cn}^2(K + iK') - \operatorname{sn}^2(K + iK')\operatorname{dn}^2(K + iK')}{1 - k^2 \operatorname{sn}^4(K + iK')} = 1$

$(c)\quad \operatorname{dn}(2K + 2iK') = \dfrac{\operatorname{dn}^2(K + iK') - k^2 \operatorname{sn}^2(K + iK')\operatorname{cn}^2(K + iK')}{1 - k^2 \operatorname{sn}^4(K + iK')} = -1$

43. Prove that $\quad (a)\ \operatorname{sn}(z + 2iK') = \operatorname{sn} z, \quad (b)\ \operatorname{cn}(z + 2K + 2iK') = \operatorname{cn} z, \quad (c)\ \operatorname{dn}(z + 4iK') = \operatorname{dn} z.$

Using Problems 39, 42, 170 and the addition formulas, we have

(a) $\operatorname{sn}(z + 2iK') = \operatorname{sn}(z - 2K + 2K + 2iK')$

$$= \frac{\operatorname{sn}(z - 2K)\operatorname{cn}(2K + 2iK')\operatorname{dn}(2K + 2iK') + \operatorname{sn}(2K + 2iK')\operatorname{cn}(z - 2K)\operatorname{dn}(z - 2K)}{1 - k^2 \operatorname{sn}^2(z - 2K)\operatorname{sn}^2(2K + 2iK')}$$

$$= \operatorname{sn} z$$

(b) $\operatorname{cn}(z + 2K + 2iK') = \dfrac{\operatorname{cn} z \operatorname{cn}(2K + 2iK') - \operatorname{sn} z \operatorname{sn}(2K + 2iK')\operatorname{dn} z \operatorname{dn}(2K + 2iK')}{1 - k^2 \operatorname{sn}^2 z \operatorname{sn}^2(2K + 2iK')}$

$$= \operatorname{cn} z$$

(c) $\operatorname{dn}(z + 4iK') = \operatorname{dn}(z - 4K + 4K + 4iK')$

$$= \frac{\operatorname{dn}(z - 4K)\operatorname{dn}(4K + 4iK') - k^2 \operatorname{sn}(z - 4K)\operatorname{sn}(4K + 4iK')\operatorname{cn}(z - 4K)\operatorname{cn}(4K + 4iK')}{1 - k^2 \operatorname{sn}^2(z - 4K)\operatorname{sn}^2(4K + 4iK')}$$

$$= \operatorname{dn} z$$

44. Construct period parallelograms or cells for the functions   (a) $\operatorname{sn} z$,   (b) $\operatorname{cn} z$,   (c) $\operatorname{dn} z$.

The results are shown in Figures 10-12, 10-13 and 10-14 respectively.

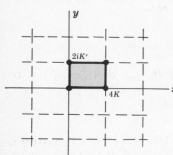

Period Parallelograms
for sn z
Fig. 10-12

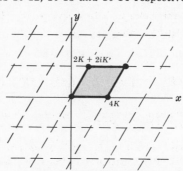

Period Parallelograms
for cn z
Fig. 10-13

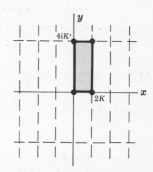

Period Parallelograms
for dn z
Fig. 10-14

MISCELLANEOUS PROBLEMS

45. Prove that $P_n(z) = F\left(-n, n+1; 1; \dfrac{1-z}{2}\right)$, $n = 0, 1, 2, 3, \ldots$.

The Legendre polynomials $P_n(z)$ are of degree n and have the value 1 for $z = 1$. Similarly from (29), Page 273, it is seen that

$$F\left(-n, n+1; 1; \frac{1-z}{2}\right) = 1 - \frac{n(n+1)}{2}(1-z) + \frac{n(n-1)(n+1)(n+2)}{16}(1-z)^2 + \cdots$$

is a polynomial of degree n having the value 1 for $z = 1$.

The required result follows if we show that P_n and F satisfy the same differential equation. To do this, let $\dfrac{1-z}{2} = u$, i.e. $z = 1 - 2u$, in Legendre's equation (25), Page 272, to obtain

$$u(1-u)\frac{d^2Y}{du^2} + (1-2u)\frac{dY}{du} + n(n+1)Y = 0$$

But this is the hypergeometric equation (30), Page 273, with $a = -n$, $b = n+1$, $c = 1$ and $u = (1-z)/2$. Hence the result is proved.

46. Prove that for $m = 1, 2, 3, \ldots$,

$$\Gamma\left(\frac{1}{m}\right)\Gamma\left(\frac{2}{m}\right)\Gamma\left(\frac{3}{m}\right)\cdots\Gamma\left(\frac{m-1}{m}\right) = \frac{(2\pi)^{(m-1)/2}}{\sqrt{m}}$$

We have

$$P = \Gamma\left(\frac{1}{m}\right)\Gamma\left(\frac{2}{m}\right)\cdots\Gamma\left(1 - \frac{1}{m}\right) = \Gamma\left(1 - \frac{1}{m}\right)\Gamma\left(1 - \frac{2}{m}\right)\cdots\Gamma\left(\frac{1}{m}\right)$$

Then multiplying these products term by term and using Problem 13 and Problem 52, Page 25, we find

$$P^2 = \left\{ \Gamma\left(\frac{1}{m}\right) \Gamma\left(1 - \frac{1}{m}\right) \right\} \left\{ \Gamma\left(\frac{2}{m}\right) \Gamma\left(1 - \frac{2}{m}\right) \right\} \cdots \left\{ \Gamma\left(1 - \frac{1}{m}\right) \Gamma\left(\frac{1}{m}\right) \right\}$$

$$= \frac{\pi}{\sin(\pi/m)} \cdot \frac{\pi}{\sin(2\pi/m)} \cdots \frac{\pi}{\sin(m-1)\pi/m}$$

$$= \frac{\pi^{m-1}}{\sin(\pi/m)\sin(2\pi/m)\cdots\sin(m-1)\pi/m} = \frac{\pi^{m-1}}{m/2^{m-1}} = \frac{(2\pi)^{m-1}}{m}$$

or $P = (2\pi)^{(m-1)/2}/\sqrt{m}$, as required.

47. Show that for large positive values of z,

$$J_n(z) \sim \sqrt{\frac{2}{\pi z}} \cos\left(z - \frac{n\pi}{2} - \frac{\pi}{4}\right)$$

By Problem 33, Chapter 6, we have

$$J_n(z) = \frac{1}{\pi} \int_0^\pi \cos(nt - z\sin t)\, dt = \text{Re}\left\{ \frac{1}{\pi} \int_0^\pi e^{-int}\, e^{iz\sin t}\, dt \right\}$$

Let $F(t) = i\sin t$. Then $F'(t) = i\cos t = 0$ where $t = \pi/2$. If we let $t = \pi/2 + v$, the integral in braces becomes

$$\frac{1}{\pi} \int_{-\pi/2}^{\pi/2} e^{-in(\pi/2+v)}\, e^{iz\sin(\pi/2+v)}\, dv = \frac{e^{-in\pi/2}}{\pi} \int_{-\pi/2}^{\pi/2} e^{-inv}\, e^{iz\cos v}\, dv$$

$$= \frac{e^{-in\pi/2}}{\pi} \int_{-\pi/2}^{\pi/2} e^{-inv}\, e^{iz(1 - v^2/2 + v^4/24 - \cdots)}\, dv$$

$$= \frac{e^{i(z-n\pi/2)}}{\pi} \int_{-\pi/2}^{\pi/2} e^{-inv}\, e^{-izv^2/2 + izv^4/24 - \cdots}\, dv$$

Let $v^2 = -2i\,u^2/z$ or $v = (1-i)u/\sqrt{z}$, i.e. $u = \frac{1}{2}(1+i)\sqrt{z}\, v$. Then the integral can be approximated by

$$\frac{(1-i)\, e^{i(z-n\pi/2)}}{\pi\sqrt{z}} \int_{-\infty}^\infty e^{-(1+i)nu/\sqrt{z}}\, e^{-u^2 - iu^4/6z - \cdots}\, du$$

or for large positive values of z,

$$\frac{(1-i)\, e^{i(z-n\pi/2)}}{\pi\sqrt{z}} \int_{-\infty}^\infty e^{-u^2}\, du = \frac{(1-i)\, e^{i(z-n\pi/2)}}{\sqrt{\pi z}}$$

and the real part is

$$\frac{1}{\sqrt{\pi z}} \left\{ \cos\left(z - \frac{n\pi}{2}\right) + \sin\left(z - \frac{n\pi}{2}\right) \right\} = \sqrt{\frac{2}{\pi z}} \cos\left(z - \frac{n\pi}{2} - \frac{\pi}{4}\right)$$

Higher order terms can also be obtained [see Problem 162].

48. If C is the contour of Fig. 10-15, prove that for all values of z

$$\Gamma(z) = \frac{1}{e^{2\pi i z} - 1} \oint_C t^{z-1}\, e^{-t}\, dt$$

Referring to Fig. 10-15 below, we see that along AB, $t = x$; along BDE, $t = \epsilon e^{i\theta}$; and along EF, $t = x e^{2\pi i}$. Then

$$\int_{ABDEF} t^{z-1}\, e^{-t}\, dt = \int_R^\epsilon x^{z-1}\, e^{-x}\, dx + \int_0^{2\pi} (\epsilon e^{i\theta})^{z-1}\, e^{-\epsilon e^{i\theta}}\, i\epsilon e^{i\theta}\, d\theta$$

$$+ \int_\epsilon^R x^{z-1}\, e^{2\pi i(z-1)}\, e^{-x}\, dx$$

$$= (e^{2\pi i z} - 1) \int_\epsilon^R x^{z-1}\, e^{-x}\, dx + i \int_0^{2\pi} \epsilon^z\, e^{i\theta z}\, e^{-\epsilon e^{i\theta}}\, d\theta$$

Now if $\text{Re}\,\{z\} > 0$, we have on taking the limit as $\epsilon \to 0$ and $R \to \infty$,

$$\int_C t^{z-1}\,e^{-t}\,dt \;=\; (e^{2\pi i z} - 1)\int_0^\infty x^{z-1}\,e^{-x}\,dx$$

$$\qquad\qquad\quad =\; (e^{2\pi i z} - 1)\,\Gamma(z)$$

But the functions on both sides are analytic for all z. Hence for all z,

$$\Gamma(z) \;=\; \frac{1}{e^{2\pi i z} - 1}\oint_C t^{z-1}\,e^{-t}\,dt$$

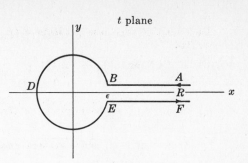

Fig. 10-15

---

49. Prove that $\quad \text{sn}\,(z_1 + z_2) \;=\; \dfrac{\text{sn}\,z_1\,\text{cn}\,z_2\,\text{dn}\,z_2 \;+\; \text{cn}\,z_1\,\text{sn}\,z_2\,\text{dn}\,z_1}{1 \;-\; k^2\,\text{sn}^2\,z_1\,\text{sn}^2\,z_2}$.

Let $z_1 + z_2 = \alpha$, a constant. Then $dz_2/dz_1 = -1$. Let us define $U = \text{sn}\,z_1$, $V = \text{sn}\,z_2$. It follows that

$$\frac{dU}{dz_1} = \dot{U} = \text{cn}\,z_1\,\text{dn}\,z_1, \qquad \frac{dV}{dz_1} = \dot{V} = \frac{dV}{dz_2}\frac{dz_2}{dz_1} = -\,\text{cn}\,z_2\,\text{dn}\,z_2$$

where dots denote differentiation with respect to z_1. Then

$$\dot{U}^2 = (1 - U^2)(1 - k^2 U^2) \qquad\text{and}\qquad \dot{V}^2 = (1 - V^2)(1 - k^2 V^2)$$

Differentiating and simplifying, we find

$$(1)\quad \ddot{U} = 2k^2 U^3 - (1 + k^2)U, \qquad\qquad (2)\quad \ddot{V} = 2k^2 V^3 - (1 + k^2)V$$

Multiplying (1) by V, (2) by U, and subtracting, we have

$$\ddot{U}V - U\ddot{V} \;=\; 2k^2 UV(U^2 - V^2) \tag{3}$$

It is easy to verify that $\qquad \dot{U}^2 V^2 - U^2 \dot{V}^2 \;=\; (1 - k^2 U^2 V^2)(V^2 - U^2) \tag{4}$

or $\qquad\qquad\qquad\qquad \dot{U}V - U\dot{V} \;=\; \dfrac{(1 - k^2 U^2 V^2)(V^2 - U^2)}{\dot{U}V + U\dot{V}} \tag{5}$

Dividing equations (3) and (5), we have

$$\frac{\ddot{U}V - U\ddot{V}}{\dot{U}V - U\dot{V}} \;=\; \frac{-\,2k^2 UV(\dot{U}V + U\dot{V})}{1 - k^2 U^2 V^2} \tag{6}$$

But $\quad \ddot{U}V - U\ddot{V} = \dfrac{d}{dz_1}(\dot{U}V - U\dot{V})\quad$ and $\quad -2k^2 UV(\dot{U}V + U\dot{V}) = \dfrac{d}{dz_1}(1 - k^2 U^2 V^2),\quad$ so that (6) becomes

$$\frac{d(\dot{U}V - U\dot{V})}{\dot{U}V - U\dot{V}} \;=\; \frac{d(1 - k^2 U^2 V^2)}{1 - k^2 U^2 V^2}$$

An integration yields $\dfrac{\dot{U}V - U\dot{V}}{1 - k^2 U^2 V^2} = c$ (a constant), i.e.,

$$\frac{\text{sn}\,z_1\,\text{cn}\,z_2\,\text{dn}\,z_2 \;+\; \text{cn}\,z_1\,\text{sn}\,z_2\,\text{dn}\,z_1}{1 \;-\; k^2\,\text{sn}^2\,z_1\,\text{sn}^2\,z_2} \;=\; c$$

is a solution of the differential equation. It is also clear that $z_1 + z_2 = \alpha$ is a solution. The two solutions must be related as follows:

$$\frac{\text{sn}\,z_1\,\text{cn}\,z_2\,\text{dn}\,z_2 \;+\; \text{cn}\,z_1\,\text{sn}\,z_2\,\text{dn}\,z_1}{1 \;-\; k^2\,\text{sn}^2\,z_1\,\text{sn}^2\,z_2} \;=\; F(z_1 + z_2)$$

Putting $z_2 = 0$, we see that $F(z_1) = \text{sn}\,z_1$. Then $F(z_1 + z_2) = \text{sn}\,(z_1 + z_2)$ and the required result follows.

Supplementary Problems

ANALYTIC CONTINUATION

50. (a) Show that $F_1(z) = z + \frac{1}{2}z^2 + \frac{1}{3}z^3 + \frac{1}{4}z^4 + \cdots$ converges for $|z| < 1$.

(b) Show that $F_2(z) = \frac{1}{4}\pi i - \frac{1}{2}\ln 2 + \left(\dfrac{z-i}{1-i}\right) + \frac{1}{2}\left(\dfrac{z-i}{1-i}\right)^2 + \frac{1}{3}\left(\dfrac{z-i}{1-i}\right)^3 + \cdots$ converges for $|z-i| < \sqrt{2}$.

(c) Show that $F_1(z)$ and $F_2(z)$ are analytic continuations of each other.

(d) Can you find a function which represents all possible analytic continuations of $F_1(z)$? Justify your answer.

Ans. (d) $-\ln(1-z)$

51. A function $F(z)$ is represented in $|z-1| < 2$ by the series

$$\sum_{n=0}^{\infty} \frac{(-1)^n (z-1)^{2n}}{2^{2n+1}}$$

Prove that the value of the function at $z = 5$ is $1/16$.

52. (a) Show that $F_1(z) = \displaystyle\int_0^{\infty} (1+t)e^{-zt}\,dt$ converges only if $\mathrm{Re}\,\{z\} > 0$.

(b) Find a function which is the analytic continuation of $F_1(z)$ into the left hand plane.
Ans. (b) $(z+1)/z^2$

53. (a) Find the region of convergence of $F_1(z) = \displaystyle\int_0^{\infty} e^{-(z+1)^2 t}\,dt$ and graph this region.

(b) Find the value of the analytic continuation of $F_1(z)$ corresponding to $z = 2 - 4i$.
Ans. (a) $\mathrm{Re}\,\{z+1\}^2 > 0$, (b) $(-7 + 24i)/625$

54. (a) Prove that $\dfrac{z}{1-z^2} + \dfrac{z^2}{1-z^4} + \dfrac{z^4}{1-z^8} + \cdots = \begin{cases} z/(1-z) & \text{if } |z| < 1 \\ 1/(1-z) & \text{if } |z| > 1 \end{cases}$

(b) Discuss these results from the point of view of analytic continuation.

55. Show that the series $\displaystyle\sum_{n=0}^{\infty} z^{3^n}$ cannot be continued analytically beyond the circle $|z| = 1$.

56. If $\displaystyle\sum_{n=1}^{\infty} a_n z^{\beta_n}$ has $|z| = 1$ as a natural barrier, would you expect $\displaystyle\sum_{n=1}^{\infty} (-1)^n a_n z^{\beta_n}$ to have $|z| = 1$ as natural barrier also? Justify your conclusion.

57. Let $\{z_n\}$, $n = 1, 2, 3, \ldots$ be a sequence such that $\displaystyle\lim_{n \to \infty} z_n = a$, and suppose that for all n, $z_n \neq a$. Let $F(z)$ and $G(z)$ be analytic at a and such that $F(z_n) = G(z_n)$, $n = 1, 2, 3, \ldots$. (a) Prove that $F(z) = G(z)$. (b) Explain the relationship of the result in (a) with analytic continuation. [*Hint.* Consider the expansion of $F(z) - G(z)$ in a Taylor series about $z = a$.]

SCHWARZ'S REFLECTION PRINCIPLE

58. Work Problem 2 using Schwarz's reflection principle.

59. (a) Given that $\sin 2z = 2 \sin z \cos z$ holds for all real values of z, prove that it also holds for all complex values of z.

(b) Can you use the Schwarz reflection principle to prove that $\tan 2z = (2 \tan z)/(1 - \tan^2 z)$? Justify your conclusion.

60. Does the Schwarz reflection principle apply if reflection takes place in the imaginary rather than the real axis? Prove your statements.

61. Can you extend the Schwarz reflection principle to apply to reflection in a curve C?

INFINITE PRODUCTS

62. Investigate the convergence of the infinite products

$$(a) \ \prod_{k=1}^{\infty}\left(1+\frac{1}{k^3}\right), \qquad (b) \ \prod_{k=1}^{\infty}\left(1-\frac{1}{\sqrt{k+1}}\right), \qquad (c) \ \prod_{k=1}^{\infty}\left(1+\frac{\cos k\pi}{k^2+1}\right)$$

Ans. (a) conv., (b) div., (c) conv.

63. Prove that a necessary condition for $\prod_{k=1}^{\infty}(1+w_k)$ to converge is that $\lim_{n\to\infty} w_n = 0$.

64. Investigate the convergence of (a) $\prod_{k=1}^{\infty}\left(1+\frac{1}{k}\right)$, (b) $\prod_{k=1}^{\infty}\left(1+\frac{k}{\sqrt{k^2+1}}\right)$, (c) $\prod_{k=1}^{\infty}(1+\cot^{-1}k^2)$.
    Ans. (a) div., (b) div., (c) conv.

65. If an infinite product is absolutely convergent, prove that it is convergent.

66. Prove that $\cos z = \prod_{k=1}^{\infty}\left(1-\frac{4z^2}{(2k-1)^2\pi^2}\right)$.

67. Show that $\prod_{k=1}^{\infty}\left(1+\frac{e^{-kz}}{k^2}\right)$ (a) converges absolutely and uniformly in the right half plane Re $\{z\} \geqq 0$ and (b) represents an analytic function of z for Re $\{z\} \geqq 0$.

68. Prove that $\left(1-\frac{1}{2^2}\right)\left(1-\frac{1}{3^2}\right)\left(1-\frac{1}{4^2}\right)\cdots = \frac{1}{2}$.

69. Prove that $\left(1-\frac{1}{2}\right)\left(1+\frac{1}{3}\right)\left(1-\frac{1}{4}\right)\cdots = \frac{1}{2}$.

70. Prove that (a) $\sinh z = \prod_{k=1}^{\infty}\left(1+\frac{z^2}{k^2\pi^2}\right)$

    (b) $\cosh z = \prod_{k=1}^{\infty}\left(1+\frac{4z^2}{(2k-1)^2\pi^2}\right)$.

71. Use infinite products to show that $\sin 2z = 2\sin z\cos z$. Justify all steps.

72. Prove that $\prod_{k=1}^{\infty}\left(1+\frac{1}{k}\sin\frac{z}{k}\right)$ (a) converges absolutely and uniformly for all z and (b) represents an analytic function.

73. Prove that $\prod_{k=1}^{\infty}\left(1+\frac{z}{k}\right)e^{-z/k}$ converges.

THE GAMMA FUNCTION

74. Evaluate each of the following by use of the gamma function.

$$(a) \ \int_0^{\infty} y^3 e^{-2y}\,dy \qquad (c) \ \int_0^{\infty} y^2 e^{-2y^2}\,dy$$

$$(e) \ \int_0^{\infty}\{ye^{-y^2}\}^{1/4}\,dy$$

$$(b) \ \int_0^{\infty} u^{3/2}e^{-3u}\,du \qquad (d) \ \int_0^1\{\ln(1/t)\}^{-1/2}\,dt$$

Ans. (a) 3/8, (b) $\sqrt{3\pi}/36$, (c) $\sqrt{2\pi}/16$, (d) $\sqrt{\pi}$, (e) $\Gamma(5/8)/\sqrt[4]{2}$

75. Prove that $\Gamma(z) = \int_0^1\{\ln(1/t)\}^{z-1}\,dt$ for Re $\{z\} > 0$.

76. Show that $\int_1^{\infty}\frac{(x-1)^p}{x^2}\,dx = \Gamma(1+p)\,\Gamma(1-p)$, $-1 < p < 1$.

77. If m, n and a are positive constants, show that

$$\int_0^{\infty} x^m\,e^{-ax^n}\,dx = \frac{1}{n}a^{-(m+1)/n}\,\Gamma\left(\frac{m+1}{n}\right)$$

78. Show that $\displaystyle\int_0^\infty \frac{e^{-zt}}{\sqrt{t}}\,dt \;=\; \sqrt{\frac{\pi}{z}}$ if   Re $\{z\} > 0$.

79. Evaluate $\displaystyle\int_0^1 (x\ln x)^4\,dx$.   *Ans.* 24/3125

80. Evaluate   (a) $\Gamma(-7/2)$, (b) $\Gamma(-1/3)$.   *Ans.* (a) $16\sqrt{\pi}/105$, (b) $-3\,\Gamma(2/3)$

81. Show that $\displaystyle\Gamma(-\tfrac{1}{2}-m) \;=\; \frac{(-1)^{m+1}\sqrt{\pi}\,2^{m+1}}{1\cdot 3\cdot 5\cdots(2m+1)}$, $m = 0,1,2,\ldots$.

82. Prove that the residue of $\Gamma(z)$ at $z = -m$, $m = 0,1,2,3,\ldots$, is $(-1)^m/m!$ where $0! = 1$ by definition.

83. Use the infinite product representation of the gamma function to prove that

    (a) $\displaystyle\Gamma(z)\,\Gamma(1-z) \;=\; \frac{\pi}{\sin \pi z}$

    (b) $\displaystyle 2^{2z-1}\,\Gamma(z)\,\Gamma(z+\tfrac{1}{2}) \;=\; \sqrt{\pi}\,\Gamma(2z)$

84. Prove that if $y > 0$, $\displaystyle|\Gamma(iy)| \;=\; \sqrt{\frac{\pi}{y\sinh \pi y}}$.

85. Discuss Problem 84 if $y < 0$.

86. Prove (a) Property 6, (b) Property 7, (c) Property 9   on Pages 268 and 269.

87. Prove that $\Gamma(\tfrac{1}{5})\,\Gamma(\tfrac{2}{5}) \;=\; 4\pi^2/\sqrt{5}$.

88. (a) By using the infinite product representation of the gamma function, prove that for any positive integer m,
$$\frac{m^{mz}\,\Gamma(z)\,\Gamma(z+1/m)\,\Gamma(z+2/m)\cdots\Gamma(z+[m-1]/m)}{\Gamma(mz)}$$
    is a constant independent of z.

    (b) By letting $z \to 0$ in the result of (a), evaluate the constant and thus establish Property 5, Page 268.

THE BETA FUNCTION

89. Evaluate   (a) $B(3,5/2)$, (b) $B(1/3,2/3)$.   *Ans.* (a) 16/315, (b) $2\pi/\sqrt{3}$

90. Evaluate each of the following using the beta function.

    (a) $\displaystyle\int_0^1 t^{-1/3}(1-t)^{2/3}\,dt$,   (b) $\displaystyle\int_0^1 u^2(1-u^2)^{-1/2}\,du$,   (c) $\displaystyle\int_0^3 (9-t^2)^{3/2}\,dt$,   (d) $\displaystyle\int_0^4 \frac{dt}{\sqrt{4t-t^2}}$.
    Ans. (a) $4\pi/3\sqrt{3}$, (b) $\pi/4$, (c) $243\pi/16$, (d) π

91. Prove that $\displaystyle\frac{B(m+1,n)}{B(m,n+1)} \;=\; \frac{m}{n}$.

92. If $a > 0$, prove that $\displaystyle\int_0^a \frac{dy}{\sqrt{a^4-y^4}} \;=\; \frac{\{\Gamma(1/4)\}^2}{4a\sqrt{2\pi}}$.

93. Prove that $\displaystyle\frac{B\left(\dfrac{p+1}{2},\dfrac{1}{2}\right)}{B\left(\dfrac{p+1}{2},\dfrac{p+1}{2}\right)} \;=\; 2^p$ stating any restrictions on p.

94. Evaluate   (a) $\displaystyle\int_0^{\pi/2} \sin^6\theta\cos^4\theta\,d\theta$, (b) $\displaystyle\int_0^{\pi/2} \sqrt{\tan\theta}\,d\theta$.   *Ans.* (a) $3\pi/512$, (b) $\pi/\sqrt{2}$

95. Prove that $\quad B(m, n) = \dfrac{1}{2} \displaystyle\int_0^1 \dfrac{x^{m-1} + x^{n-1}}{(1+x)^{m+n}}\, dx \quad$ where $\ \text{Re}\,\{m\} > 0\ $ and $\ \text{Re}\,\{n\} > 0$.

[*Hint.* Let $y = x/(1+x)$.]

96. Prove that $\quad \displaystyle\int_0^\infty \dfrac{x^3\, dx}{1 + x^6} = \dfrac{\pi}{3\sqrt{3}}$.

97. (*a*) Show that if either m or n (but not both) is a negative integer and if $\ m + n < 0$, then $B(m, n)$ is infinite. (*b*) Investigate $B(m, n)$ when both m and n are negative integers.

DIFFERENTIAL EQUATIONS

98. Determine the singular points of each of the following differential equations and state whether they are regular or irregular.

(*a*) $(1 - z^2)Y'' - 2Y' + 6Y = 0$

(*b*) $(2z^4 - z^5)Y'' + zY' + (z^2 + 1)Y = 0$

(*c*) $z^2(1 - z)^2\, Y'' + (2 - z)Y' + 4z^2 Y = 0$.

Ans. (*a*) $z = \pm 1$, regular.   (*b*) $z = 2$, regular; $z = 0$, irregular.   (*c*) $z = 0, 1$, irregular.

99. Solve each of the following differential equations using power series and find the region of convergence. If possible, sum the series and show that the sum satisfies the differential equation.

(*a*) $Y'' + 2Y' + Y = 0$, (*b*) $Y'' + zY = 0$, (*c*) $zY'' + 2Y' + zY = 0$.

Ans. (*a*) $Y = Ae^{-z} + Bze^{-z}$

(*b*) $Y = A\left(1 - \dfrac{z^3}{3!} + \dfrac{1 \cdot 4}{6!}z^6 - \dfrac{1 \cdot 4 \cdot 7}{9!}z^9 + \cdots\right) + B\left(z - \dfrac{2z^4}{4!} + \dfrac{2 \cdot 5}{7!}z^7 - \dfrac{2 \cdot 5 \cdot 8}{10!}z^{10} + \cdots\right)$

(*c*) $Y = \dfrac{A \sin z + B \cos z}{z}$

100. (*a*) If you solved $\ (1 - z^2)Y'' + 2Y = 0\ $ by substituting the assumed solution $\ Y = \Sigma a_n z^n$, what region of convergence would you expect? Explain.

(*b*) Determine whether your expectations in (*a*) are correct by actually finding the series solution.

Ans. (*b*) $Y = A(1 - z^2) + B\left(z - \dfrac{z^3}{1 \cdot 3} - \dfrac{z^5}{3 \cdot 5} - \dfrac{z^7}{5 \cdot 7} - \cdots\right)$

101. (*a*) Solve $Y'' + z^2 Y = 0$ subject to $Y(0) = 1$, $Y'(0) = -1$ and (*b*) determine the region of convergence.

Ans. (*a*) $Y = 1 - z - \dfrac{z^4}{3 \cdot 4} + \dfrac{z^5}{4 \cdot 5} + \dfrac{z^8}{3 \cdot 4 \cdot 7 \cdot 8} - \dfrac{z^9}{4 \cdot 5 \cdot 8 \cdot 9} - \cdots$ (*b*) $|z| < \infty$

102. If $Y = Y_1(z)$ is a solution of $\ Y'' + p(z)\,Y' + q(z)\,Y = 0$, show that the general solution is

$$Y = A\, Y_1(z) + B\, Y_1(z) \int \dfrac{e^{-\int p(z)\, dz}}{\{Y_1(z)\}^2}\, dz$$

103. (*a*) Solve $zY'' + (1 - z)\,Y' - Y = 0$ and (*b*) determine the region of convergence.

Ans. (*a*) $Y = (A + B \ln z)e^z - B\left\{z + \dfrac{z^2}{2!}(1 + \tfrac{1}{2}) + \dfrac{z^3}{3!}(1 + \tfrac{1}{2} + \tfrac{1}{3}) + \cdots\right\}$ (*b*) $|z| > 0$

104. (*a*) Use Problem 102 to show that the solution to the differential equation of Problem 103 can be written as

$$Y = Ae^z + Be^z \int \dfrac{e^{-z}}{z}\, dz$$

(*b*) Reconcile the result of (*a*) with the series solution obtained in Problem 103.

105. (*a*) Solve $zY'' + Y' - Y = 0$ and (*b*) determine the region of convergence.

Ans. (*a*) $Y = (A + B \ln z)\left\{\dfrac{z}{(1!)^2} + \dfrac{z^2}{(2!)^2} + \dfrac{z^3}{(3!)^2} + \cdots\right\}$

$- 2B\left\{\dfrac{z}{(1!)^2} + \dfrac{z^2}{(2!)^2}(1 + \tfrac{1}{2}) + \dfrac{z^3}{(3!)^2}(1 + \tfrac{1}{2} + \tfrac{1}{3}) + \cdots\right\}$

106. Prove that $Y = V e^{-\frac{1}{2}\int p(z)\,dz}$ transforms the differential equation $Y'' + p(z)\,Y' + q(z)\,Y = 0$ into

$$V'' + \{q(z) - \tfrac{1}{2}p'(z) - \tfrac{1}{4}[p(z)]^2\}V = 0$$

107. Use the method of Problem 106 to find the general solution of $zY'' + 2Y' + zY = 0$ [see Prob. 99(c)].

SOLUTION OF DIFFERENTIAL EQUATIONS BY CONTOUR INTEGRALS

108. Use the method of contour integrals to solve each of the following.

(a) $Y'' - Y' - 2Y = 0$, (b) $Y'' + 4Y' + 4Y = 0$, (c) $Y'' + 2Y' + 2Y = 0$.

Ans. (a) $Y = Ae^{2z} + Be^{-z}$, (b) $Y = Ae^{-2z} + Bze^{-2z}$, (c) $Y = e^{-z}(A\sin z + B\cos z)$.

109. Prove that a solution of $zY'' + (a - z)Y' - bY = 0$, where Re $\{a\} > 0$, Re $\{b\} > 0$, is given by

$$Y = \int_0^1 e^{zt}\, t^{b-1}\,(1-t)^{a-b-1}\,dt$$

BESSEL FUNCTIONS

110. Prove that $J_{-n}(z) = (-1)^n\,J_n(z)$ for $n = 0, 1, 2, 3, \ldots$.

111. Prove (a) $\dfrac{d}{dz}\{z^n\,J_n(z)\} = z^n\,J_{n-1}(z)$, (b) $\dfrac{d}{dz}\{z^{-n}\,J_n(z)\} = -z^{-n}\,J_{n+1}(z)$.

112. Show that (a) $J_0'(z) = -J_1(z)$, (b) $\displaystyle\int z^3\,J_2(z)\,dz = z^3\,J_3(z) + c$, (c) $\displaystyle\int z^3\,J_0(z)\,dz = z^3\,J_1(z) - 2z^2\,J_2(z) + c$.

113. Show that (a) $J_{1/2}(z) = \sqrt{2/\pi z}\,\sin z$, (b) $J_{-1/2}(z) = \sqrt{2/\pi z}\,\cos z$.

114. Prove the result of Problem 27 for non-integral values of n.

115. Show that $J_{3/2}(z)\sin z - J_{-3/2}\cos z = \sqrt{2/\pi z^3}$.

116. Prove that $J_n'(z) = \tfrac{1}{2}\{J_{n-1}(z) - J_{n+1}(z)\}$.

117. Prove that (a) $J_n''(z) = \tfrac{1}{4}\{J_{n-2}(z) - 2J_n(z) + J_{n+2}(z)\}$

(b) $J_n'''(z) = \tfrac{1}{8}\{J_{n-3}(z) - 3J_{n-1}(z) + 3J_{n+1}(z) - J_{n+3}(z)\}$.

118. Generalize the results in Problems 116 and 117.

119. By direct substitution prove that $J_0(z) = \dfrac{1}{\pi}\displaystyle\int_0^\pi \cos(z\sin\theta)\,d\theta$ satisfies the equation

$$zY'' + Y' + zY = 0$$

120. If Re $\{z\} > 0$, prove that $\displaystyle\int_0^\infty e^{-zt}\,J_0(t)\,dt = \dfrac{1}{\sqrt{z^2+1}}$.

121. Prove that: (a) $\cos(\alpha\cos\theta) = J_0(\alpha) - 2J_2(\alpha)\cos 2\theta + 2J_4(\alpha)\cos 4\theta + \cdots$

(b) $\sin(\alpha\cos\theta) = 2J_1(\alpha)\cos\theta - 2J_3(\alpha)\cos 3\theta + 2J_5(\alpha)\cos 5\theta - \cdots$.

122. If p is an integer, prove that

$$J_p(x+y) = \sum_{n=-\infty}^{\infty} J_n(x)\,J_{p-n}(y)$$

[*Hint.* Use the generating function.]

123. Establish Property 8, Page 271.

124. If $\text{Re}\,\{z\} > 0$, prove that $\quad J_n(z) = \dfrac{z^n}{2\pi i} \oint_C e^{\frac{1}{2}(t - z^2/t)}\, t^{-n-1}\, dt \quad$ where C is the contour of Fig. 10-5, Page 268.

125. If $\text{Re}\,\{z\} > 0$, prove that

$$J_n(z) \;=\; \frac{1}{\pi} \int_0^{\pi} \cos{(n\phi - z \sin \phi)}\, d\phi \;-\; \frac{\sin n\pi}{\pi} \int_0^{\infty} e^{-n\phi - z \sinh \phi}\, d\phi$$

126. (a) Verify that $Y_0(z)$, given by equation (23) on Page 272, is a solution to Bessel's equation of order zero. (b) Verify that $Y_n(z)$ given by equation (22) on Page 271 is a solution to Bessel's equation of order n.

127. Show that: (a) $z\,Y_{n-1}(z) \;-\; 2n\,Y_n(z) \;+\; z\,Y_{n+1}(z) \;=\; 0$

       (b) $\dfrac{d}{dz}\{z^n\,Y_n(z)\} \;=\; z^n\,Y_{n-1}(z)$ (c) $\dfrac{d}{dz}\{z^{-n}\,Y_n(z)\} \;=\; -z^{-n}\,Y_{n+1}(z).$

128. Prove that the general solution of

$$V'' + \left\{1 - \frac{(n^2 - 1/4)}{z^2}\right\} V \;=\; 0$$

is $V \;=\; \sqrt{z}\,\{A\,J_n(z) + B\,Y_n(z)\}.$

129. Prove that $\quad J_{n+1}(z)\,Y_n(z) \;-\; J_n(z)\,Y_{n+1}(z) \;=\; 1/z.$

130. Show that the general solution of $\quad V'' + z^{m-2}\,V = 0 \quad$ is

$$V \;=\; \sqrt{z}\left\{A\,J_{1/m}\left(\frac{2}{m}z^{m/2}\right) + B\,Y_{1/m}\left(\frac{2}{m}z^{m/2}\right)\right\}$$

131. (a) Show that the general solution to Bessel's equation $\quad z^2\,Y'' + z\,Y' + (z^2 - n^2)Y = 0 \quad$ is

$$Y \;=\; A\,J_n(z) \;+\; B\,J_n(z)\int \frac{dz}{z\,J_n^2(z)}$$

(b) Reconcile this result with that of equation (24), Page 272.

LEGENDRE FUNCTIONS

132. Obtain the Legendre polynomials   (a) $P_3(z)$,   (b) $P_4(z)$,   (c) $P_5(z)$.

   Ans.   (a) $\frac{1}{2}(5z^3 - 3z)$,   (b) $\frac{1}{8}(35z^4 - 30z^2 + 3)$,   (c) $\frac{1}{8}(63z^5 - 70z^3 + 15z)$

133. Prove   (a) $P'_{n+1}(z) - P'_{n-1}(z) = (2n+1)P_n(z)$,   (b) $(n+1)P_n(z) = P'_{n+1}(z) - z\,P'_n(z)$.

134. Prove that $\quad nP'_{n+1}(z) - (2n+1)zP'_n(z) + (n+1)P'_{n-1}(z) = 0.$

135. Prove that   (a) $P_n(-1) = (-1)^n$,   (b) $P_{2n+1}(0) = 0$.

136. Prove that $\quad P_{2n}(0) \;=\; \dfrac{(-1)^n}{n!}\left(\dfrac{1}{2}\right)\left(\dfrac{3}{2}\right)\left(\dfrac{5}{2}\right)\cdots\left(\dfrac{2n-1}{2}\right) \;=\; (-1)^n\,\dfrac{1\cdot 3\cdot 5\cdots(2n-1)}{2\cdot 4\cdot 6\cdots(2n)}.$

137. Verify Property 2, Page 272.

138. If $[n/2]$ denotes the greatest integer $\leqq n/2$, show that

$$P_n(z) \;=\; \sum_{k=0}^{[n/2]} \frac{(-1)^k\,(2n - 2k)!}{2^n\,k!\,(n-k)!\,(n-2k)!}\, z^{n-2k}$$

139. Prove that the general solution of Legendre's equation $\quad (1-z^2)Y'' - 2zY' + n(n+1)Y = 0 \quad$ for $n = 0, 1, 2, 3, \ldots$ is

$$Y \;=\; A\,P_n(z) \;+\; B\,Q_n(z) \qquad \text{where} \qquad Q_n(z) \;=\; P_n(z)\int_z^{\infty} \frac{dt}{(t^2 - 1)\{P_n(t)\}^2}$$

140. Use Problem 139 to find the general solution of the differential equation $\quad (1-z^2)Y'' - 2zY' + 2Y = 0.$

   Ans. $Y \;=\; Az + B\left\{1 + \frac{1}{2}z \ln\left(\dfrac{z-1}{z+1}\right)\right\}$

THE ZETA FUNCTION

141. If $\text{Re}\{z\} > 0$, prove that

$$\zeta(z) \;=\; \frac{1}{1^z} + \frac{1}{2^z} + \frac{1}{3^z} + \cdots \;=\; \frac{1}{\Gamma(z)} \int_0^\infty \frac{t^{z-1}\,dt}{e^t - 1}$$

142. Prove that $\left(1 - \dfrac{1}{2^2}\right)\left(1 - \dfrac{1}{3^2}\right)\left(1 - \dfrac{1}{5^2}\right)\left(1 - \dfrac{1}{7^2}\right)\cdots \;=\; \dfrac{\pi^2}{6}$ where $2, 3, 5, 7, \ldots$ represent prime numbers.

143. Prove that the only singularity of $\zeta(z)$ is a simple pole at $z = 1$ whose residue is equal to 1.

144. Use the analytic continuation of $\zeta(z)$ given by equation *(33)*, Page 273, to show that *(a)* $\zeta(-1) = -1/12$, *(b)* $\zeta(-3) = 1/120$.

145. Show that if z is replaced by $1 - z$ in equation *(33)*, Page 273, the equation remains the same.

THE HYPERGEOMETRIC FUNCTION

146. Prove that: *(a)* $\ln(1 + z) \;=\; z\,F(1, 1; 2; -z)$

             (b) $\dfrac{\tan^{-1} z}{z} \;=\; F(1/2, 1; 3/2; -z^2).$

147. Prove that $\cos 2\alpha z \;=\; F(\alpha, -\alpha; 1/2; \sin^2 z).$

148. Prove that $\dfrac{d}{dz}\,F(a, b; c; z) \;=\; \dfrac{ab}{c}\,F(a+1, b+1; c+1; z).$

149. If $\text{Re}\{c - a - b\} > 0$ and $c \neq 0, -1, -2, \ldots,$ prove that

$$F(a, b; c; 1) \;=\; \frac{\Gamma(c)\,\Gamma(c - a - b)}{\Gamma(c - a)\,\Gamma(c - b)}$$

150. Prove the result *(31)*, Page 273.

151. Prove that: *(a)* $F(a, b; c; z) \;=\; (1 - z)^{c - a - b}\,F(c - a, c - b; c; z)$

             (b) $F(a, b; c; z) \;=\; (1 - z)^{-a}\,F(a, c - b; c; z/[z-1]).$

152. Show that for $|z - 1| < 1$, the equation $z(1 - z)Y'' + \{c - (a + b + 1)z\}Y' - abY = 0$ has the solution $F(a, b; a + b - c + 1; 1 - z)$.

ASYMPTOTIC EXPANSIONS AND THE METHOD OF STEEPEST DESCENTS

153. Prove that

$$\int_p^\infty e^{-zt^2}\,dt \;=\; \frac{e^{-zp^2}}{2pz}\left\{1 - \frac{1}{2p^2 z} + \frac{1 \cdot 3}{(2p^2 z)^2} - \cdots (-1)^n \frac{1 \cdot 3 \cdot 5 \cdots (2n - 1)}{(2p^2 z)^n}\right\}$$

$$\cdot (-1)^{n+1} \frac{1 \cdot 3 \cdot 5 \cdots (2n + 1)}{(2z)^{n+1}} \int_p^\infty \frac{e^{-zt^2}}{t^{2n+2}}\,dt$$

and thus obtain an asymptotic expansion for the integral on the left.

154. Use Problem 153 to verify the result *(48)* on Page 275.

155. Evaluate $50!$.   *Ans.* 3.04×10^{64}

156. Show that for large values of n, $\dfrac{1 \cdot 3 \cdot 5 \cdots (2n - 1)}{2 \cdot 4 \cdot 6 \cdots (2n)} \;\sim\; \dfrac{1}{\sqrt{\pi n}}.$

157. Obtain the asymptotic expansions:

(a) $\displaystyle\int_0^\infty \frac{e^{-zt^2}}{1+t^2}\,dt \;\sim\; \frac{1}{2}\sqrt{\frac{\pi}{z}}\left\{1 - \frac{1}{2z} + \frac{1\cdot 3}{(2z)^2} - \frac{1\cdot 3\cdot 5}{(2z)^3} + \cdots\right\}$

(b) $\displaystyle\int_0^\infty \frac{e^{-zt}}{1+t}\,dt \;\sim\; \frac{1}{z} - \frac{1!}{z^2} + \frac{2!}{z^3} - \frac{3!}{z^4} + \cdots$

158. Verify the asymptotic expansion (*49*) on Page 275.

159. Use asymptotic series to evaluate $\displaystyle\int_{10}^\infty \frac{e^{-t}}{t}\,dt.$ *Ans.* .915, approx.

160. Under suitable conditions on $F(t)$, prove that

$$\int_0^\infty e^{-zt}\,F(t)\,dt \;\sim\; \frac{F(0)}{z} + \frac{F'(0)}{z^2} + \frac{F''(0)}{z^3} + \cdots$$

161. Perform the steps needed in order to go from (*4*) to (*5*) of Problem 36.

162. Prove the asymptotic expansion (*46*), Page 275, for the Bessel function.

163. If $F(z) \sim \displaystyle\sum_{n=0}^\infty \frac{a_n}{z^n}$ and $G(z) \sim \displaystyle\sum_{n=0}^\infty \frac{b_n}{z^n},$ prove that:

(a) $F(z) + G(z) \;\sim\; \displaystyle\sum_{n=0}^\infty \frac{a_n + b_n}{z^n}$

(b) $F(z)\,G(z) \;\sim\; \displaystyle\sum_{n=0}^\infty \frac{c_n}{z^n}$ where $c_n = \displaystyle\sum_{k=0}^n a_k b_{n-k}.$

164. If $F(z) \sim \displaystyle\sum_{n=2}^\infty \frac{a_n}{z^n},$ prove that $\displaystyle\int_z^\infty F(z)\,dz \;\sim\; \sum_{n=2}^\infty \frac{a_n}{(n-1)\,z^{n-1}}.$

165. Show that for large values of z,

$$\int_0^\infty \frac{dt}{(1+t^2)^z} \;\sim\; \frac{\sqrt{\pi}}{2}\left\{\frac{1}{z^{1/2}} + \frac{3}{8z^{3/2}} + \frac{25}{128z^{5/2}} + \cdots\right\}$$

ELLIPTIC FUNCTIONS

166. If $0 < k < 1$, prove that

$$K = \int_0^{\pi/2} \frac{d\theta}{\sqrt{1 - k^2 \sin^2\theta}} = \frac{\pi}{2}\left\{1 + \left(\frac{1}{2}\right)^2 k^2 + \left(\frac{1\cdot 3}{2\cdot 4}\right)^2 k^4 + \cdots\right\}$$

167. Prove: (a) $\operatorname{sn} 2z = \dfrac{2\operatorname{sn} z\,\operatorname{cn} z\,\operatorname{dn} z}{1 - k^2\operatorname{sn}^4 z},$ (b) $\operatorname{cn} 2z = \dfrac{1 - 2\operatorname{sn}^2 z + k^2\operatorname{sn}^4 z}{1 - k^2\operatorname{sn}^4 z}.$

168. If $k = \sqrt{3}/2$, show that   (a) $\operatorname{sn}(K/2) = \sqrt{2/3},$ (b) $\operatorname{cn}(K/2) = \sqrt{1/3},$ (c) $\operatorname{dn}(K/2) = \sqrt{1/2}.$

169. Prove that $\dfrac{\operatorname{sn} A + \operatorname{sn} B}{\operatorname{cn} A + \operatorname{cn} B} = \operatorname{tn}\tfrac{1}{2}(A+B)\,\operatorname{dn}\tfrac{1}{2}(A-B).$

170. Prove that   (a) $\operatorname{sn}(4K + 4iK') = 0,$ (b) $\operatorname{cn}(4K + 4iK') = 1,$ (c) $\operatorname{dn}(4K + 4iK') = 1.$

171. Prove: (a) $\operatorname{sn} z = z - \tfrac{1}{6}(1+k^2)z^3 + \tfrac{1}{120}(1+14k+k^4)z^5 + \cdots$

(b) $\operatorname{cn} z = 1 - \tfrac{1}{2}z^2 + \tfrac{1}{24}(1+4k^2)z^4 + \cdots$

(c) $\operatorname{dn} z = 1 - \tfrac{1}{2}k^2 z^2 + \tfrac{1}{24}k^2(k^2+4)z^4 + \cdots$

172. Prove that $\displaystyle\int_1^\infty \frac{dt}{\sqrt{t^4 - 1}} = \frac{1}{\sqrt{2}}K\left(\frac{1}{\sqrt{2}}\right).$

173. Use contour integration to prove the results of Problem 40 (b) and (c).

174. (a) Show that $\displaystyle\int_0^\phi \frac{d\phi}{\sqrt{1 - k^2 \sin^2 \phi}} = \frac{2}{1+k} \int_0^{\phi_1} \frac{d\phi_1}{\sqrt{1 - k^2 \sin^2 \phi_1}}$ where $k_1 = 2\sqrt{k}/(1+k)$ by using

Landen's transformation, $\tan \phi = (\sin 2\phi_1)/(k + \cos 2\phi_1)$.

(b) If $0 < k < 1$, prove that $k < k_1 < 1$.

(c) Show that by successive applications of Landen's transformation a sequence of moduli k_n, $n = 1, 2, 3, \ldots$ is obtained such that $\lim\limits_{n \to \infty} k_n = 1$. Hence show that if $\Phi = \lim\limits_{n \to \infty} \phi_n$,

$$\int_0^\phi \frac{d\phi}{\sqrt{1 - k^2 \sin^2 \phi}} = \sqrt{\frac{k_1 k_2 k_3 \cdots}{k}} \ln \tan\left(\frac{\pi}{4} + \frac{\Phi}{2}\right)$$

(d) Explain how the result in (c) can be used in the evaluation of elliptic integrals.

175. Is $\operatorname{tn} z = (\operatorname{sn} z)/(\operatorname{cn} z)$ a doubly periodic function? Explain.

176. Derive the addition formulas for (a) $\operatorname{cn}(z_1 + z_2)$, (b) $\operatorname{dn}(z_1 + z_2)$ given on Page 276.

MISCELLANEOUS PROBLEMS

177. If $|p| < 1$, show that $\displaystyle\int_0^{\pi/2} \tan^p \theta \, d\theta = \tfrac{1}{2}\pi \sec(p\pi/2)$.

178. If $0 < n < 2$, show that $\displaystyle\int_0^\infty \frac{\sin t}{t^n} dt = \frac{\pi \csc(n\pi/2)}{2\,\Gamma(n)}$.

179. If $0 < n < 1$, show that $\displaystyle\int_0^\infty \frac{\cos t}{t^n} dt = \frac{\pi \sec(n\pi/2)}{2\,\Gamma(n)}$.

180. Prove that the general solution of $(1 - z^2)Y'' - 4zY' + 10Y = 0$ is given by
$$Y = A\,F(5/2, -1; 1/2; z^2) + Bz\,F(3, -1/2; 3/2; z^2)$$

181. Show that: (a) $\displaystyle\int_0^\infty \sin t^3 \, dt = \tfrac{1}{6}\Gamma(1/3)$

(b) $\displaystyle\int_0^\infty \cos t^3 \, dt = \frac{\sqrt{3}}{6}\Gamma(1/3)$.

182. (a) Find a solution of $zY'' + Y' + zY = 0$ having the form $(\ln z)\left(\displaystyle\sum_{k=0}^\infty a_k z^k\right)$, and thus verify the result (23) given on Page 272. (b) What is the general solution?

183. Use the method of Problem 182 to find the general solution of $z^2Y'' + zY' + (z^2 - n^2)Y = 0$. [See equation (22), Page 271.]

184. Show that the general solution of $zU'' + (2m+1)U' + zU = 0$ is
$$U = z^{-m}\{A\,J_m(z) + B\,Y_m(z)\}$$

185. (a) Prove that $z^{1/2}J_1(2i z^{1/2})$ is a solution of $zU'' - U = 0$. (b) What is the general solution?
Ans. (b) $Y = z^{1/2}\{A\,J_1(2i z^{1/2}) + B\,Y_1(2i z^{1/2})\}$

186. Prove that $\{J_0(z)\}^2 + 2\{J_1(z)\}^2 + 2\{J_2(z)\}^2 + \cdots = 1$.

187. Prove that $e^{z \cos \alpha} J_0(z \sin \alpha) = \displaystyle\sum_{n=0}^\infty \frac{P_n(\cos \alpha)}{n!} z^n$.

188. Prove that $\Gamma'(\tfrac{1}{2}) = -\sqrt{\pi}\,(\gamma + 2\ln 2)$.

189. (a) Show that $\displaystyle\int_z^\infty \frac{e^{-t}}{t} dt = -\gamma - \ln z + z - \frac{z^2}{2 \cdot 2!} + \frac{z^3}{3 \cdot 3!} - \cdots$.

(b) Is the result in (a) suitable for finding the value of $\displaystyle\int_{10}^\infty \frac{e^{-t}}{t} dt$? Explain. [Compare with Problem 159.]

190. If m is a positive integer, show that $\quad F(\frac{1}{2}, -m; \frac{1}{2} - m; 1) = \dfrac{2 \cdot 4 \cdot 6 \cdots 2m}{1 \cdot 3 \cdot 5 \cdots (2m-1)}$.

191. Prove that $\quad (1 + z)\left(1 - \dfrac{z}{2}\right)\left(1 + \dfrac{z}{3}\right)\left(1 - \dfrac{z}{4}\right) \cdots = \dfrac{\sqrt{\pi}}{\Gamma\left(\dfrac{1+z}{2}\right)\Gamma\left(\dfrac{2-z}{2}\right)}$.

192. Prove that $\quad \displaystyle\int_0^{\pi/2} \dfrac{d\phi}{\sqrt{1 - k^2 \sin^2 \phi}} = \dfrac{\pi}{2} F(\frac{1}{2}, \frac{1}{2}; 1; k^2)$.

193. The *associated Legendre functions* are defined by
$$P_n^{(m)}(z) = (1 - z^2)^{m/2} \frac{d^m}{dz^m} P_n(z)$$

(a) Determine $P_3^{(2)}(z)$.

(b) Prove that $P_n^{(m)}(z)$ satisfies the differential equation
$$(1 - z^2)Y'' - 2zY' + \left\{n(n+1) - \frac{m^2}{1 - z^2}\right\} Y = 0$$

(c) Prove that $\displaystyle\int_{-1}^1 P_n^{(m)}(z) P_l^{(m)}(z)\, dz = 0 \quad$ if $n \neq l$.

This is called the *orthogonality property for the associated Legendre functions*.
Ans. (a) $15z(1 - z^2)$

194. Prove that if m, n and r are positive constants,
$$\int_0^1 \frac{x^{m-1}(1 - x)^{n-1}}{(x + r)^{m+n}}\, dx = \frac{B(m, n)}{r^m(1 + r)^{m+n}}$$
[*Hint.* Let $x = (r+1)y/(r+y)$.]

195. Prove that if m, n, a and b are positive constants,
$$\int_0^{\pi/2} \frac{\sin^{2m-1}\theta \cos^{2n-1}\theta\, d\theta}{(a \sin^2\theta + b \cos^2\theta)^{m+n}} = \frac{B(m, n)}{2a^n b^m}$$
[*Hint.* Let $x = \sin^2\theta$ in Problem 194 and choose r appropriately.]

196. Prove that: (a) $\dfrac{z}{2} = J_1(z) + 3 J_3(z) + 5 J_5(z) + \cdots$

(b) $\dfrac{z^2}{8} = 1^2 J_2(z) + 2^2 J_4(z) + 3^2 J_6(z) + \cdots$

197. If m is a positive integer, prove that:

(a) $P_{2m}(z) = \dfrac{(-1)^m (2m)!}{2^{2m}(m!)^2} F(-m, m + \frac{1}{2}; \frac{1}{2}; z^2)$

(b) $P_{2m+1}(z) = \dfrac{(-1)^m (2m+1)!}{2^{2m}(m!)^2} z\, F(-m, m + \frac{3}{2}; \frac{3}{2}; z^2)$

198. (a) Prove that $1/(\text{sn } z)$ has a simple pole at $z = 0$ and (b) find the residue at this pole. *Ans.* 1

199. Prove that $\quad \{\Gamma(\frac{1}{4})\}^2 = 8\sqrt{\pi}\, \dfrac{4 \cdot 6 \cdot 8 \cdot 10 \cdot 12 \cdot 14 \cdot 16 \cdot 18 \cdots}{5 \cdot 5 \cdot 9 \cdot 9 \cdot 13 \cdot 13 \cdot 17 \cdot 17 \cdots}$.

200. If $|z| < 1$, prove *Euler's identity*: $\quad (1 + z)(1 + z^2)(1 + z^3) \cdots = \dfrac{1}{(1 - z)(1 - z^3)(1 - z^5) \cdots}$.

201. If $|z| < 1$, prove that $\quad (1 - z)(1 - z^2)(1 - z^3) \cdots = 1 + \displaystyle\sum_{n=1}^{\infty} (-1)^n \{z^{n(3n-1)/2} + z^{n(3n+1)/2}\}$.

202. (a) Prove that $\dfrac{z}{1 + z} + \dfrac{z^2}{(1 + z)(1 + z^2)} + \dfrac{z^4}{(1 + z)(1 + z^2)(1 + z^4)} + \cdots$ converges for $|z| < 1$ and $|z| > 1$.

(b) Show that in each region the series represents an analytic function, say $F_1(z)$ and $F_2(z)$ respectively.

(c) Are $F_1(z)$ and $F_2(z)$ analytic continuations of each other? Is $F_1(z) = F_2(z)$ identically? Justify your answers.

203. (*a*) Show that the series $\displaystyle\sum_{n=1}^{\infty} \frac{z^n}{n^2}$ converges at all points of the region $|z| \leqq 1$.

(*b*) Show that the function represented by all analytic continuations of the series in (*a*) has a singularity at $z = 1$ and reconcile this with the result in (*a*).

204. Let $\Sigma\, a_n z^n$ have a finite circle of convergence C and let $F(z)$ be the function represented by all analytic continuations of this series. Prove that $F(z)$ has at least one singularity on C.

205. Prove that $\displaystyle\frac{\text{cn } 2z + \text{dn } 2z}{1 + \text{cn } 2z} = \text{dn}^2 z$.

206. Prove that a function which is not identically constant cannot have two periods whose ratio is a real irrational number.

207. Prove that a function, not identically constant, cannot have three or more independent periods.

208. (*a*) If a doubly-periodic function is analytic everywhere in a cell [period parallelogram], prove that it must be a constant. (*b*) Deduce that a doubly-periodic function, not identically constant, has at least one singularity in a cell.

209. Let $F(z)$ be a doubly-periodic function. (*a*) Prove that if C is the boundary of its period parallelogram, then $\displaystyle\oint_C F(z)\, dz = 0$. (*b*) Prove that the number of poles inside a period parallelogram equals the number of zeros, due attention being paid to their multiplicities.

210. Prove that the Jacobian elliptic functions $\text{sn } z$, $\text{cn } z$ and $\text{dn } z$ (*a*) have exactly two zeros and two poles in each cell and that (*b*) each function assumes any given value exactly twice in each cell.

211. Prove that $\displaystyle\left(1 + \frac{1}{1^2}\right)\left(1 + \frac{1}{4^2}\right)\left(1 + \frac{1}{7^2}\right) \cdots = \frac{\{\Gamma(1/3)\}^2}{\left\{\Gamma\left(\dfrac{1+i}{3}\right)\right\}^2 \left\{\Gamma\left(\dfrac{1-i}{3}\right)\right\}^2}.$

212. Prove that $\displaystyle\int_0^{\pi/2} e^{-z\tan\theta}\, d\theta \sim \frac{1}{z} - \frac{2!}{z^3} + \frac{4!}{z^5} - \frac{6!}{z^7} + \cdots$

213. Prove that $\displaystyle P_n(\cos\theta) = 2\left\{\frac{1\cdot 3\cdot 5\cdots(2n-1)}{2\cdot 4\cdot 6\cdots(2n)}\right\}\left\{\cos n\theta + \frac{1\cdot 2n}{2\cdot(2n-1)}\cos(n-2)\theta\right.$

$$\left. + \frac{1\cdot 3\cdot 2n(2n-2)}{2\cdot 4\cdot(2n-1)(2n-3)}\cos(n-4)\theta + \cdots\right\}$$

[*Hint.* $1 - 2t\cos\theta + t^2 = (1 - te^{i\theta})(1 - te^{-i\theta})$.]

214. (*a*) Prove that $\Gamma(z)$ is a meromorphic function and (*b*) determine the principal part at each of its poles.

215. If $\text{Re}\,\{n\} > -1/2$, prove that

$$J_n(z) = \frac{z^n}{2^n\sqrt{\pi}\,\Gamma(n+\frac{1}{2})}\int_{-1}^{1} e^{izt}(1-t^2)^{n-1/2}\, dt$$

$$= \frac{z^n}{2^n\sqrt{\pi}\,\Gamma(n+\frac{1}{2})}\int_0^{\pi} \cos(z\cos\theta)\sin^{2n}\theta\, d\theta$$

216. Prove that $\displaystyle\int_0^{\infty} t^n J_m(t)\, dt = \frac{2^n\,\Gamma\left(\dfrac{m+n+1}{2}\right)}{\Gamma\left(\dfrac{m-n+1}{2}\right)}.$

217. Prove that $\displaystyle\int_0^{\pi/2} \cos^p\theta \cos q\theta\, d\theta = \frac{\pi\,\Gamma(p+1)}{2^{p+1}\,\Gamma\left(\dfrac{2+p+q}{2}\right)\Gamma\left(\dfrac{2+p-q}{2}\right)}.$

218. Prove that $\displaystyle\{\Gamma(\tfrac{1}{4})\}^2 = 4\sqrt{\pi}\int_0^{\pi/2} \frac{d\theta}{\sqrt{1 - \frac{1}{2}\sin^2\theta}}.$

INDEX

Abel's theorem, 142
Absolute convergence, 140, 147, 148
  convergence and, 141, 159
  of infinite products, 266, 267
  of power series, 152
Absolutely convergent series,
    140, 141, 147, 148, 152, 159
    (*see also* Absolute convergence)
  theorems on, 141
Absolute value, of a complex number,
    2, 4
  of a real number, 2
Acceleration, along a curve, 69, 82, 90
  centripetal, 82
Addition, associative law of, 3, 8
  commutative law of, 3, 8
  identity with respect to, 3
  inverse with respect to, 1, 3
  of complex numbers, 2, 6
  of real numbers, 1
  of vectors, 10, 11, 15
Addition formulas for elliptic
    functions, 276, 295
Aerodynamic theory, 224, 234
Airfoil, 224
  flow pattern about, 238
  Joukowski, 224
Airplane wing, 224 (*see also* Airfoil)
Algebra, fundamental theorem of
    (*see* Fundamental theorem of
    algebra)
Algebraic functions, 36
Algebraic numbers, 23
Alternating currents, 91
Alternating series test, 142
Amplitude, 4
Analytic continuation, 146, 159, 265,
    277-279
  branch points and, 265
  Cauchy's integral formulas and,
    278
  Morera's theorem and, 278
  of gamma function, 268, 269, 281
  of integrals, 265, 279
  of series, 146, 159, 265, 279
  of zeta function, 273
  singularities and, 146
  uniqueness theorem for, 265
Analytic extension (*see* Analytic
    continuation)
Analytic functions, 63, 66
  and conformal mapping (*see*
    Conformal mapping)
  and continuity, 63, 71, 72
  elements of, 146, 265
  harmonic functions and, 131
  imaginary part of, 84
  in neighborhood of a singularity,
    160
  necessary and sufficient conditions
    for, 63, 72-74
  real part of, 84
Analytic part, of Laurent series, 144
Angle, between vectors, 21
Angles, preservation of, under map-
    ping, 201, 213, 214, 229 (*see also*
    Conformal mapping)
Annular region or annulus, 143, 211
  mapping of, 211
Anti-derivatives, 95
Arc, 68 (*see also* Curves)
Area, bounded by a simple closed
    curve, 113, 114
  of ellipse, 113, 230
  of parallelogram, 6, 21
  of region, 113, 114
  of triangle, 22
Area magnification factor, 201, 214

Argand diagram, 3
Argument, 4, 18
Argument theorem, 119
  generalization of, 137
  proof of, 127, 128
Arg z (*see* Argument)
Associated Legendre functions, 305
Associative law, of addition, 3, 8
  of multiplication, 3, 8
Asymptotic expansions, 274, 275,
    288-290, 294
  of Bessel functions, 275, 294
  of error function, 275
  of exponential integral, 275
  of gamma function, 275, 289, 290
  special, 275
  Stirling's, 275
Asymptotic series, 274 (*see also*
    Asymptotic expansions)
Attraction, of electric charges, 238
Axiomatic foundations of complex
    number system, 3, 13, 14
Axis, imaginary, 4
  real, 4
  x and y, 3

Base of logarithms, 35
Bernoulli numbers, 171, 273
Bernoulli's theorem, 238
Bessel functions, 161, 270, 271
  asymptotic expansion for, 275, 294
  generating function for, 271
  of first and second kinds, 271
  recursion formula for, 271, 286
Bessel's differential equation, 270,
    271
  general solution of, 271, 272, 284
Beta function, 222, 269, 282
  relation of, to gamma function,
    269, 282
Biharmonic equation, 263
Bilinear transformation, 35, 203, 216,
    217
  cross ratio of, 203, 216
  transformation of circles into
    circles using, 217
  use of, in mapping circle on to half
    plane, 203, 204, 216, 217
Binomial coefficients, 16
Binomial formula or theorem, 16, 143
  use of, in obtaining Laurent series,
    157, 158
Blasius, theorems of, 238, 250-252
Bolzano-Weierstrass theorem, 8, 23
Borel-Heine theorem, 8
Boundary conditions, 232
Boundary, natural, 146, 159, 265
  points, 7, 22, 23
  transformation of, on to real axis,
    205, 221, 228
Boundary-value problems, 232, 243,
    244
  Dirichlet and Neumann (*see*
    Dirichlet problem and
    Neumann problem)
  existence of solutions to, 232
Bounded functions, 39
  sequences, 141
  sets, 7, 22, 23
Bound, least upper, 62, 93
Branch cut, 37, 44
Branch lines, 37, 44, 48-50
Branch, of a function, 33
Branch points, 37, 44, 48-50, 55, 56,
    67, 76, 80, 145
  and analytic continuation, 265
  integration involving, 185, 186,
    193, 194
  of logarithmic functions, 46, 76, 80

Branch, principal, 33, 44, 46, 48

Cables, transmission line, 256
Capacitance, 240, 255, 256
Capacitor, 240
Cardioid, mapping of, 210, 229
Casorati-Weierstrass theorem, 145
Cauchy-Goursat theorem, 95, 103-106
    (*see also* Cauchy's theorem)
  converse of (*see* Morera's theorem)
  proof of, for closed polygon, 104,
    105
  proof of, for multiply-connected
    regions, 106
  proof of, for simple closed curve,
    105, 106
  proof of, for triangle, 103, 104
Cauchy principal value of integrals,
    174
Cauchy-Riemann equations, 63, 72-74
  gradient and, 70
  harmonic functions obtained from,
    73, 74
  polar form of, 83, 84
  proof of, 72, 73
Cauchy's convergence criterion, 141
Cauchy's inequality, 118, 124, 169
  proof of, 124
Cauchy's integral formulas, 118,
    120-123
  and related theorems, 118-138
  for multiply-connected regions, 123
  proof of, 120-122
  use of, in analytic continuation, 278
Cauchy's integral theorem (*see*
    Cauchy's theorem)
Cauchy's theorem, 95, 103-106 (*see
    also* Cauchy-Goursat theorem)
  consequences of, 96, 97, 106-108
  converse of (*see* Morera's theorem)
  proof of, 103
Cavitation, 259
Cell, 276, 293
Centripetal acceleration, 82
Centroid, 114
Chain rule, for differentiation, 65, 77,
    84, 85
  proof of, 84, 85
Change of variables, in integration,
    93
Channel, mapping of, 208, 211
Charge distribution, 239
Charge, electric, 238
  line (*see* Line charge)
  potential due to (*see* Potential)
Christoffel-Schwarz transformation,
    204, 218-223
Circle, harmonic functions for, 119,
    120
  mappings of, 203, 204, 207, 216, 217
  of convergence, 140, 143, 150
  Poisson's integral formulas for,
    119, 129, 130, 233
  unit, 5, 201
Circular obstacle, flow around, 237,
    238, 246, 250
Circulation, 234, 249
  about a vortex, 249
  flow with, 236, 249, 250
Cis θ, 4
Closed curves, 68 (*see also* Simple
    closed curve)
  interval, 2
  regions, 7
  sets, 7, 22, 23 (*see also* Closure)
Closure, law or property, 1, 3
  of a set, 7, 22, 23
Coefficients, binomial, 16
Commutative law, of addition, 3, 8
  of multiplication, 3, 8

Compact set, 7, 22, 23
Comparison test, 141
  proof of, 148
Complementary modulus of elliptic
    functions, 276
Complement, of a region, 94
  of a set, 8, 22, 23
Complex, conjugate, 2, 9
  coordinates, 7 (see also Conjugate
      coordinates)
  differentiation, 63-91 (see also
      Differentiation)
  integration (see Integration)
  line integrals, 92 (see also Line
      integrals)
  number system, 2 (see also Com-
      plex numbers)
Complex numbers, 2
  absolute value of, 2, 4
  addition of, 2, 6
  as ordered pairs of real numbers, 3
  axiomatic foundations of, 3, 13, 14
  conjugate of, 2, 9
  division of, 2
  equality of, 2, 3
  fundamental operations with,
      2, 8, 9
  graphical representation of, 3, 4,
      10-13
  imaginary part of, 2
  multiplication of, 2
  polar form of, 4, 14, 15
  product of, 2-4
  quotient of, 2, 4
  real part of, 2
  roots of, 4, 18, 19
  spherical representation of, 6
  subtraction of, 2
  vector interpretation of, 5
Complex plane, 3
  entire or extended, 6
Complex potential, due to line charge,
      252
  in electrostatics, 238, 239, 252-254
  in fluid flow, 235, 245-250
Complex temperature, 240, 241
Complex variable, 2, 33
  functions of a, 33
Complex velocity, 235
Components, of a vector, 10
Composite functions, 39, 65
Condenser, 240
Conditional convergence, of infinite
      products, 266, 267
  of infinite series, 140, 141
Conductivity, thermal, 240
Conductors, perfect, 240
Conformal mapping, 200-231
    (see also Mapping)
  conditions for, 201
  definition of, 201
  Dirichlet and Neumann problems
      and, 233, 234 (see also Dirichlet
      problem and Neumann prob-
      lem)
  harmonic functions under, 242
  proof of, for analytic functions,
      213, 214
  solution of boundary-value prob-
      lems by, 232-264
Conformal transformation (see Con-
      formal mapping)
Conjugate coordinates, 7, 22, 69
  and del operator, 69, 82, 83
  equation of circle in, 22
  Green's theorem expressed in, 95,
      102, 103
Conjugate, functions, 63, 232
  of a complex number, 2, 9
  pairs, 20
Connected regions, 94
  sets, 7, 22, 23
  simply- (see Simply-connected
      regions)
Constant of integration, 95, 96

Continuation, analytic (see Analytic
      continuation)
Continuity, 38, 39, 53-54
  and analyticity, 63, 71, 72
  and uniform convergence, 142, 151
  equation of, 234
  in a region, 39, 40
  theorems on, 39
  uniform, 39, 40, 54
Continuous, curve or arc, 68
  function (see Continuity)
Contour, 69
Contour integrals, 94
  solution of differential equations
      by, 270, 284, 285
Convergence, absolute (see Absolute
      convergence)
  circle of, 140, 143, 150
  conditional (see Conditional con-
      vergence)
  criterion of Cauchy, 141
  of infinite products (see Infinite
      products)
  of power series, 143
  of sequences, 40, 54, 55, 139, 146,
      147
  of series, 41, 54, 55, 139, 147
  radius of, 140, 150
  region of, 139, 149, 150
  tests for, 141, 142, 148-150
  uniform (see Uniform convergence)
Convergent sequences, 40, 54, 55
Convergent series, 41, 54, 55, 139, 147
  necessary condition for, 41, 55, 139,
      141
Coordinate curves, 34
Coordinates, conjugate, 7 (see also
      Conjugate coordinates)
  curvilinear, 34, 42
  polar, 4
  rectangular, 3, 34
Coulomb's law, 238
Countability of a set, 8, 22, 23
Counterclockwise direction or sense,
      69, 94
Critical points, 201, 214
Cross cut, 101
Cross product, 6, 20, 21
  and curl, 70
Cross ratio, 203, 216
  invariance of, 203
Curl, 70, 82, 83, 85
  and cross product, 70
  identities involving, 70
Currents, alternating, 91
Curves, 68, 69
  acceleration along, 69, 82, 90
  continuous, 68
  coordinate, 34
  direction or sense of, 69, 94
  families of, 68
  integrals along, 92 (see also
      Line integrals)
  Jordan, 94
  normal vector to, 70, 83
  of infinite length, 111, 112
  orthogonal, 68, 81
  piecewise smooth, 69
  rectifiable, 92
  simple closed (see Simple closed
      curve)
  smooth, 68, 69
  stream, 235
  tangents to, 69, 83
  velocity along, 69, 82
Curvilinear coordinates, 34, 42
Cut, branch, 37, 44
  cross, 101
Cycloid, 113
  mapping of, 228
Cylindrical obstacles, force on, 251,
      252

Definite integrals, 92
  evaluation of, by residues, 173, 174,
      179-188, 193

Definite integrals (cont.)
  special theorems used in evaluat-
      ing, 174, 179, 182, 183
Degree, of polynomial equation, 5
  of polynomial function, 34
Del, 69, 82, 83
  and complex conjugate
      coordinates, 69, 82, 83
Deleted neighborhood, 7, 22
Delta neighborhood, 7, 23
De Moivre's theorem, 4, 15-18
  in terms of Euler's formula, 5, 16
  proof of, 16
Denominator, 1
Density, fluid, 238
Dependent variable, 33
Derivative operators, 65
Derivatives, 63, 71, 72
  anti-, 95
  geometric interpretation of, 64
  higher order, 66
  of elementary functions, 65, 66,
      74-78
  of multiple-valued functions, 65,
      66, 76, 77
  of power series, 142, 143, 152, 153
Determinant, Jacobian (see Jacobian)
Determinants, multiplication rule
      for, 215, 230
Diagonals of parallelogram, 12
Diagram, Argand, 3
Dielectric constant, 238, 256
Differentiability, 63 (see also
      Analytic functions)
  continuity and, 63, 71, 72
Differential equation, 91, 269, 270,
      283-285
  Bessel's, 270-272, 284
  Gauss', 273
  general solution of (see General
      solution)
  hypergeometric, 273
  Legendre's, 272, 273
  ordinary points of, 269
  partial (see Partial differential
      equations)
  regular singular points of, 269,
      283, 284
  singular points of, 269, 283
  solution of, by contour integrals,
      270, 284, 285
Differential operators, complex, 69
Differentials, 64, 65, 74
  principal part in, 64
  rules for finding, 65
Differentiation, chain rule for, 65, 77,
      84, 85
  complex, 63-91
  of series, 142, 143, 152, 153
  rules for, 65, 74-78
  under integral sign, 174, 175, 182
Dipole, 237, 240
  moment, 237
Direction, of vector, 5
Direction or sense, of curve, 69
  convention regarding, 94
Dirichlet problem, 232, 233, 243-245
    (see also Neumann problem)
  for the half plane, 233
  for the unit circle, 233
  solution of, by conformal mapping,
      233, 234
  uniqueness of solution to, 256, 257
Discontinuities, 39 (see also
      Singularities)
  removable, 39, 53
Discontinuous functions, 39 (see also
      Continuity)
Disjoint sets, 8
Disk, unit, 201
Distance, between two points, 2, 4, 12
  in complex plane, 4
Distributive law, 3, 9, 14
  for sets, 23
Divergence, identities involving, 70
  of functions, 70, 82, 83

Divergence (cont.)
of sequences and series, 40, 41, 139
(see also Convergence)
Divergent sequences, 40
series, 41, 139 (see also
Convergence)
Division, of complex numbers, 2
of real numbers, 1
Domain, 7, 22, 23
Dot product, 6, 21
divergence in terms of, 70, 82, 83
Double pole, 158
Doublet, 237, 240
Doubly-periodic functions, 276
Dummy symbol, 97
Dummy variable, 97
Duplication formula for the gamma
function, 268
Dynamics, fluid, 234 (see also Fluid
flow)

Elasticity, theory of, 263
Electric charges, 238
potential due to (see Potential)
Electric field intensity, 238, 239
Electricity, theory of alternating
currents in, 91
Electrostatic potential, 238, 239,
252-254
complex, 239, 252-254
sources and sinks in, 240
Electrostatics, applications to,
238-240
complex potential in, 239, 252-254
Gauss' theorem in, 239
Elementary functions, 34-37, 44-48
derivatives of, 65, 66, 74-78
Element, of an analytic function,
146, 265
of a set, 7
Elements, function, 265
Ellipse, area of, 113, 230
mapping of, 209, 224
transformation of, on to real axis,
221
Elliptic functions, 276, 277, 290-293
addition formulas for, 276, 295
Jacobian, 276
periods of, 276, 292, 293
Elliptic integral, 276, 277 (see also
Elliptic functions)
of first kind, 276
of second and third kinds, 277
Entire functions, 145
infinite product representation of,
267
Entire or extended complex plane, 6
Equality, of complex numbers, 2, 3
of ordered pairs of real numbers, 3
of vectors, 5
Equation, biharmonic, 263
differential (see Differential
equation)
indicial, 270, 284
Laplace's (see Laplace's equation)
of circle in conjugate coordinates,
22
of continuity, 234
of straight line in parametric,
standard and symmetric
form, 13
polynomial (see Polynomial
equations)
product of roots of, 20
quadratic, 19
roots of, 5, 18, 20
solutions of, 18
sum of roots of, 20
Equations, parametric, 41, 62
transformation, 200
Equipotential lines or curves, 235,
239, 252
Error function, 275
asymptotic expansion of, 275
Essential singularities, 67, 80, 157

Essential singularities (cont.)
behavior of analytic function near,
160
defined from Laurent series, 144
theorems involving, 145
Euler's constant, 268
Euler's formula, 5
and De Moivre's theorem, 5, 16
Euler's identity, 305
Even functions, 45
Existence, of solutions to boundary-
value problems, 232
Expansions, asymptotic
(see Asymptotic series)
series (see Series)
Exponential functions, 35
relation of, to trigonometric functions, 16, 17, 35
Exponential integral, 275
asymptotic expansion of, 275
Extended complex plane, 6
Extension, analytic (see Analytic
continuation)
Exterior, of a closed curve, 94
Exterior points, 7

Factored form, of polynomial equation, 5
Factorial function, 268 (see also
Gamma function)
Factor theorems of Weierstrass,
267, 282
Families of curves, 68
orthogonal, 68, 81
Field, 3
force, 111
intensity, 238, 239
Finite sequence, 40
Fixed or invariant points of a transformation, 202, 217
Flow pattern, about an airfoil, 238
Flows (see also Fluid flow)
around circular obstacle, 237, 238,
246, 250
around obstacles, 237, 238, 250
due to source or sink, 237
special, 236, 237
stationary, 234
steady, 234
superposition of, 237
uniform, 236, 245
vortex, 249
with circulation, 236, 249, 250
Fluid, density, 238
dynamics, 234 (see also Fluid flow)
ideal, 234
incompressible, 234
pressure in, 238, 251
real, 234
viscous, 234
Fluid flow, applications to, 234-238,
245-250 (see also Flows)
complex potential in, 235, 245-250
sources and sinks in, 235-237, 247,
248
Flux, heat, 240
lines, 239, 241
Force field, 111
Force, on a cylindrical obstacle, 251,
252
on an obstacle in a fluid, 238,
250-252
Fourier series, 170
Fractional linear transformation, 35,
203
cross ratio of, 203
Fractions, 1
Function elements, 265
Functions, 33, 41-50
algebraic, 36
analytic (see Analytic functions)
Bessel (see Bessel functions)
Beta (see Beta function)
bounded, 39
branches of, 33

Functions (cont.)
composite, 39, 65
conjugate, 63, 232
continuous (see Continuity)
divergence of, 70, 82, 83
doubly-periodic, 276
elementary (see Elementary
functions)
elliptic (see Elliptic functions)
entire, 145, 267
error, 275
even, 45
expansion of, in Laurent series,
143, 144, 155-158
exponential, 35
factorial, 268 (see also Gamma
function)
gamma (see Gamma function)
generating (see Generating
function)
harmonic (see Harmonic functions)
hyperbolic (see Hyperbolic
functions)
hypergeometric, 273, 288, 293
inverse, 33
lacunary, 146
Legendre (see Legendre functions)
limits of, 37, 38, 50-53
logarithmic (see Logarithmic
functions)
multiple-valued (see Multiple-
valued functions)
Neumann's, 271, 272
odd, 45
of a complex variable, 33
of a function, 39
polynomial, 34
rational algebraic, 35
sequences of, 139, 146, 147
series of, 146, 147
single-valued, 33
stream, 235, 245
transcendental, 36, 37
trigonometric (see Trigonometric
functions)
uniformly continuous, 39, 40
value of, 33
Fundamental theorem of algebra, 5,
119
proof of, using Liouville's theorem,
125
proof of, using Rouché's theorem,
128, 129

Gamma function, 222, 267-269,
280-282, 294
analytic continuation of, 268, 269,
281
asymptotic expansion of, 275, 289,
290
duplication formula for, 268
recursion formula for, 268, 280
Gauss' differential equation, 273
mean value theorem, 119, 125
Π function, 268
test, 142
theorem on electrostatics, 239
General solution, of a differential
equation, 269
of Bessel's differential equation,
271, 272, 284
Generating function, for Bessel
functions, 271
for Legendre polynomials, 272
Geometric interpretations, of
derivatives, 64
of limits, 50
Geometric series, 148, 149
Geometry, applications to, 69, 81, 82
Goursat-Cauchy theorem
(see Cauchy-Goursat theorem)
Gradient, 69, 70, 82, 83, 85
as a vector normal to a curve, 70, 83
Cauchy-Riemann equations and, 70
identities involving, 70

Graphical representation, of complex
    numbers, 3, 4, 10-13
  of real numbers, 1
  of roots, 18, 19
Greater, 1
Green's first and second identities,
    117
Green's theorem in the plane, 95,
    99-102
  complex form of, 95, 102, 103
  generalization of, 114
  in conjugate coordinates, 95, 102,
    103
  proof of Cauchy's theorem using,
    103
  proof of, for multiply-connected
    regions, 101
  proof of, for simply-connected
    regions, 99, 100

Half plane, Dirichlet problem for,
    233
  harmonic functions for, 120
  mapping of, on to unit circle, 203,
    204, 216, 217
  Poisson's integral formulas for,
    120, 130, 233
Hardy, 274
Harmonic functions, 63, 64, 70, 73,
    74, 85, 232, 233, 241-243
  and Poisson's integral formulas,
    119, 120
  for a circle, 119, 120
  for a half plane, 120
  obtained from Cauchy-Riemann
    equations, 73, 74
  relation of, to analytic functions,
    131
  under conformal mappings or
    transformations, 242
Harmonic motion, simple, 82
Heat flow, applications to, 240, 241,
    254, 255
Heat flux, 240
Heine-Borel theorem, 8
Hexagon, mapping of, 230
Higher order derivatives, 66
Holomorphic, 63 (see also Analytic)
Hydrodynamics, 234 (see also Fluid
    flow)
Hyperbola, mapping of, 228
Hyperbolic functions, 35
  inverse, 36, 48
  properties of, 35, 46
  relation of, to trigonometric
    functions, 36
Hypergeometric differential
    equation, 273
Hypergeometric function, 273, 288,
    293
Hypergeometric series, 169
Hypocycloid, 113, 114
  mapping of, 228

Ideal fluid, 234
Identities involving gradient,
    divergence and curl, 70
Identity with respect to addition, 3
  with respect to multiplication, 3
Image, 33, 34, 200
Imaginary axis, 4
Imaginary numbers, pure, 2
Imaginary part, of analytic function,
    84
  of complex number, 2
Imaginary unit, 2
  as ordered pair of real numbers, 3
Incompressible fluids, 234
Increment, 64
Indefinite integrals, 95, 107
Independence of path, 96, 102, 106
  necessary and sufficient condition
    for, 102
Independent variable, 33

Indeterminate forms, 67 (see also
    L'Hospital's rule)
Indicial equation, 270, 284
Induction, mathematical, 16
Infinite length, of curves, 111, 112
Infinite products, 266, 267, 279, 280,
    293, 294
  absolute, conditional and uniform
    convergence of, 266, 267
  important theorems on, 267
  rearrangement of terms in, 266
  special, 267, 280
Infinite sequence, 40 (see also
    Sequences)
Infinite series, 40, 41 (see also Series)
Infinitesimal, 74
Infinity, 38
  point at, 6, 38, 47, 80, 81
  singularities at, 68, 145
Initial point, of vector, 5
Inside, of simple closed curve, 94
Integers, 1
Integrable, 92
Integral, along a curve, 92 (see also
    Line integrals)
  differentiation under, 174, 175, 182
  elliptic, 276, 277 (see also Elliptic
    functions)
  exponential, 275
  formulas of Cauchy (see Cauchy's
    integral formulas)
  formulas of Poisson (see Poisson's
    integral formulas)
  line, 92 (see also Line integrals)
  test, 141
Integrals, analytic continuation of,
    265, 279
  Cauchy principal value of, 174
  contour (see Contour integrals)
  definite (see Definite integrals)
  evaluation of, by residues, 173, 174,
    179-188, 193
  of special functions, 96, 109, 110
Integration, around a singularity,
    184
  branch points and, 185, 186, 193,
    194
  by parts, 109, 110
  change of variables in, 93
  complex, 92-117
  constant of, 95, 96
  contour, 94
  of series, 142, 152
Intensity, electric field, 238, 239
Interior, of a simple closed curve, 94
Interior points, 7, 22, 23
Intersection of sets, 8, 23
Interval, closed, 2
  open, 2
Invariance, of cross ratio, 203
Invariant or fixed points of a trans-
    formation, 202, 217
Inverse, with respect to addition, 1, 3
  with respect to multiplication, 1, 3
Inverse functions, 33
Inverse hyperbolic functions, 36, 48
  relation of, to logarithmic
    functions, 36
Inverse transformation, 200
Inverse trigonometric functions, 36,
    48
  relation of, to logarithmic
    functions, 36
Inversion, 202, 217
Involutory transformation, 230
Irrational numbers, 1
Irregular singular point, 269, 283,
    284
Irrotational, 234
Isogonal mapping, 201
Isolated singularity, 67, 79, 80, 144
  behavior of analytic function near,
    160
Isothermal lines, 241

Jacobian elliptic function, 276
    (see also Elliptic functions)
Jacobian, of a transformation, 200,
    214-216
  of conformal transformation, 201,
    214-216
Jensen's theorem, 138
Jordan curve, 94
Jordan curve theorem, 94
Joukowski airfoils or profiles, 224
Joukowski, transformation, 224, 229

Kernel, 270

Lacunary function, 146
Lagrange's expansion, 145, 159
  proof of, 159
Landen's transformation, 304
Laplace's equation, 63, 64, 232, 233
  and Dirichlet or Neumann prob-
    lems, 232, 233 (see also
    Dirichlet problem and
    Neumann problem)
  in polar form, 84
Laplace's method, 275 (see also
    Steepest descents, method of)
Laplacian operator, 63, 70, 82, 83
Laurent's series, 143, 144 (see also
    Laurent's theorem)
  analytic part of, 144
  classification of singularities from,
    144, 145, 157, 158
  expansion of functions in, 143, 144,
    155-158
  principal part of, 144
  residues and, 172
Laurent's theorem, 143, 144, 155-157
    (see also Laurent's series)
  proof of, 155-157
Least upper bound, 62, 93
Legendre functions, 272, 273, 287,
    288 (see also Legendre
    polynomials)
  associated, 305
  properties of, 272
Legendre polynomials, 161, 162, 272,
    293 (see also Legendre functions)
  generating function for, 272
  orthogonality of, 287
  recursion formula for, 272, 287, 288
  relation of, to hypergeometric
    function, 293
  Schlaefli's formula for, 161, 162
Legendre's differential equation, 272
  general solution of, 272, 273
Leibnitz's rule, 174, 175, 182
Lemniscate, mapping of, 229
Length or magnitude of a vector, 5
Lens-shaped region, mapping of, 207
Less, 1
L'Hospital's rule, 67, 78, 79
  proof of, 78
  use of, in evaluating residues, 177
Limit points, 7, 22, 23
Limits, 37, 38, 50-53
  geometric interpretation of, 50
  of functions, 37, 38, 50-53
  of sequences, 40, 51-53, 55
    (see also Sequences)
  theorems on, 38, 51-53
  uniqueness of, 38, 40, 51, 140
Linearly independent solutions of
    differential equations, 269, 270
Linear magnification factor, 201
Linear transformation, 34, 203
  fractional, 35, 203
Line (see Straight line)
Line, branch, 37, 44, 48-50
  equipotential, 235, 239, 252
  isothermal, 241
  stream, 235
Line charge, 239, 240
  complex potential due to, 252
Line integrals, 92, 98, 99
  complex, 92, 98

Line integrals (cont.)
  connection between real and
    complex, 93
  properties of, 93, 99
  real, 92, 98
Line sinks, 235 (*see also* Line sources)
Line sources, 235
  complex potential due to, 247, 252
Liouville's theorem, 119, 124, 125, 145
  fundamental theorem of algebra
    and, 125
  proof of, 124, 125
Logarithmic functions, 36, 46
  branch points of, 46, 76, 80
  principal branch of, 36, 46
  relation of, to hyperbolic
    functions 36
Logarithm, natural, 35, 36 (*see also*
  Logarithmic functions)

Maclaurin series, 143
Magnification factor, 201, 214
Magnitude or length of a vector, 5
Many-valued function, 33 (*see also*
  Multiple-valued functions)
Mapping, 33, 41–43, 200 (*see also*
  Transformations)
  conformal (*see* Conformal
    mapping)
  isogonal, 201
  of annular region, 211
  of cardioid, 210, 229
  of channel, 208, 211
  of cycloid, 228
  of ellipse, 209, 221, 224
  of half plane on to a circle, 203,
    204, 216, 217
  of hexagon, 230
  of hyperbola, 228
  of hypocycloid, 228
  of lemniscate, 229
  of lens-shaped region, 207
  of parabola, 208, 210, 228
  of polygon, 204, 218–221, 223
  of rectangle, 209, 211
  of sector, 205, 207
  of semi-circle, 206, 207
  of strip, 205, 206, 210, 211, 219
  of triangle, 209, 221, 222
  of unit circle, 203, 204, 207, 216,
    217
  of wedge-shaped regions, 225
  one to one, 200
Mapping functions, 34, 41, 42, 57
  complex, 200, 201
  conformal (*see* Conformal
    mapping)
  special, 205–211, 218
Mapping theorem of Riemann, 201,
  202, 233
Mathematical induction, 16
  proof of De Moivre's theorem by, 16
Mathematical model, 234
Maximum modulus theorem, 119, 125,
  126
  proof of, 125, 126, 135
Mean value theorem of Gauss, 119,
  125
Mechanics, applications to, 69, 81, 82
Member, of a set, 7
Meromorphic functions, 145
Minimum modulus theorem, 119, 126,
  127
Mittag-Leffler's expansion theorem,
  175, 191, 192
  proof of, 191, 192
Model, mathematical, 234
Modulus, complementary, 276
Modulus, of complex numbers, 4
  of elliptic functions, 276
Modulus theorems (*see* Maximum or
  Minimum modulus theorems)
Mod *z* (*see* Modulus of complex
  numbers)

Moment, dipole, 237
Moment, of pressure forces, 238, 251
Monotonic increasing or decreasing
  sequences, 141
Morera's theorem, 95, 110, 111, 118,
  160
  proof of, 110, 111, 124
  use of, in analytic continuation, 278
M test of Weierstrass, 142, 150, 151,
  267, 282
Multiple-valued functions, 33, 37, 43,
  44, 67, 76
  derivatives of, 65, 66, 76, 77
Multiplication, associative law of,
  3, 8
  commutative law of, 3, 8
  identity with respect to, 3
  inverse with respect to, 1, 3
  of determinants, 215, 230
  of real numbers, 1
Multiply-connected regions, 93, 94
  Cauchy-Goursat theorem for, 106
  Green's theorem for, 101
Mutually exclusive sets, 8

Natural base of logarithms, 35
Natural boundary, 146, 159, 265
Natural logarithm, 35, 36 (*see also*
  Logarithmic functions)
Natural numbers, 1
Negative integers, 1
Neighborhood, deleted, 7, 22
  delta, 7, 23
Nested triangles, 104
Neumann problem, 232, 233, 243–245
  (*see also* Dirichlet problem)
  solution of, by conformal mapping,
    233, 234
  solution of, in terms of Dirichlet
    problem, 262
  uniqueness of solution to, 262
Neumann's function, 271, 272
Non-analytic functions, 71
Non-isolated singularity, 67, 80
Normal vector to curves, 70, 83
North pole, 6
*n*th roots of unity, 5, 21
*n*th root test, 141
Null set, 8, 23
Numbers, 1
  algebraic, 23
  Bernoulli, 171, 273
  complex (*see* Complex numbers)
  irrational, 1
  natural, 1
  prime, 274
  real, 1
  theory of, 273
  transcendental, 23
Numerator, 1

Obstacles, flow around, 237, 238, 250
  force on cylindrical, 251, 252
Odd functions, 45
One to one mapping or transforma-
  tion, 200
Open interval, 2
Open regions, 7, 22, 23
Open sets, 7, 22, 23
Operators, 17, 82
  derivative, 65
  Laplacian, 63, 70, 82, 83
Order, of a pole, 67, 144
  of a zero, 67
Ordered pairs, of real numbers, 3
  graphical representation of, 3, 4
Ordinary point, 67, 269
  of a differential equation, 269
Origin, 1
Orthogonal families, 68, 81
  set, 287
  trajectories, 81
Orthogonality, of associated
  Legendre functions, 305
  of Legendre polynomials, 287

Outside, of a simple closed curve, 94

Parabola, mappings of, 208, 210, 228
Parallelogram, area of, 6, 21
  diagonals of, 12
  law, 6, 10, 11
  period, 276, 293
Parallel vectors, 6
Parametric equations, of a curve,
  41, 68
  of a line, 13
  use of, in mapping, 205, 221, 228
Partial differential equations, 85, 86,
  232
  Dirichlet and Neumann problem of
    (*see* Dirichlet problem and
    Neumann problem)
Partial sums, of infinite series, 40,
  139
Path, independence of, 96, 102, 106
Perfect conductors, 240
Period, of exponential function, 45
  of simple harmonic motion, 82
Period parallelograms of elliptic
  functions, 276, 293
Periods, of elliptic functions, 276,
  292, 293
Perpendicularity of vectors, 6
Picard's theorem, 145
Piecewise smooth curve, 69
Plane, complex, 3, 6
  z, 4
Planets, motion of, 82
Point at infinity, 6, 38, 47, 80, 81
Points, boundary, 7, 22, 23
  branch (*see* Branch points)
  critical, 201, 214
  distance between, 2, 4, 12
  exterior, 7
  in complex plane, 3
  initial, 5
  interior, 7, 22, 23
  limit, 7, 22, 23
  on real axis, 1
  ordinary, 67, 269
  saddle, 274
  singular (*see* Singular points)
  stagnation, 235
  terminal, 5
Point sets, 7, 8, 22, 23
Poisson's integral formulas, for a
  circle, 119, 129, 130, 233
  for a half plane, 120, 130, 233
Polar coordinates, 4 (*see also* Polar
  form)
Polar form, of Cauchy-Riemann
  equations, 83, 84
  of complex numbers, 4, 14, 15
  of Laplace's equation, 84
Pole, north and south, 6
Poles, 67, 79, 157, 158
  defined from Laurent series, 144
  double, 158
  number of, 119
  order of, 67, 144
  series expansion in terms of, 175,
    191, 192
  simple, 67, 80
Polygon, mapping of, on to half
  plane, 204, 218, 219, 223
  mapping of, on to unit circle, 220,
    221
Polynomial equations, 5, 19, 20, 23, 36
  degree of, 5
  factored form of, 5
  fundamental theorem of algebra
    for, 5, 119, 125, 128, 129
Polynomial functions, 34
  degree of, 34
Polynomials, Legendre (*see* Legendre
  polynomials)
  zeros of, 5, 20
Position vector, 5, 69
Positive integers, 1
Positive sense or direction, 94

Potential, 238, 252, 253
  complex, 235, 245-250
  due to charge and plane, 253
  due to charge between two parallel
      planes, 253, 254
  due to two cylinders, 255
  electrostatic, 238, 239, 252-254
  velocity, 234, 235
Power series, 140
  absolute convergence of, 152
  analytic continuation of
      (see Analytic continuation)
  circle of convergence of, 140, 143
  continuity of, 142
  differentiation of, 142, 143, 152, 153
  integration of, 142, 152
  radius of convergence of, 140, 150
  singularities and, 143, 146
  theorems on, 142, 152, 153
  uniform convergence of, 142, 152,
      153
Pressure, in a fluid, 238, 251
Prime numbers, 274
Principal branch, 33, 44, 46
  of inverse hyperbolic functions, 48
  of logarithmic functions, 36
  of trigonometric functions, 48
Principal part, in differentials, 64
  of Laurent series, 144
Principal range, 4, 44
Principal value, 4, 33
  of integrals, 174
  of logarithms, 36, 46
Product, cross (see Cross product)
  dot (see Dot product)
  infinite (see Infinite product)
  of complex numbers, 2-4
  of natural numbers, 1
  of ordered pairs of real numbers, 3
  of roots of an equation, 20
  scalar, 6 (see also Dot product)
Profiles, Joukowski, 224
Projection, of vectors, 6
  stereographic, 6, 229
Proper subset, 8
p-series, 148, 149
Pure imaginary numbers, 2

Quadratic equation, 19
Quotient, 1
  of complex numbers, 2, 4

Raabe's test, 141
Radius of convergence, 140, 150
Range, principal, 4, 44
Ratio, cross, 203, 216
  test, 141, 149
Rational algebraic functions, 35
Rational numbers, 1
Rational roots of an equation, 20
Rational transformation, 35
Ratio test, 141
  proof of, 149
Real axis, 1, 4
  points on, 1
Real fluid, 234
Real line integrals, 92 (see also Line
      integrals)
Real numbers, 1
  absolute value of, 2
  addition of, 1
  division of, 1
  graphical representation of, 1
  operations with, 1
  ordered pairs of, 3
Real number system, 1
Real part, of analytic function, 84
  of complex number, 2
Real variable, 2
Rearrangement of terms, in infinite
      series, 141
  of infinite product, 266
Rectangle, mapping of, on to annulus,
      211
  mapping of, on to half plane, 209

Rectangular coordinates, 3, 34
Rectifiable curve, 92
Recursion formulas, for Bessel
      functions, 271, 286
  for gamma function, 268, 280
  for Legendre polynomials, 272,
      287, 288
Reflection principle, Schwarz's, 266,
      279
Region, 7, 22, 23
  annular, 143, 211
  area of, 113, 114
  closed, 7
  complement of, 94
  connected, 94
  continuity in, 39, 40
  multiply-connected, 93, 94
  of convergence, 139, 149, 150
  open, 7, 22, 23
  simply-connected (see Simply-
      connected regions)
Regular, 63 (see also Analytic)
Regular singular point, 269, 283, 284
Removable discontinuity, 39, 53
Removable singularity, 67, 80, 144,
      157
  defined from Laurent series, 144
Repulsion, of electric charges, 238
Residues, 172, 173, 176-179
    (see also Residue theorem)
  calculation of, 172, 173, 176-179
  evaluation of integrals by, 133, 173,
      174, 179-188, 193
  relation of, to Laurent series, 172
  summation of series using, 175,
      188-191, 194, 195
  use of L'Hospital's rule in
      evaluating, 177
  use of series in finding, 178
Residue theorem, 173, 176-179
    (see also Residues)
  proof of, 176
Resultant, of vectors, 10
Riemann-Cauchy equations (see
      Cauchy-Riemann equations)
Riemann's conjecture, 274
Riemann sphere, 6
Riemann's mapping theorem, 201,
      202, 233
Riemann surfaces, 37, 48-50
  and analytic continuation, 265
  sheets of, 37
Riemann's zeta function, 273, 274, 288
Rodrigue's formula, 161
Roots, graphical representation of,
      18, 19
  number of, 129
  of complex numbers, 4, 18, 19
  of equations, 5, 18, 20
  of unity, 5, 21
  product of, 20
  rational, 20
  sum of, 20
Root test, 141
Rotation, of vector, 14, 15, 17
  transformation, 202, 212-214, 217
Rouché's theorem, 119, 128
  fundamental theorem of algebra
      proved by, 128, 129
  proof of, 128

Saddle point method, 274
Saddle points, 274
Scalar, 69
Scalar product, 6, 21, 70, 82, 83
    (see also Dot product)
Schlaefli's formula for Legendre
      polynomials, 161, 162
Schwarz-Christoffel transformation,
      204, 218-223
  proof of, 218, 219, 223
Schwarz's inequality, 32
Schwarz's reflection principle, 266,
      279

Schwarz's theorem, 132
Sectionally smooth curve, 69
Sector, mapping of, 206, 207
Sense or direction of curve, 69
  convention regarding, 94
Sequences, 40, 54, 55, 57, 139, 146, 147
  bounded, 141
  convergence of, 40, 54, 55
  divergent, 40
  finite, 40
  important theorems on, 140-144
  limits of, 40, 51-53, 55
  monotonic, 141
  of functions, 139, 146, 147
  terms of, 40
  uniform convergence of, 140, 147,
      148
Series, 40, 41, 54, 55, 57, 139-171
  absolute convergence of (see Abso-
      lutely convergent series)
  alternating, 142
  analytic continuation of, 146, 159,
      265, 279 (see also Analytic
      continuation)
  asymptotic, 274 (see also
      Asymptotic expansions)
  conditional convergence of, 140,
      141
  convergence of, 41, 54, 55, 139, 147
  differentiation of, 142, 143, 152, 153
  divergent, 41, 139
      (see also Convergence)
  Fourier, 170
  geometric, 148, 149
  hypergeometric, 169
  important theorems on, 140-144
  integration of, 142, 152
  Laurent's (see Laurent's series)
  Maclaurin, 143
  Mittag-Leffler, 175, 191, 192
  of functions, 146, 147
  p-, 148, 149
  partial sums of, 40, 139
  power (see Power series)
  rearrangement of terms in, 141
  residues obtained by, 178
  special, 143, 175
  sum of, 41, 139
  summation of, 175, 188-191, 194,
      195
  Taylor's, 143, 153-155
  uniform convergence of, 140, 147,
      148
Sets, 1
  boundary points of, 7, 22, 23
  bounded, 7, 22, 23
  closed, 7, 22, 23
  closure of, 7, 22, 23
  compact, 7, 22, 23
  complement of, 8, 22, 23
  connected, 7, 22, 23
  countability of, 8, 22, 23
  disjoint or mutually exclusive, 8
  distributive law for, 23
  elements or members of, 7
  exterior points of, 7
  interior points of, 7, 22, 23
  intersection of, 8, 23
  null, 8, 23
  of points in the complex plane, 3, 6
  open, 7, 22, 23
  orthogonal, 287
  point, 7, 8, 22, 23
  subset of, 1, 2, 8
  unbounded, 7
  union of, 8, 23
Sheets, of Riemann surface, 37
Simple closed curve, 68, 93
  area bounded by, 113, 114
  exterior of, 94
  interior of, 94
Simple harmonic motion, 82
Simple pole, 67, 80
Simple zero, 67
Simply-connected regions, 93, 94

Simply-connected regions (cont.)
Cauchy-Goursat theorem for, 105, 106
Cauchy's theorem for, 103
Green's theorem for, 95, 99, 100
Single-valued function, 33
Singularities, 67, 68 (see also Singular points)
and analytic continuation, 146
and convergence of power series, 143, 146
behavior of analytic functions near, 160
classification of, by Laurent series, 144, 145, 157, 158
essential (see Essential singularities)
integration around, 184
isolated, 67, 79, 80, 144, 160
removable, 67, 80, 144, 157
Singular points, 67, 68, 79-81 (see also Singularities)
of a differential equation, 269, 283
regular and irregular, 269, 283, 284
Sinks, 235, 236, 240 (see also Sources)
in electrostatics, 240
in fluid flow, 235, 236
line, 235
Smooth curve or arc, 68, 69
Solar system, motion of planets in, 82
Solutions, linearly independent, 269, 270
of an equation, 18
of differential equations, 269, 270 (see also Differential equation)
Sources, in electrostatics, 240
in fluid flow, 235-237, 247, 248
line, 235, 247, 252
strength of, 236, 247
South pole, 6
Speed, 82 (see also Velocity)
Sphere, Riemann, 6
unit, 6
Sperical representation of complex numbers, 6
Square roots, determination of, 19
Stagnation points, 235
Standard form of equation of line, 13
Stationary flow, 234
Stationary phase, method of, 275 (see also Steepest descents, method of)
Steady flow, 234
Steady-state temperature, 240, 254, 255, 257, 258
Steepest descents, method of, 274, 275, 288-290
Stereographic projection, 6, 229
Stirling's asymptotic formula, 275
Straight line, equation of, 13
mapping of curve into, 205, 221, 228
Stream function, 235, 245
Streamlines or stream curves, 235
Strength, of a source, 236, 247
of a vortex, 237
Stretching, 202, 212, 217
Strip, mapping of, 205, 206, 210, 211, 219
Subset, 1, 2, 8
Subtraction, of complex numbers, 2
of real numbers, 1
Successive transformations, 203
Sum, of infinite series, 41, 139
of natural numbers, 1
of ordered pairs of real numbers, 3
of roots of an equation, 20
Sum function, discontinuity of, 148
of series, 139, 148
Summation of series, 175, 188-191, 194, 195
Superposition of flows, 237
Surfaces, Riemann, 37, 48-50, 265

Symmetric form of equation of line, 13
Tangent, to a curve, 69, 83
Taylor's series, 143, 153-155
Taylor's theorem, 143, 153-155
proof of, 153, 154
Temperature, complex, 240, 241
steady-state, 240, 254, 255, 257, 258
Terminal point, of vector, 5
Terms, of a sequence, 40
rearrangement of, in a series, 141
removal or addition of, to a series, 141
Tests for convergence, 141, 142, 148-150
alternating series, 142
comparison, 141, 148
Gauss', 142
integral, 141
nth root, 141
Raabe's, 141
ratio, 141, 149
Thermal conductivity, 240
Trajectories, orthogonal, 81
Transcendental functions, 36, 37
Transcendental numbers, 23
Transformation equations, 200
Transformation, Joukowski, 224, 229
Transformations, 33, 34, 41-43, 200, 211-213 (see also Mapping)
bilinear (see Bilinear transformation)
conformal (see Conformal mapping)
fixed or invariant points of, 202, 217
fractional linear, 35, 203 (see also Bilinear transformation)
general, 202
inversion, 202, 217
involutory, 230
Jacobian of, 200, 201, 214-216
Joukowski, 224, 229
Landen's, 304
linear, 34, 203
of boundaries in parametric form, 205, 221, 228
rational, 35
rotation, 202, 212-214, 217
Schwarz-Christoffel, 204, 218-223
special, 205-211
stretching, 202, 212, 217
successive, 203
Translation, 202, 212, 217
Transmission line cables, 256
Triangle, mapping of, 209, 221, 222
Triangles, areas of, 22
nested, 104
Trigonometric functions, 35
in terms of exponential functions, 16, 17, 35
principal branch of, 48
properties of, 35, 45
relation of, to hyperbolic functions, 36
zeros of, 45

Unbounded sets, 7
Uniform continuity, 39, 40, 54
Uniform convergence, 140, 142, 147, 148, 150-153, 160
and continuity, 142, 151
of infinite products, 266, 267
of power series, 142, 152, 153
of sequences, 140, 147, 148
of series, 140, 147, 148
theorems on, 142, 150-152
Uniform flow, 236, 245
Uniform continuity, 39, 40, 54
Uniformly continuous functions, 39, 40, 54
Uniformly convergent series (see Uniform convergence)

Union of sets, 8, 23
Uniqueness, of limits, 38, 40, 51, 140
of solution to Dirichlet problem, 256, 257
of solution to Neumann problem, 262
theorem for analytic continuation, 265
Unit cell, 276, 293
Unit circle, 5, 201
Dirichlet problem for, 233
mappings on to, 203, 204, 207, 216, 217
Poisson's integral formula for, 119, 129, 130, 233
Unit disk, 201
Unit sphere, 6
Unity, nth roots of, 5, 21
Upper bound, 62, 93
Value, absolute (see Absolute value)
of a function, 33
principal (see Principal value)
Variables, 2, 33
change of, in integration, 93
complex, 2, 33
dependent, 33
dummy, 97
independent, 33
real, 2
Vectors, 5, 10-13
addition of, 10, 11, 15
angle between, 21
components of, 10
equality of, 5
initial point of, 5
interpretation of complex numbers as, 5
length or magnitude of, 5
normal, to a curve, 70, 83
parallel, 6
perpendicularity of, 6
projection of, 6
resultant of, 10
rotation of, 14, 15, 17
terminal point of, 5
Velocity, along curve, 69, 82
Velocity, complex, 235
Velocity potential, 234, 245
Viscous fluid, 234
Vortex, 237, 249
circulation about, 249
strength of, 237
Vortex flow, 249

Wedge-shaped region, mapping of, 225 (see also Sector)
Weierstrass-Bolzano theorem, 8, 23
Weierstrass-Casorati theorem, 145
Weierstrass factor theorems, 267, 282
Weierstrass M test, 142, 150, 151
analog of, for infinite products, 267
proof of, 150
Weierstrass' theorem for infinite products, 267, 282
Wing, of airplane, 224
(see also Airfoil)
Work, 111

x axis, 3

y axis, 3

Zero, 1
Zeros, number of, 119
of polynomials, 5, 20
of trigonometric functions, 45
order of, 67
simple, 67
Zeta function, of Riemann, 273, 274, 288
analytic continuation of, 273
z plane, 4
entire, 6